Springer Series in Solid and Structural Mechanics

Volume 13

The Springer Series in Solid and Structural Mechanics (SSSSM) publishes new developments and advances dealing with any aspect of mechanics of materials and structures, with a high quality. It features original works dealing with mechanical, mathematical, numerical and experimental analysis of structures and structural materials, both taken in the broadest sense. The series covers multi-scale, multi-field and multiple-media problems, including static and dynamic interaction. It also illustrates advanced and innovative applications to structural problems from science and engineering, including aerospace, civil, materials, mechanical engineering and living materials and structures. Within the scope of the series are monographs, lectures notes, references, textbooks and selected contributions from specialized conferences and workshops.

More information about this series at http://www.springer.com/series/10616

Friedel Hartmann · Peter Jahn

Statics and Influence Functions

From a Modern Perspective

Second Edition

Springer

Friedel Hartmann
Institute of Structural Mechanics
University of Kassel
Kassel, Germany

Peter Jahn
Institute of Structural Mechanics
University of Kassel
Kassel, Germany

ISSN 2195-3511 ISSN 2195-352X (electronic)
Springer Series in Solid and Structural Mechanics
ISBN 978-3-030-55891-8 ISBN 978-3-030-55889-5 (eBook)
https://doi.org/10.1007/978-3-030-55889-5

This Springer imprint is published by the registered company Springer Nature Switzerland AG
The registered company address is: Gewerbestrasse 11, 6330 Cham, Switzerland

Preface to the Second Edition

This second edition of *Statics and Influence Functions—from a Modern Perspective* has been completely revised and new material has been added to extend the scope of the book.

The list of the differential equations in the first chapter has been more than doubled and the second chapter on Betti's theorem includes new sections on adjoint operators and St. Venant's principle.

The equivalent stress transformation in chapter three is a novel approach to couple diverse structural elements.

In the fourth chapter, sections on the patch test, pollution and super convergence were added.

In the fifth chapter on reanalysis, new sections on adjoint sensitivity analysis via duality techniques investigate the close connection between reanalysis with Green's functions, optimization and parameter identification.

A new direct approach to calculate the modified displacement vector without solving the full system is introduced in chapter five. Practical application of this technique is *one-click-reanalysis*, which allows modifications of a structure to be made with single mouse clicks.

The mixed and nonlinear problems each have been given separate chapters, seven and eight, to allow for more specifics.

In the ninth chapter, a section on Sobolev's Embedding Theorem was added to allow a more detailed exposition of the practical implications of this important theorem.

The new edition offers four programs for download, see Chap. 10: Solution of the Poisson equation, 2-D elasticity, Kirchhoff plates and planar frames. These are general purpose programs, but with a particular emphasis on influence functions to put the theory into practice. The frame program also demonstrates *one-click-reanalysis*.

Kassel, Germany
June 2020

Friedel Hartmann
Peter Jahn

Preface to the First Edition

The new is the old and the old is mightier than ever before.

The subject of this book are influence functions and the role they play in finite element analysis of structures. Influence functions are a classical tool of structural analysis and are dearly loved by "old-school" engineers since with some clever sketches—if need be on a beer mat—it is easy to understand the behavior of a structure or to find the weak spots in a design.

Unfortunately, with the advent of finite element programs though, the application of influence functions has faded into the background. When in doubt, one rather studies variants in a design with the computer than strive for a deeper understanding which the study of influence functions can provide so easily and so well.

But new results have rekindled the interest in influence functions because we know today that in linear analysis, finite elements compute "everything"—as we are tempted to say—with influence functions. This equals a loop backward. Classical hand methods seemed outdated and old fashioned but in finite elements, they have risen like Phoenix from the ashes. FE-analysis is more classical than we ever imagined.

In the olden days, the subject of influence functions mostly focused on the analysis of frames with the Müller-Breslau principle but in FE-analysis, the concept of influence functions has a much broader and wider scope.

The key word is *functionals*.

The deflection at the midspan of a beam, the bending moment at a fixed edge, the force in a pier, all these are functionals. Anything you can calculate with finite elements is considered a functional. And to each linear functional belongs a Green's function, an influence function.

Influence functions, too, are displacements; they are the reaction of a structure to special point loads, to Dirac deltas, but normally a mesh is not that flexible enough to generate the exact intricate shape and exact peaks of these influence functions. This is the reason why FE-results are not exact. FE-programs operate with approximate, substitute influence functions.

The influence functions are the "real" shape functions, the *physical shape functions*. A good approximation of these functions is the key to good FE-results.

So, in structural analysis—and we dare say all of linear computational mechanics—influence functions play a dominant role. This is why we have written this book.

It is not a book for a first course in structural mechanics; the reader should be acquainted with the basic principles and the reader should have seen influence functions in action.

We treat the subject also, as it seems, with a rather sharp pencil but this is more or less self-defense because in the advent of time, many things in structural analysis have so much settled in that it is hard to discover the mathematics behind the formulas—all too often the ubiquitous $\delta W_e = \delta W_i$ is considered sufficient proof.

Kassel, Germany
October 2016

Friedel Hartmann
Peter Jahn

PS. The original title of the book was *Influence Functions—from a Modern Perspective* but in the time of search engines, it seemed better to extend the title to *Statics and Influence Functions—from a Modern Perspective*. In doing so, it was not our intention to construct a (non-existing) contrast between old and new statics. Only the view on influence functions has changed with the advent of the computer.

Contents

Chapter 1
Foundation

1.1 Introduction

We start with the variational principles of statics and Betti's theorem,

- the principle of virtual displacements
- the principle of conservation of energy
- the principle of virtual forces
- Betti's theorem

Since they form the core of modern structural analysis. As diverse as these principles and theorems are, technically they are all based on one equation: Integration by parts.

1.1.1 Integration by Parts

The integration by parts formula

$$\int_0^l u' \, \delta u \, dx = [u \, \delta u]_0^l - \int_0^l u \, \delta u' \, dx \,, \tag{1.1}$$

derived from

$$\int_0^l (u \, \delta u)' \, dx = \int_0^l (u' \, \delta u + u \, \delta u') \, dx = [u \, \delta u]_0^l \,, \tag{1.2}$$

can be written as a **"zero sum"**,

F. Hartmann and P. Jahn, *Statics and Influence Functions*,
Springer Series in Solid and Structural Mechanics 13,
https://doi.org/10.1007/978-3-030-55889-5_1

$$\mathscr{I}(u, \delta u) = \int_0^l u' \, \delta u \, dx - [u \, \delta u]_0^l + \int_0^l u \, \delta u' \, dx = 0 \,, \tag{1.3}$$

since this invariant form better expresses the built-in **duality**: The equation is zero for ***all pairs*** of functions in $C^1(0, l)$, like $u = \sin(x)$ and $\delta u = \cos(x)$.

1.1.2 Principle of Virtual Displacements

Zero sums are easy to deal with. When two forces $\pm f$ pull at the two ends of a bar as in Fig. 1.1a, *the zero sum* of the two forces,

$$-f + f = 0 \,, \tag{1.4}$$

can be multiplied with any number δu without changing the mathematics

$$\delta u \cdot (-f + f) = -\delta u \cdot f + \delta u \cdot f = 0 \,, \tag{1.5}$$

which means we can slide the bar forwards and backwards on the table and each time the work done by the two forces is zero. This is an application of the ***principle of virtual displacements***.

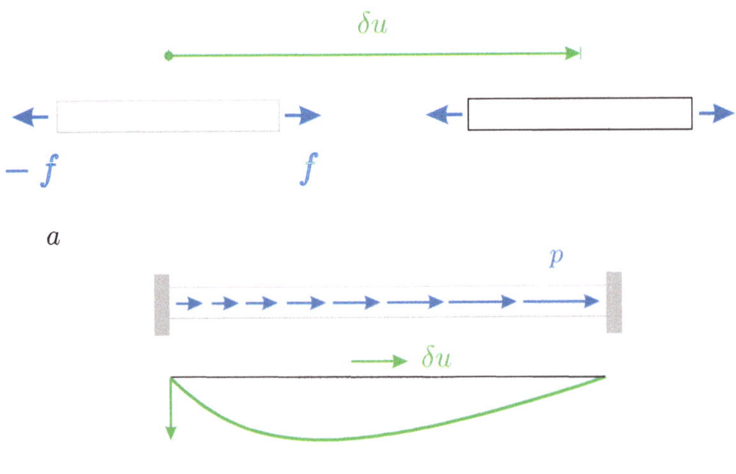

$$\int_0^l \frac{N \delta N}{EA} \, dx = \int_0^l p \, \delta u \, dx$$

Fig. 1.1 Virtual displacement, **a** of an unrestrained rigid bar, and **b** an elastic bar fixed at both ends. For each admissible δu the two integrals have the same value

Granted, the logic is remarkably simple. If an equation is zero

$$Eq = 0\,, \tag{1.6}$$

the product of the equation with any number δu is zero as well

$$\delta u \cdot Eq = 0\,, \tag{1.7}$$

and this holds also true for functions, see Fig. 1.1b, since if $u(x)$ satisfies the differential equation

$$-EA\,u''(x) - p(x) = 0 \qquad 0 < x < l\,, \tag{1.8}$$

then

$$\int_0^l (-EA\,u'' - p)\,\delta u\,dx = 0\,, \tag{1.9}$$

or after integration by parts, if $\delta u(0) = \delta u(l) = 0$,

$$\int_0^l \frac{N\,\delta N}{EA}\,dx = \int_0^l p\,\delta u\,dx \tag{1.10}$$

where $N = EA\,u'$ is the normal force.

1.1.3 Betti's Theorem

When two numbers u_1 and u_2 solve the two **"twin"** equations (the 3 makes them twins)

$$3 \cdot u_1 = 12 \qquad 3 \cdot u_2 = 18\,, \tag{1.11}$$

and we multiply each twin with the other solution,

$$u_2 \cdot 3 \cdot u_1 = 12 \cdot u_2 \qquad u_1 \cdot 3 \cdot u_2 = 18 \cdot u_1\,, \tag{1.12}$$

then the left sides are the same, and so also the right sides must be the same, see Fig. 1.2,

$$W_{1,2} = 12 \cdot x_2 = 18 \cdot x_1 = W_{2,1}\,. \tag{1.13}$$

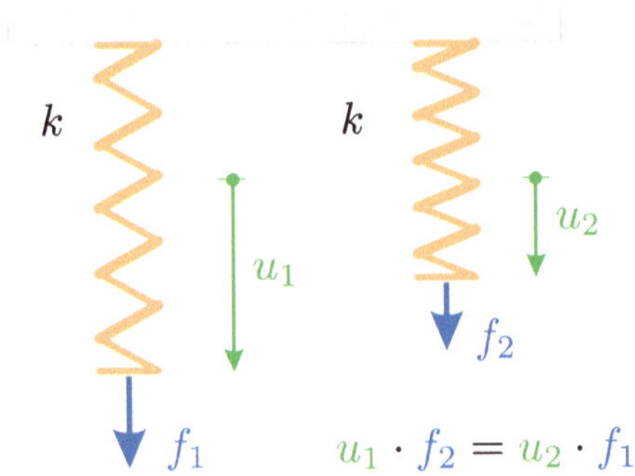

Fig. 1.2 Two springs and application of *Betti's theorem*, stiffness $k = 3$

This is *Betti's theorem*: *The reciprocal external work of two equilibrium states is the same.* It is a universal property of self-adjoint operators and symmetric matrices. If $\boldsymbol{u}_1$ and $\boldsymbol{u}_2$ are the nodal displacements of a truss under two different loads

$$\boldsymbol{K}\boldsymbol{u}_1 = \boldsymbol{f}_1 \qquad \boldsymbol{K}\boldsymbol{u}_2 = \boldsymbol{f}_2\,, \tag{1.14}$$

the left sides can be made the same

$$\boldsymbol{u}_2^T\boldsymbol{K}\boldsymbol{u}_1 = \boldsymbol{u}_2^T\boldsymbol{f}_1 \qquad \boldsymbol{u}_1^T\boldsymbol{K}\boldsymbol{u}_2 = \boldsymbol{u}_1^T\boldsymbol{f}_2\,, \tag{1.15}$$

and so also the right sides must be the same

$$\boldsymbol{u}_2^T\boldsymbol{f}_1 = \boldsymbol{u}_1^T\boldsymbol{f}_2\,. \tag{1.16}$$

1.1.4 Influence Functions

To solve the equation

$$3 \cdot x = 12 \tag{1.17}$$

we divide the right side by the number 3, or—as we could say as well—we multiply the right side with the inverse $g = 1/3$, the response to a "point load", a "Dirac delta",

$$3 \cdot g = 1\,, \tag{1.18}$$

since this equation implies that

$$x = g \cdot 12 = \frac{1}{3} \cdot 12 = 4 \tag{1.19}$$

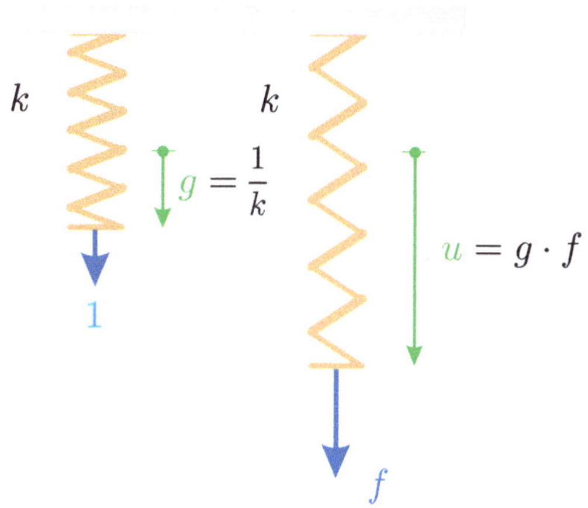

Fig. 1.3 If a force $f = 1$ stretches the spring by $g = 1/k$ units, a force f will stretch the spring by $u = f \cdot g$ units

is the solution to (1.17), see Fig. 1.3. *This is the technique of influence functions or Green's functions.*

To calculate the nodal displacement u_i of a truss in this way, we apply a single force $f_i = 1$ at the node, determine the corresponding nodal displacements of the truss, the vector g_i,

$$\boldsymbol{K}\, g_i = \boldsymbol{e}_i \qquad (i\text{-th unit vector})\,, \tag{1.20}$$

and form the scalar product of g_i and $\boldsymbol{f}$

$$u_i = \boldsymbol{e}_i^T \boldsymbol{u} = \boldsymbol{e}_i^T \boldsymbol{K}^{-1} \boldsymbol{f} = g_i^T \boldsymbol{f}\,. \tag{1.21}$$

In a beam, instead, we would place a **single force** $P = 1$ at the **source point** x, determine the response $G(y, x)$ of the beam, see Fig. 1.4, and integrate

$$w(x) = \int_0^l G(y, x)\, p(y)\, dy\,. \tag{1.22}$$

1.1.5 Identities

In statics we solve scalar equations

$$k\, u = f\,, \tag{1.23}$$

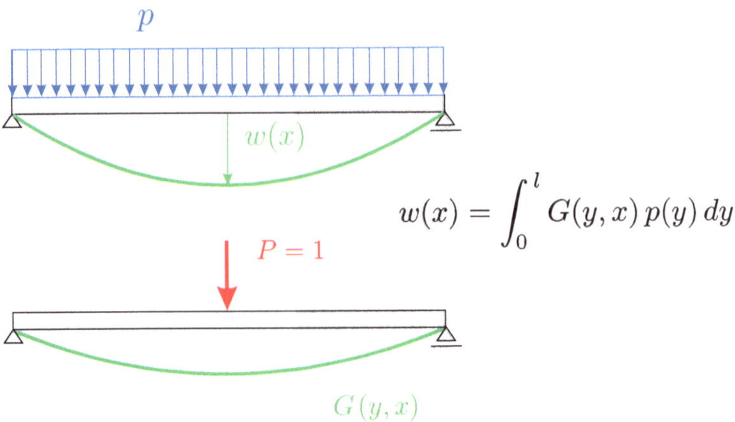

Fig. 1.4 Beam and influence function for the deflection at the midpoint

or systems of equations

$$\boldsymbol{K}\boldsymbol{u} = \boldsymbol{f}\,, \tag{1.24}$$

or differential equations

$$EI\,w^{IV}(x) = p(x)\,. \tag{1.25}$$

To each of the operators on the left belongs a simple identity

$$\mathscr{B}(u, \delta u) = \delta u\,k\,u - u\,k\,\delta u = 0 \tag{1.26}$$

$$\mathscr{B}(\boldsymbol{u}, \boldsymbol{\delta u}) = \boldsymbol{\delta u}^T\,\boldsymbol{K}\,\boldsymbol{u} - \boldsymbol{u}^T\,\boldsymbol{K}\,\boldsymbol{\delta u} = 0 \tag{1.27}$$

$$\mathscr{G}(w, \delta w) = \int_0^l EI\,w^{IV}\,\delta w\,dx + [V\,\delta w - M\,\delta w']_0^l - \int_0^l \frac{M\,\delta M}{EI}\,dx = 0\,. \tag{1.28}$$

Only this last identity requires a warning, since it is based on integration by parts, and so the functions w and δw must be from $C^4(0, l)$ and $C^2(0, l)$ respectively for it to be true, since in (1.28) we integrate two times.

Fig. 1.5 Check of the equilibrium of the beam with Green's first identity by applying the virtual displacement $\delta w = 1$, $\mathcal{G}(w, 1) = p \cdot l + V(l) - V(0) = 0$

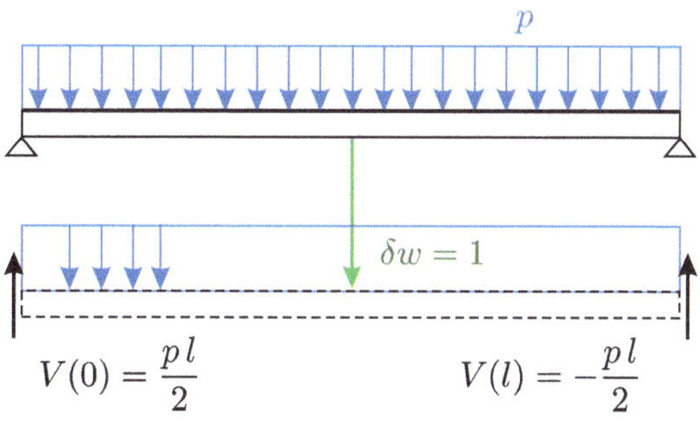

The energy principles of statics are based on Green's first and second identity (integration by parts)

This is why the essential formulations of statics and mechanics are ***dual formulations***, are "stereo", not "mono". The prime mathematical operation in mechanics is the **scalar product**. The ubiquitous *for all* δu of the variational principles has its root in the scalar product, see Fig. 1.5.

1.2 Green's Identities

To continue in a more systematic and orderly fashion we will list in the following the major differential equations of frame analysis, and formulate the associated identities as for example the identity of a bar

$$\int_0^l -EA\, u''\, \delta u\, dx = [(-EA\, u')\, \delta u]_0^l - \int_0^l -EA\, u'\, \delta u'\, dx\,. \tag{1.29}$$

The principal tool, integration by parts, is like climbing stairs ($\delta u = 1$)

$$\int_a^b f'(x)\, dx = f(b) - f(a)\,, \tag{1.30}$$

since with each step dx in horizontal direction the gain in height is $df = f'(x)\, dx$ and so at the end the total is $f(b) - f(a)$, see Fig. 1.6.

The staircase formula (1.30) is the ***fundamental theorem of calculus***. It implies that the integral of the normal force $N(x) = EA\, u'(x)$ in a bar, fixed at both ends, is zero, see Fig. 1.7a,

$$\int_0^l EA\, u'(x)\, dx = [EA\, u]_0^l = EA\,(u(l) - u(0)) = 0\,, \tag{1.31}$$

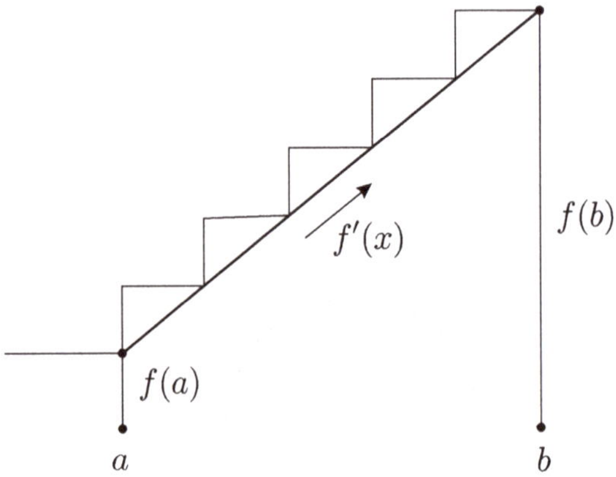

Fig. 1.6 On climbing stairs we sense the fundamental theorem of differential and integral calculus

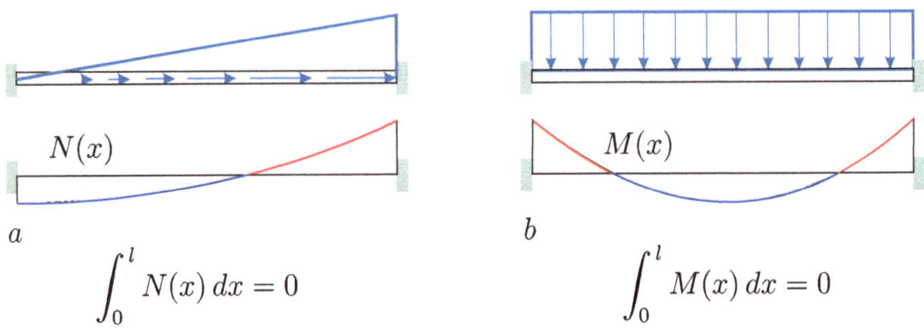

Fig. 1.7 The integrals of the normal force N and the bending moment M are zero

and the mean value of the bending moment $M(x) = -EI\,w''(x)$ in a beam, clamped on both sides, see Fig. 1.7b, is zero as well

$$\int_0^l -EI\,w''(x)\,dx = -EI\,(w'(l) - w'(0)) = 0\,. \tag{1.32}$$

For **partial derivatives** integration by parts reads

$$\int_\Omega u_{,i}\;\delta v\,d\Omega = \int_\Gamma u\,n_i\,\delta v\,ds - \int_\Omega u\,\delta v_{,i}\;d\Omega\,, \tag{1.33}$$

where Γ is the edge of the domain Ω, n_i is the i-th component of the normal vector $\boldsymbol{n}$ (length $|\boldsymbol{n}| = 1$) on Γ and $u_{,i} = \partial u/\partial x_i$ is the derivative with respect to x_i. If we let $\delta v = 1$ Eq. (1.33) becomes

$$\int_\Omega u_{,i}\, d\Omega = \int_\Gamma u\, n_i\, ds\,, \tag{1.34}$$

and so, if a plate Ω is fixed along its edge Γ, $u_x = u_y = 0$, the mean of the stress

$$\sigma_{xx} = E\,(\varepsilon_{xx} + \nu\,\varepsilon_{yy}) = E\,(u_{x,x} + \nu\, u_{y,y}) \tag{1.35}$$

(and of σ_{yy} as well) is zero, since

$$\int_\Omega E\,(u_{x,x} + \nu\, u_{y,y})\, d\Omega = \int_\Gamma E\,(u_x\, n_x + \nu\, u_y\, n_y)\, ds = 0\,. \tag{1.36}$$

1.2.1 Longitudinal Displacement $u(x)$ of a Bar

$$-EA\, u''(x) = p(x) \tag{1.37}$$

$$\mathcal{G}(u, \delta u) = \underbrace{\int_0^l -EA\, u''(x)\, \delta u(x)\, dx + [N\, \delta u]_0^l}_{\text{external virt. work}} - \underbrace{\int_0^l \frac{N\, \delta N}{EA}\, dx}_{\text{internal virt. work}} = 0\,, \tag{1.38}$$

where $N = EA\, u'$ is the normal force, see Fig. 1.8.

If $EA(x)$ is a function of x, the differential equation of the bar is $-(EA(x)\, u')' = p(x)$ and then

$$\int_0^l -(EA(x)\, u')'\, \delta u\, dx = [(-EA(x)\, u')\, \delta u]_0^l - \int_0^l -EA(x)\, u'\, \delta u'\, dx \tag{1.39}$$

which repeats the identity (1.38), because the definition of $N = EA(x)\, u'$ does not change

$$\mathcal{G}(u, \delta u) = \int_0^l -(EA(x)\, u')'\, \delta u(x)\, dx + [N\, \delta u]_0^l - \int_0^l \frac{N\, \delta N}{EA}\, dx = 0\,. \tag{1.40}$$

Since $-N' = p$ is the same as $-(EA(x)\, u')' = p$, we can also write

$$\int_0^l -N'\, \delta u\, dx = [N\, \delta u]_0^l - \int_0^l -N\, \delta u'\, dx\,. \tag{1.41}$$

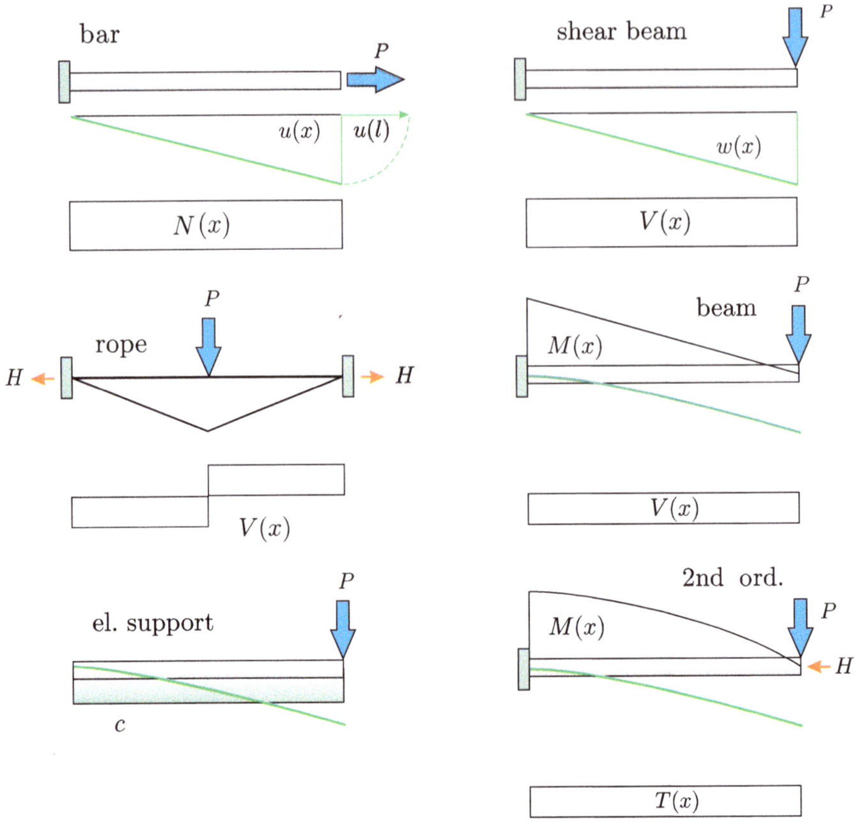

Fig. 1.8 Structural elements

When the longitudinal displacement must work against some friction (c),

$$-EA\,u''(x) + c\,u(x) = p(x)\,, \tag{1.42}$$

the identity reads

$$\mathscr{G}(u, \delta u) = \underbrace{\int_0^l (-EA\,u''(x) + c\,u(x))\,\delta u(x)\,dx + [N\,\delta u]_0^l}_{\delta W_e}$$

$$-\underbrace{\int_0^l \Big(\frac{N\,\delta N}{EA} + c\,u\,\delta u\Big)\,dx}_{\delta W_i} = 0\,. \tag{1.43}$$

1.2.2 Shear Deformation $w_S(x)$ of a Beam

$$-GA\,w_s''(x) = p(x) \tag{1.44}$$

$$\mathcal{G}(w_s, \delta w_s) = \underbrace{\int_0^l -GA\,w_s''(x)\,\delta w_s(x)\,dx + [V\,\delta w_s]_0^l}_{\delta W_e} - \underbrace{\int_0^l \frac{V\,\delta V}{GA}\,dx}_{\delta W_i} = 0\,, \tag{1.45}$$

with $V = GA\,w_s'$ as the shear force, and GA as the shear modulus.

When the beam sits on an elastic foundation (c),

$$-GA\,w_s''(x) + c\,w_s(x) = p(x)\,, \tag{1.46}$$

the identity has the form

$$\mathcal{G}(w_s, \delta w_s) = \underbrace{\int_0^l (-GA\,w_s''(x) + c\,w_s(x))\,\delta w_s(x)\,dx + [V\,\delta w_s]_0^l}_{\delta W_e}$$
$$- \underbrace{\int_0^l (\frac{V\,\delta V}{GA} + c\,w_s\,\delta w_s)dx}_{\delta W_i} = 0\,. \tag{1.47}$$

1.2.3 Deflection w of a Rope

$$-H\,w''(x) = p(x) \qquad H = \text{horizontal prestress in the rope} \tag{1.48}$$

with $V(x) = H\,w'(x)$ as the shear force in the rope

$$\mathcal{G}(w, \delta w) = \underbrace{\int_0^l -H\,w''(x)\,\delta w(x)\,dx + [V\,\delta w]_0^l}_{\delta W_e} - \underbrace{\int_0^l \frac{V\,\delta V}{H}\,dx}_{\delta W_i} = 0\,. \tag{1.49}$$

1.2.4 Deflection w of a Beam

$$EI\, w^{IV}(x) = p(x) \tag{1.50}$$

$$\mathcal{G}(w, \delta w) = \underbrace{\int_0^l EI\, w^{IV}(x)\, \delta w\, dx + [V\, \delta w - M\, \delta w']_0^l}_{\delta W_e} - \underbrace{\int_0^l \frac{M\, \delta M}{EI}\, dx}_{\delta W_i} = 0\,, \tag{1.51}$$

with $M(x) = -EI\, w''(x)$ and $V(x) = -EI\, w'''(x)$.

If $EI(x)$ is a function of x, the beam equation is $(EI(x)\, w'')'' = p(x)$, and so

$$\int_0^l (EI(x)\, w'')''\, \delta w\, dx = [(EI(x)\, w'')'\, \delta w - EI(x)\, w''\, \delta w']_0^l + \int_0^l EI(x)\, w''\, \delta w''\, dx\,, \tag{1.52}$$

where $M = -EI(x)\, w''$ and $V = -(EI(x)\, w'')'$ and this means

$$\mathcal{G}(w, \delta w) = \int_0^l (EI(x)\, w'')''\, \delta w\, dx + [V\, \delta w - M\, \delta w']_0^l - \int_0^l \frac{M\, \delta M}{EI}\, dx = 0\,. \tag{1.53}$$

Here too we can, since $-M'' = p$ is the same as $(EI(x)w'')'' = p$, write

$$\int_0^l -M''\, \delta w\, dx = -[M'\, \delta w]_0^l + \int_0^l M'\, \delta w'\, dx\,. \tag{1.54}$$

1.2.5 Deflection w of a Beam, 2nd Order Theory

$$EI\, w^{IV}(x) + (D(x)\, w'(x))' = p_z(x) \qquad D(x) = P + \int_0^x p_x(y)\, dy \tag{1.55}$$

$$\mathcal{G}(w, \delta w) = \underbrace{\int_0^l (EI\, w^{IV}(x) + (D(x)\, w'(x))')\, \delta w\, dx + [T\, \delta w - M\, \delta w']_0^l}_{\delta W_e}$$
$$- \underbrace{\int_0^l \left(\frac{M\, \delta M}{EI} - D(x)\, w'(x)\, \delta w'(x)\right) dx}_{\delta W_i} = 0 \tag{1.56}$$

where T is the *transverse force*

$$T(x) = -EI\, w'''(x) - D(x)\, w'(x) = V(x) - D(x)\, w'(x)\,. \tag{1.57}$$

The D is the compressive force and $p_x(x)$ and $p_z(x)$ are distributed forces in axial direction and perpendicular to it.

1.2.6 Beam on an Elastic foundation

$$EI\, w^{IV}(x) + c\, w(x) = p(x) \tag{1.58}$$

and

$$\mathscr{G}(w, \delta w) = \underbrace{\int_0^l (EI\, w^{IV}(x) + c\, w(x))\, \delta w(x)\, dx + [V\, \delta w - M\, \delta w']_0^l}_{\delta W_e} - \underbrace{\int_0^l (\frac{M\, \delta M}{EI} + c\, w(x)\, \delta w(x))\, dx}_{\delta W_i} = 0\,. \tag{1.59}$$

1.2.7 Tensile Chord Bridge

Imagine a beam through which runs a prestressed rope (tendon), so that the beam and the rope jointly carry the distributed load p

$$EI\, w^{IV}(x) - H\, w''(x) = p(x) \qquad H = \text{prestress}\,, \tag{1.60}$$

and Green's first identity reads

$$\mathscr{G}(w, \delta w) = \underbrace{\int_0^l (EI\, w^{IV}(x) - H\, w''(x))\, \delta w(x)\, dx + [V\, \delta w - M\, \delta w']_0^l}_{\delta W_e} - \underbrace{\int_0^l (\frac{M\, \delta M}{EI} + H\, w'(x)\, \delta w'(x))\, dx}_{\delta W_i} = 0\,, \tag{1.61}$$

with $V = -EI\, w'''(x) + H\, w'(x)$.

1.2.8 Torsion

The differential equation of ***St. Venant's torsion***

$$-G\,I_T\,\vartheta'' = m_x \tag{1.62}$$

$$\mathcal{G}(\vartheta, \delta\vartheta) = \underbrace{\int_0^l -G\,I_T\,\vartheta''(x)\,\delta\vartheta(x)\,dx + [M_T\,\delta\vartheta]_0^l}_{\delta W_e} - \underbrace{\int_0^l \frac{M_T\,\delta M_T}{G\,I_T}\,dx}_{\delta W_i} = 0\,, \tag{1.63}$$

and ***warping torsion***

$$E I_\omega\,\vartheta^{IV} - G\,I_T\,\vartheta'' = m_x \tag{1.64}$$

$$\mathcal{G}(\vartheta, \delta\vartheta) = \underbrace{\int_0^l (E I_\omega\,\vartheta^{IV}(x) - G\,I_T\,\vartheta''(x))\,\delta\vartheta(x)\,dx + [M_T\,\delta\vartheta - M_\omega\,\delta\vartheta']_0^l}_{\delta W_e}$$
$$- \underbrace{\int_0^l (\frac{M_\omega\,\delta M_\omega}{E I_\omega} + G\,I_T\,\vartheta'(x)\,\delta\vartheta'(x))\,dx}_{\delta W_i} = 0\,, \tag{1.65}$$

with

$$M_\omega = -E I_\omega\,\vartheta''(x) \qquad M_T = -E I_\omega\,\vartheta'''(x) + G\,I_T\,\vartheta'(x) \tag{1.66}$$

repeat the patterns from above.

1.3 Variational Principles of Structural Analysis

In all identities, as for example the identity of a rope,

$$\mathcal{G}(w, \delta w) = \int_0^l -H\,w''(x)\,\delta w(x)\,dx + [V\,\delta w]_0^l - \int_0^l \frac{V\,\delta V}{H}\,dx = 0\,, \tag{1.67}$$

we equate the external virtual work with the internal strain energy. So, each term is an energy, is the scalar product of a **force** [N] and a **displacement** [m] and the bottom line is: The sum is always zero, the account is always balanced. This simple observation is the key to the variational and energy principles of statics.

Principle of Virtual Displacements

$$\boxed{\mathscr{G}(w, \delta w) = \delta W_e - \delta W_i = 0} \tag{1.68}$$

Conservation of Energy

If the second entry is identical with the first, $\delta w = w$, the identities formulate the *principle of conservation of energy*

$$\boxed{\frac{1}{2}\,\mathscr{G}(w, w) = W_e - W_i = 0} \tag{1.69}$$

The external eigenwork is stored as internal energy (this needs the factor $1/2$).

Principle of Virtual Forces

If the test function δw^* takes the first spot, and—as is tradition—is then written with an asterisk, it is the *principle of virtual forces*

$$\boxed{\mathscr{G}(\delta w^*, w) = \delta W_e^* - \delta W_i^* = 0} \tag{1.70}$$

Betti's Theorem

When we formulate Green's first identity twice, but switch the positions of w and $\hat{w}$ in the second round, and subtract the two equations, $0 - 0 = 0$, we arrive at Green's second identity

$$\mathscr{B}(w, \hat{w}) = \mathscr{G}(w, \hat{w}) - \mathscr{G}(\hat{w}, w) = \underbrace{\int_0^l EI\, w^{IV}(x)\, \hat{w}(x)\, dx + [V\, \hat{w} - M\, \hat{w}']_0^l}_{W_{1,2}}$$

$$\underbrace{- [w\, \hat{V} - w' \hat{M}]_0^l - \int_0^l w(x)\, EI\, \hat{w}^{IV}(x)\, dx}_{W_{2,1}} = 0\,, \tag{1.71}$$

or *Betti's theorem*: The reciprocal external work of two deflections w and $\hat{w}$ is the same.

$$\boxed{\mathscr{B}(w, \hat{w}) = W_{1,2} - W_{2,1} = 0} \tag{1.72}$$

Principle of Minimum Potential Energy

The **potential energy** of a hinged beam—in classical and modern notation side by side—is the expression

$$\begin{aligned}\Pi(w) &= \frac{1}{2}\int_0^l \frac{M^2}{EI}\,dx - \int_0^l p(x)\,w(x)\,dx = \frac{1}{2}\,a(w,w) - (p,w)\\ &= \frac{1}{2}\,a(w,w) - \frac{1}{2}\,(p,w) - \frac{1}{2}\,(p,w)\,.\end{aligned} \tag{1.73}$$

If w is the deflection of the beam, $EI\,w^{IV} = p$, Green's first identity implies $\mathcal{G}(w,w) = a(w,w) - (p,w) = 0$, and $\Pi(w)$ reduces to

$$\boxed{\Pi(w) = -\frac{1}{2}\,(p,w) = \text{half the eigenwork} \times (-1)} \tag{1.74}$$

which means: The potential energy in the equilibrium position is negative, since ***eigenwork*** (p,w) is always positive.

If we add an admissible virtual displacement δw, i.e. $\delta w(0) = \delta w(l) = 0$, to the equilibrium position w, the potential energy will increase

$$\Pi(w + \delta w) = \Pi(w) + \underbrace{\mathcal{G}(w,\delta w)}_{=0} + \underbrace{a(\delta w,\delta w)}_{>0}\,, \tag{1.75}$$

and so $\Pi(w)$ must be **the deepest point**.

1.4 Zero Sums

Green's first identity resembles the game played by the **desert wind** $(= \delta u)$ with the dried-out ***tumbleweed*** $(= u)$, see Fig. 1.9. No matter how strong the wind blows and how big the capers are, at the end the result always comes out zero, $\mathcal{G}(u,\delta u) = 0$.

If we read the identity as a **variational statement**

$$\mathcal{G}(u,\delta u) = 0 \qquad \text{for all } \delta u\,, \tag{1.76}$$

it reminds of the **path-independence** of the work integral of a point mass m in the gravitational field of the Earth, see Fig. 1.10. Near the Earth's surface the potential energy has the form $\Pi = m \cdot g \cdot y$, and if the point mass m moves on a closed path $C = \{x(s), y(s)\}^T$ its total work is zero[1]

[1] Int. by parts and $y(0) = y(L)$ with $L =$ length of the path.

Fig. 1.9 Tumbleweed

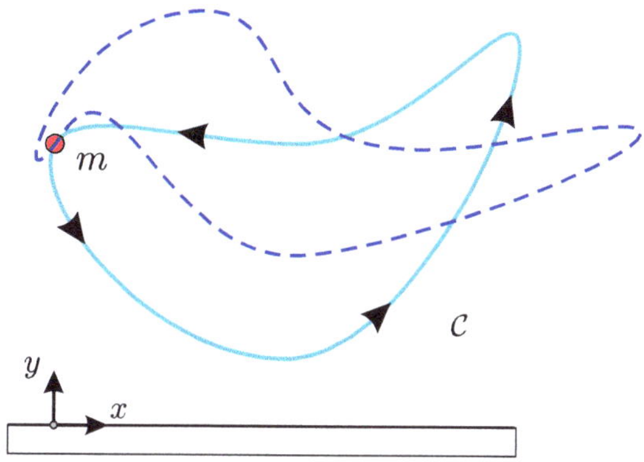

Fig. 1.10 Closed path

$$\int_C \cdot\nabla\Pi \cdot \boldsymbol{ds} = m \cdot g \int_0^L \begin{bmatrix} 0 \\ 1 \end{bmatrix} \cdot \begin{bmatrix} x' \\ y' \end{bmatrix} ds = m \cdot g \int_0^L y'\, ds$$
$$= m \cdot g \cdot (y(L) - y(0)) = 0\,, \tag{1.77}$$

regardless of the shape of the curve C—the path δu so to speak.

It all starts with the **scalar product**[2] **of two conjugated quantities**, of a force and a displacement,

$$\int_0^l -EA\, u''\, \delta u\, dx = \textit{force} \times \textit{displacement} \tag{1.78}$$

and integration by parts

$$\mathscr{I}(u, \delta u) = \int_0^l u'\, \delta u\, dx - [u\, \delta u]_0^l + \int_0^l u\, \delta u'\, dx = 0\,, \tag{1.79}$$

[2]The superposition of two functions is called an L_2-scalar product.

allows to link infinitely many pairs of functions u and δu in a "zero-sum game"

$$\mathcal{G}(u, \delta u) = \left\{ \begin{array}{l} \displaystyle\int_0^l -EA\, u''\, \delta u\, dx + \cdots \\ \displaystyle\int_0^l EI\, u^{IV}\, \delta u\, dx + \cdots \\ \displaystyle\int_\Omega -\Delta u\, \delta u\, d\Omega + \cdots \\ \ldots \end{array} \right\} = 0\,. \tag{1.80}$$

And that these are zero sums, are ***invariant***—think of the desert wind—makes the variational principles so valuable.

1.5 Examples

With the following examples we will try to clarify the connection between Green's first identity and the variational principles of mechanics.

1.5.1 The Principle of Virtual Displacements

The **deflection of the cantilever beam** in Fig. 1.11

$$EI\, w^{IV}(x) = 10 \qquad w(0) = w'(0) = 0 \qquad M(l) = V(l) = 0 \tag{1.81}$$

is

$$w(x) = \frac{1}{EI}\left(\frac{10}{24}\,x^4 - \frac{50}{6}\,x^3 + \frac{125}{2}\,x^2\right), \tag{1.82}$$

and the **internal forces** are

$$M(x) = -5\,x^2 + 50\,x - 125 \qquad V(x) = -10\,x + 50\,. \tag{1.83}$$

Green's first identity

$$\mathcal{G}(w, \delta w) = \int_0^l EI\, w^{IV}(x)\, \delta w\, dx + [V\, \delta w - M\, \delta w']_0^l - \int_0^l \frac{M\, \delta M}{EI}\, dx = 0\,, \tag{1.84}$$

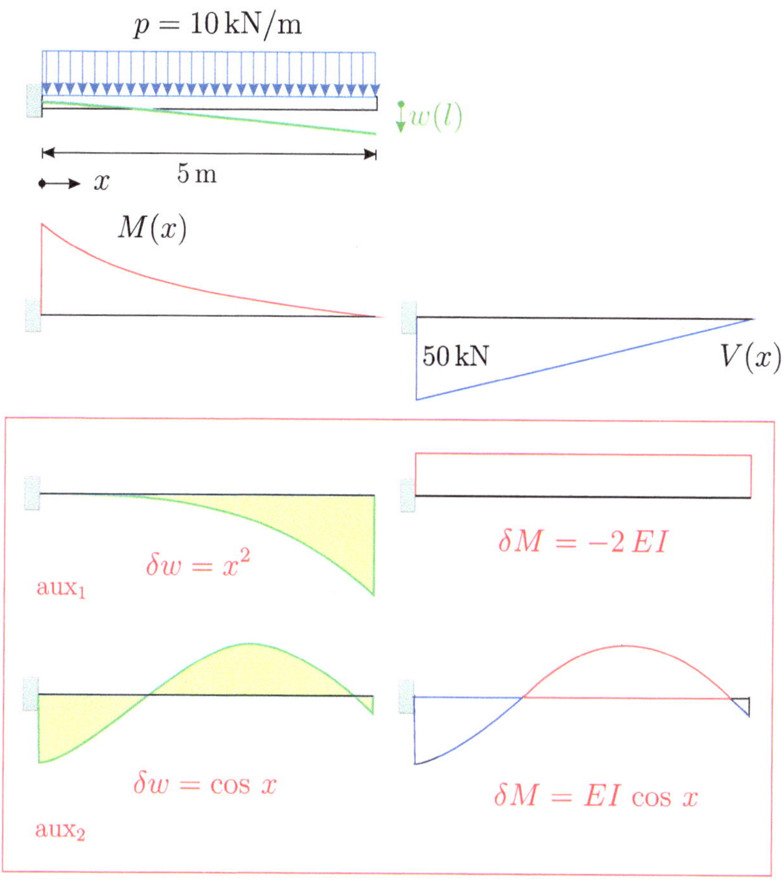

Fig. 1.11 Cantilever beam, aux = auxiliary system

becomes, if we pair w with an **admissible virtual displacement** $\delta w(x)$

$$\delta w(0) = 0 \qquad \delta w'(0) = 0\,, \tag{1.85}$$

the expression

$$\mathscr{G}(w, \delta w) = \int_0^l 10 \cdot \delta w\,dx - \int_0^l \frac{M\,\delta M}{EI}\,dx = 0\,, \tag{1.86}$$

which is the statement $\delta W_e - \delta W_i = 0$.

A **numerical test** with $\delta w(x) = x^2$ confirms this result

$$\mathcal{G}(w, x^2) = \int_0^5 10 \cdot x^2\, dx - \int_0^5 (-5\,x^2 + 50\,x - 125) \cdot (-2)\, dx$$
$$= \frac{1250}{3} - \frac{1250}{3} = \delta W_e - \delta W_i = 0\,. \tag{1.87}$$

But we can substitute for δw also displacements, which, technically, are not admissible as for example the function

$$\delta w(x) = \cos x\,, \tag{1.88}$$

since the fixed left support gets displaced, $\delta w(0) = \cos 0 = 1$. But this does not invalidate the statement $\mathcal{G}(w, \delta w) = 0$. **You just have to count correctly** and add the work done by the shear force $V(0) = 50$ to the sum, and so with

$$\delta M(x) = -EI\, \delta w''(x) = EI\ \cos x \tag{1.89}$$

the result is correct (it is $-50 \cdot 1 = -V(0) \cdot \cos 0$)

$$\mathcal{G}(w, \cos x) = \int_0^5 10 \cos x\, dx - 50 \cdot 1 - \int_0^5 (-5\,x^2 + 50\,x - 125) \cos x\, dx$$
$$= \underbrace{-9.59 - 50}_{\delta W_e} + \underbrace{59.59}_{\delta W_i} = 0\,. \tag{1.90}$$

Also the **rigid body displacements** $\delta w = a + b\,x$ disregard the fixed supports, but the two displacements $\delta w(x) = 1$ and $\delta w(x) = x$ control the **equilibrium** of the beam, namely the sum of the vertical forces and the sum of the moments about the left support

$$\mathcal{G}(w, 1) = \int_0^5 10 \cdot 1\, dx - V(0) \cdot 1 = 50 - 50 = 0 \qquad \delta w = 1\,, \tag{1.91}$$
$$\mathcal{G}(w, x) = \int_0^5 10 \cdot x\, dx - M(0) \cdot 1 = 125 - 125 = 0 \qquad \delta w = x\,. \tag{1.92}$$

$(M(0) \cdot 1 = M(0) \cdot x')$. Only if the deflection w of the beam is orthogonal to these two displacements, is the balance maintained.

1.5.2 Conservation of Energy

It is also easy to prove that the energy is conserved

$$\begin{aligned}\frac{1}{2}\,\mathcal{G}(w,w) &= \frac{1}{2}\int_0^l p(x)\,w(x)\,dx - \frac{1}{2}\int_0^l \frac{M^2}{EI}\,dx = W_e - W_i\\ &= \frac{1}{2}\,\frac{1}{EI}(1562.5 - 1562.5) = 0\,,\end{aligned} \tag{1.93}$$

since the external eigenwork W_e equals the internal energy W_i.

1.5.3 The Principle of Virtual Forces

In this principle the **order** of w and δw **is reversed**, and to indicate this switch we write δw^* instead of δw

$$\begin{aligned}\mathcal{G}(\delta w^*, w) = &\int_0^l EI\,\delta w^{*IV}(x)\,w(x)\,dx + [\delta V^*\,w - \delta M^*\,w']_0^l\\ &- \int_0^l \frac{\delta M^*\,M}{EI}\,dx = 0\,.\end{aligned} \tag{1.94}$$

Mohr's equation is based on this equation.

To calculate the deflection at the end of the cantilever beam in Fig. 1.12a in this way, we place a single force $P^* = 1$ at the end point, and we determine the bending moment δM^* under this load

$$EI\,\delta w^{*IV} = 0 \qquad \delta V^*(l) = 1 \qquad \delta M^*(l) = 0\,. \tag{1.95}$$

Since $w(0) = w'(0) = 0$, Green's first identity becomes

$$\mathcal{G}(\delta w^*, w) = P^* \cdot w(l) - \int_0^l \frac{\delta M^*\,M}{EI}\,dx = 0\,, \tag{1.96}$$

or

$$1 \cdot w(l) = \int_0^l \frac{\delta M^*\,M}{EI}\,dx\,, \tag{1.97}$$

and this is Mohr's equation.

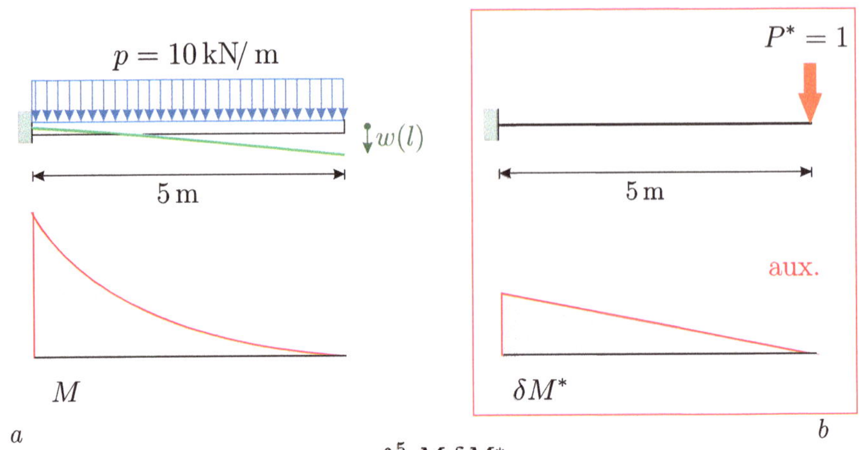

$$w(l) = \int_0^5 \frac{M\,\delta M^*}{EI}\,dx$$

Fig. 1.12 Cantilever beam

1.5.3.1 Principle of Virtual Forces

Let δF^* denote the forces, which we associate with a function δw^*, that is the distributed load $EI\,\delta\, w^{*IV}(x)$, the moments $\delta M^* = -EI\,\delta w^{*\prime\prime}$, and the shear forces $\delta V^* = -EI\,\delta w^{*\prime\prime\prime}$ at the beam ends.

If a system of external forces δF^* is in equilibrium, the external work δW_e^* of these forces on acting through a displacement w

$$\delta W_e^* = \int_0^l EI\,\delta\, w^{*IV}\, w(x)\,dx + [\delta\, V^* w - \delta\, M^* w']_0^l\,, \tag{1.98}$$

equals the virtual internal energy

$$\delta W_i^* = \int_0^l \frac{\delta\, M^* M}{EI}\,dx\,. \tag{1.99}$$

Sometimes the claim is that the forces δF^* must be *infinitesimal small* for this to apply, but there is no objective reason for this, since integration by parts does not distinguish between large and small.

The equilibrium of the virtual forces is guaranteed, because they are the derivatives of δw^* (the "mother lode"), and each function $\delta w^* \in C^4(0, l)$ automatically satisfies the equilibrium conditions

$$\mathcal{G}(\delta w^*, \delta w) = 0 \qquad \delta w = a + b\,x\,. \tag{1.100}$$

It would be different if $EI\,\delta\,w^{*IV}$ and the beam end forces $\delta\,V^*$ and δM^* would not match, if they were picked at random, then $\delta W_e^* - \delta W_i^*$ would probably not be zero.

1.6 Frames

The extension of these identities to frame-like structures as in Fig. 1.13 is easy, since zeros can be added arbitrarily often, $0 + 0 + \cdots + 0 = 0$.

The frame may consist of n frame elements with corresponding longitudinal and bending displacements u_i and w_i, and each element will experience certain virtual displacements δu_i and δw_i. With each pair $\{u_i, \delta u_i\}$ or $\{w_i, \delta w_i\}$ we formulate the associated identity, and we add all these identities

$$0 + 0 + \cdots + 0 = 0\,. \tag{1.101}$$

Then we split the sum into **external** and **internal work**. What is external work remains on the left, and what is internal work is placed on the right

$$\delta W_e = \delta W_i\,. \tag{1.102}$$

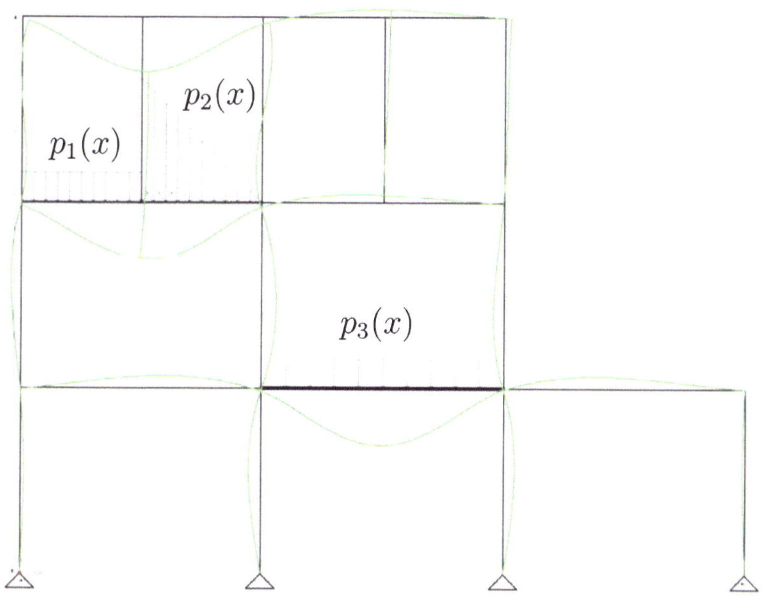

Fig. 1.13 Frame, load and displacement

Often the expression δW_e can be simplified further. The identities of two functions u_i and w_i,

$$\mathcal{G}(u_i, \delta u_i) = 0 \quad \text{(long.)} \qquad \mathcal{G}(w_i, \delta w_i) = 0 \quad \text{(trans.)} \tag{1.103}$$

add an expression such as

$$\int_0^{l_i} p_x\, \delta u_i\, dx + \int_0^{l_i} p_z\, \delta w_i\, dx + \underbrace{[N_i\, \delta u_i]_0^{l_i} + [V_i\, \delta w_i - M_i\, \delta w_i']_0^{l_i}}_{\textit{work at the boundary}} \tag{1.104}$$

to δW_e, which is the virtual external work of the distributed load p_x (longitudinal) and p_z (transverse) in between the nodes, and the work done at the beam ends by the section forces, N_i, V_i and M_i on acting through the virtual displacements.

If a node carries no nodal forces, the end forces of the single frame elements attached to a node, balance and since the virtual displacements of the beam ends are the same on each side of a node, (1.102) reduces to

$$\begin{aligned} \delta W_e &= \sum_i [\int_0^{l_i} p_x\, \delta u_i\, dx + \int_0^{l_i} p_z\, \delta w_i\, dx] \\ &= \sum_i [\int_0^{l_i} \frac{N_i\, \delta N_i}{EA_i}\, dx + \int_0^{l_i} \frac{M_i\, \delta M_i}{EI_i}\, dx] = \delta W_i\,. \end{aligned} \tag{1.105}$$

In the presence of nodal forces, the work done at the beam ends results in contributions like $P \cdot \delta w(x)$ at the nodes, see Sect. 1.8.

The expression (1.105) can be simplified further by omitting the integration limits and by dropping the indices on u_i and w_i and EA_i and EI_i etc., since we know, which part of the frame is meant

$$\delta W_e = \int p_z\, \delta w\, dx + \int p_x\, \delta u\, dx \tag{1.106}$$

$$\int \frac{N\, \delta N}{EA}\, dx + \int \frac{M\, \delta M}{EI}\, dx = \delta W_i\,, \tag{1.107}$$

and so the many identities become at the end the convenient expression

$$\delta W_e = \int p_z\, \delta w\, dx + \int p_x\, \delta u\, dx = \int \frac{N\, \delta N}{EA}\, dx + \int \frac{M\, \delta M}{EI}\, dx = \delta W_i\,. \tag{1.108}$$

1.7 Spring Support

A spring support poses no additional problems, see Fig. 1.14. Given $EI\, w^{IV} = p$ and the boundary conditions $V(l) = -k\, w(l)$ and $w(0) = w'(0) = M(l) = 0$ and $\delta w(0) = \delta w'(0) = 0$, we find

$$\mathcal{G}(w, \delta w) = \int_0^l p\, \delta w\, dx - k\, w(l)\, \delta w(l) - \int_0^l \frac{M\, \delta M}{EI}\, dx = 0\,, \qquad (1.109)$$

which is the **first variation** $\delta \Pi(w, \delta w)$ of the potential energy

$$\Pi(w) = \frac{1}{2}\int_0^l \frac{M^2}{EI}\, dx + \frac{1}{2}\, k\, w^2 - \int_0^l p\, w\, dx\,. \qquad (1.110)$$

1.8 Single Forces and Moments

We have yet to explain how single forces and single moments, which means terms like $P \cdot \delta w(x)$, get into the picture.

These terms are the "residues" of the **square brackets**, the work done on the boundary, because single forces and single moments in between the nodes necessitate a **split** of the deflection curve into two parts $w_L(x)$ and $w_R(x)$, since one cannot simply integrate over a single force. The boundary forms where the two halves meet.

Therefore, we integrate from the left end of the beam to the foot $\bar{x}$ of the force, stop there, and continue the integration behind the force

$$\mathcal{G}(w, \delta w) = \mathcal{G}(w_L, \delta w)_{(0,\bar{x})} + \mathcal{G}(w_R, \delta w)_{(\bar{x},l)} = 0 + 0 = 0\,. \qquad (1.111)$$

The two parts $w_L(x)$ and $w_R(x)$ are homogeneous solutions of the beam equation, since we assume here, for simplicity, that the distributed loads are zero

$$EI\, w_L(x) = 0 \qquad 0 < x < \bar{x} \qquad EI\, w_R(x) = 0 \qquad \bar{x} < x < l\,, \qquad (1.112)$$

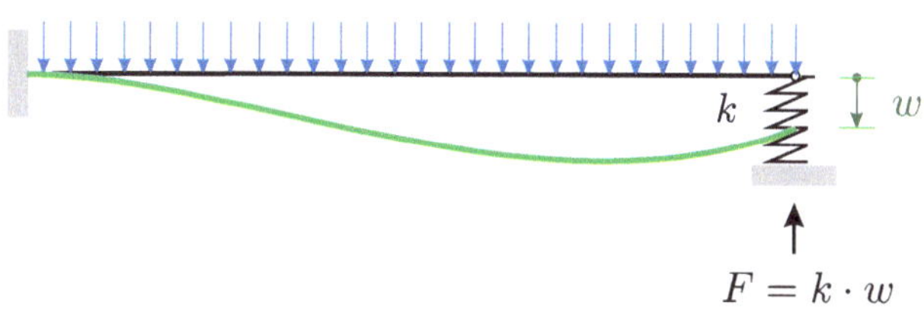

Fig. 1.14 Elastic support at the end of a beam

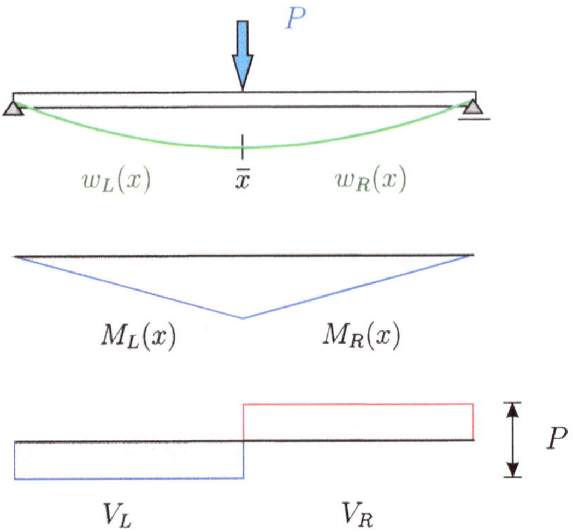

Fig. 1.15 Single forces require a split of the beam

and at the point $\bar{x}$ the two solutions match, except for the two transverse forces V_L and V_R, which differ by the size of the single force, see Fig. 1.15,

$$M_R(\bar{x}) - M_L(\bar{x}) = 0 \qquad V_L(\bar{x}) - V_R(\bar{x}) = P \tag{1.113}$$

$$[V_L\,\delta w - M_L\,\delta w']_0^{\bar{x}} + [V_R\,\delta w - M_R\,\delta w']_{\bar{x}}^{l} = P \cdot \delta w(\bar{x}) \tag{1.114}$$

and so

$$\delta W_e = P \cdot \delta w(\bar{x}) = \int_0^l \frac{M\,\delta M}{EI}\,dx = \delta W_i\,. \tag{1.115}$$

1.9 Support Settlements

Now let us see how we handle the settlement of a support as in Fig. 1.16a. In this situation we split the deflection of the beam

$$EI\,w^{IV} = 0 \qquad w(0) = w'(0) = 0 \qquad M(l) = 0 \quad w(l) = w_\Delta\,, \tag{1.116}$$

into **two functions**, a curve $w_1(x)$ which matches the geometric boundary conditions

$$w_1(0) = w_1'(0) = 0 \qquad w_1(l) = w_\Delta\,, \tag{1.117}$$

and a second curve $w_2(x)$, zero at the supports, which supplies the necessary fit

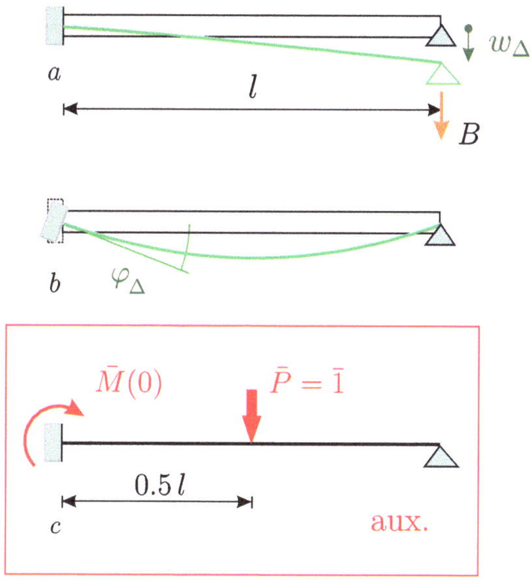

Fig. 1.16 Settlement and rotation of supports

$$EI\, w_2^{IV}(x) = -EI\, w_1^{IV}(x) \quad w_2(0) = w_2'(0) = w_2(l) = 0 \quad M_1(l) + M_2(l) = 0\,, \tag{1.118}$$

so that the sum $w(x) = w_1(x) + w_2(x)$ satisfies (1.116).

With finite elements we let $w(x) = w_\Delta \cdot \varphi_3(x)$, and we place the column $\boldsymbol{f}_3$ of $\boldsymbol{K}$ on the right side, $\boldsymbol{K}\,\boldsymbol{w} = -w_\Delta\,\boldsymbol{f}_3$, and we determine with these $n-1$ equations (row 3 is dropped) the remaining nodal values w_i.

Conservation of Energy

The law of conservation of energy also applies, of course, to such load cases. When we superpose w with itself

$$\begin{aligned} \mathcal{G}(w, w) &= \int_0^l EI\, w^{IV}\, w\, dx + [V\, w - M\, w']_0^l - \int_0^l \frac{M^2}{EI}\, dx \\ &= V(l) \cdot w_\Delta - \int_0^l \frac{M^2}{EI}\, dx = 0\,, \end{aligned} \tag{1.119}$$

and multiply the identity with 1/2

$$\frac{1}{2}\, \mathcal{G}(w, w) = \frac{1}{2}\, V(l) \cdot w_\Delta - \frac{1}{2} \int_0^l \frac{M^2}{EI}\, dx = 0\,, \tag{1.120}$$

we see that the energy is conserved, $W_e - W_i = 0$.

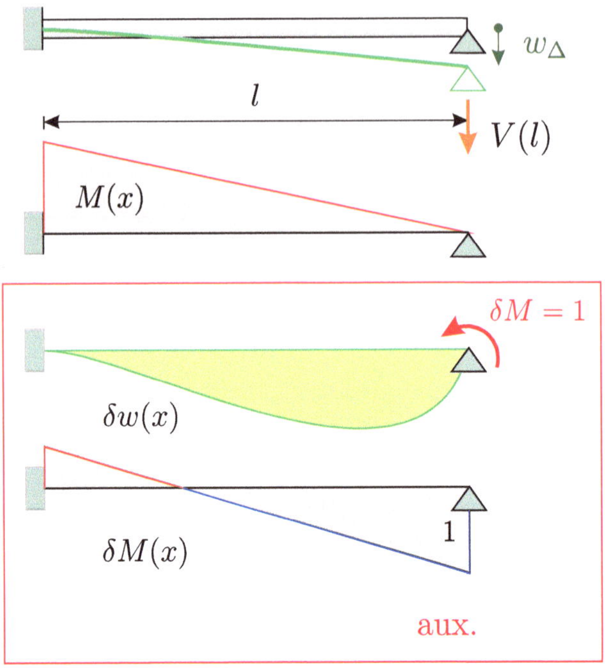

Fig. 1.17 Virtual displacement, $\delta W_e = \delta W_i = 0$

Principle of Virtual Displacements

The "trick" with the principle of virtual displacements is that the admissible virtual displacements δw of the beam in Fig. 1.17 know nothing of the settlement of the support

$$\delta w(0) = \delta w'(0) = \delta w(l) = 0\,, \tag{1.121}$$

and so, since the distributed load is zero, $EI\, w^{IV} = 0$, no work is done, $\delta W_e = 0$, and therefore the strain energy δW_i must be zero as well

$$\mathscr{G}(w, \delta w) = -\int_0^l \frac{M \delta M}{EI}\, dx = -\delta W_i = 0\,. \tag{1.122}$$

To understand the zero: Suppose we calculate with Mohr's equation the deflection $\delta w(l) = 0$ by applying $V(l)$ as "virtual" force (dummy load) at the free end of the beam (free because the *Reduction principle*, see Sect. 1.36, allows to remove the support), the result would be this equation, $\delta w(l) = 0$.

Principle of Virtual Forces

Now, we swap the places of w and $\delta w \to \delta w^*$, and we formulate the *principle of virtual forces*

$$\mathscr{G}(\delta w^*, w) = \delta W_e^* - \delta W_i^* = 0\,. \tag{1.123}$$

For an application we calculate the deflection in the middle of the beam in Fig. 1.16, by applying a single force $P^* = 1$ in the direction of $w(x)$, and we formulate with the two parts

$$w^* = w_L^* + w_R^* \tag{1.124}$$

and w Green's first identity

$$\begin{aligned}\mathscr{G}(w_L^*, w)_{(0,0.5\,l)} + \mathscr{G}(w_R^*, w)_{(0.5\,l,l)}\\ = 1 \cdot w(0.5\,l) + V^*(l) \cdot w_\Delta - \int_0^l \frac{M^* M}{EI}\,dx = 0\,,\end{aligned} \tag{1.125}$$

which gives

$$w(0.5\,l) = \int_0^l \frac{M^* M}{EI}\,dx - V^*(l) \cdot w_\Delta\,. \tag{1.126}$$

1.10 Springs

In matrix notation the coupling between the displacements and nodal forces of a spring, see Fig. 1.18, reads

$$\begin{bmatrix} k & -k \\ -k & k \end{bmatrix} \begin{bmatrix} u_1 \\ u_2 \end{bmatrix} = \begin{bmatrix} f_1 \\ f_2 \end{bmatrix}, \tag{1.127}$$

or $\boldsymbol{K}\boldsymbol{u} = \boldsymbol{f}$ for short. To this matrix $\boldsymbol{K}$ belongs the identity

$$\mathscr{G}(\boldsymbol{u}, \boldsymbol{\delta u}) = \boldsymbol{\delta u}^T \boldsymbol{K}\boldsymbol{u} - \boldsymbol{u}^T \boldsymbol{K}\,\boldsymbol{\delta u} = 0\,. \tag{1.128}$$

Fig. 1.18 Coil spring

If $\boldsymbol{u}$ is the equilibrium position of the spring, $\boldsymbol{K}\,\boldsymbol{u} = \boldsymbol{f}$, the identity formulates the ***principle of virtual displacements***

$$\mathscr{G}(\boldsymbol{u}, \delta\boldsymbol{u}) = \delta\boldsymbol{u}^T \boldsymbol{f} - \boldsymbol{u}^T \boldsymbol{K}\,\delta\boldsymbol{u} = \delta W_e - \delta W_i = 0 \tag{1.129}$$

and

$$\mathscr{G}(\delta\boldsymbol{u}^*, \boldsymbol{u}) = \boldsymbol{u}^T \boldsymbol{f}^* - \boldsymbol{u}^{*T} \boldsymbol{K}\,\boldsymbol{u} = \delta W_e^* - \delta W_i^* = 0 \tag{1.130}$$

is the ***principle of virtual forces***.

1.11 Temperature

In linear statics we can superpose results, and this allows to split off temperature effects as a separate load case

$$_w(x) = w_{LF\,1} + w_{LF\,2} + \cdots + w_T\,. \tag{1.131}$$

We can further assume w_T to be calculated on the statically determinate structure,

$$w_T(x) = \alpha_T \frac{\Delta T}{h}\, x^2 + a\,x + b\,, \qquad (a, b \text{ are constants}) \tag{1.132}$$

since any necessary corrections can be treated separately.

The ***principle of virtual forces*** for a deflection w_T, we place a point load $P^* = 1$ at the source point x, reads

$$\mathscr{G}(\delta w^*, w_T) = 1 \cdot w_T(x) - \int_0^l EI\,\delta w^{*\prime\prime}\, w_T^{\prime\prime}\, dx = 0\,, \tag{1.133}$$

or with $w_T^{\prime\prime} = \alpha_T\,\Delta T/h$

$$w_T(x) = \int_0^l \delta M^*\, \alpha_T \frac{\Delta T}{h}\, dx\,. \tag{1.134}$$

where $\alpha_T \sim 10^{-5}$ (steel, concrete) is the **temperature coefficient** of the material, ΔT is the temperature difference between the top and bottom fibers of the beam and h is the beam height, see Fig. 1.19.

In the same way, the temperature displacement in longitudinal direction is

$$u_T(x) = \int_0^l \delta N^*\, \alpha_T\, T\, dx\,, \tag{1.135}$$

where T is the increase in temperature.

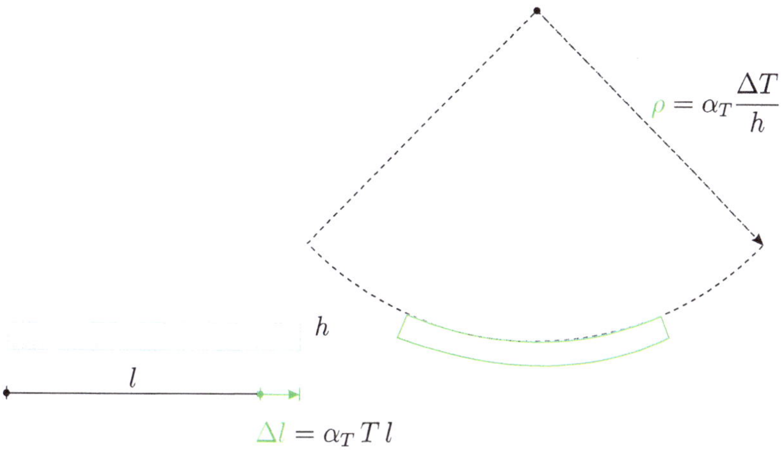

Fig. 1.19 Temperature effects

1.12 The Complete Equation

We can now assemble all these single parts in one formula, ***Mohr's equation***,

$$
\begin{aligned}
1 \cdot \delta = &\int \frac{M^* M}{EI}\, dx + \int \frac{N^* N}{EA}\, dx + \int M^* \alpha_T \frac{\Delta T}{h}\, dx + \int N^* \alpha_T\, T\, dx \\
&+ \underbrace{\sum_i \frac{F_i^* F_i}{k_i}}_{\text{force springs}} + \underbrace{\sum_j \frac{M_j^* M_j}{k_{\varphi j}}}_{\text{moment springs}} \\
&- \underbrace{\sum_k F_k^* w_{\Delta k}}_{\text{sup. settlements}} - \underbrace{\sum_l M_l^* \tan \varphi_{\Delta l}}_{\text{sup. rotations}} . \qquad (1.136)
\end{aligned}
$$

The δ on the left side stands for a displacement or a rotation.

If it is a rotation, it is the slope w', because in Green's first identity the moment is paired with the slope $w' = \tan \varphi$

$$\cdots + [V\, w - M\, w'] + \cdots \qquad (1.137)$$

and not with the rotation φ. This is consistent with the requirement that the left side of Mohr's equation must equate to an expression of work

$$M^* \cdot \delta = 1\,\text{Nm} \cdot \tan \varphi = [\text{Nm}] \cdot [\ ] = \int_0^l \ldots \qquad (1.138)$$

1.13 Shortcuts

It would certainly be too tedious a task to diligently add the identities of the individual members to formulate

$$\delta W_e = \delta W_i \,. \tag{1.139}$$

No engineer does it this way. With a little practice the engineer automatically knows what counts as virtual external work δW_e, namely the work of the distributed loads

$$\int_0^l p_z \, \delta w \, dx \qquad \int_0^l p_z \, \delta u \, dx \,, \tag{1.140}$$

and of the point loads

$$P_z \, \delta w(x) \qquad P_x \, \delta u(x) \qquad M \, \delta w'(x) \qquad \text{etc.} \tag{1.141}$$

and the contributions to δW_i are also easy to guess

$$\begin{aligned} \delta W_i = & \int \frac{M \, \delta M}{EI} \, dx + \int \frac{N \, \delta N}{EA} \, dx + \int \delta M \, \alpha_T \, \frac{\Delta T}{h} \, dx + \int \delta N \, \alpha_T \, T \, dx \\ & + \sum_i \frac{\delta F_i \, F_i}{k_i} + \sum_j \frac{\delta M_j \, M_j}{k_{\varphi j}} - \sum_k \delta F_k \, w_{\Delta k} - \sum_l \delta M_l \, \tan \varphi_{\Delta l} \,, \end{aligned} \tag{1.142}$$

and they usually reduce to

$$\delta W_i = \int_0^l \frac{M \, \delta M}{EI} \, dx + \int_0^l \frac{N \, \delta N}{EA} \, dx \tag{1.143}$$

or often even simpler to

$$\delta W_i = \int_0^l \frac{M \, \delta M}{EI} \, dx \,. \tag{1.144}$$

1.14 Duality

The scalar product of a force vector and a displacement vector, $\delta W_e = \boldsymbol{f}^T \, \boldsymbol{\delta u}$ becomes an L_2-scalar product when both are functions

$$\int_0^l EI \, w^{IV}(x) \, \delta w(x) \, dx = \text{force} \times \text{displacement} \,, \tag{1.145}$$

and integration by parts leads to the energy principles—it is a logical step.

Force and displacement are the two poles around which statics revolves. Work is the central quantity and the basic operation is the scalar product.

The skill in dealing with Green's identities consists in choosing the proper **auxiliary function** which returns the information—either a force or a displacement—we are looking for.

Principle of virtual displacements	$\mathcal{G}(w, \delta w) = 0$	if forces
Principle of virtual forces	$\mathcal{G}(\delta w^*, w) = 0$	if displacements
Betti's theorem	$\mathcal{B}(w_1, w_2) = 0$	both .

In the *principle of virtual displacements* the statement is[3]

$$\mathcal{G}(w, \delta w) = \text{real forces} \times \text{virtual disp.} - a(w, \delta w) = 0 \tag{1.146}$$

and in the *principle of virtual forces*

$$\mathcal{G}(\delta w^*, w) = \text{virtual forces} \times \text{real disp.} - a(\delta w^*, w) = 0\,. \tag{1.147}$$

To determine the support reaction R_A of the beam in Fig. 1.20, we choose for δw a rotation of the beam about its right support, $\delta w(x) = 1 - x/l$,

$$\mathcal{G}(w, \delta w) = \int_0^l p(x)\,\delta w(x)\,dx - V(0)\,\delta w(0) = 0\,, \tag{1.148}$$

and solve for $R_A = V(0)$

$$R_A \cdot 1 = \int_0^l p \cdot (1 - \frac{x}{l})\,dx\,. \tag{1.149}$$

Note, in the case of a slanted beam, a rafter as in Fig. 1.21, only the part of $\delta w(x)$, which points in the direction of the traveling load counts.

To calculate section forces $M(x)$ or $V(x)$ with the principle of virtual displacements we install a corresponding hinge at the source point and spread it by one unit apart, see Fig. 1.20. Technically, though, this is the same as if we would apply *Betti's theorem*

$$\mathcal{B}(w_1, w_2) = W_{1,2} - W_{2,1} = W_{1,2} = 0 \qquad (W_{2,1} = 0)\,, \tag{1.150}$$

---

[3]The functions δw and δw^* are of course just as real as w. They are simply cleverly chosen auxiliary functions.

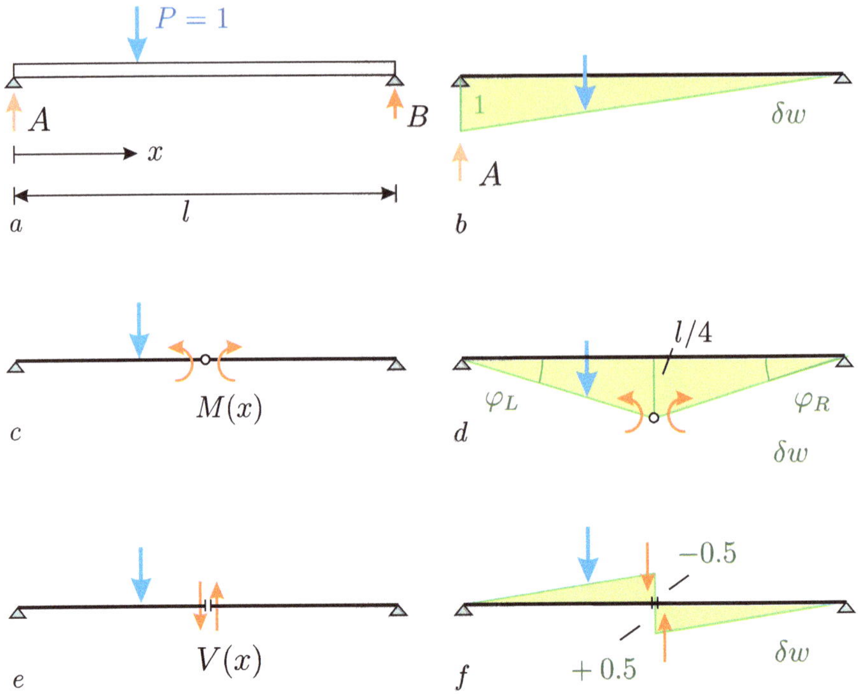

Fig. 1.20 Influence functions are based on *Betti's theorem*, $W_{1,2} = W_{2,1}$

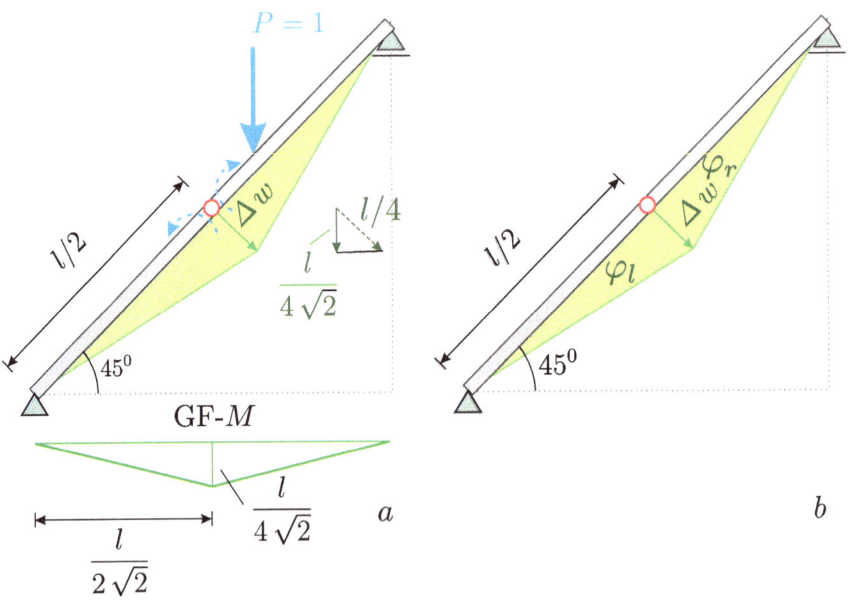

Fig. 1.21 The influence function for the bending moment M in the rafter corresponds to a unit-spread, $\tan\varphi_l + \tan\varphi_r = 1$, of the added hinge

where w_2 is the rigid-body motion, the rotation of the beam about its support or the spread of the shear hinge, and w_1 is the deflection curve of the beam. The equations are the same.

1.15 Mohr Versus Betti

The deflection $w(x)$ of a beam can either be calculated with the ***principle of virtual forces*** (M^* is the moment of G_0)

$$w(x) = \int_0^l \frac{M(y)\, M^*(y,x)}{EI}\, dy \qquad \mathscr{G}(G_0, w) = 0\,, \tag{1.151}$$

or with ***Betti's theorem***

$$w(x) = \int_0^l G_0(y,x)\, p(y)\, dy \qquad \mathscr{B}(G_0, w) = 0\,. \tag{1.152}$$

Engineers prefer the first formula since it is easier to determine the bending moment distribution $M^* = -EI\, G_0''$ of the point load P^* than the deflection $G_0(y,x)$, see Fig. 1.22.

On the other hand, we eventually must do more. Even if the load P happens to act only at one point, the product of M and M^* must be integrated over all frame elements, see Fig. 1.23, while Betti requires only the evaluation at one point.

With regard to finite elements one might suspect that the *principle of virtual forces* provides more accurate results, since on average the local errors in M_h and M_h^* (the FE-solutions) should cancel, while Betti must get the value $G_0(y,x)$ at one point, the foot y of the single force, right, but *Mohr and Betti are two sides of the same coin!* Via integration by parts we can always switch from one formula to the other.

1.16 Weak and Strong Influence Functions

We call

$$w(x) = \int_0^l \frac{M(y)\, \bar{M}(y,x)}{EI}\, dy \tag{1.153}$$

a **weak influence function** and

$$w(x) = \int_0^l G_0(y,x)\, p(y)\, dy \tag{1.154}$$

a **strong influence function**. Both equations have the same result $w(x)$.

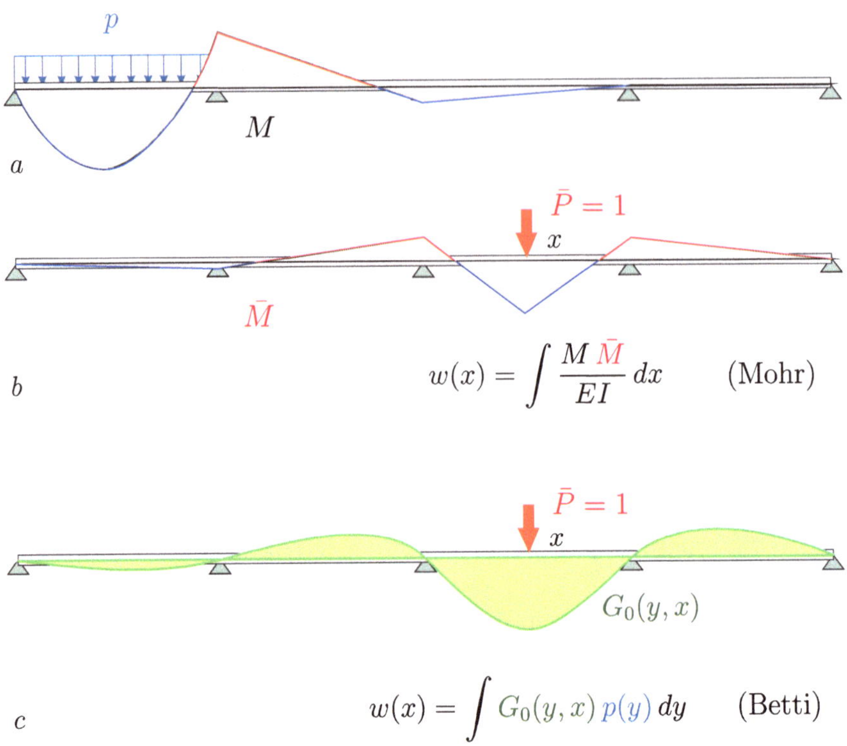

Fig. 1.22 Mohr and *Betti's theorem*, it is easier to determine $\bar{M}$ than to find the deflection curve $G_0(y, x)$

Weak influence functions are based on the *principle of virtual forces*, i.e. Green's first identity

$$\mathscr{G}(G_0, w) = 0\,, \tag{1.155}$$

where the Green's function $G_0(y, x)$ is the δw^* and strong influence functions are based on *Betti's theorem*, Green's second identity,

$$\mathscr{B}(G_0, w) = 0\,. \tag{1.156}$$

The standard example of a weak influence function is Mohr's Eq. (1.153).

Weak influence functions can only calculate displacements but not forces, which seems obvious, since: Which virtual force would you apply to compute the shear force $V(x)$ in a beam?

To see what goes wrong, we try to compute the shear force in the middle of the beam in Fig. 1.24 with Mohr's equation

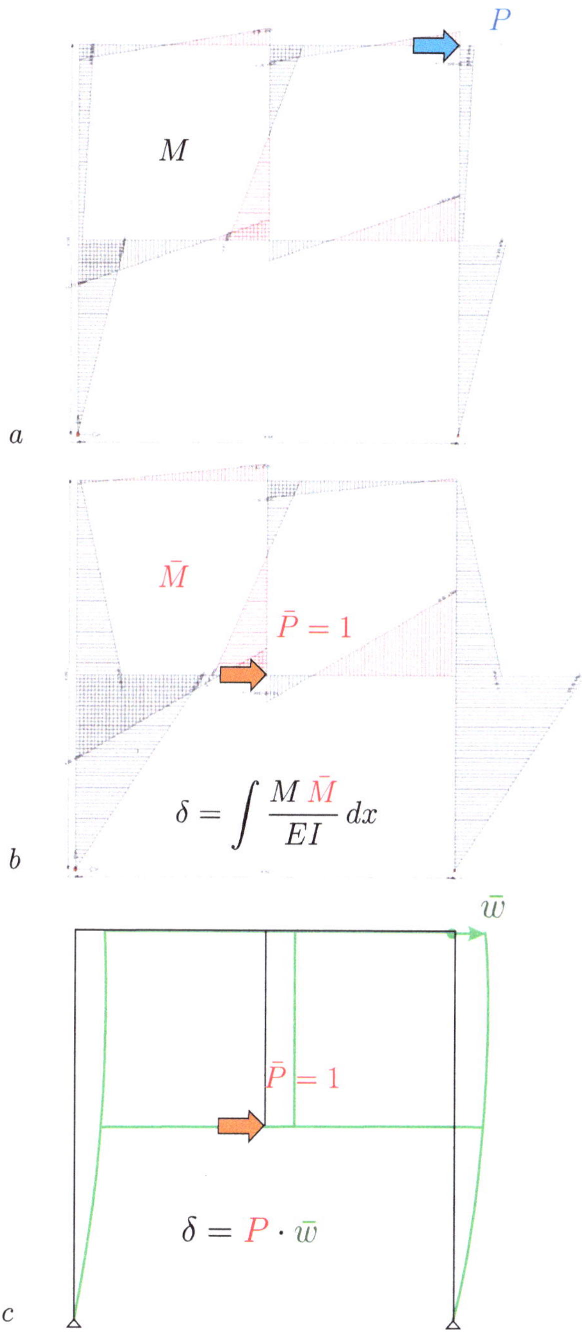

Fig. 1.23 Calculating the horizontal displacement of a node with Mohr and with *Betti's theorem*, **a** load moment, and **b** moment from $\bar{P} = 1$, **c** displacements due to $\bar{P} = 1$ (=influence function for the horizontal displacement)

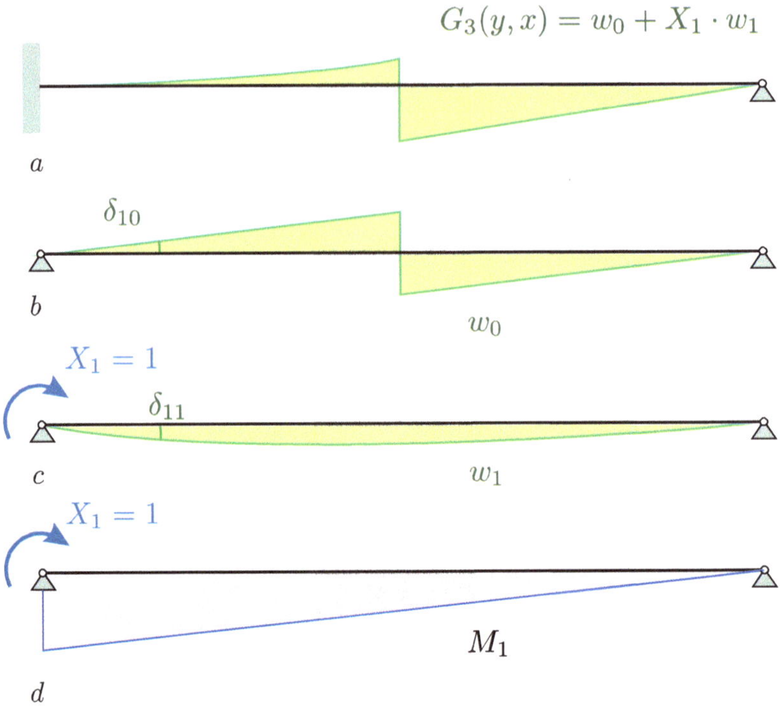

Fig. 1.24 Trying to calculate $V(x)$ with Mohr, **a** influence function for $V(x)$, **b** for $V(x)$ in the statically determinate beam, **c** auxiliary moment X_1, **d** moment of X_1

$$V(x) \stackrel{?}{=} \int_0^l \frac{M(y)\, M^*(y,x)}{EI}\, dy\,, \tag{1.157}$$

where M^* would be the bending moment of the influence function $G_3(y,x)$ for $V(x)$.

This influence function $G_3(y,x) = w_0 + X_1 \cdot w_1$ is a sum of two functions, the scissor-like motion w_0 in Fig. 1.24b, and an auxiliary function $X_1 \cdot w_1$, see Fig. 1.24c, which makes that the slope is zero at the fixed edge. So, M^* has the form

$$M^* = X_1 \cdot M_1 + 0\,, \tag{1.158}$$

where the zero is the moment of the piecewise linear shearing motion w_0 (the 0 at the source point is somewhat problematic[4]). But the scalar product of M and M^* is zero and not $V(x)$,

$$V(x) \stackrel{?}{=} \int_0^l \frac{M\, M^*}{EI}\, dy = X_1 \int_0^l \frac{M\, M_1}{EI}\, dy = 0\,. \tag{1.159}$$

[4]Setting $\bar{M} = X_1 \cdot M_1 + \delta_1$ would save the situation, since (we let $EI = 1$) $(M, \bar{M}) = X_1 \cdot (M, M_1) + (M, \delta_1) = X_1 \cdot 0 + V(x) = V(x)$, see Sect. 9.22.

In the force method the zero result for the scalar product of M and M_1, the moment of the redundant $X_1 = 1$, is a check that the rotation of the beam at the left clamped edge is zero—as it must be. So, the result is not really a surprise.

Remark 1.1 To give a graphical description of the difference between the two type of influence functions imagine the load to be a single point load P. In *Betti's theorem*

$$w(x) = \int_0^l G_0(y, x)\, p(y)\, dy = G_0(y, x) \cdot P \tag{1.160}$$

we—figuratively speaking—let a stone (a Dirac delta) drop into the water at the source point x and we watch how the wave $G_0(y, x)$ spreads over the beam, by how much the point load is buoyed.

In the *principle of virtual forces (Mohr)*

$$w(x) = \int_0^l \frac{M(y)\, M^*(y, x)}{EI}\, dy \tag{1.161}$$

instead, we study the interaction of two waves. The first wave, $M(y)$, is the bending moment of the point load and the second wave $M^*(y, x)$ is the bending moment of the Dirac delta. Only the segments where both moments are large (and not orthogonal) deserve out attention.

1.17 The Canonical Boundary Values

Green's first identity consists of a string of scalar products of conjugated terms, and in particular are the terms in the brackets

$$[N\, u]_0^l = N(l)\, u(l) - N(0)\, u(0) = f_2\, u_2 + f_1\, u_1 \tag{1.162}$$

$$\begin{aligned}[V\, w - M\, w']_0^l &= V(l)\, w(l) - M(l)\, w'(l) - V(0)\, w(0) + M(0)\, w'(0) \\ &= f_3\, u_3 + f_4\, u_4 + f_1\, u_1 + f_2\, u_2\,,\end{aligned} \tag{1.163}$$

the ***canonical boundary values***, see Fig. 1.25. The stiffness matrix of a bar

$$\frac{EA}{l}\begin{bmatrix} 1 & -1 \\ -1 & 1 \end{bmatrix}\begin{bmatrix} u_1 \\ u_2 \end{bmatrix} = \begin{bmatrix} f_1 \\ f_2 \end{bmatrix} \tag{1.164}$$

and a beam

$$\frac{EI}{l^3}\begin{bmatrix} 12 & -6l & -12 & -6l \\ -6l & 4l^2 & 6l & 2l^2 \\ -12 & 6l & 12 & 6l \\ -6l & 2l^2 & 6l & 4l^2 \end{bmatrix}\begin{bmatrix} u_1 \\ u_2 \\ u_3 \\ u_4 \end{bmatrix} = \begin{bmatrix} f_1 \\ f_2 \\ f_3 \\ f_4 \end{bmatrix} \tag{1.165}$$

Fig. 1.25 The canonical boundary values of a bar and a beam

formulate ***coupling conditions*** $\boldsymbol{K}\,\boldsymbol{u} = \boldsymbol{f}$ between these values, more precisely, between the boundary values of homogeneous solutions, $-EAu'' = 0$ and $EI\,w^{IV} = 0$ respectively.

Distributed loads, $EIw^{IV} = p_z$ and $-EAu'' = p_x$, add a further vector $\boldsymbol{d}$

$$\boldsymbol{K}\,\boldsymbol{u} = \boldsymbol{f} + \boldsymbol{d}\,, \tag{1.166}$$

whose entries are the equivalent nodal forces of the load

$$d_i = \int_0^l p_x(x)\,\varphi_i^e(x)\,dx \qquad i = 1, 2 \qquad \text{(bar)} \tag{1.167}$$

$$d_i = \int_0^l p_z(x)\,\varphi_i^e(x)\,dx \qquad i = 1, 2, 3, 4 \qquad \text{(beam)} \tag{1.168}$$

where the $\varphi_i^e(x)$ are the unit displacements, see Fig. 3.7.

The FEM only operators with one vector $\boldsymbol{f} \equiv \boldsymbol{f} + \boldsymbol{d}$, the forces f_i which act directly at the nodes plus the equivalent nodal forces d_i of the distributed load.

Each of the n equations of the system $\boldsymbol{K}\boldsymbol{u} = \boldsymbol{f} + \boldsymbol{d}$ formulates Green's first identity. To see this, we split the displacement $u(x) = u_h(x) + u_p(x)$ of a bar into a **homogeneous solution**

$$u_h(x) = u_1\,\varphi_1(x) + u_2\,\varphi_2(x) \tag{1.169}$$

and a **particular solution** $u_p(x)$

$$-EA\,u_p''(x) = p(x) \qquad u_p(0) = u_p(l) = 0\,, \tag{1.170}$$

and so

$$\mathscr{G}(u_h + u_p, \varphi_i^e) = \int_0^l p\, \varphi_i^e\, dx + [N\, \varphi_i^e]_0^l - a(u_h + u_p, \varphi_i^e)$$

$$= d_i + f_i - a(u_h, \varphi_i^e) - \underbrace{a(u_p, \varphi_i^e)}_{=0} = d_i + f_i - \sum_{j=1}^{2} k_{ij}\, u_j = 0\,, \tag{1.171}$$

since

$$\mathscr{G}(\varphi_i^e, u_p) = \int_0^l 0 \cdot u_p\, dx - a(\varphi_i^e, u_p) = a(\varphi_i^e, u_p) = 0\,, \tag{1.172}$$

because the boundary values of the particular solution u_p are zero.

Some rules apply to the system $\boldsymbol{K}\boldsymbol{u} = \boldsymbol{f} + \boldsymbol{d}$. In each row either u_i or $f_i + d_i$ can be specified but not both at the same time. You cannot pull on a bar with a force of 10 kN and simultaneously require the longitudinal displacement $u(l)$ to be 1 cm. So, each equation must contain one and only one unknown, as for example in the case of a hinged one-element beam

$$\frac{EI}{l^3}\begin{bmatrix} 12 & -6l & -12 & -6l \\ -6l & 4l^2 & 6l & 2l^2 \\ -12 & 6l & 12 & 6l \\ -6l & 2l^2 & 6l & 4l^2 \end{bmatrix}\begin{bmatrix} 0 \\ u_2\,? \\ 0 \\ u_4\,? \end{bmatrix} = \begin{bmatrix} f_1\,? \\ 0 \\ f_3\,? \\ 0 \end{bmatrix} + \begin{bmatrix} d_1 \\ d_2 \\ d_3 \\ d_4 \end{bmatrix}, \tag{1.173}$$

where the end rotations u_2, u_4 and the support reactions f_1, f_3 are the unknowns.

1.18 The Reduction of the Dimension

In the slope deflection method the *degree of kinematic indeterminacy* is the number of unknown nodal displacements in the equation

$$\boldsymbol{K}\boldsymbol{u} = \boldsymbol{f} + \boldsymbol{d}\,, \tag{1.174}$$

and once these have been determined we say the structure is **kinematically determinate**.

In frame analysis it is obviously enough to know the nodal displacements, the displacements on “the boundary”, see Fig. 1.26.

Finitely many nodal values determine infinitely many numbers, the displacements $u(x)$, $w(x)$, $w'(x)$ and section forces $N(x)$, $M(x)$, $V(x)$ at the infinitely many points of the frame.

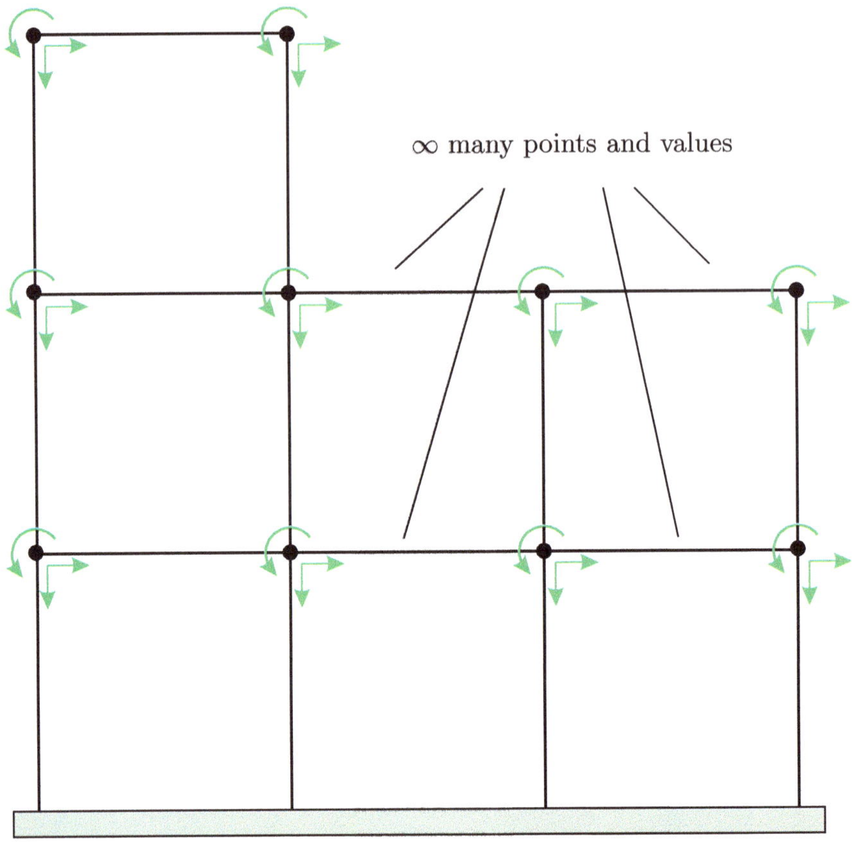

Fig. 1.26 The nodes are the "boundary" of a frame. The nodal displacements (+the distributed load) determine the section forces N, M and V in between the nodes

But this is a reduction by one. The $n = 1$-dimensional frame members shrink to a $n - 1 = 0$-dimensional set of points, the nodes. This reduction makes the slope deflection method possible, it is enough to deal with the nodes!

The slope deflection method is a genuine boundary element method.

All linear, self-adjoint differential equations allow such a **reduction of the dimension** of a problem by one, $n \to (n - 1)$. The practical value of this reduction cannot be overestimated.

Given the displacements and forces on the boundary, the displacements and forces in the interior can be calculated with influence functions.

To calculate the stresses inside a dam ($n = 3$), see Fig. 1.27, only the displacements and stresses on the surface of the dam ($n = 2$) are needed. To analyze a plate ($n = 2$)

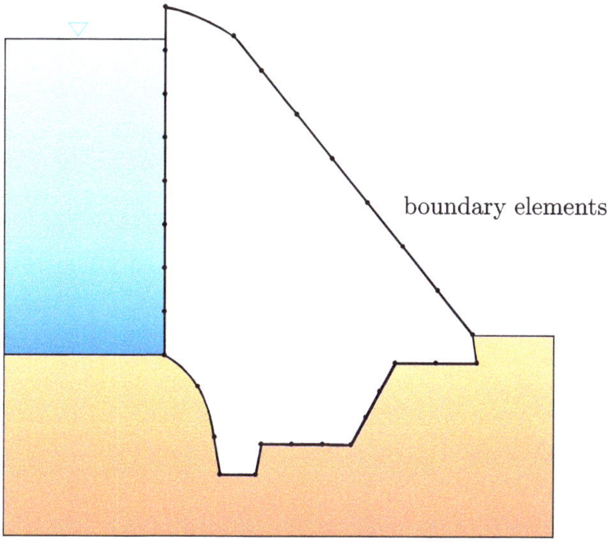

Fig. 1.27 Water dam, on the airside and on the waterside the traction vector $\boldsymbol{t} = \boldsymbol{S}\boldsymbol{n}$ is known and in the rock the displacement vector $\boldsymbol{u} = \boldsymbol{0}$. The missing values, the traction vector $\boldsymbol{t}$ in the rock and the displacement vector $\boldsymbol{u}$ of the upper part, can be determined by solving an integral equation on the surface of the dam. With these boundary values any value in the interior can be calculated

we only must know the displacements and forces on the edge ($n = 1$) and in the case of a beam ($n = 1$) only the nodal values.

The simplest application of this idea is the **ruler**. A straight line (the solution of the differential equation $u'' = 0$) is uniquely determined by its two boundary values, and therefore you can draw the line when you hold the ruler to the endpoints. *The ruler is the universal influence function of the straight lines.*

Exterior problems are often solved by such methods. Just by discretizing the surface of a motor block one can calculate the noise—the sound pressure—at 3 m, 30 m or 300 m distance.

1.19 Boundary Element Method

The boundary element method is the application of these ideas to 2-D and 3-D problems. The BEM is a mixture of the slope deflection method and influence functions. The edge of a plate is divided into boundary elements, see Figs. 1.28 and 1.29, which allow to model the edge displacements and edge forces with piecewise polynomials, and the unknown nodal values are determined by solving a linear system of equations. Influence functions allow then to compute the displacements and internal forces inside the plate.

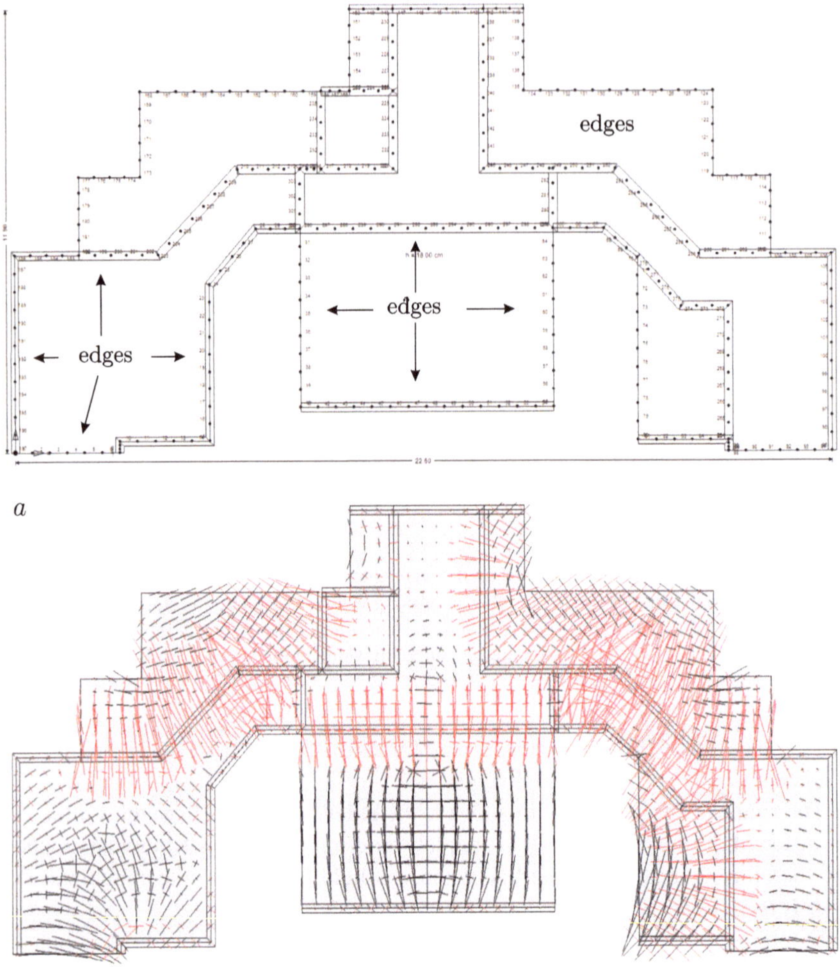

Fig. 1.28 Plate, **a** the nodes on the boundary, and **b** the principal moments under gravity load

And what may come as a surprise: Also the finite element method is in a sense a "boundary element method" since the kernel $G_h(\boldsymbol{y}, \boldsymbol{x})$ in the influence function[5]

$$u_h(\boldsymbol{x}) = \int_{\Omega} G_h(\boldsymbol{y}, \boldsymbol{x})\, p(\boldsymbol{y})\, d\Omega_{\boldsymbol{y}} \tag{1.175}$$

is itself based on an influence function

[5] $d\Omega_{\boldsymbol{y}}$ and $ds_{\boldsymbol{y}}$ means we integrate over $\boldsymbol{y}$.

a

b

c

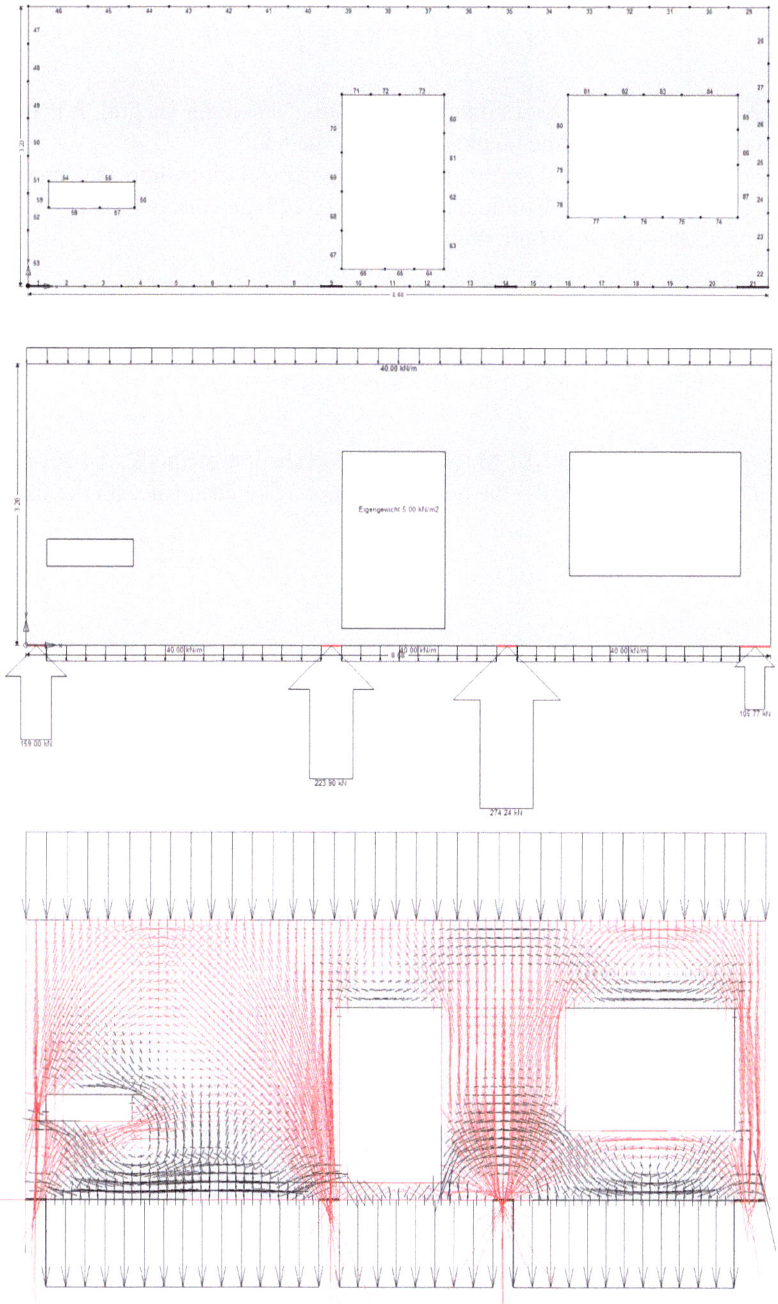

Fig. 1.29 Plate, **a** nodes on the boundary, **b** support reactions, **c** principal stresses

$$G_h(\boldsymbol{y}, \boldsymbol{x}) = \int_\Gamma \ldots ds_{\boldsymbol{y}} + \int_\Omega \ldots \, d\Omega_{\boldsymbol{y}} \,, \tag{1.176}$$

which consists of a string of boundary integrals + a domain integral. It is the same influence function as in the boundary element method.

The boundary integrals propagate the edge singularities into the interior, see Sect. 6.9, make that the FE-influence function (1.175) and therewith the FE-solution loses accuracy—**in the whole domain**!

1.20 Finite Elements and Boundary Elements

Both methods, FEM and BEM, are influence function methods. An FE-program calculates—quite classical—the displacement of a stretched bar with the formula

$$u_h(x) = \int_0^l G_h(y, x)\, p(y)\, dy \,, \tag{1.177}$$

only that the influence function $G_h(y, x)$ is an approximation, see Fig. 1.30b, since the program cannot produce the necessary kink in $G(y, x)$ at the single force (at least, if the source point x lies between the nodes).

The boundary element method does it the same way, but it uses a so-called ***fundamental solution*** $g(y, x)$. This is a function, which has the correct kink at the source point, but which is not zero at the end points, it violates the boundary conditions.

The consequence is that, when we formulate *Betti's theorem*,

$$\lim_{\varepsilon \to 0} \mathscr{B}(g, u)_{\Omega_\varepsilon} = 0 \,, \tag{1.178}$$

also the normal forces $N(0)$ and $N(l)$ at the end of the bar, see Fig. 1.30a, get displaced and contribute to the Dirac energy, **the influence function gets "longer"**

$$1 \cdot u(x) = \underbrace{\int_0^l g(y, x)\, p(y)\, dy + N(l)\, g(l, x) - N(0)\, g(0, x)}_{Dirac\ energy} \tag{1.179}$$

compared to the original formulation with $G(y, x)$

$$1 \cdot u(x) = \underbrace{\int_0^l G(y, x)\, p(y)\, dy}_{Dirac\ energy} \,. \tag{1.180}$$

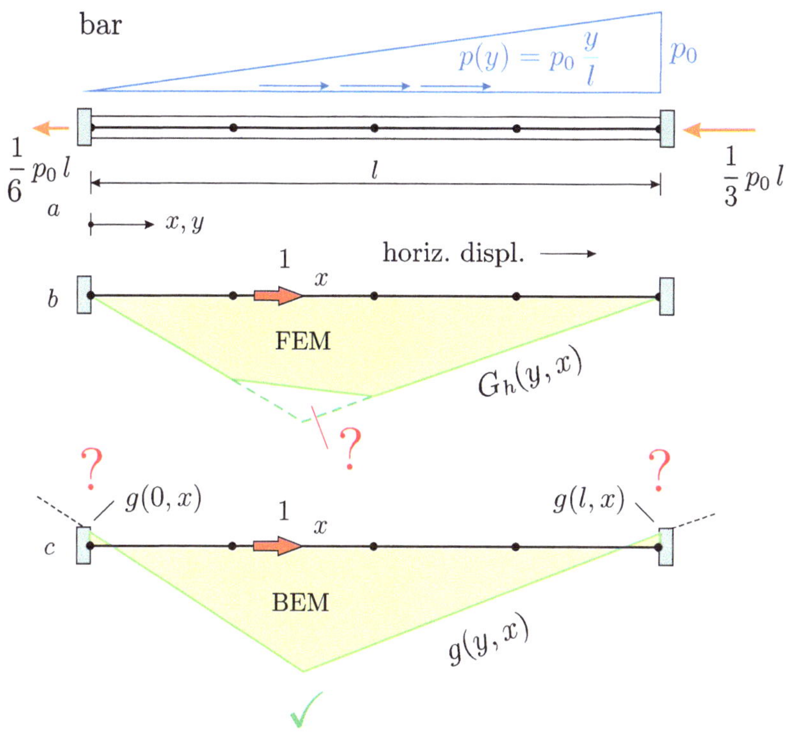

Fig. 1.30 Stretched bar, **a** system and load, **b** approximate influence function (no kink), **c** fundamental solution with the correct kink, but the boundary values are not zero; note, the horizontal displacements are plotted downward

Note, both formulas provide the same value for the Dirac energy. But the $g(y, x)$ approach must also count the work at the supports, the contribution of the domain load, the integral (p, g) alone is "too little". This is the difference.

Fundamental solutions are like ***passe-partouts*** which fit everywhere. With one fundamental solution we can analyze all plates. In BE programs the fundamental solutions are "hardwired". *One solution suffices to rule them all.*

The downside is that also the edge forces and edge displacements contribute to the ***Dirac energy***, and so these boundary values, if unknown, must be determined by solving a linear system. Technically, this poses no problem. It is just that in 2-D and 3-D these auxiliary problems can only be solved approximately, and so the *Dirac energy* is only an approximation, as in the finite element method.

Operating with fundamental solutions has the further advantage that an approximation of the influence functions for the internal forces, as with finite elements, is not necessary. BE-programs use **exact influence functions**, only the marks on the edge to which we hold—so to speak—the curved rulers, the influence functions, are slightly displaced.

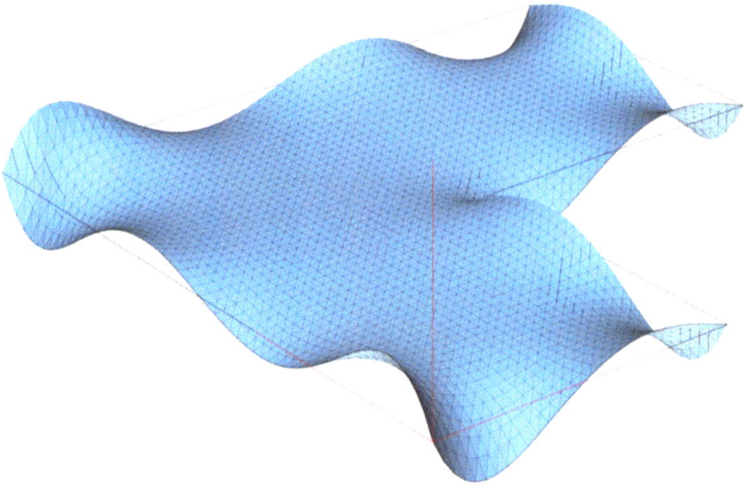

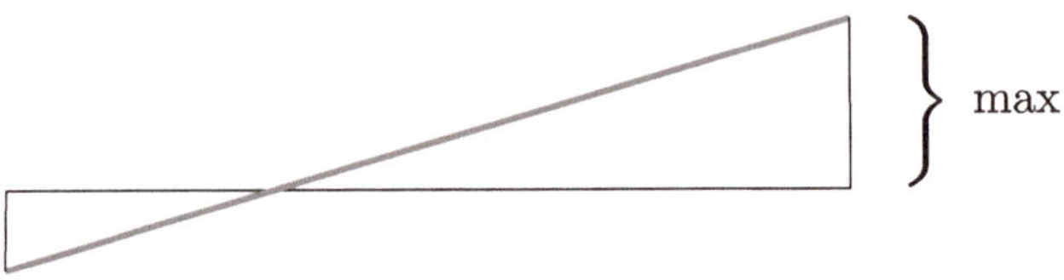

Fig. 1.31 The maximum deflection occurs on the edge. The same is true for a straight line, $u'' = 0$

We do not want to let the subject of boundary elements pass by, without mentioning the ***maximum principle***, since every homogeneous solution of the Laplace equation, $\Delta u = 0$, attains its maximum on the edge, see Fig. 1.31. In 1-D these are the straight lines, $u'' = 0$.

Because $EI\, w^{IV} = p$ is identical to $-M'' = p$, in any zero-load span, $p = 0$, the maximum moment occurs at the nodes.

1.21 Test Functions

The energy principles are based on Green's first identity

$$\mathcal{G}(w, \delta w) = 0\,, \tag{1.181}$$

and the notation $\mathcal{G}(w, \delta w)$ is suitable to make clear that there is nothing special about a virtual displacement δw. Mathematically, it is simply a function like w.

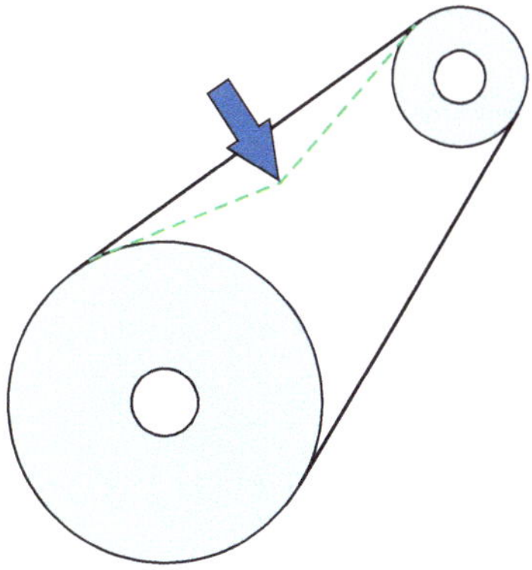

Fig. 1.32 Test of a transmission belt

Unfortunately, however, the concept of virtual displacements is historically so burdened that one is tempted to replace it with a more harmless term like ***test function***. After all, testing, wiggling, deflecting from the equilibrium position is an everyday maneuver.

To determine the weight of a suitcase, we lift it up. Since *force* = *mass* × *acceleration* we conclude from the acceleration a and the force in the arm, the mass m of the suitcase.

To determine the tension in a transmission-belt or the pressure in a football, we press our thumb against it, see Fig. 1.32. Likewise, the draftsman projects the structure onto three orthogonal planes, and the drafts are the scalar product between the structure and the two unit vectors $\boldsymbol{e}_i, \boldsymbol{e}_j$ of each of the three planes. From the more sides we look at something, the more we learn about the object. In a figurative sense it means, the more often we wiggle an object, the more tests φ_i we run, the clearer the picture becomes.

1.22 Do Virtual Displacements Have to Be Small?

No, virtual displacements need not be small, see Fig. 1.33. The equation

$$\delta W_e = \int_0^l p\, \delta w\, dx = \int_0^l \frac{M\, \delta M}{EI}\, dx = \delta W_i , \tag{1.182}$$

is not true because $\delta W_e = \delta W_i$ is true. Attaching labels to an equation proves nothing.

Sure, the variational principles form the core of mechanics and at no point are we as close to mechanics as in formulations like $\delta W_e = \delta W_i$. It is also true that the static interpretation of an equation is the best check. But this tempts engineers to prove mathematics with mechanics, which is not possible.

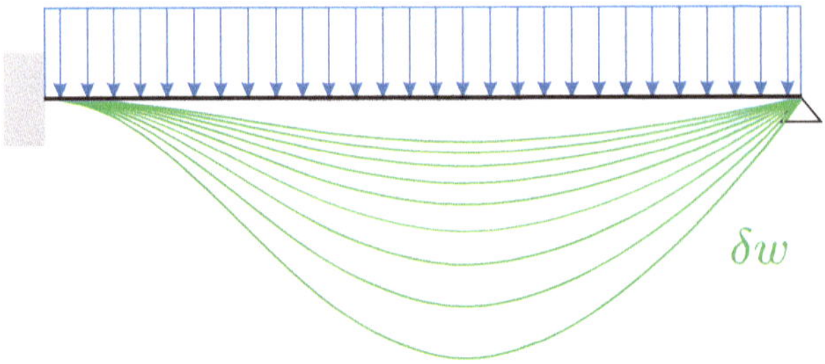

$$\int_0^l \frac{M\delta M}{EI}\,dx = \int_0^l p\,\delta w\,dx$$

Fig. 1.33 Whatever the size of δw, the formula $\delta W_i = \delta W_e$ holds true

Not to be misunderstood: We consider the energy principles of mechanics an extraordinarily successful concept. From a didactic point of view there is no better approach to statics—generations of engineers have successfully learned statics in this way.

It is not our intention to transform statics into a rigorous system of axioms and theorems. Such an attempt would do more harm than good. Statics cannot and should not be treated as in a mathematical textbook. We only believe that at one point we should explain to students why the central Eq. (1.182) is correct. It is true because

- $w \in C^4(0, l)$ (as we assume) is a solution to the boundary value problem

$$EI\,w^{IV} = p \qquad w(0) = w(l) = M(0) = M(l) = 0\,, \tag{1.183}$$

- $\delta w \in C^2(0, l)$ (as we suppose) is an admissible virtual displacement, $\delta w(0) = \delta w(l) = 0$
- and integration by parts (twice) of the scalar product $(EI\,w^{IV}, \delta w)$ results in

$$\mathscr{G}(w, \delta w) = \underbrace{\int_0^l p\,\delta w\,dx}_{\delta W_e} - \underbrace{\int_0^l \frac{M\,\delta M}{EI}\,dx}_{\delta W_i} = 0\,. \tag{1.184}$$

Since integration by parts makes no difference between "large" and "small", δw can have any size. Which mechanical principle can guarantee that two numbers

$$0.56789\ldots = 0.56789\ldots \tag{1.185}$$

agree in all digits? Or if we turn the argument on its head, which *mathematical law* prohibits δw to be large? Has a mathematician ever proved an equation by referring to a law of nature?

To reduce the beam curvature to $\kappa \simeq w''$ when $w'' \ll 1$ is a legitimate argument to linearize the beam equation, but it is a surprise when an engineer claims—as has happened to us multiple times—that Eq. (1.182) is only correct if δw is small.[6]

1.23 Only When in Equilibrium?

Since the identities are based on integration by parts the expression

$$\mathscr{G}(w, \delta w) = 0 \tag{1.186}$$

is guaranteed to be zero, while in the literature the claim is: *A system must be an equilibrium state for the energy principles to apply*. Why this restriction?

To prove, that the deflection of a hinged beam

$$EI\, w^{IV}(x) = p(x) \qquad w(0) = w(l) = 0 \qquad M(0) = M(l) = 0 \tag{1.187}$$

satisfies the *principle of virtual displacements*, we substitute w and an admissible virtual displacement δw, with zero boundary values,

$$\delta w(0) = \delta w(l) = 0\,, \tag{1.188}$$

into Green's first identity

$$\mathscr{G}(w, \delta w) = \int_0^l EI\, w^{IV}(x)\, \delta w\, dx + [V\, \delta w - M\, \delta w']_0^l - \int_0^l \frac{M\, \delta M}{EI}\, dx = 0\,, \tag{1.189}$$

and obtain the well known result

$$\mathscr{G}(w, \delta w) = \int_0^l p(x)\, \delta w(x)\, dx - \int_0^l \frac{M\, \delta M}{EI}\, dx = \delta W_e - \delta W_i = 0\,. \tag{1.190}$$

In the literature instead, Green's first identity (1.189) is never mentioned explicitly, the authors start directly with the shortened identity (1.190), but then the equilibrium conditions (1.187) and the boundary conditions (1.188) become **necessary side conditions**.

[6]In such conversations you wish mathematicians were present, they would listen, leave their *splendid isolation* and understand what engineering mathematics is. *"Epsilontic"* is easy, the hard part is the interpretation, the application! Many misunderstandings are based on this *twin nature* of engineering mathematics.

1.24 What Counts as Displacement and What as Force?

When we integrate the integral

$$\int_0^l EI\, w^{IV}(x)\, w(x)\, dx = -[-EI\, w''' \cdot w - (-EI\, w'') \cdot w']_0^l + a(w, w) \tag{1.191}$$

by parts (twice) the inner structure of the differential equation unfolds before our eyes. We discover the strain energy product

$$a(w, w) = \int_0^l EI\, (w'')^2\, dx = \int_0^l \frac{M^2}{EI}\, dx\,, \tag{1.192}$$

and we see how the displacements and forces are defined, since the square bracket is the work done at the ends of the beam

$$[V\, w - M\, w']_0^l = V(l)\, w(l) - M(l)\, w'(l) - V(0)\, w(0) + M(0)\, w'(0)\,, \tag{1.193}$$

and so the expression $V = -EI\, w'''$ must be the shear force, since it is paired with the deflection w, and $M = -EI\, w''$ must be the moment since it is paired with the slope w'. This seems clear, because we have learnt it this way, but here is the place where it is made official.

Integration by parts of the **second-order beam equation** results in

$$\int_0^l (EI\, w^{IV}(x) + P\, w''(x))\, w(x)\, dx = -[\underbrace{(-EI\, w'''(x) - P\, w'(x))}_{T(x)}\, w - M\, w']_0^l$$
$$+ \int_0^l (\frac{M^2}{EI} - P\, (w'(x))^2)\, dx\,, \tag{1.194}$$

and so the force which is adjoint to $w(x)$ is now

$$T(x) = -EI\, w'''(x) - P\, w'(x) = V(x) - P\, w'(x)\,, \tag{1.195}$$

but $M(x)$ is the same as in first-order theory, since it remains adjoint to w'.

1.25 The Number of Force and Displacement Terms

The second-order differential equation of a bar, $-EA\, u'' = p$, evolves from a displacement and a force

$$u(x) \quad \text{0-th derivative} \qquad N(x) = EA\, u'(x) \quad \text{1st derivative}\,, \tag{1.196}$$

and in a rope, $-H\,w''(x)$,

$$w(x) \quad \text{0-th derivative} \qquad V(x) = H\,w'(x) \quad \text{1st derivative}\,, \tag{1.197}$$

and a shear beam, $-GA\,w_s''(x)$,

$$w(x) \quad \text{0-th derivative} \qquad V(x) = GA\,w'(x) \quad \text{1st derivative}\,, \tag{1.198}$$

the situation is the same, while the fourth-order beam equation splits into $2 + 2$ such terms

$$w(x),\ w'(x) \qquad \text{0-th and 1st derivative} \tag{1.199}$$

$$M(x) = -EI\,w''(x),\ V(x) = -EI\,w'''(x) \quad \text{2nd and 3rd derivative}\,. \tag{1.200}$$

The lower order derivatives are the displacements and the higher order derivatives are the forces. The distinction is not entirely unimportant, since influence functions for displacements can be formulated with the *principle of virtual forces* and *Betti's theorem*, while influence functions for force terms require *Betti's theorem*.

1.26 Why the Minus in $-H\,w'' = p$?

Why do all second-order differential equations start with a minus sign?

A mathematician will point to the trigonometric functions and will explain: If the load $p(x)$ on a rope has a wave-like shape, $p(x) = \sin(x)$, the deflection $w(x)$ of the rope follows this rhythm, $w(x) = 1/H\ \sin(x)$, see Fig. 1.34, but since the second derivative of $\sin(x)$ is negative, a minus sign up front must correct this

$$-H\,w''(x) = -H\,(-\frac{1}{H}\,\sin(x)) = \sin(x)\,. \tag{1.201}$$

The engineer instead would say, the minus sign comes from the equilibrium condition $-V + V + dV + p\,dx = 0$, see Fig. 9.8. This is also true, but what do the trigonometric functions know about equilibrium?

1.27 The Virtual Internal Energy

Green's first identity includes a symmetric integral, the virtual internal energy,

$$a(w, \delta w) = \int_0^l \frac{M\,\delta M}{EI}\,dx\,, \tag{1.202}$$

which we call the ***strain energy product*** of w and δw.

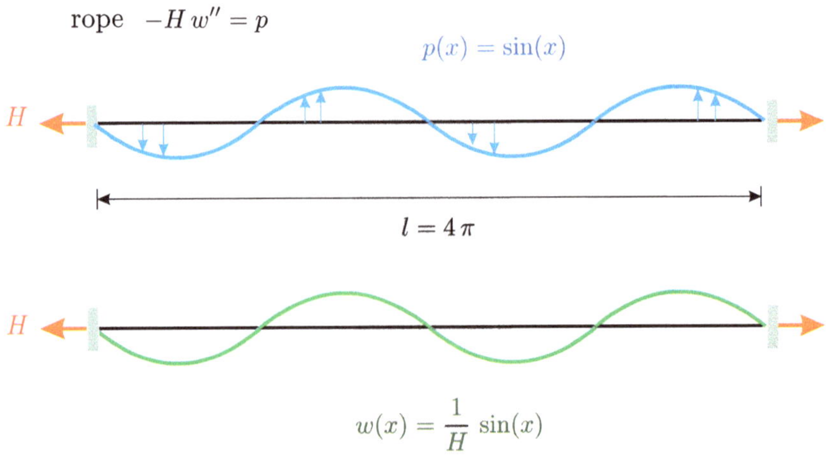

Fig. 1.34 Oscillating load on a rope and response of the rope

On the diagonal, $\delta w = w$, the strain energy product is the internal energy

$$\frac{1}{2}\,a(w, w) = \frac{1}{2}\int_0^l \frac{M^2}{EI}\,dx\,, \tag{1.203}$$

and if $w(x)$ and $\delta w(x)$ are compound functions, let c_i and d_i be any real numbers,

$$w(x) = c_1\,w_1(x) + c_2\,w_2(x) \qquad \delta w(x) = d_1\,\delta w_1(x) + d_2\,\delta w(x)\,, \tag{1.204}$$

the sum can be split into the single contributions

$$\begin{aligned} a(w, \delta w) = c_1\,d_1\,a(w_1, \delta w_1) + c_1\,d_2\,a(w_1, \delta w_2) \\ + c_2\,d_1\,a(w_2, \delta w_1) + c_2\,d_2\,a(w_2, \delta w_2)\,, \end{aligned} \tag{1.205}$$

which is why $a(w, \delta w)$ is called a ***symmetric bilinear form***.

1.28 Castigliano's Theorem(s)

Castigliano's Theorem is pure algebra: It is based on the fact that the strain energy product is a bilinear form. The derivative of the strain energy of an FE-solution

$$\frac{1}{2}\,a(u_h, u_h) = \frac{1}{2}\boldsymbol{u}^T\boldsymbol{K}\boldsymbol{u} = \frac{1}{2}\sum_{i,j} k_{ij}u_i u_j \tag{1.206}$$

with respect to a nodal value u_l (chain rule and $k_{lj} = k_{jl}$)

$$\frac{\partial}{\partial u_l}\frac{1}{2}a(u_h, u_h) = \frac{1}{2}\sum_j k_{lj}u_j + \frac{1}{2}\sum_j k_{jl}u_j = \sum_j k_{lj}u_j = f_l \tag{1.207}$$

is the nodal force f_l.

Castigliano's first theorem *The derivative of the strain energy in a structure with respect to any particular displacement u_l is equal to the force f_l applied at that point in the direction of this displacement (if the strain energy is finite).*

We can also express the strain energy as

$$\frac{1}{2}a(u_h, u_h) = \frac{1}{2}\boldsymbol{u}^T\boldsymbol{K}\boldsymbol{u} = \frac{1}{2}\boldsymbol{u}^T\boldsymbol{f} = \frac{1}{2}\boldsymbol{f}^T\boldsymbol{K}^{-1}\boldsymbol{f} \tag{1.208}$$

and so

$$\frac{\partial}{\partial f_l}\frac{1}{2}a(u_h, u_h) = \sum_j k_{lj}^{(-1)}f_j = u_l\,. \tag{1.209}$$

Castigliano's second theorem *The derivative of the strain energy in a structure with respect to any particular force f_l is equal to the displacement u_l in the direction of this force (if the strain energy is finite).*

If we allow also point solutions

$$\boldsymbol{u} = \sum_i \boldsymbol{G}(\boldsymbol{y}_i, \boldsymbol{x})\boldsymbol{f}_i \tag{1.210}$$

the strain energy of one or all displacement fields $\boldsymbol{G}_i$ might be infinite, $a(\boldsymbol{G}_i, \boldsymbol{G}_i) = \infty$, see Fig. 1.35, and then the derivative

$$\frac{\partial}{\partial u_l}a(\boldsymbol{u}, \boldsymbol{u}) \tag{1.211}$$

does not exist. This happens in elasticity theory, where point solutions have infinite energy, see Sect. 1.34. So, we must add a warning to Castigliano's Theorem—***true if the strain energy is finite***.

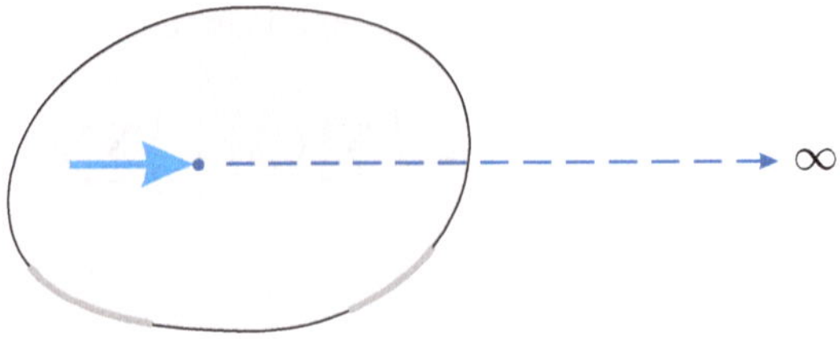

Fig. 1.35 Castigliano's theorem, *and the force was never seen again...*

1.29 Equilibrium

The functions δw with the property $a(w, \delta w) = 0$ *for all* w implicitly define the equilibrium conditions. In a beam these are the straight lines $\delta w = a\,x + b$ since, $\delta M = -EI\,\delta w'' = 0$, so that Green's identity reduces to

$$\mathcal{G}(w, \delta w) = \int_0^l p\,\delta w\,dx + [V\,\delta w - M\,\delta w']_0^l = 0\,. \tag{1.212}$$

The simplest function of this type is the translation $\delta w = 1$, which establishes that the sum of the vertical forces is zero

$$\mathcal{G}(w, 1) = \int_0^l p\,dx + V(l) - V(0) = 0\,, \tag{1.213}$$

and the function $\delta w = x$, a pseudorotation about the left end of the beam, implies that the moment about the left end is zero as well

$$\mathcal{G}(w, x) = \int_0^l p \cdot x\,dx + V(l) \cdot l - V(0) \cdot 0 - M(l) \cdot 1 + M(0) \cdot 1 = 0\,. \tag{1.214}$$

We call $\delta w = x$ a ***pseudorotation***, since the points of the beam move on the tangents to the circles, see Fig. 1.36,

$$\tan\varphi = \frac{y}{x}\,. \tag{1.215}$$

Only such pseudorotations leave the beam stress free, $\delta M = 0$, and thus $a(w, \delta w) = 0$. All influence functions of statically determinate structures are piecewise rigid-body motions (displ. of mechanisms) and the rotations are pseudorotations. *True rotations would give incorrect results.*

In linear mechanics virtual displacements δw consist of translations plus linearized rotations (following the tangent). This is what we mean when we speak of infinitesimal displacements. If this is agreed upon we have no problem with the **infinitesimal**

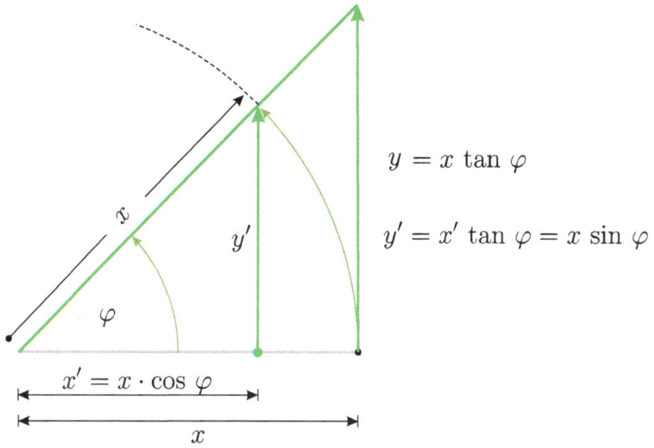

Fig. 1.36 True vertical displacement y' of a point x under a rotation φ and engineering displacement y (pseudorotation)

small since we can make the virtual displacements—as in mathematics—arbitrarily large.

infinitesimal $\equiv$ linearized rigid body rotations

May we recall the definition of the **moment of a force about a point**? The line of action of the force is congruent with the tangent to the turning circle and in the first semester we learned that we can shift the force on its line of action *any distance* without changing the moment. There are—fortunately(!)—no "infinitesimal" moments.

It is this concept, which requires forces to move on the tangent. Tangent and moment are complementary terms.

Beam on an Elastic Foundation

In the case of a beam on an elastic foundation (c)

$$EI\, w^{IV}(x) + c\, w(x) = p(x)\,, \tag{1.216}$$

the situation is different. To this differential equation belongs the identity

$$\begin{aligned}\mathcal{G}(w, \delta w) &= \int_0^l p(x)\, \delta w\, dx + [V\, \delta w - M\, \delta w']_0^l \\ &\quad - \int_0^l (\frac{M\, \delta M}{EI} + c\, w(x)\, \delta w(x))\, dx = 0\,,\end{aligned} \tag{1.217}$$

and suddenly there is no function δw, other than $\delta w = 0$, which reduces the virtual internal energy

$$a(w, \delta w) = \int_0^l (\frac{M\,\delta M}{EI} + c\,w(x)\,\delta w(x))\,dx \qquad (1.218)$$

to zero. Are there no equilibrium conditions? The terms in Green's first identity must be orthogonal to all functions $\delta w(x) \in C^2$, this we know. So, if we choose the function $\delta w(x) = 1$, it follows

$$\mathcal{G}(w, 1) = \int_0^l p(x)\,dx + V(l) - V(0) - \int_0^l c\,w(x)\,dx = 0\,, \qquad (1.219)$$

or when we substitute for $p(x)$ the expression $p(x) = EI\,w^{IV}(x) + c\,w(x)$ this results in

$$\mathcal{G}(w, 1) = \int_0^l EI\,w^{IV}\,dx + V(l) - V(0) = 0\,, \qquad (1.220)$$

which looks familiar. On the same track the pseudorotation $\delta w(x) = x$ formulates the condition $M = 0$ about the left end of the beam

$$\mathcal{G}(w, x) = \int_0^l EI\,w^{IV}\,x\,dx + V(l)\,x - M(l)\cdot 1 + M(0)\cdot 1 = 0\,. \qquad (1.221)$$

But note, only the part $EI\,w^{IV}$ of the total load $p(x) = EI\,w^{IV}(x) + c\,w(x)$ appears in these equations.

When we place a single force $P = V(l)$ at the end of such a beam, then must hold

$$\int_0^l EI\,w^{IV}(x)\,dx + P = 0\,. \qquad (1.222)$$

Because of $EI\,w^{IV}(x) + c\,w(x) = 0$ (no distributed load), we can interchange $EI\,w^{IV}(x)$ with $-c\,w(x)$, and therefore we must have as well

$$P = \int_0^l c\,w(x)\,dx\,, \qquad (1.223)$$

and so the soil pressure balances the vertical force P.

1.30 The Mathematics Behind the Equilibrium Conditions

The equilibrium conditions are based on the ***fundamental theorem of calculus***

$$\int_0^l f'(x)\,dx = f(l) - f(0)\,. \tag{1.224}$$

Since $EI\,w^{IV}(x) = -V'(x) = p(x)$ equilibrium is automatically established

$$\int_0^l -V'(x)\,dx = -V(l) + V(0) \quad \Rightarrow \quad \int_0^l p(x)\,dx + V(l) - V(0) = 0\,. \tag{1.225}$$

To verify that the moment about the left end vanishes, $M = 0$, we calculate the moment of the distributed load $p(x) = -V'(x)$ and apply integration by parts

$$\int_0^l -V'(x)\,x\,dx = [-V\,x]_0^l - \int_0^l -V(x)\cdot 1\,dx\,. \tag{1.226}$$

Since $V(x) = M'(x)$ this is the same as

$$\int_0^l -V'(x)\,x\,dx = [-V\,x]_0^l + M(l) - M(0)\,, \tag{1.227}$$

or

$$\int_0^l p(x)\,x\,dx + V(l)\cdot l - M(l) + M(0) = 0\,. \tag{1.228}$$

1.31 Balance and Second-Order Theory

In first-order theory we formulate the equilibrium conditions with the undeformed structure and in second-order theory with the deformed structure—this is how we have learned it. But this is, we think, not true: Second-order theory is actually **a mixture of both**, see Fig. 1.37.

The contribution of the horizontal force H to the equilibrium about the fixed end enters the equation, but the shortening of the lever arm (the beam gets compressed) of the vertical force P does not.

It is therefore not possible to check the equilibrium of a frame, since the nodal displacements contain contributions from first- and second-order theory. That this goes unnoticed is, because the axial deformations of the members are usually exceedingly small, and so one is inclined to attribute deviations to rounding errors.

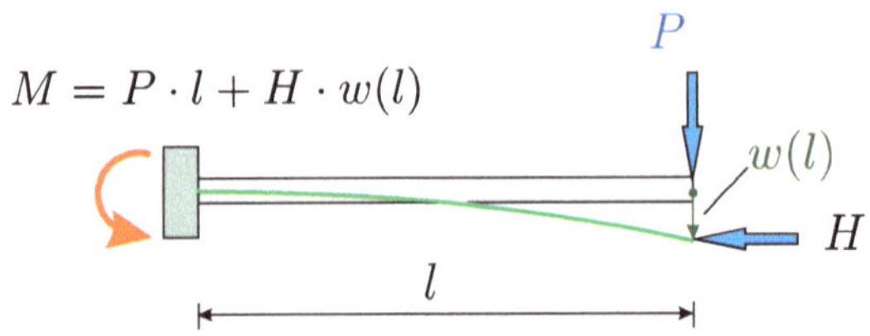

Fig. 1.37 Second-order theory

1.32 Sources and Sinks

From a physical point of view an equilibrium condition formulates a **conservation principle**. The support reactions v_n (Kirchhoff shear) at the edge of a plate Ω, the flow out so to speak, must match the load p[7]

$$\mathcal{G}(w, 1) = \int_\Omega p\, d\Omega + \int_\Gamma v_n\, ds = 0\,. \tag{1.229}$$

Or, the temperature distribution $T(\boldsymbol{x})$ in a floor plate, which is kept warm by a floor heating p (=sources), while the edge is kept at zero Celsius, satisfies the equations

$$-\Delta T = p \qquad T = 0 \qquad \text{on } \Gamma\,, \tag{1.230}$$

and so the identity

$$\mathcal{G}(T, \delta T) = \int_\Omega -\Delta T\, \delta T\; d\Omega + \int_\Gamma \frac{\partial T}{\partial n}\, \delta T\, ds - \int_\Omega \nabla T \bullet \nabla \delta T\, d\Omega = 0\,, \tag{1.231}$$

implies ***flow in = flow out***,

$$\mathcal{G}(T, 1) = \int_\Omega p\, d\Omega + \int_\Gamma \frac{\partial T}{\partial n}\, ds = 0\,. \tag{1.232}$$

For the same reason the normal forces at the two ends of a truss element must balance

$$\mathcal{G}(u, 1) = [N \cdot 1]_0^l = N(l) - N(0) = 0\,, \tag{1.233}$$

since there are no sources in the interior, $-EA\, u'' = 0$.

[7] We neglect the corner forces.

1.33 The Principle of Minimum Potential Energy

In a spring,

$$k\,u = f\,, \tag{1.234}$$

the displacement u is proportional to the force f, see Fig. 1.38, and the eigenwork of f

$$W_e = \frac{1}{2}\,f\,u\,, \tag{1.235}$$

is stored as internal energy in the spring

$$W_i = \frac{1}{2}\,k\,u^2\,, \tag{1.236}$$

and we expect these two energies to be the same

$$W_e = \frac{1}{2}\,f\,u = \frac{1}{2}\,k\,u^2 = W_i\,. \tag{1.237}$$

But this is guaranteed by the identity

$$\mathscr{G}(u, \delta u) = \delta u\,k\,u - u\,k\,\delta u = 0\,, \tag{1.238}$$

since

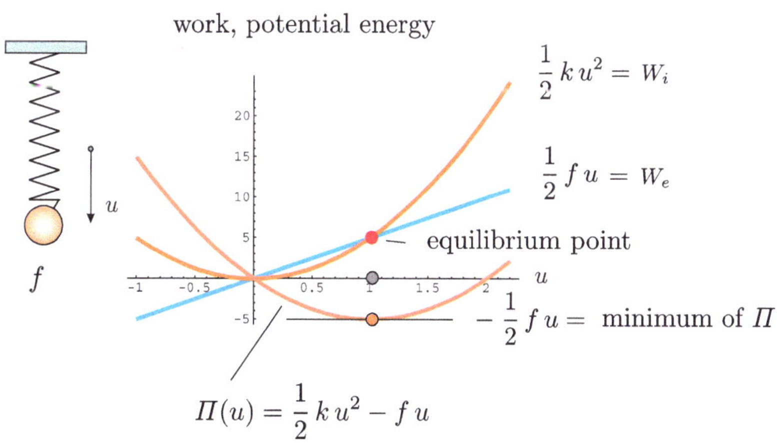

Fig. 1.38 At $W_i = W_e$ lies the equilibrium point u of the spring. Since the internal energy W_i is quadratic and the external work W_e linear, will W_i always pass W_e, will there always be an equilibrium point

$$\frac{1}{2}\,\mathscr{G}(u,u) = \frac{1}{2}\,u\,k\,u - \frac{1}{2}\,u\,k\,u = \frac{1}{2}\,u\,f - \frac{1}{2}\,k\,u^2 = W_e - W_i = 0\,. \quad (1.239)$$

The point at which the two curves $1/2\, f\, u$ and $1/2\, k\, u^2$ intersect, is exactly the equilibrium position u of the spring, see Fig. 1.38.

The third curve in Fig. 1.38 is the potential energy Π of the spring

$$\Pi(u) = \frac{1}{2}\,k\,u^2 - f\,u\,, \quad (1.240)$$

which has zero slope at the equilibrium position

$$\Pi'(u) = k\,u - f = 0\,, \quad (1.241)$$

since $k\,u = f$.

The point where the external and internal work intersect, is also the point at which the potential energy attains its minimum value, see Fig. 1.38.

At the start the external work grows faster, than the internal energy, but it soon passes a point, after which the internal energy has the upper hand. This **cross-over point** is the equilibrium point. It is the point at which the two energies coincide, $W_e = W_i$.

If right from the start the internal energy would grow faster than the external energy, the spring would not move one *iota*.

If we set al.l values equal to one, $k = 1$ and $f = 1$, the equilibrium point is exactly the point $u = 1$. This means:

Statics is based on the fact that in the interval (0, 1) the inequality $u > u^2$ holds true and that beyond this point the opposite is true, $u^2 > u$.

1.33.1 Minimum or Maximum?

The potential energy can either be positive or negative, depending on the type p (forces) or d (displ.) of the load case (LC):

- In a LC p loads are applied (but no supports get displaced)
- In a LC d supports get displaced (but no load is applied, $p = 0$)

In a LC p the potential energy is negative, since $k\,u = f$ implies a negative value for $\Pi(u)$

$$\Pi(u) = \frac{1}{2}\,k\,u^2 - f\,u = \frac{1}{2}\,f\,u - f\,u = -\frac{1}{2}\,f\,u\,, \quad (1.242)$$

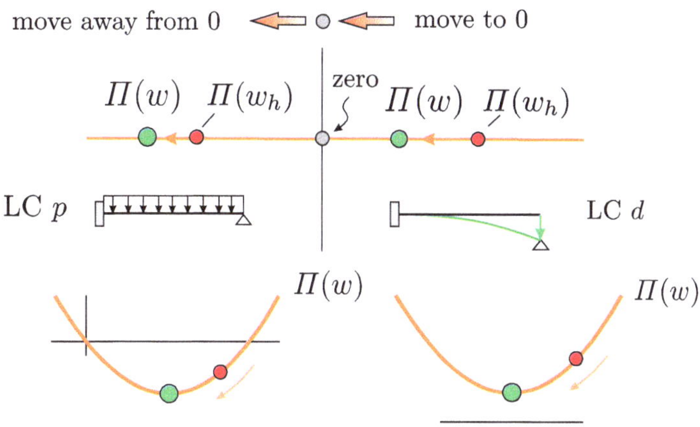

Fig. 1.39 The potential energy $\Pi(w_h)$ of the FE-solution always lies to the right of the exact potential energy, it trails $\Pi(w)$

and so in the equilibrium position u the absolute value $|\Pi(u)|$ has the ***maximum possible distance*** from zero, see Fig. 1.39.

In a LC p the *principle of minimum potential energy* is effectively a maximum principle: The absolute value $|\Pi(u)|$ gets maximized. Only because the potential energy is negative at the equilibrium point, is this the same as to say the potential energy gets minimized.

But many engineers understand the principle in the sense of *"make the effort as small as possible"* while the true meaning ist just the opposite. The force f aims to extract as much energy as possible from the spring, to make $|\Pi(u)|$ as large as possible, and the larger the space $\mathcal{V}$ is, the less constraints such as $w(x_0) = 0$, are placed on w, the more negative the potential energy can become, the larger $|\Pi(w)|$ will be, see Fig. 1.40.

Perhaps the following will help: Imagine a beam carrying a load p, and we search on $\mathcal{V}$ for the deflection w which maximizes the functional

$$J(w) = \int_0^l p\, w\, dx\,. \tag{1.243}$$

For the match to be fair we require all w to be normalized, $w/\|w\|$, to have the same "length" $\|w\|/\|w\| = 1$, since otherwise you could multiply w with a big number. The winner is the deflection w of the beam, see Fig. 1.41, since it gets ***the most mileage*** out of the integral (1.243). It maximizes $J(w)$ and it minimizes $\Pi(w) = -0.5 \cdot J(w)$.

Next, let us look at an LC d (a settlement) as in Fig. 1.42. Since $p = 0$, the potential energy is positive, see Sect. 9.12,

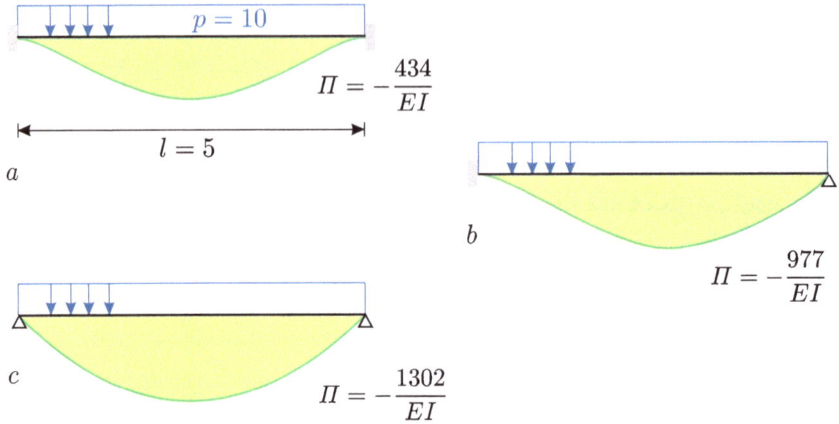

Fig. 1.40 The fewer the number of supports the larger $\mathcal{V}$ is, and the larger is the absolute value of the potential energy, $|\Pi|$, at the equilibrium point

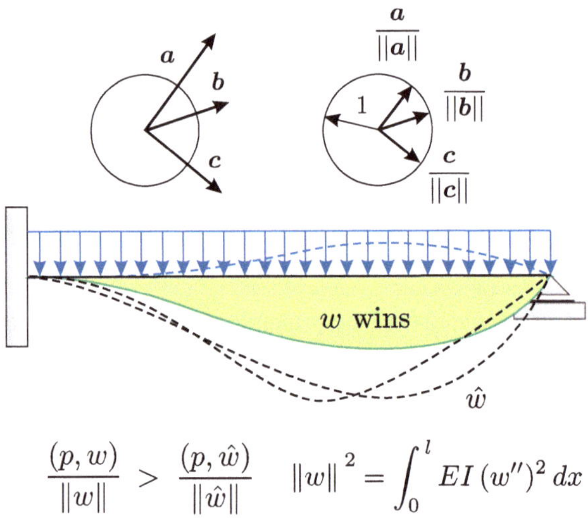

Fig. 1.41 Among all normalized deflections $w/\|w\|$ in $\mathcal{V}$, the deflection curve w of the beam is the function, which gets the most mileage out of p [1]

$$\Pi(w) = \frac{1}{2}\,a(w, w) = \frac{1}{2}\int_0^l \frac{M^2}{EI}\,dx\,, \tag{1.244}$$

and in this situation the principle of minimum potential energy retains its original meaning—it aims to minimize the strains, see Fig. 1.39, to lessen the burden on the beam, to make $\Pi(w)$ as small as possible.

With a ***spline interpolation*** it is the same logic: We leave it to the material to find the optimal curve w through the slalom of the pegs, optimal in the sense that the bending energy

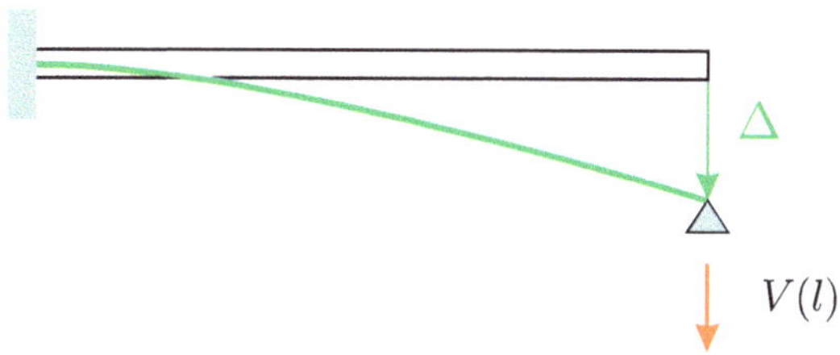

Fig. 1.42 Settlement of a support

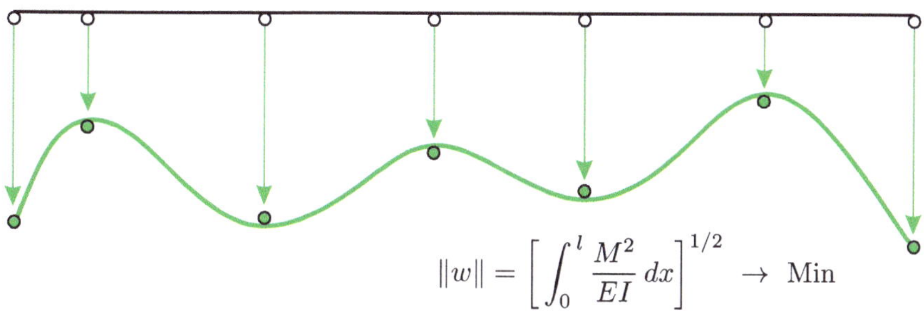

Fig. 1.43 Spline interpolation as settlement of supports

$$\|w\| = \sqrt{a(w, w)} = \left[\int_0^l \frac{M^2}{EI}\, dx \right]^{\frac{1}{2}} \tag{1.245}$$

is as small as possible, is as close to zero as possible, see Fig. 1.43.

1.33.2 Cracks

The ***principle of minimum potential energy*** is also the reason, why the strain energy in a structure increases when it starts to crack, see Fig. 1.44.

Once a crack appears, the displacements on the two faces of a crack no longer need to be the same, and so the trial space $\mathcal{V}$ gets larger, more functions can compete for the minimum value of the potential energy, and thus the potential energy $\Pi(\boldsymbol{u})$ moves further away from zero, further into the direction $-\infty$, its absolute value increases.

When the strains are caused by the settlement of a support, cracks will have the opposite effect—they lessen the strains in the structure. As before, the trial space $\mathcal{V}$ increases and so $\Pi(u)$ decreases, but since $\Pi(u) > 0$, see Sect. 9.12, the potential energy effectively gets smaller.

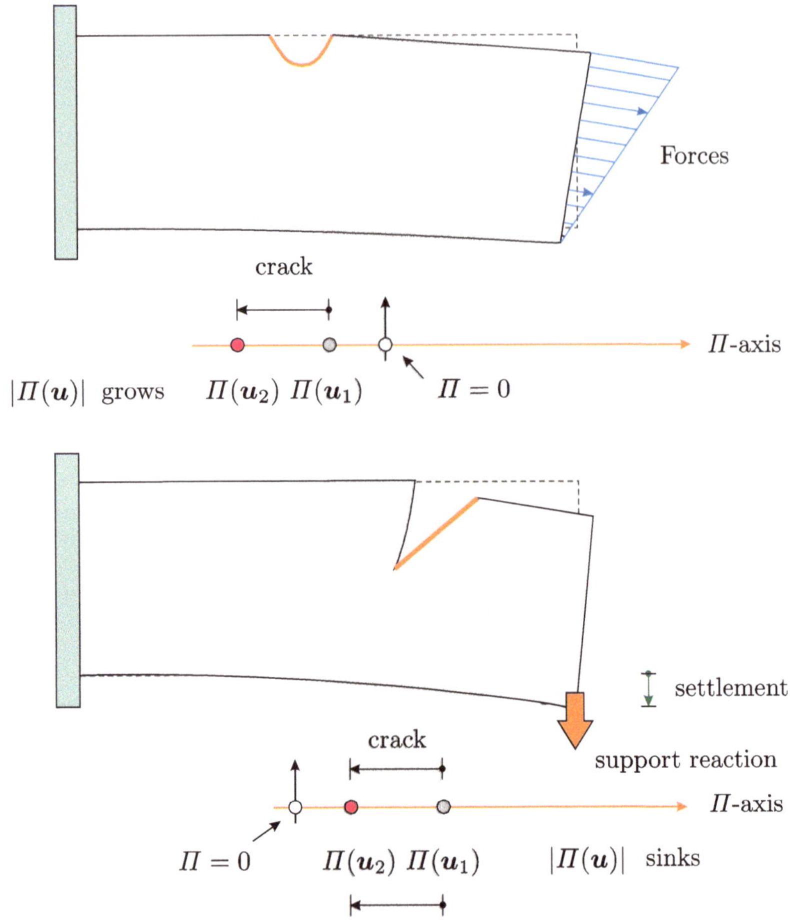

Fig. 1.44 In a LC p (forces) the stresses increase if the plate cracks, $\boldsymbol{u}_1 \rightarrow \boldsymbol{u}_2$, while they shrink in a LC d (settlements)

1.33.3 The Size of the Trial Space $\mathcal{V}$

In structures the size of the trial space $\mathcal{V}$, see Figs. 1.45 and 1.46, is directly linked to the number of constraints, since additional supports, $w(x) = 0$, let the number of possible trial functions decrease, the space $\mathcal{V}$ shrinks, while the removal of a support has the opposite effect, has a liberating effect.

One would assume $\mathcal{V}$ to shrink only when rigid constraints like $w(x) = 0$ are added, but also "soft" constraints let $\mathcal{V}$ shrink. A pier ($EA < \infty$) placed under the end of the cantilever beam as in Fig. 1.47 would be such a ***soft constraint***.

Adding frame elements to a structure lets $\mathcal{V}$ shrink, the structure gets stiffer.

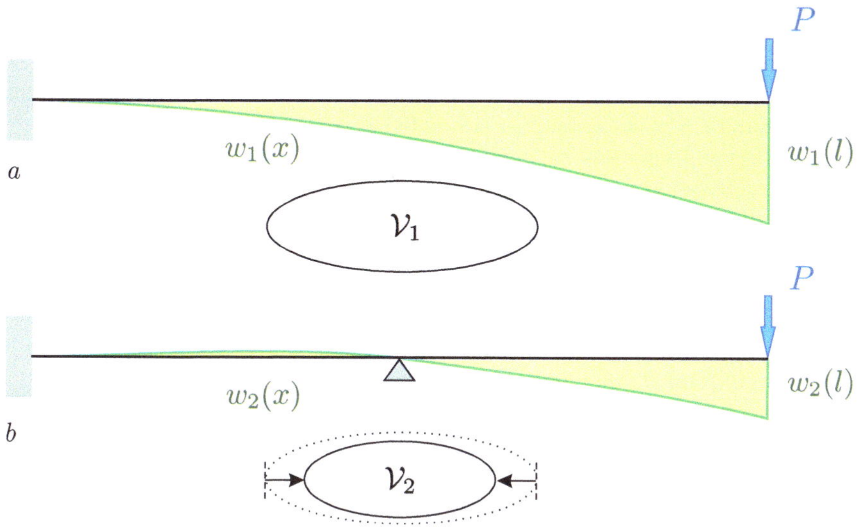

Fig. 1.45 The more supports, the smaller $|\Pi(w)|$, since the trial space $\mathcal{V}$ shrinks with each additional support

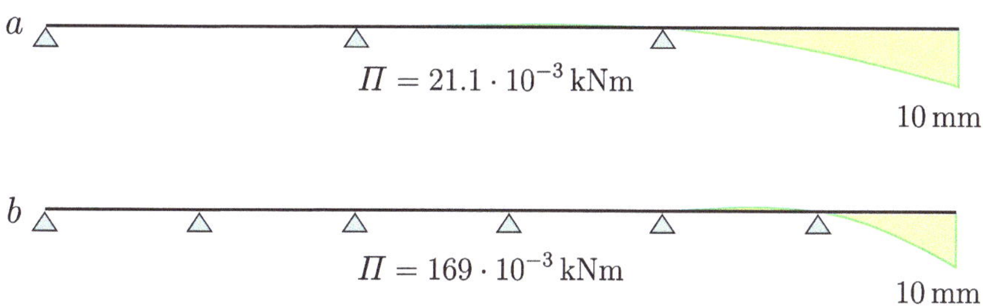

Fig. 1.46 Adding supports to a beam already stressed by a displaced support lets the potential energy increase further

Fig. 1.47 Primary structure

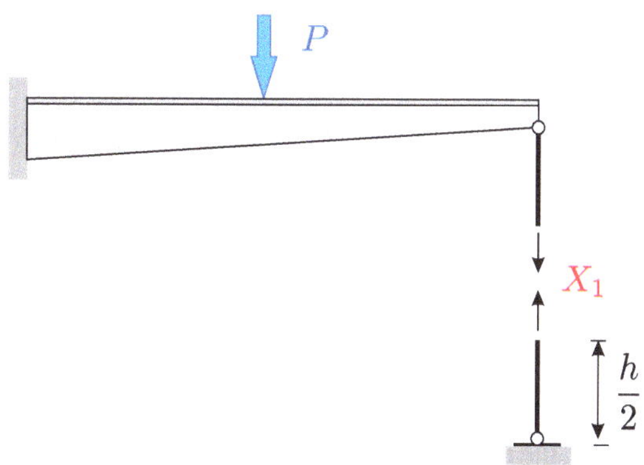

1.34 Infinite Energy

Influence functions are the response to point loads or dislocations and these actions (usually) produce infinitely large strains in the overstressed structure.

To understand why this happens, we look at a circular membrane with radius $R = 1$, which acts as a seal for a circular opening Ω. Under a uniform pressure p the membrane will take a round bowl like shape $u(\boldsymbol{x})$, which is the solution to the boundary value problem

$$-H\,\Delta u = p \quad \text{on } \Omega \qquad u = 0 \quad \text{on } \Gamma\,, \tag{1.246}$$

where the constant H is the prestress.

The internal forces in the membrane are the shear forces v_1, v_2 (N/m),

$$\begin{bmatrix} v_1 \\ v_2 \end{bmatrix} = H \begin{bmatrix} u_{,1} \\ u_{,2} \end{bmatrix} = H\,\nabla u\,, \tag{1.247}$$

which are proportional to the slope of the surface in x_1- and x_2-direction, respectively.

The energy principles of the membrane are based on the identity (we set $H = 1$)

$$\mathcal{G}(u, \delta u) = \underbrace{\int_\Omega -\Delta u\,\delta u\,d\Omega + \int_\Gamma \nabla u \bullet \boldsymbol{n}\,\delta u\,ds}_{\delta W_e} - \underbrace{\int_\Omega \nabla u \bullet \nabla \delta u\,d\Omega}_{\delta W_i} = 0\,, \tag{1.248}$$

where

$$\nabla u \bullet \boldsymbol{n} = u_{,1}\,n_1 + u_{,2}\,n_2 = \frac{\partial u}{\partial n} \tag{1.249}$$

is the slope at the edge. Since the deflection gets smaller towards the edge, its slope is negative and so the downward pressure $p \downarrow$ and the support reactions $\partial u/\partial n \uparrow$ balance

$$\mathcal{G}(u, 1) = \int_\Omega p \cdot 1\,d\Omega + \int_\Gamma \nabla u \bullet \boldsymbol{n} \cdot 1\,ds = 0\,. \tag{1.250}$$

If we place a single force $P = 1$ at the center $\boldsymbol{x} = \boldsymbol{0}$ of the membrane, see Fig. 1.48, it will produce a funnel-like shape

$$G(\boldsymbol{y}, \boldsymbol{x}) = -\frac{1}{2\pi}\ln r \qquad r = |\boldsymbol{y} - \boldsymbol{x}|\,, \tag{1.251}$$

and so the force sinks infinitely deep, $G(\boldsymbol{0}, \boldsymbol{0}) = \infty$. But even in this extreme situation the *energy balance* between external and internal eigenwork holds true

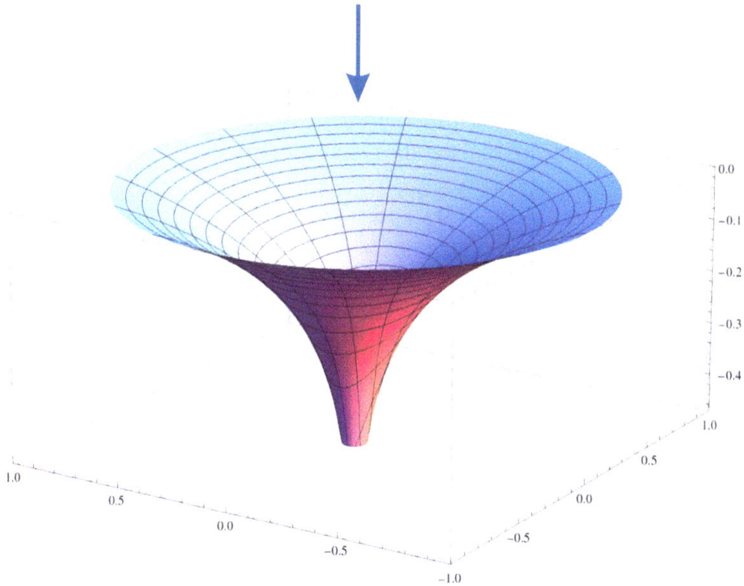

Fig. 1.48 Membrane and point load

$$\frac{1}{2}\,\mathscr{G}(G, G) = W_e - W_i = 0\,. \tag{1.252}$$

We leave off the factor $1/2$ in the following since it is not essential.

The **gradient** of the funnel-like surface is of the order $1/r$

$$\nabla G = \frac{1}{2\,\pi}\,\frac{1}{r}\begin{bmatrix}\cos\varphi\\ \sin\varphi\end{bmatrix}, \tag{1.253}$$

and the scalar product of the gradient with itself is of the order $1/r^2$

$$\nabla G \cdot \nabla G = \frac{1}{4\,\pi^2}\,\frac{1}{r^2}(\cos^2\varphi + \sin^2\varphi) = \frac{1}{4\,\pi^2}\,\frac{1}{r^2}\,, \tag{1.254}$$

and so the internal energy is infinitely large

$$W_i = \int_\Omega \nabla G \cdot \nabla G\, d\Omega = \int_0^{2\pi}\int_0^1 \frac{1}{4\,\pi^2}\frac{1}{r^2}\, r\, dr\, d\varphi = \infty\,, \tag{1.255}$$

because

$$\int_0^1 \frac{1}{r}\, dr = \infty\,. \tag{1.256}$$

This agrees with the infinite value of the external work

$$W_e = P \cdot \infty\,, \tag{1.257}$$

since P sinks infinitely deep. In itself the result is correct

$$W_e - W_i = \infty - \infty = 0\,, \tag{1.258}$$

but we cannot calculate with infinite, infinite is simply "infinitely many".

The reason for the infinite energy is that we are on the ***diagonal*** of Green's first identity when we formulate the ***principle of conservation of energy***

$$\frac{1}{2}\,\mathcal{G}(G, G) = W_e - W_i = 0\,, \tag{1.259}$$

and so the singularity doubles. The product of $1/r \cdot 1/r$ is $1/r^2$ and this funnel-like shape has an **infinite volume**

$$\int_\Omega \nabla G \bullet \nabla G\, d\Omega \simeq \int_0^{2\pi} \int_0^1 \frac{1}{r^2}\, r\, dr\, d\varphi = \infty\,. \tag{1.260}$$

In plates the situation is the same. Near a point force placed (in tangential direction) on a circular plate with radius $R = 1$, the strains and stresses increase as $1/r$ and so the internal energy W_i is infinitely large,

$$W_i = \int_\Omega \sigma_{ij}\, \varepsilon_{ij}\, d\Omega \simeq \int_0^{2\pi} \int_0^{R=1} \frac{1}{r^2}\, r\, dr\, d\varphi = \infty\,. \tag{1.261}$$

In 3-D the strains and stresses are of the order $1/r^2$ and the volume element $d\Omega = r^2\, dr\, d\varphi\, \sin\theta\, d\theta$ cannot counter the $1/r^4 = 1/r^2 \cdot 1/r^2$ of two such singularities, which means that also in 3-D the internal energy is infinite.

The situation is different in a Kirchhoff plate, which is governed by a fourth-order equation, $K\, \Delta\Delta w = p$.

The deflection of a circular plate with radius $R = 1$ has the form

$$w(\boldsymbol{x}) = c \cdot \frac{1}{8\,\pi\, K}\, r^2 \ln r + f(r) \tag{1.262}$$

—here $f(r)$ is a bounded function and c a constant—and so the deflection is finite at the source point since

$$\lim_{r \to 0} r^2 \cdot \ln r = 0\,. \tag{1.263}$$

The moments m_{ij} and curvatures κ_{ij} produced by the point load have a logarithmic singularity $m_{ij} \sim \ln r$, and this is why the internal energy

$$W_i = \int_{\Omega} m_{ij}\, \kappa_{ij}\, d\Omega \sim \int_0^{2\pi} \int_0^{R=1} \ln^2 r \underbrace{r\, dr\, d\varphi}_{d\Omega} \tag{1.264}$$

is bounded.

Remark 1.2 As in any 2-D problem, the point load $P = 1$ is balanced by $1/r$ forces, here the shear force v_n (the "third" derivative),

$$\lim_{\varepsilon \to 0} \int_{\Gamma_{N_\varepsilon}} \frac{1}{2\pi\, r}\, ds_y = 1\,, \tag{1.265}$$

but since the equation is of fourth order it takes **three steps** from v_n to w

$$w = \int \int \int v_n (d\Omega)^3 \simeq r^2 \ln r \qquad \text{(three steps)} \tag{1.266}$$

and this is enough to dampen the singularity. In second-order problems only one integration separates $\sigma \simeq 1/r$ from u and this is too little

$$u \simeq \int \frac{1}{r}\, dr = \ln r \qquad \text{(one step)}\,. \tag{1.267}$$

1.35 Sobolev's Embedding Theorem

It is *Sobolev's Embedding Theorem*, which allows to classify the singularities of point solutions in a more systematic way. A point solution has finite energy[8] if

$$\boxed{m - i > \frac{n}{2}} \tag{1.268}$$

Here, $2\,m$ is the order of the differential operator, i is the order of the singularity, which in our notation is identical with the index i on the Dirac delta δ_i, and n is the dimension of the space. In plates, $n = 2$, the system of equations is of order $2\,m = 2$ and therefore the displacement field generated by a single force, $i = 0$, has infinite energy, since the inequality

$$1 - 0 > \frac{2}{2} = 1 \quad ? \tag{1.269}$$

[8]The table on page 456 contains a systematic evaluation of the inequality.

is not true, while for a Kirchhoff plate, $2m = 4$, it is true

$$2 - 0 > \frac{2}{2} = 1\,. \tag{1.270}$$

We can state it also differently:

The internal energy W_i is infinite, if the external work W_e is infinite, and this is the case if in the expression

$$W_e = force \times displacement \tag{1.271}$$

one of the two terms is infinite.

The force placed on the membrane, $P = 1$, is finite, but the deflection is not—it is ∞. In a plate the situation is the same. But in the case of a Kirchhoff plate the deflection produced by a unit force $P = 1$ is finite and therefore the external work $W_e = force \times deflection < \infty$ is bounded.

The energy is always infinite in influence functions for forces since these are triggered by dislocations or kinks and such actions overstress any material. In plates and shells also influence functions for displacements have infinite energy—with the exception of the deflection $w(\boldsymbol{x})$ of a Kirchhoff plate.

But: "If the energy of an influence function is infinite, how can we calculate with it?" The difference is that when we apply the *principle of virtual displacements* or the *principle of virtual forces*, we are on the **off-diagonal**

$$\mathscr{G}(G, u) = \delta W_e - \delta W_i = 0\,, \tag{1.272}$$

and therefore the singularity does not double, the virtual internal energy δW_i remains finite

$$\delta W_e = 1 \cdot u(\boldsymbol{x}) = \int_\Omega \nabla G \bullet \nabla u \, d\Omega = \delta W_i\,, \tag{1.273}$$

since the $1/r$ of the gradient ∇G is balanced by the r of the domain element $d\Omega = r\,dr\,d\varphi$. Strictly speaking, we must also require that the gradient of u is bounded, $|\nabla u| \leq \infty$.

Betti's theorem is also affected by the singularities of the influence functions, but because we see only the final result

$$\lim_{\varepsilon \to 0} \mathscr{B}(G, u)_{\Omega_\varepsilon} = u(\boldsymbol{x}) - \int_\Omega G(\boldsymbol{y}, \boldsymbol{x})\, p(\boldsymbol{y})\, d\Omega_{\boldsymbol{y}} = 0\,, \tag{1.274}$$

everything looks smooth. What was singular has led to the term $u(\boldsymbol{x})$, and the rest are all integrals whose evaluation can be left to the computer (numerical quadrature).

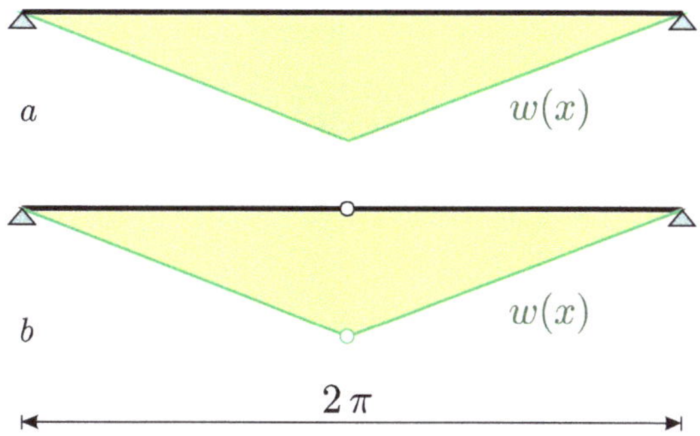

Fig. 1.49 Deflection curves, **a** with infinite energy, **b** with finite energy, here even zero

Remark 1.3 The characteristic singularities of the different influence functions G_i and the limits

$$\lim_{\varepsilon \to 0} \mathscr{G}(G_i, u)_{\Omega_\varepsilon} = 0 \qquad \lim_{\varepsilon \to 0} \mathscr{B}(G_i, u)_{\Omega_\varepsilon} = 0\,, \tag{1.275}$$

are discussed in detail in [2].

Remark 1.4 *Engineer versus mathematician* A mathematician will not fail to remark that an infinite amount of energy is needed to create a kink in a beam, and to support his claim, he will expand the deflection w, see Fig. 1.49a, into a Fourier series

$$w(x) = \frac{\pi}{2} - \frac{4}{\pi}(\cos x + \frac{1}{3^2}\cos 3x + \frac{1}{5^2}\cos 5x + \cdots) \tag{1.276}$$

and calculate its energy

$$\frac{1}{2} \cdot EI \int_0^{2\pi} (w''(x))^2\, dx = \frac{1}{2} \cdot \frac{16}{\pi}(1 + 1 + 1 \ldots) = \infty\,. \tag{1.277}$$

But an engineer does it differently: The engineer installs a hinge and rotates both sides of the hinge so that $\tan \varphi_l + \tan \varphi_r = 1$. The bending energy is then simply the energy necessary to rotate the beam ends, and this energy is finite. In the case of the statically determinate beam in Fig. 1.49b it is even zero, since $w'' = 0$.

Remark 1.5 For further remarks about Sobolev's embedding theorem, see Sect. 9.23.

1.36 Reduction Principle

The *Reduction principle* is a special version of Mohr's equation. Mohr would calculate the horizontal displacement u_i of the frame in Fig. 1.50 by applying a unit point load $X_1 = 1$ in the direction of u_i, and then evaluate the integral

$$1 \cdot u_i = \sum_e \int_0^{l_e} (\frac{M\,M_1}{EI} + \frac{N\,N_1}{EA})\,dx \tag{1.278}$$

by integrating over all frame elements.

But according to the *Reduction principle* it suffices to apply the point load $X_1 = 1$ to a substructure of the original structure, where substructure means any statically determinate system **contained** in the original system, as for example the single post in Fig. 1.50d.

This surprising result is better understood when we realize that the node, which carries the load, can be reached from different starting points, see Fig. 1.50c, and that the sum of the horizontal displacement increments along each path must be the same, must be u_i.

This means, we can compute u_i by integrating for example only over the end post in Fig. 1.50d

$$1 \cdot u_i = \int_0^{l} (\frac{M\,M_1}{EI} + \frac{N\,(N_1 = 0)}{EA})\,dx\,. \tag{1.279}$$

The *Reduction principle* essentially states that the Dirac delta, the X_1, need not be applied to the original system, but it can be any substructure, which is "contained" in the original system, see Fig. 1.50d. Contained means, that in the transition to the substructure nodes of the original structure can be released, but no nodes may additionally be fixed. In the language of mathematics this means the Dirichlet boundary may shrink, but it may not grow [1], p. 149. The admissible substructures are typically the structures, which we would eventually choose in the force method. The single post is such a substructure.

The *Reduction principle* is a clever application of Green's first identity. The sum of all the single identities of the frame elements is zero

$$\sum_e \mathcal{G}(u, u_1)_e = 0\,, \tag{1.280}$$

and this holds also true for any substructure or for any statically determinate system contained in the original system. The "trick" is to choose substructures which are easier to analyze than the original structure. Then one needs to integrate over fewer frame elements, or the moment distribution M_1 and the normal forces N_1 are easier to determine, because the substructure is statically determinate.

But in picking a substructure care must be taken to avoid systems which require unknown nodal displacements or forces (coupling forces), when $\mathcal{G}(u, u_1) = 0$

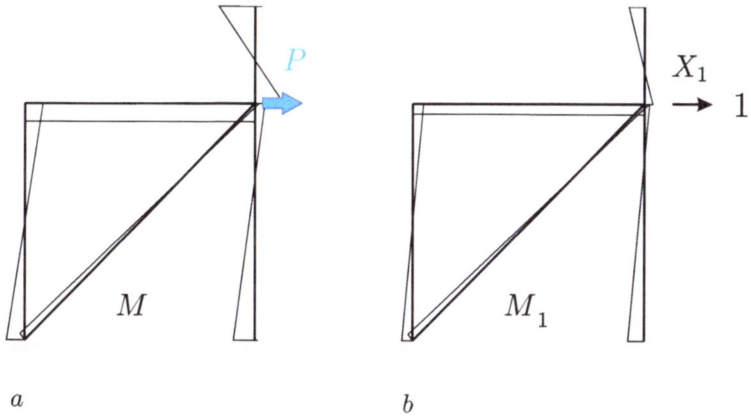

$$1 \cdot u_i = \sum_e \int_0^{l_e} \left(\frac{M\,M_1}{EI} + \frac{N\,N_1}{EA}\right) dx$$

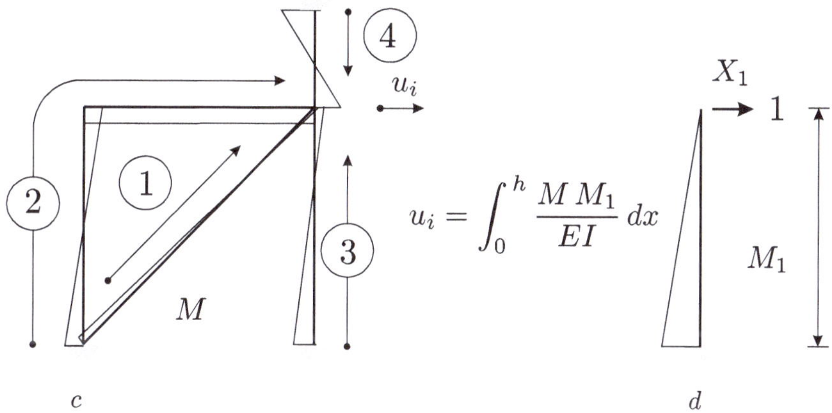

Fig. 1.50 Application of the *Reduction principle*, **a** and **b** moment distribution and Mohr's approach, **c** different paths have the same result, **d** *Reduction principle*

is formulated. This is the essence of the "Dirichlet constraint" mentioned above. Choosing a substructure "contained" in the original system, avoids this trap. So, the *Reduction principle* and the force method go hand in hand.

The idea of substructures is of course applicable to any FE-mesh. Anytime one can choose a sub-mesh and formulate a zero-pairing between two stress states on this sub-mesh via Green's first identity, see Fig. 1.51.

The path-independent integral in fracture mechanics is the result of such a pairing [3].

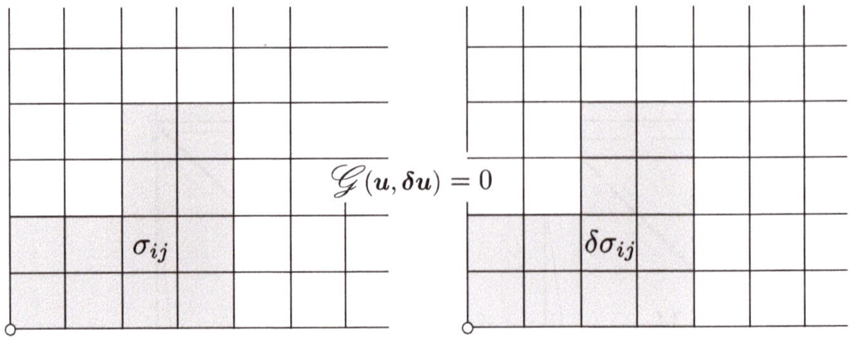

Fig. 1.51 Linking two stress states in a sub-region

1.37 The Force Method

In the force method a structure is made statically determinate by installing M-, V- or N-joints, see Fig. 1.52.

The **redundants** X_i acting at these joints are subject to the condition that the gaps at the joints close

$$\begin{bmatrix} \delta_{11} & \delta_{12} \\ \delta_{21} & \delta_{22} \end{bmatrix} \begin{bmatrix} X_1 \\ X_2 \end{bmatrix} - \begin{bmatrix} \delta_{10} \\ \delta_{20} \end{bmatrix} = \begin{bmatrix} 0 \\ 0 \end{bmatrix} . \tag{1.281}$$

The δ_{ij} are the relative rotations/displacements between the X_i (left and right side of the joint) and the δ_{i0} are the same values, but due to the load. The values are calculated with Mohr's equation

$$\delta_{ij} = \int_0^l \frac{M_i\, M_j}{EI}\, dx\,, \tag{1.282}$$

which is identical with Green's first identity.

$$\mathscr{G}(w_i, w_j) = \delta_{ij} - a(w_i, w_j) = \delta_{ij} - \int_0^l \frac{M_i\, M_j}{EI}\, dx = 0\,. \tag{1.283}$$

The deflections w_i and w_j are the deflections, which belong to the X_i and X_j (fortunately they are not needed—only their moments) and the single term

$$\delta_{ij} = [V_i\, w_j - M_i\, w_j']_0^l \tag{1.284}$$

is, so to speak, the "rest" of the square brackets.

For to calculate the δ_{i0} in the above formulas the moment M_j is replaced by the load moment M_0 in the statically determinate primary structure.

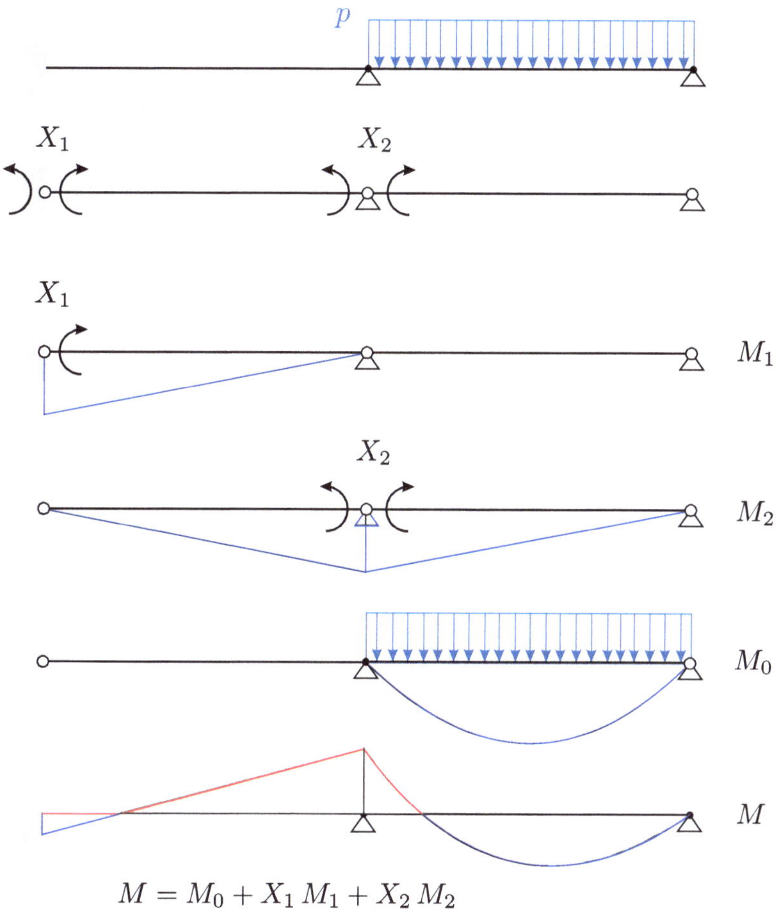

Fig. 1.52 Force method

1.38 Where Does It Run To?

The zero sum of Green's first identity also has a direct static relevance, as the following two examples will illustrate.

A fixed point is not enough for Archimedes to unhinge the Earth. He also needs a lever with an infinite bending stiffness $EI = \infty$, see Fig. 1.53a, since otherwise his whole effort will only serve to give the lever a nicely rounded shape.

Because of $\mathscr{G}(w, w) = W_e - W_i = 0$ the external work and the bending energy in the beam are at any moment the same (we omit the factor 1/2)

$$W_e = P_r \cdot w_r - P_l \cdot w_l = a(w, w) = \int_0^l \frac{M^2}{EI}\, dx = W_i\,, \tag{1.285}$$

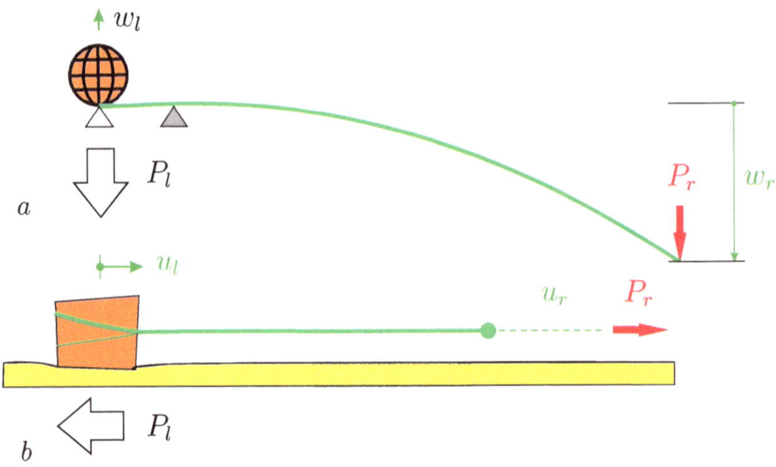

Fig. 1.53 Archimedes' dilemma, **a** his efforts let the strain energy increase, but the Earth will not move one *iota*, **b** and the rubber band will stretch and stretch...

or solved for the wished-for effect

$$P_l \cdot w_l = a(w, w) - P_r \cdot w_r \simeq 0\,, \tag{1.286}$$

which is practically zero, because Archimedes' whole effort, $P_r \cdot w_r$, flows into the curvature of the beam, into the bending energy $a(w, w)$, and practically nothing arrives on the left side of the equation, nothing is left to lift the globe.

The same happens when a heavy weight is dragged over the wet sand on the beach, see Fig. 1.53b,

$$W_e = \underbrace{P_r \cdot u_r}_{effort} - P_l \cdot u_l = a(u, u) = \int_0^l \frac{N^2}{EA}\, dx = W_i\,. \tag{1.287}$$

The rubber band (EA) will stretch and stretch, the strain energy $a(u, u)$ will continue to grow, but the weight will hardly move, $u_l \simeq 0$,

$$P_l \cdot u_l = a(u, u) - P_r \cdot u_r \simeq 0\,. \tag{1.288}$$

Since Green's first identity is a sum of different energies, we have no control over where our effort or energy flows to. It must be in the system, but where exactly at the end?

A similar situation occurs, when the stiffness of a structure changes. How does a structure react if a support is removed or an element cracks? The load does not change, only the distribution of the internal forces—but how? We only know, see Chap. 5, that the new equilibrium position $\boldsymbol{u}_c$ is tied to the previous position $\boldsymbol{u}$

$$(\boldsymbol{I} + \boldsymbol{K}^{-1}\boldsymbol{\Delta K})\boldsymbol{u}_c = \boldsymbol{u}\,, \tag{1.289}$$

since $\boldsymbol{f} = \boldsymbol{K}\,\boldsymbol{u} = (\boldsymbol{K} + \boldsymbol{\Delta K})\,\boldsymbol{u}_c$ does not change.

1.39 Finite Elements and Green's Identity

We want to end this chapter with a warning, point at a possible misunderstanding: Green's first identity is identical with $\delta\Pi = 0$, but it is not identical with the FE-system $\boldsymbol{K}\boldsymbol{u} = \boldsymbol{f}$. (The exception are the standard 1-D problems, $EI\,w^{IV} = p_z$ and $-EA\,u'' = p_x$, see Sect. 1.17).

If u is the deflection of a membrane

$$-\Delta u = p \qquad u = 0 \quad \text{on } \Gamma \tag{1.290}$$

and δu is an admissible virtual displacement then holds

$$\mathcal{G}(u, \delta u) = (p, \delta u) - a(u, \delta u) = 0\,, \tag{1.291}$$

and therefore also

$$\mathcal{G}(u, \varphi_i) = (p, \varphi_i) - a(u, \varphi_i) = 0\,. \tag{1.292}$$

This equation is the template for the FE-solution $u_h(x) = \sum_j u_j\,\varphi_j(x)$

$$(p, \varphi_i) - a(u_h, \varphi_i) = 0 \qquad i = 1, 2, \ldots, n\,, \tag{1.293}$$

and so one is tempted to think this is identical to

$$\mathcal{G}(u_h, \varphi_i) = (p_h, \varphi_i) - a(u_h, \varphi_i) = 0\,, \tag{1.294}$$

but the load in this identity is $p_h = -\Delta\,u_h''$ and not p,

$$\text{FEM} = (\underset{\uparrow}{p}, \varphi_i) - a(u_h, \varphi_i) \neq (\underset{\uparrow}{p_h}, \varphi_i) - a(u_h, \varphi_i) = \mathcal{G}(u_h, \varphi_i)\,. \tag{1.295}$$

So, the FE-equation (1.293) looks like Green's first identity, (1.294), but it is not. Only in the end, $h \to 0$, will the two sides be the same, since then $p_h = p$.

References

1. Hartmann F (2013) Green's functions and finite elements. Springer, Heidelberg
2. Hartmann F (1989) Introduction to boundary elements. Springer, Heidelberg
3. Szabó B, Babuška I (1991) Finite element analysis. Wiley

Chapter 2
Betti's Theorem

The subject of this chapter is the formulation of **influence functions** with *Betti's theorem*. Influence functions are central to finite elements since in linear static analysis any FE-solution is an expansion in terms of the nodal influence functions $\boldsymbol{g}_i$ of a structure, see Fig. 2.1.

$$\boldsymbol{u} = f_1\, \boldsymbol{g}_1 + f_2\, \boldsymbol{g}_2 + \cdots + f_n \boldsymbol{g}_n \tag{2.1}$$

2.1 Basics

According to ***Betti's theorem*** the reciprocal external work of two equilibrium states is the same

$$W_{1,2} = W_{2,1}\,. \tag{2.2}$$

The work done by the load on system 1 on acting through the displacements of system 2, $W_{1,2}$, is the same as the work done by the load on system 2 on acting through the displacements of system 1, $W_{2,1}$.

This theorem is based on Green's second identity

$$\mathscr{B}(w, \hat{w}) = \mathscr{G}(w, \hat{w}) - \mathscr{G}(\hat{w}, w) = 0 - 0 = 0\,, \tag{2.3}$$

which for a beam is the expression

F. Hartmann and P. Jahn, *Statics and Influence Functions*, Springer Series in Solid and Structural Mechanics 13,
https://doi.org/10.1007/978-3-030-55889-5_2

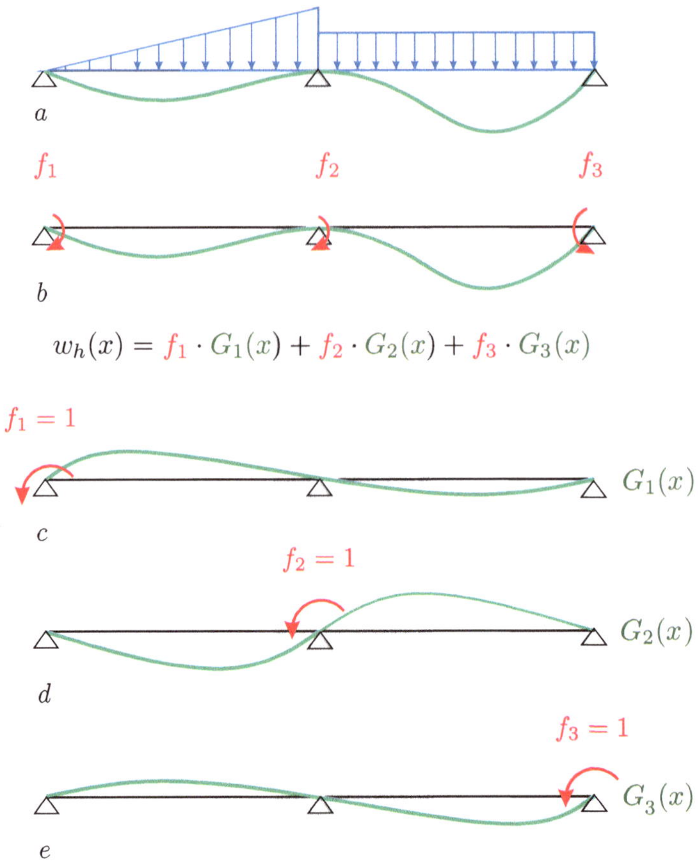

Fig. 2.1 The FE-solution $w_h(x)$ is the sum of the influence functions G_1, G_2, G_3 for the three nodal displacements and the nodal moments f_i are the weights

$$\mathscr{B}(w_1, w_2) = \underbrace{\int_0^l EI\, w_1^{IV}\, w_2\, dx + [V_1\, w_2 - M_1\, w_2{}']_0^l}_{W_{1,2}}$$
$$\underbrace{- [V_2\, w_1 - M_2\, w_1']_0^l - \int_0^l w_1\, EI\, w_2^{IV}\, dx}_{W_{2,1}} = 0\,, \tag{2.4}$$

and a numerical test with the two beams in Fig. 2.2

$$p_1 = 10 \qquad w_1(x) = \frac{10 \cdot 5^4}{24\, EI}\, (\xi - 2\,\xi^3 + \xi^4) \qquad \xi = \frac{x}{l} \tag{2.5}$$

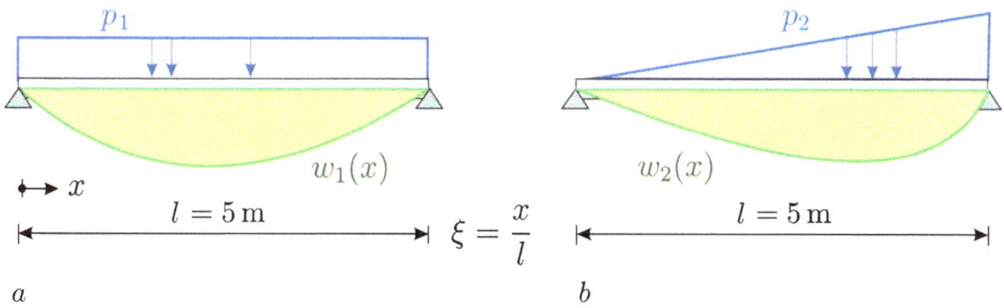

Fig. 2.2 *Betti's theorem*

$$p_2 = 7\,\xi \qquad w_2(x) = \frac{7 \cdot 5^3\,x}{360\,EI}\,(7 - 10\,\xi^2 + 3\,\xi^4)\,, \tag{2.6}$$

confirms it: The work done **"crosswise"**, is the same

$$\begin{aligned}\mathscr{B}(w_1, w_2) &= \int_0^l p_1(x)\,w_2(x)\,dx - \int_0^l p_2(x)\,w_1(x)\,dx \\ &= \frac{1}{EI} \cdot 911.46 - \frac{1}{EI} \cdot 911.46 = 0\,. \end{aligned} \tag{2.7}$$

2.2 Influence Functions for Displacements

The influence function $G_0(y, x)$ for a displacement $u(x)$ is the shape of the structure if a force $P = 1$ pushes the source point x in the direction of $u(x)$; the y in $G_0(y, x)$ are the points y where we register the effects caused by the single force. Since the influence function is symmetric, $G_0(y, x) = G_0(x, y)$, (Maxwell), we could also imagine that a force $P = 1$ acts at the point y, and we register the effect at x.

For our purposes it will be best to think of x as the observation point and the load to be acting at y.

What is perhaps a bit confusing at the beginning is: To ***generate*** the influence function $G_0(y, x)$ we place a unit force $P = 1$ at x, at the source point, but when we ***evaluate*** the influence function

$$u(x) = \int_0^l G_0(y, x)\,p(y)\,dy\,, \tag{2.8}$$

the force at x is gone and the load acts at the points y

$$x = \begin{cases} \text{source point when } G_0(y, x) \text{ is generated} \\ \text{observation point when } G_0(y, x) \text{ is evaluated.} \end{cases} \tag{2.9}$$

A distributed load p can be compared to a series of small single forces

$$dP(y) = p(y)\,dy\,, \tag{2.10}$$

each of which adds a part

$$dw = G_0(y, x)\,dP(y) \tag{2.11}$$

to the deflection at x, and so the total is the superposition of the influence function with the load

$$w(x) = \int_0^l dw = \int_0^l G_0(y, x)\,p(y)\,dy\,. \tag{2.12}$$

If the load is a single force P at a point y then

$$w(x) = G_0(y, x) \cdot P\,. \tag{2.13}$$

And just as the slope w' and the section forces respectively are the derivatives of w

$$w'(x) = \frac{d}{dx}\,w(x) \qquad M(x) = -EI\,\frac{d^2}{dx^2}\,w(x) \qquad V(x) = -EI\,\frac{d^3}{dx^3}\,w(x)\,, \tag{2.14}$$

so their influence functions are the derivatives of $G_0(y, x)$ with respect to the source point x, see Fig. 2.3,

$$G_0(y, x) \qquad = \text{influence function for } w(x) \tag{2.15a}$$

$$G_1(y, x) = \frac{d}{dx} G_0(y, x) \qquad = \text{influence function for } w'(x) \tag{2.15b}$$

$$G_2(y, x) = -EI\,\frac{d^2}{dx^2} G_0(y, x) \qquad = \text{influence function for } M(x) \tag{2.15c}$$

$$G_3(y, x) = -EI\,\frac{d^3}{dx^3} G_0(y, x) \qquad = \text{influence function for } V(x)\,. \tag{2.15d}$$

We write the influence functions with a capital G, because in mathematics influence functions are called ***Green's functions***, and since the deflection is the zero-order derivative, we write its influence function G_0 with an index 0.

Note, we use the term influence function in two ways: It can either mean the integral

$$w(x) = \int_0^l G_0(y, x)\,p(y)\,dy\,, \tag{2.16}$$

or the ***kernel*** $G_0(y, x)$ of the influence function.

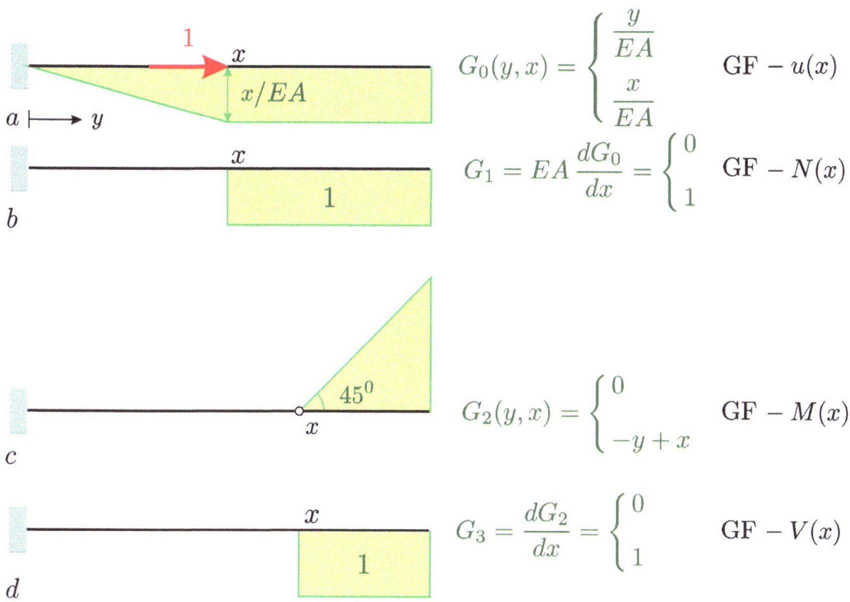

Fig. 2.3 Formulating influence functions by differentiating the kernel G_0

The central role of the Green's functions cannot be overestimated, since all solutions are the superposition of the load with point solutions, and the key question therefore is:

How does the medium react to a single force?

2.2.1 Formulation

The single steps which lead to the influence function (2.16) are the following: We apply at the source point x a unit force $P = 1$, see Fig. 2.4b, determine the deflection $G_0(y, x)$, and formulate *Betti's theorem* (Green's second identity) with the two deflections $G_0(y, x)$ and $w(y)$.

This cannot be done in one step, since the third derivative (the shear force) of the influence function **jumps at the source point**. So, we integrate from the left end up to the source point x, pause there, and continue the integration behind the source point

$$\mathscr{B}(G_0, w) = \mathscr{B}(G_0^l, w)_{(0,x)} + \mathscr{B}(G_0^R, w)_{(x,l)}\,. \tag{2.17}$$

Since w and M are both zero at the beam ends, this gives

$$
\begin{aligned}
\mathscr{B}(G_0, w) &= \mathscr{B}(G_0^l, w)_{(0,x)} + \mathscr{B}(G_0^R, w)_{(x,l)} \\
&= V_0^l(x)\, w(x) - M_0^l(x)\, w'(x) - \int_0^x G_0^l(y,x)\, p(y)\, dy \\
&\quad - V_0^R(x)\, w(x) + M_0^R(x)\, w'(x) - \int_x^l G_0^R(y,x)\, p(y)\, dy \\
&= \underbrace{(V_0^l(x) - V_0^R(x))}_{=1}\, w(x) - \underbrace{(M_0^l(x) - M_0^R(x))}_{=0}\, w'(x) \\
&\quad - \int_0^l G_0(y,x)\, p(y)\, dy = 1 \cdot w(x) - \int_0^l G_0(y,x)\, p(y)\, dy = 0\,.
\end{aligned} \tag{2.18}
$$

or

$$
w(x) = \int_0^l G_0(y,x)\, p(y)\, dy\,. \tag{2.19}
$$

2.2.1.1 Influence Function for $w'(x)$

In this case we apply at the source point a **single moment** $M = 1$ and we formulate *Betti's theorem* with the two parts G_1^L and G_1^R, see Fig. 2.4c,

$$
\mathscr{B}(G_1, w) = \mathscr{B}(G_1^L, w)_{(0,x)} + \mathscr{B}(G_1^R, w)_{(x,l)} = 0 + 0\,. \tag{2.20}
$$

The unit jump of M at the source point produces $w'(x)$

$$
(M_L(x,x) - M_R(x,x))\, w'(x) = 1 \cdot w'(x)\,, \tag{2.21}
$$

and so

$$
1 \cdot w'(x) = \int_0^l G_1(y,x)\, p(y)\, dy\,. \tag{2.22}
$$

2.2.1.2 Influence Function for Longitudinal Displacement $u(x)$

Betti's theorem for a bar, $-EA\, u''(x) = p(x)$, is

$$
\begin{aligned}
\mathscr{B}(u, \hat{u}) &= \int_0^l -EA\, u''(x)\, \hat{u}(x)\, dx + [N\, \hat{u}]_0^l \\
&\quad - [u\, \hat{N}]_0^l - \int_0^l u(x)\, (-EA\, \hat{u}''(x))\, dx = 0\,,
\end{aligned} \tag{2.23}
$$

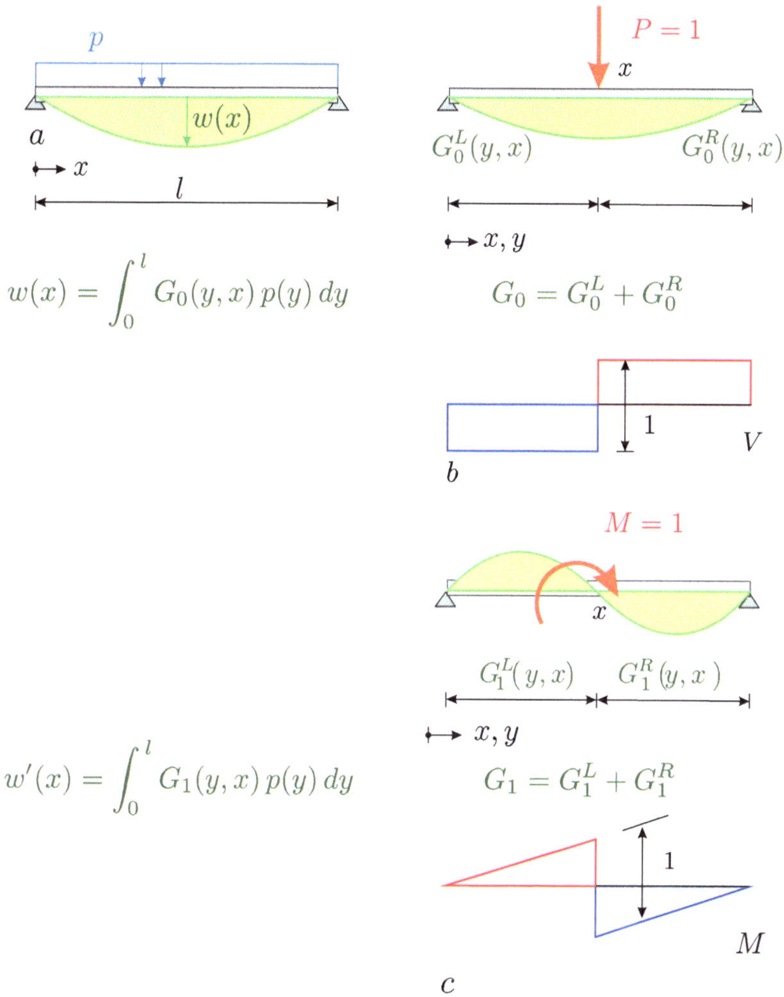

Fig. 2.4 *Betti's theorem* applied to a beam, **a** influence function $G_0(y, x)$ for $w(x)$, and **b** influence function $G_1(y, x)$ for the slope $w'(x)$ in the middle of the beam

and with a **force** $P = 1$ at the source point, see Fig. 2.5, this gives—we leave the intermediate steps off

$$1 \cdot u(x) = \int_0^l G_0(y, x)\, p(y)\, dy\,. \tag{2.24}$$

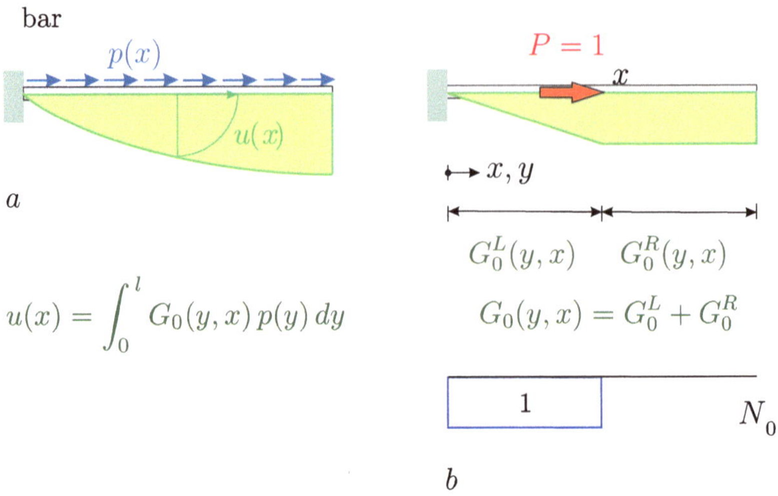

Fig. 2.5 *Betti's theorem* applied to a bar

2.3 Influence Functions for Forces

The required action is always conjugated (adjoint would mean the same) to the quantity we are looking for: So, we search in Green's second identity for the quantity, look at its partner, and we know that a **unit-spread** of the partner, as in Fig. 2.6, will produce the searched-for quantity.

The spreads, the dislocations are made possible by installing a corresponding M-, N- and V-hinge, see Fig. 2.7. If the structure is statically determinate, no forces are necessary to spread the hinge.

A statically indeterminate structure, instead, hinders the spread. So, to determine the necessary action, we apply a pair $X = \pm 1$ of forces on both sides of the hinge, register the resulting spread, and normalize the pair $\pm X$ so that a unit-spread will result.

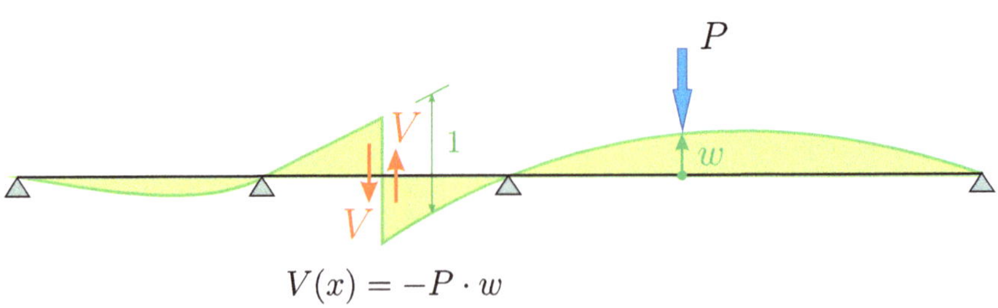

Fig. 2.6 An influence function can be compared to a seesaw

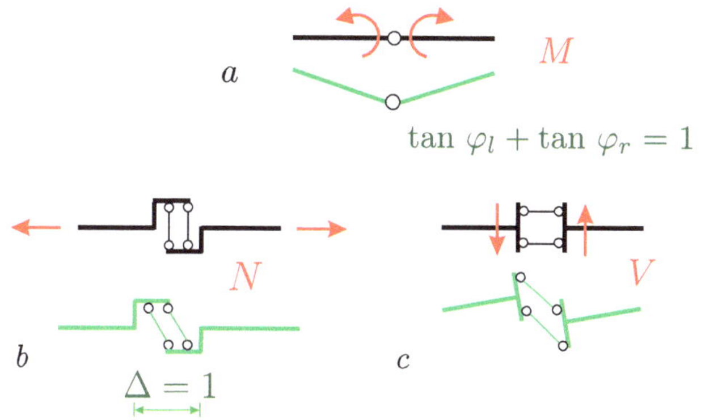

Fig. 2.7 Hinges allow the formation of influence functions, **a** M-hinge, **b** N-hinge, **c** V-hinge

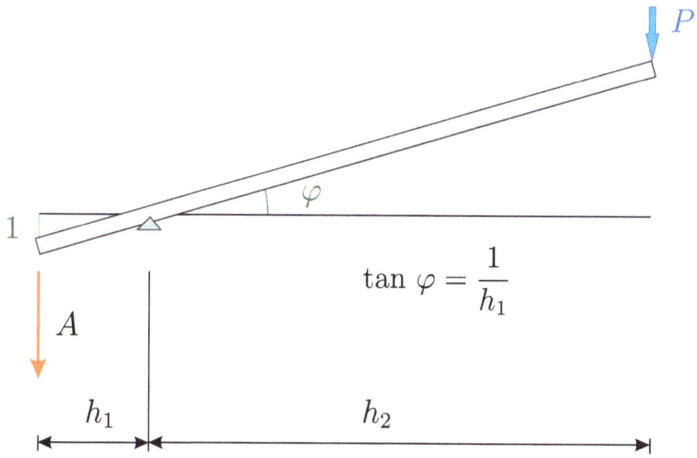

Fig. 2.8 Archimedes' lever

Archimedes

This was also **Archimedes** approach who knew: When he removed the left support of the **lever**, see Fig. 2.8, and pushed the lever down by one $\delta\alpha\kappa\tau\upsilon\lambda o\varsigma$, the work of the support reaction A and the work of the force P had to be the same, the sum had to be zero

$$W_{1,2} = A \cdot 1 - P\, h_2 \tan \varphi = 0\,, \tag{2.25}$$

and so he found for A the value

$$A = P\, h_2 \tan \varphi = P \cdot \frac{h_2}{h_1}\,. \tag{2.26}$$

All influence functions for forces are such **seesaws**, see Fig. 2.6.

Statics is not static, but statics is "**kinematics**".

To each section force belongs a hinge and the deformation, which "ripples" through the structure, ***the echo***, when the hinge is spread, determines how large the section force is at the source point.

2.3.1 Influence Function for $N(x)$

The influence function $G_1(y, x)$ for the normal force $N(x)$ is generated by a (horizontal) unit dislocation at the source point x, see Fig. 2.9b,

$$G_1(x_-) - G_1(x_+) = 1\,. \tag{2.27}$$

In Green's second identity

$$\begin{aligned}\mathcal{B}(G_1, u) &= \mathcal{B}(G_1^L, u)_{(0,x)} + \mathcal{B}(G_1^R, u)_{(x,l)} = 0 + 0\\ &= [\ldots]_0^x - \int_0^x G_1^L(y, x)\, p(y)\, dy + [\ldots]_x^l - \int_x^l G_1^R(y, x)\, p(y)\, dy\,,\end{aligned} \tag{2.28}$$

the work done at the ends of the bar is, due to the boundary conditions, zero, and so we must only be concerned with the action at the source point x.

Since G_1 has the **same slope** on both sides of the source point, see Fig. 2.9b, N_1 is continuous at x so that

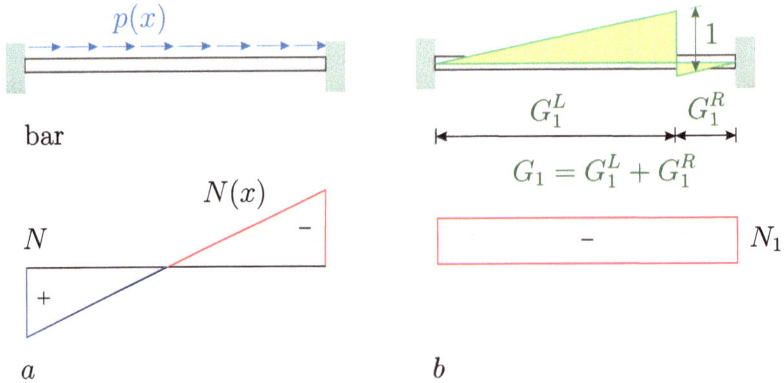

Fig. 2.9 Bar and influence function for the normal force $N(x)$

$$\underbrace{N_1(x_-)\,u(x)}_{left} - \underbrace{N_1(x_+)\,u(x)}_{right} = (N_1(x_-) - N_1(x_+))\,u(x) = 0\,, \tag{2.29}$$

and this gives

$$\begin{aligned}\mathscr{B}(G_1, u) &= N(x)(G_1(x_-) - G_1(x_+)) - \int_0^l G_1(y, x)\,p(y)\,dy \\ &= N(x) \cdot 1 - \int_0^l G_1(y, x)\,p(y)\,dy = 0\,. \end{aligned} \tag{2.30}$$

2.3.2 *Influence Function for $M(x)$*

We install a **hinge** at the source point x, and spread it by one unit, see Fig. 2.10d,

$$G_2'(x_-) - G_2'(x_+) = 1\,. \tag{2.31}$$

When *Betti's theorem* is formulated in "split" fashion

$$\mathscr{B}(G_2, w) = \mathscr{B}(G_2^L, w)_{(0,x)} + \mathscr{B}(G_2^L, w)_{(x,l)} = 0 + 0\,, \tag{2.32}$$

the work done at the beam's ends is zero and at the source point x only the virtual work of the two moments

$$G_2'(x_-)\,M(x) - G_2'(x_+)\,M(x) = 1 \cdot M(x) \tag{2.33}$$

is left, and so

$$1 \cdot M(x) = \int_0^l G_2(y, x)\,p(y)\,dy\,. \tag{2.34}$$

2.3.3 *Influence Function for $V(x)$*

The **shear hinge** at the source point is spread by one unit, see Fig. 2.10e,

$$G_3(x_-) - G_3(x_+) = 1\,, \tag{2.35}$$

and when *Betti's theorem* is formulated in two parts

$$\mathscr{B}(G_3, w) = \mathscr{B}(G_3^L, w)_{(0,x)} + \mathscr{B}(G_3^R, w)_{(x,l)} = 0 + 0\,, \tag{2.36}$$

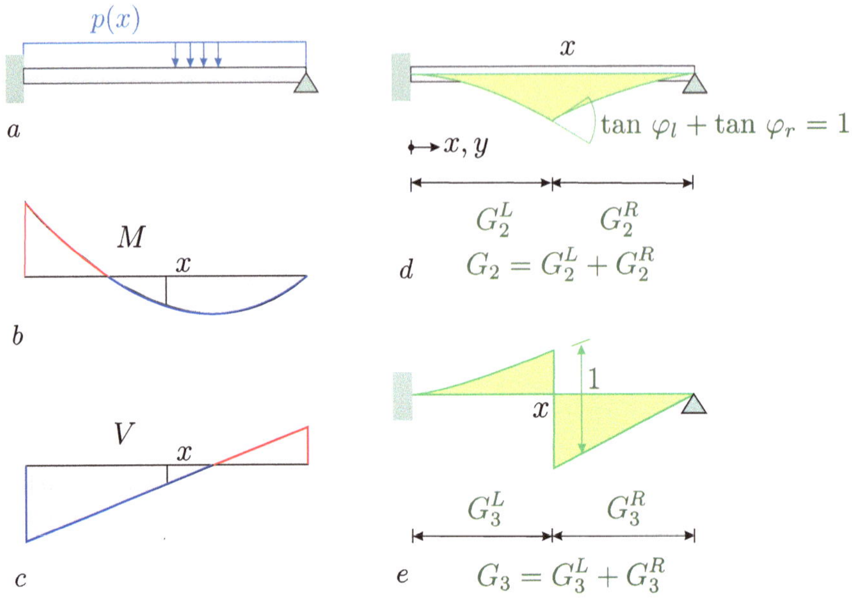

Fig. 2.10 Influence function for $M(x)$ and $V(x)$

the work at the interface is

$$G_3(x_-)\, V(x) - G_3(x_+)\, V(x) = 1 \cdot V(x)\,, \tag{2.37}$$

and so

$$1 \cdot V(x) = \int_0^l G_3(y, x)\, p(y)\, dy\,. \tag{2.38}$$

2.3.4 Settlement of a Support

If a support settles there is no distributed load, since $p = 0$, which we can superpose with the influence function

$$w(x) = \int_0^l G_0(y, x)\, p(y)\, dy \qquad ? \tag{2.39}$$

Instead we rewrite the influence function (via int. by parts) as

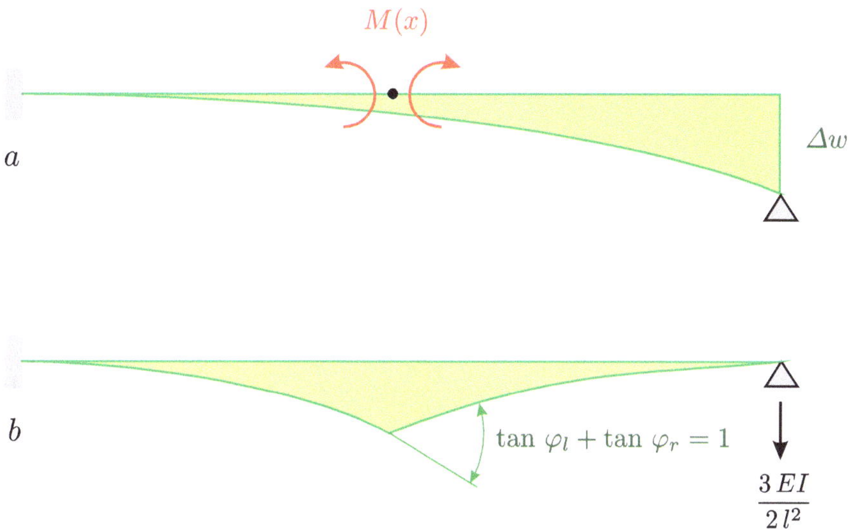

Fig. 2.11 Application of an influence function when a support settles, **a** settlement, **b** influence function $G_2(y, x)$ for $M(x)$

$$\text{virt. disp. } G \times \text{real load } p = \text{virt. force } R \times \text{real disp. } \Delta w\,, \tag{2.40}$$

which means: We multiply the support reaction R of the influence function with the settlement Δw, see Fig. 2.11, to calculate $w(x)$.

To determine $M(x)$ at the midpoint of the beam we argue as follows: The deflection w of the beam is a homogeneous solution, $EI\, w^{IV} = 0$, and the influence function is a solution of the equation $EI\, G_2^{IV}(y, x) = \delta_2(y - x)$ (kink at the source point), and so (we skip the intermediate steps)

$$\mathscr{B}(G_2, w) = -M(x) - V_2(l)\, w(l) = 0\,, \tag{2.41}$$

or

$$M(x) = -V_2(l)\, w(l) = -\frac{3\, EI}{2\, l^2} \cdot \Delta w\,. \tag{2.42}$$

The minus in $-V_2(l)\, w(l)$ has its origin in the minus in front of the second part of *Betti's theorem*

$$\mathscr{B}(G_2, w) = \int_0^l \ldots\, dy + [\ldots]_0^l \underset{\uparrow}{-} [V_2\, w + \ldots]_0^l - \int_0^l \ldots dy = 0\,. \tag{2.43}$$

Let us state this as a rule:

Given a settlement of a support an influence function is evaluated by multiplying the support reaction associated with the influence function with the support settlement $\times(-1)$.

2.3.5 Temperature Effects

An antisymmetric temperature distribution ΔT will cause a beam to curve, to deflect and **Mohr's integral** will supply the answer by how much

$$\bar{1} \cdot \delta = \ldots + \int \bar{M}\, \alpha_T \, \frac{\Delta T}{h}\, dx + \int \bar{N}\, \alpha_T \, T\, dx\,. \tag{2.44}$$

Here $\bar{M}$ and $\bar{N}$ are the bending moment and normal force respectively of the influence function.

Technically temperature effects belong in the realm of mixed formulations, and therefore the above two terms actually come from a strong influence function, see Chap. 7.

2.3.6 Single Moments Differentiate the Influence Functions

The action of the moment in Fig. 2.12 resembles the action of two single forces $P = \pm M/\Delta y$ a distance Δy apart. Seeing it this way, we can assess the effect of M on the end of the beam as in

$$w(l) = \lim_{\Delta y \to 0} (G_0(y + 0.5\,\Delta y, l) - G_0(y - 0.5\,\Delta y, l)) \cdot \frac{M}{\Delta y} = \frac{d}{dy}\, G_0(y, l) \cdot M\,, \tag{2.45}$$

and we recognize that $w(l)$ depends on the **slope of the influence function** at M.

2.4 Statically Determinate Structures

When we add a hinge to a statically determinate structure it becomes a mechanism. Therefore the influence functions for section forces or support reactions are **piecewise linear** as in Fig. 2.13.

The hinge interrupts the flow of the bending moments, and so an equivalent pair $\pm M(x)$ of external moments must compensate the loss of the previous internal moment $M(x)$. In the second figure only the hinge is spread by one unit apart, $\tan \varphi_l + \tan \varphi_r = 1$. No forces are required for this spread. It is a "noiseless" operation.

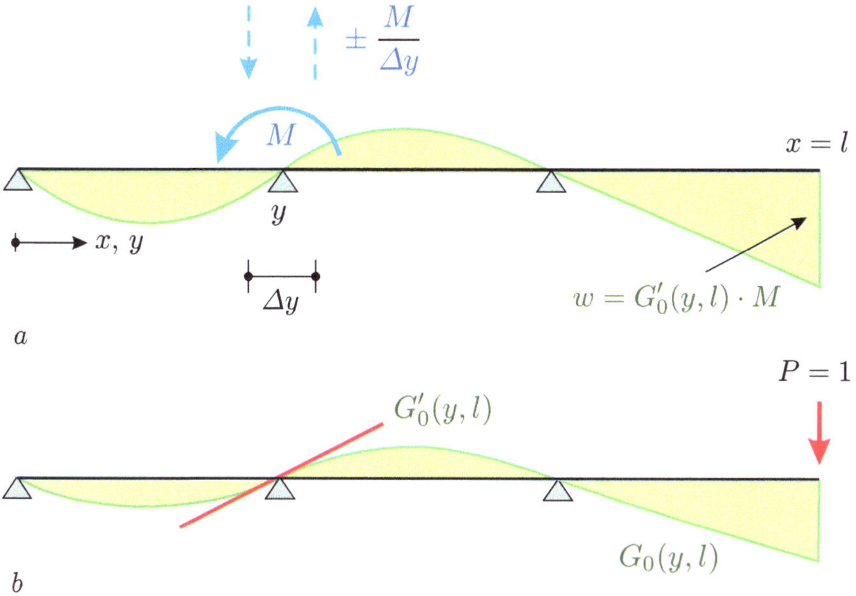

Fig. 2.12 The effect of the moment M depends on the slope of the influence function, **a** single moment, **b** influence function for the deflection of the beam's end

Fig. 2.13 *Betti's theorem* and influence function for a moment, **a** system and load, **b** hinge added and spread, $\tan \varphi_l + \tan \varphi_r = 1$

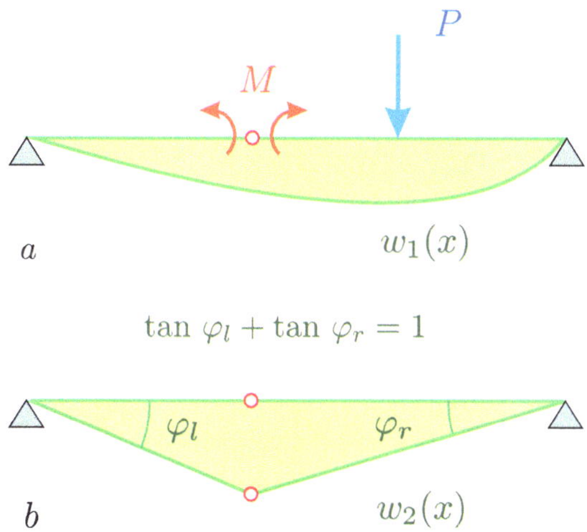

According to *Betti's theorem* the reciprocal external work of the two systems in Fig. 2.13 *a* and *b* must be the same

$$\mathscr{B}(w_1, w_2) = W_{1,2} - W_{2,1} = 0\,. \tag{2.46}$$

But $W_{2,1} = 0$ is zero, since the non-existent forces on system # 2 contribute no work on acting through $w_1(x)$, and therefore

$$W_{1,2} = -M_L \tan \varphi_l - M_R \tan \varphi_r + P\, w_2(x) = -M \cdot 1 + P\, w_2(x) = 0 \tag{2.47}$$

must be zero as well or

$$1 \cdot M = P\, w_2(x)\,, \tag{2.48}$$

and so, $w_2(x)$ is the influence function for $M(x)$.

Remark 2.1 Forces in a statically determinate structure do not change when the stiffness changes. An "academic" proof of this principle is the following:

- If EI or EA change in a member, the associated forces f^+ are equilibrium forces, see Chap. 5.
- The forces f^+ are orthogonal to all rigid-body motions
- The influence functions for N, M, V are rigid-body motions

The consequence is: The internal forces do not change.

2.4.1 Pole-Plans

To draw the displaced shape of a mechanism requires the knowledge of the ***instant centers of rotation*** or poles of the single segments. The plan of these poles we call *pole-plan*.

The following rules apply to the construction of pole-plans, see Fig. 2.14:

1. Each fixed hinge is the main pole of the attached segment.
2. Each hinge forms the secondary pole of all the segments connected with the segment.
3. The line orthogonal to a roller support forms the location of the main pole of the segment attached to the support.
4. If two segments are connected by a sliding hinge (normal force hinge or shear force hinge) their secondary pole lies on the line orthogonal to the sliding motion at infinity.
5. The main poles of two segments and their common secondary pole lie on a straight line: $(i) - (i, j) - (j)$, e.g.: $(1) - (1, 2) - (2)$.
6. The secondary poles (i, j), (j, k), (i, k) of three segments I, J, K lie on one line: $(i, j) - (j, k) - (i, k)$, e.g.: $(1, 3) - (1, 4) - (3, 4)$.

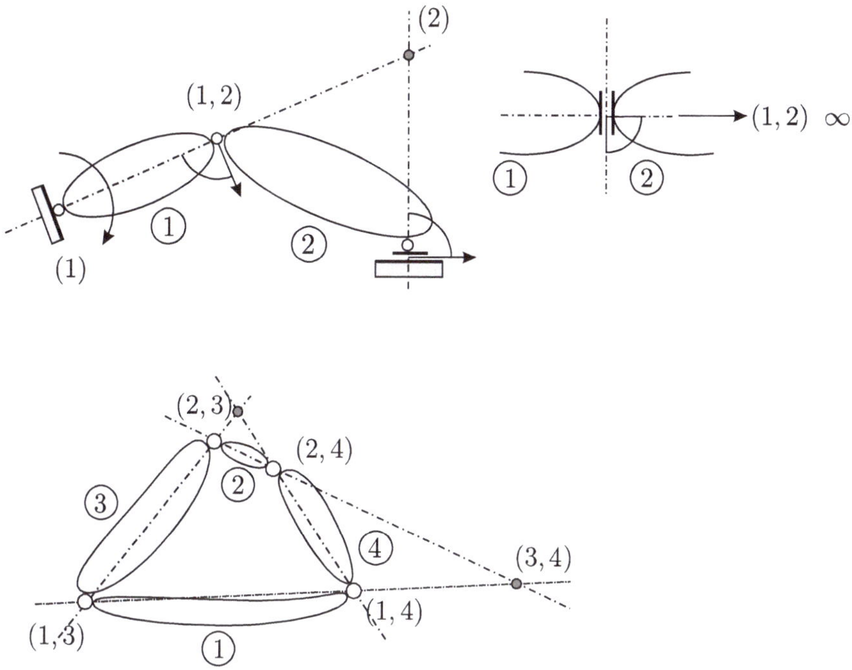

Fig. 2.14 Rules for pole-plans

2.4.2 Construction of Pole-Plans

It is best to start at hinged supports, since these are, see rule #1 the main pole of the segment attached to the hinge. Hinges in between segments are the secondary poles of the segments connected at the hinge, see rule #2.

All the other poles can be found with the help of so-called ***trace lines***. A trace line or locus is the straight line on which the pole must lie according to rules #3 to #6.

- The intersection of two trace lines of the same pole marks the location of the pole.
- If two trace lines run parallel and aim at the same pole, they intersect at infinity.
- If the main pole of a segment lies at infinity, the segment can only move parallel to itself; its rotation is zero.
- If the secondary pole of two segments lies at infinity, the two segments rotate about their main poles by equal amounts. If two segments are parallel to each other they stay parallel to each other after the motion started.

These rules allow to determine the motions when an M-, V- or N-hinge is spread by one unit. The spread is opposite to the direction of the forces at the hinge. The part of the deformed structure which has the same direction as the traveling load is the influence function. We also say: The influence function is the "projection" of the displaced shape of the structure into the direction of the traveling load.

2.4.3 How to Determine the Magnitude of Rotations

Eventually one needs the rotation angles of the single segments of a mechanism and the relation between these angles. This information is easily obtained by considering the connection between the rotations of two segments (i) and (k), connected via a secondary pole (i, k). Figure 2.15 illustrates how we proceed.

The rotation φ_i of the segment i is given and we want to find the rotation of the segment k as a function of φ_i.

Let x_i be the horizontal distance of the main pole (i) from the secondary pole (i, k), and x_k the horizontal distance of the main pole (k) from the secondary pole (i, k). Correspondingly do y_i and y_k represent the vertical distances and l_i and l_k are the distances of the main poles (i) and (k) from the secondary pole (i, k).

So, we have

$$\tan\,\varphi_i = \frac{\eta}{l_i} \qquad \tan\,\varphi_k = \frac{\eta}{l_k} \tag{2.49}$$

and

$$\boxed{l_i \cdot \tan\,\varphi_i = l_k \cdot \tan\,\varphi_k} \tag{2.50}$$

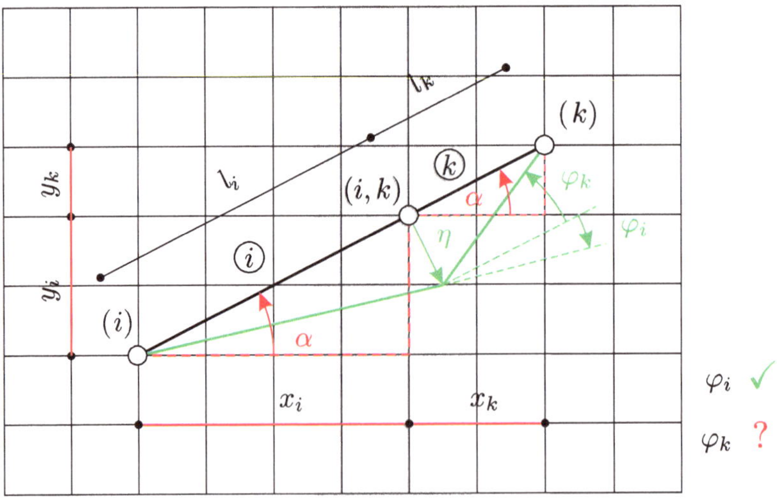

Fig. 2.15 The relative rotations between two segments

The main poles of the two segments and their secondary poles lie—as always—on a straight line which, here, is inclined at an angle α and hence

$$\sin \alpha = \frac{y_i}{l_i} = \frac{y_k}{l_k} \tag{2.51}$$

or solved for the lengths

$$l_i = \frac{y_i}{\sin \alpha} \qquad l_k = \frac{y_k}{\sin \alpha}, \tag{2.52}$$

and with (2.50) follows as well

$$\boxed{y_i \cdot \tan \varphi_i = y_k \cdot \tan \varphi_k} \tag{2.53}$$

By the same reasoning does

$$\cos \alpha = \frac{x_i}{l_i} = \frac{x_k}{l_k} \tag{2.54}$$

imply the result

$$\boxed{x_i \cdot \tan \varphi_i = x_k \cdot \tan \varphi_k} \tag{2.55}$$

A system composed of three segments, see Fig. 2.16, may illustrate the technique. We are given the angle φ_1 having the value $\tan \varphi_1 = 1/3$. We want to find the other two angles φ_2 and φ_3. Because of (2.55) the rotation of segment #2 is

$$\tan \varphi_2 = \frac{x_1 \cdot \tan \varphi_1}{x_2} = \frac{3 \cdot 1/3}{1} = 1\,. \tag{2.56}$$

The rotation of segment #3 follows from (2.53)

$$\tan \varphi_3 = \frac{y_2 \cdot \tan \varphi_2}{y_3} = \frac{2 \cdot 1}{2} = 1 \tag{2.57}$$

or with (2.55) by considering the rotation of segment #1

$$\tan \varphi_3 = \frac{\bar{x}_1 \cdot \tan \varphi_1}{\bar{x}_3} = \frac{9 \cdot 1/3}{3} = 1\,. \tag{2.58}$$

So, the rotations of the single segments are

$$\begin{bmatrix} \tan \varphi_1 \\ \tan \varphi_2 \\ \tan \varphi_3 \end{bmatrix} = \begin{bmatrix} 1 \\ 3 \\ 3 \end{bmatrix} \cdot \tan \varphi_1\,. \tag{2.59}$$

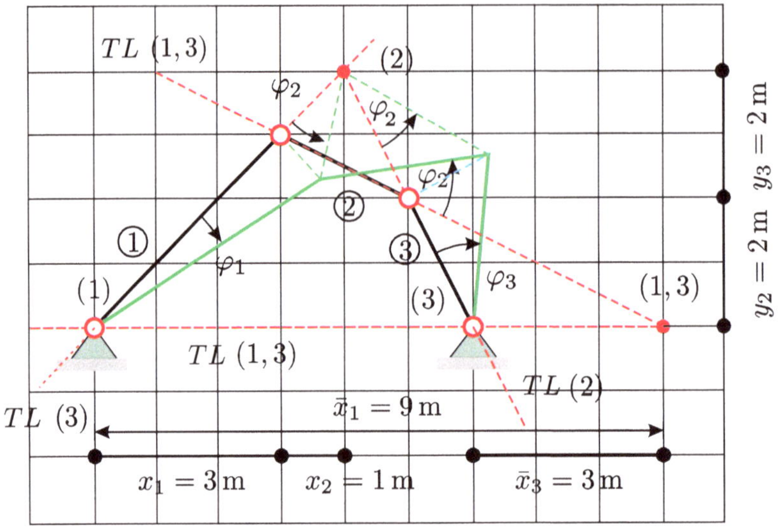

Fig. 2.16 Computing the displacements of segments 2 and 3, when the rotation of segment 1 is given, TL = trace line

2.4.4 *Influence Function for a Shear Force, Fig. 2.17*

In Fig. 2.17 we determine the influence function for the shear force in the slanted beam. The secondary pole (2,3) lies on any trace line which is orthogonal to the sliding shear hinge and therefore also on the trace line, which passes through the main pole (3). This straight line is at the same time the trace line of the main pole (2), as well as the connecting line of (1) and (1,2). The main pole (2) lies at the intersection of these two lines.

We start by spreading the shear hinge by one unit. In vertical direction, which is also the direction of the traveling load, this means a relative displacement of the segments at the hinge and also above the main poles of 0.8 m.

The relative displacement atop the main poles can be used to trace the motion of the segments in the projection. Since the main poles are fixed the displacement of segment #2 atop the main pole (3) amounts to 0.8 m and viceversa amounts the displacement of segment #3 atop the main pole (2) to the same value 0.8 m.

The vertical motion determines the influence function.

2.4.5 *Influence Function for a Normal Force, Fig. 2.18*

In Fig. 2.18 we construct the influence function for the normal force in the slanted beam. We know, the trace line of the secondary pole (2,3) is found at infinity when we follow any straight line which is orthogonal to the gliding motion of the N-hinge.

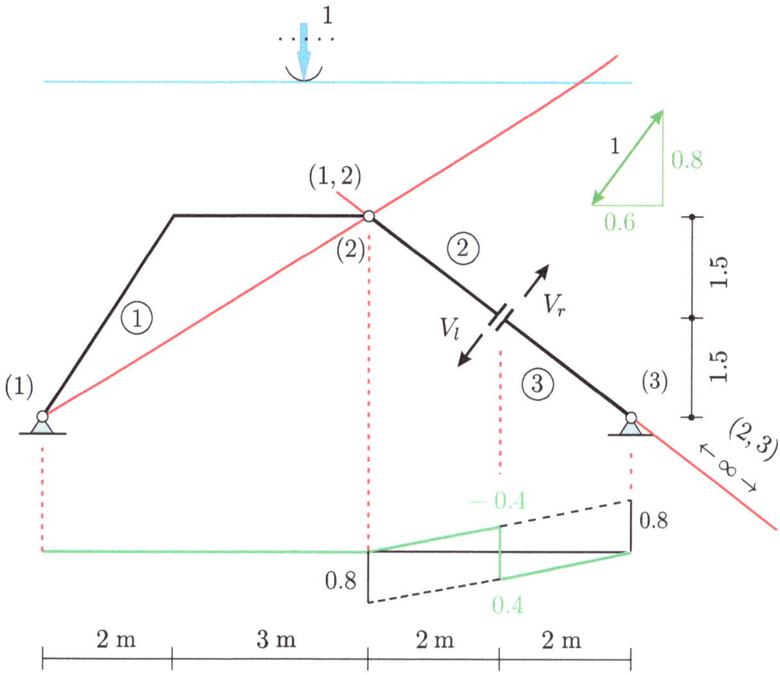

Fig. 2.17 Traveling load and influence function for a shear force, (1), (2), (3) are the main poles of the segments 1, 2 and 3 and (1, 2), (2, 3) are the secondary poles of segments 1 and 2 and segments 2 and 3 respectively

Such a line passes also through the main pole (3), and therefore this line is also the trace line for the main pole (2). The line, which passes through the main pole (1) and the secondary pole (1,2), is also a trace line for the main pole (2). The intersection of both lines must be the main pole (2).

To construct the influence function, we spread the N-hinge by one unit. In vertical direction, in the direction of the traveling load, this corresponds to a relative displacement of the segments at the hinge and atop the main poles of $0.5 \cdot \sqrt{2}$ m. The relative displacement atop the main poles can be used to construct the displacements in the projection. Since the main poles are fixed, the displacement of segment #2 atop the main pole (3) is $0.5 \cdot \sqrt{2}$ m and vice versa the displacement of segment 3 atop the main pole (2) is $0.5 \cdot \sqrt{2}$ m.

The vertical displacements form the influence function.

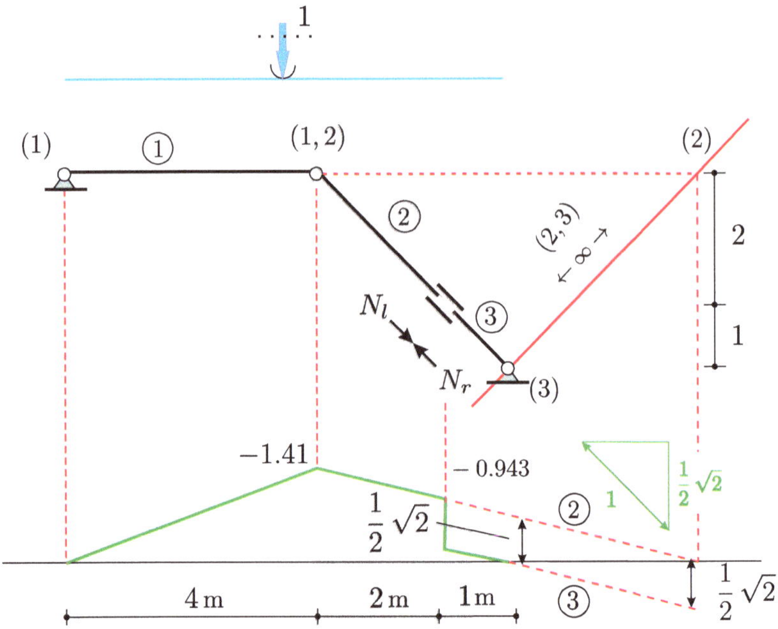

Fig. 2.18 Traveling load and influence function for a normal force

2.4.6 Influence Function for a Moment, Fig. 2.19

In Fig. 2.19 we construct the influence function for the bending moment at the point i. After we have installed a hinge at this point the previously statically determinate structure can now rotate freely about the hinge. We want a rotation where $\tan\varphi_r + \tan\varphi_l = 1$ and this requires the vertical displacement at the source point i to have the value

$$\eta = \frac{x_1 \cdot x_2}{x_1 + x_2} = \frac{3 \cdot 4}{3 + 4} = \frac{12}{7}\,. \tag{2.60}$$

In the final step the displaced structure is projected into the direction of the traveling load. The rotations of the three segments in the direction of the projections agree with the rotations of the structure. At the main poles, the displacement is zero and therewith also in the projection. With these hints it is easy to draw the displacements, which the traveling load experiences, that is the influence function.

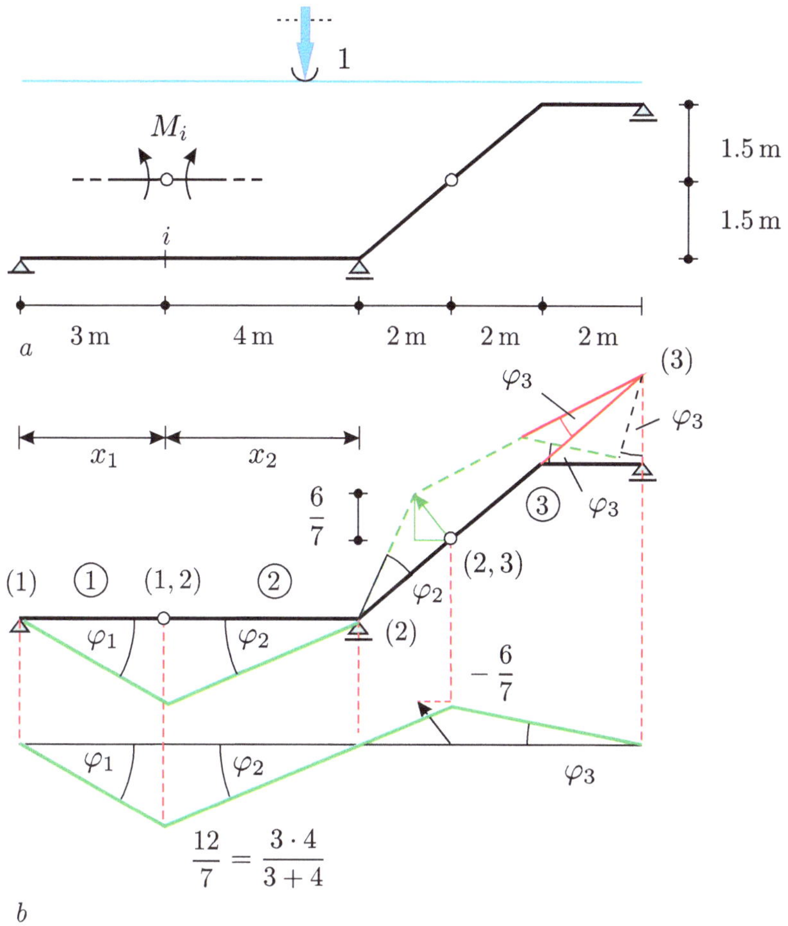

Fig. 2.19 Traveling load and influence function for the moment at point i

2.4.7 *Influence Function for a Moment, Fig. 2.20*

In Fig. 2.20 we construct the influence function for the bending moment at a point i and to this end we install a hinge at this point whereby the structure becomes a mechanism. Unlike the earlier example in Fig. 2.19 both main poles of the two segments lie on the right side of the hinge, of this secondary pole (2, 4).

We want $\tan \varphi_4 + \tan \varphi_2 = 1$. This is guaranteed if the relative rotation between the two segments #2 and #4 equals 1, and this is the case if the vertical displacement η at the source point i has the value

$$\eta = \frac{x_2 \cdot x_4}{x_2 - x_4} = \frac{3 \cdot 1}{3 - 1} = \frac{3}{2}\,. \tag{2.61}$$

2.4.8 Influence Function for a Shear Force, Fig. 2.21

In Fig. 2.21 we construct the influence function for the shear force at a point i, and we start by installing a shear-hinge at this point. As in the previous example in Fig. 2.20 both main poles of the two segments which are joined at this hinge lie to the right of the source point i.

To construct the influence function we spread the shear-hinge by one unit apart. The work is negative if segment #2 and segment #3 rotate anti-clockwise about the main poles. These rotations are also seen in the projection.

In vertical direction, the direction of the traveling load, the relative displacement of the segments atop the main poles is 1 m. The relative displacement atop the main poles can be used to construct the displacements in the projection, because the main poles are fixed, the displacement of segment #2 atop main pole (3) is 1 m and vice versa.

2.4.9 Influence Function for Two Support Reactions, Fig. 2.22

In Fig. 2.22 we construct the influence functions for two support reactions, R_A and R_B. Because the support reactions follow the direction of the projection, we only need to push the points A and B respectively by one unit in negative direction, which is downwards.

The influence function for support reaction R_A is easily found because the displaced support lies on segment #1 which is also the path of the traveling load. In the case of the support reaction R_B the displaced point belongs to segment #3 and #4, and the main poles of these two segments lie at infinity. Therefore, these segments move in parallel one unit downwards.

On these two segments we find the secondary poles (1,3) and (2,4) of the segments #1 and #2. The connection of (1) with (1,3) and of (2) with (2,4) in the projection determines the segments 1 and 2 and therewith the influence function.

2.4.10 Abutment Reaction, Fig. 2.23

The horizontal support reaction H of an arch, see Fig. 2.23,

$$H = \frac{M}{f} \tag{2.62}$$

equals the bending moment M of a beam with the same span l and load, divided by the height f of the arch.

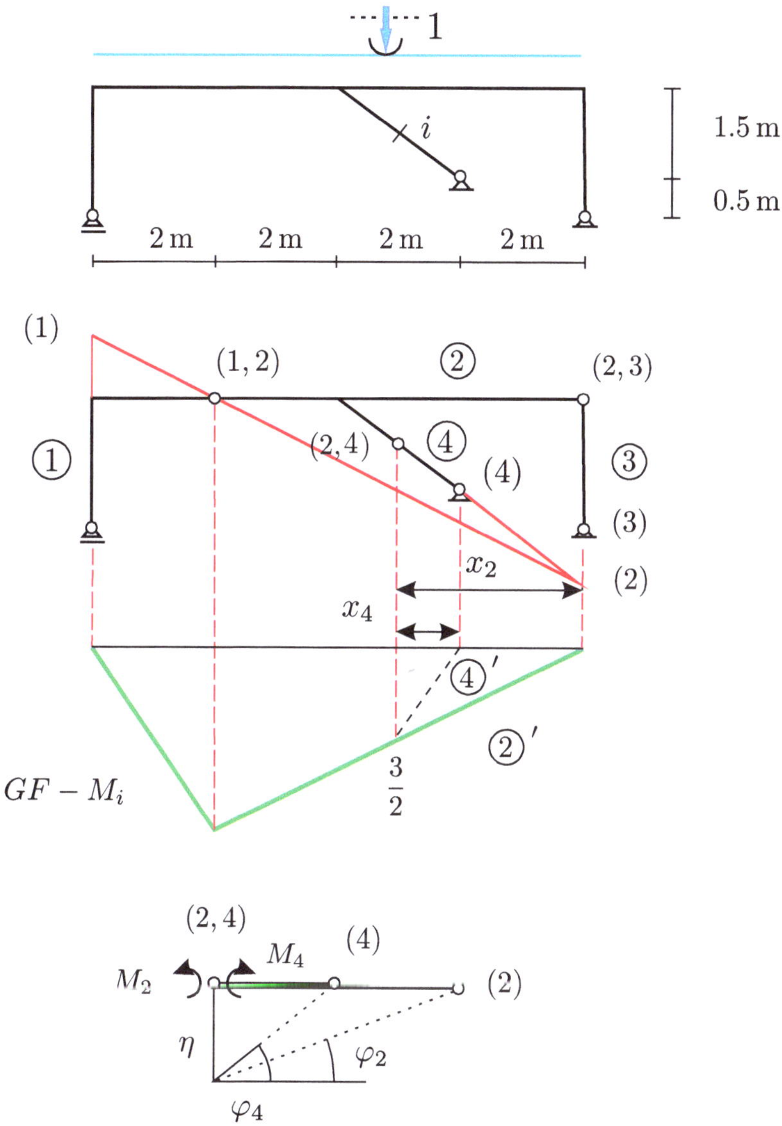

Fig. 2.20 Traveling load and influence function for the moment at point *i*

In principle a two-hinged arch is statically indeterminate, but if the load is symmetric, $V = 0$ is zero on the symmetry axis, and so we can install a shear-hinge at the top without disturbing the equilibrium of the arch. With this added shear-hinge the arch becomes statically determinate, and we can apply our technique to derive (2.62).

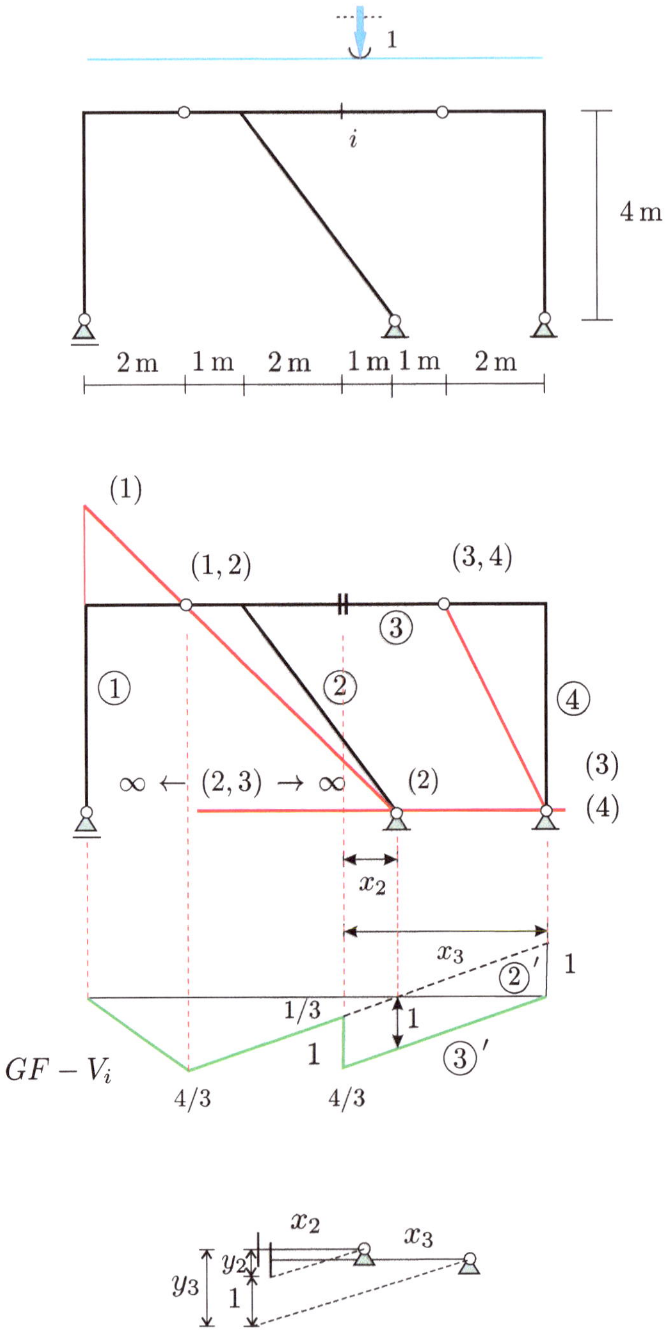

Fig. 2.21 Traveling load and influence function for the shear force at point i

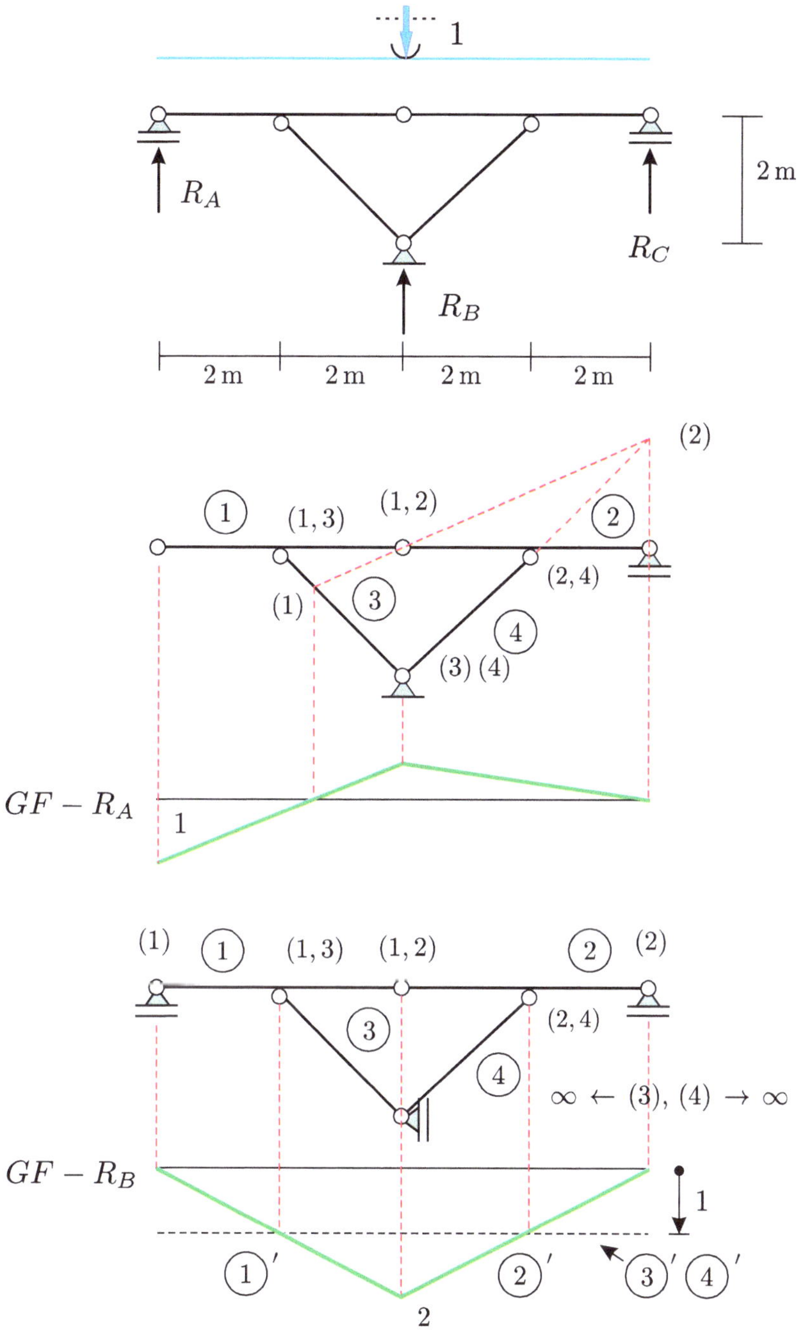

Fig. 2.22 Two influence functions for two support reactions

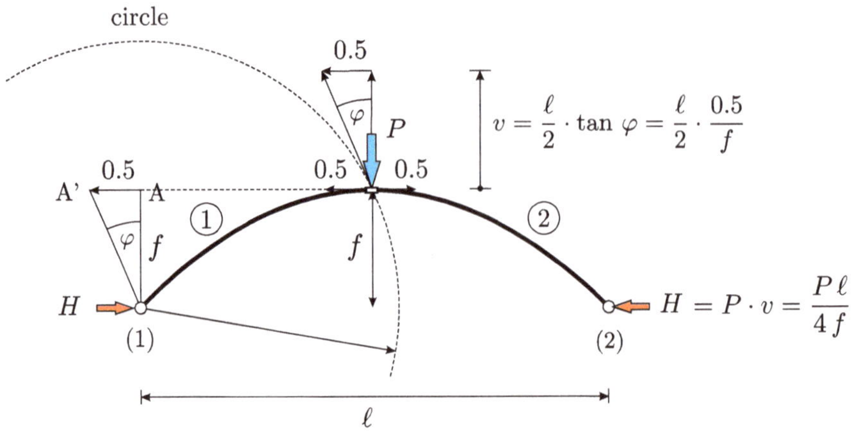

Fig. 2.23 Spread of the top of the circular arch by 1 meter. The kinematics of the arch determines the statics of masonry arches [1]

The normal force $N = H$ at the top of the arch in Fig. 2.23 turns into an external force when we install an N-hinge. So, when the hinge is spread by one meter, the left side of the arch rotates about the main pole #1 and all points, which have the same height f above the main pole, swing to the left by 0.5 m, as point A in Fig. 2.23. This allows to compute $\tan\varphi = 0.5/f$, and all other points, which have the horizontal distance $\ell/2$ from the main pole, swing by $v = \ell/2 \cdot \tan\varphi$ upwards, and so $H = P \cdot v = P \cdot l/(4 \cdot f) = M/f$. In the case of a uniform load p this results in

$$H = 2 \cdot p \int_0^{\ell/2} x \cdot \tan\varphi \, dx = \frac{p \cdot \ell^2}{8 \cdot f} = \frac{M}{f} . \tag{2.63}$$

2.5 Free Ends

One concluding remark on cantilever beams or, more generally speaking, **attached structural parts**. Why is it that only the right side, the free end, but not the left part moves, when the spread $\Delta w = 1$ or $\tan\varphi_r = -1$ is initiated, as in Fig. 2.24 a, b?

To see also a displacement on the left would require the action of a force, actually two, $\pm f$, at both sides of the joint, but then the right part would enter into a "spin", because it is not supported. So, if one side is kinematic, the other side will not move.

In the case of the angled two-span beam on "ball bearings" in Fig. 2.24c, the bearings do not hinder axial rotations and so the right part starts to rotate freely, when the joint (Green's function for M on the left side of the support) is spread, Fig. 2.24 d, and the absence of forces means absence of displacements in the first span, the Green's function is zero. The same applies in reverse order to the moment on the right of the support. The two-span beam carries any load like a series of two

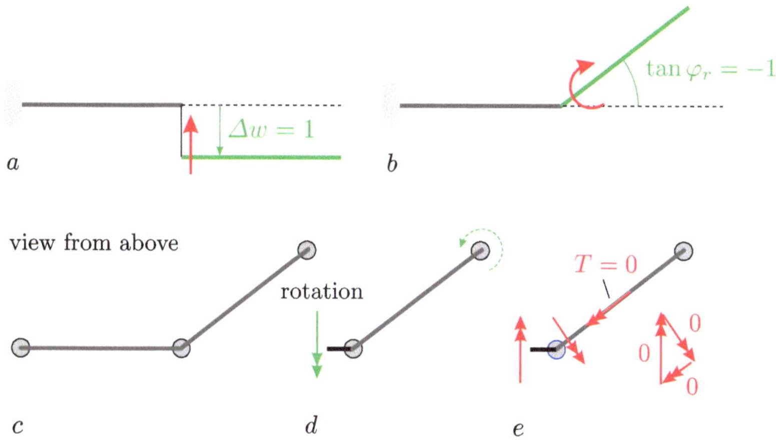

Fig. 2.24 Influence functions for a cantilever beam and an angled two-span beam

single-span hinged beams. If the beams form an angle of 90°, it is obvious, but it also true for any other angle $\neq 180°$ in between. At 180° the beam becomes unstable, since the beam will rotate about its longitudinal axis, and thus will shed off any load.

2.6 Statically Indeterminate Structures

If the structure is statically indeterminate, forces are required to spread the hinge, so system # 2 is not load free, but even in this situation the work $W_{2,1} = 0$ is zero, and *Betti's theorem* reduces as before to

$$\mathscr{B}(w_1, w_2) = W_{1,2} = 0\,. \tag{2.64}$$

A look at the beam in Fig. 2.25 will explain why. To generate the influence function for the bending moment at the midpoint of the beam, we install a hinge and rotate the two beam ends

$$\tan\,\varphi_r + \tan\,\varphi_l = 1\,. \tag{2.65}$$

In formulating *Betti's theorem* we integrate from 0 to x, and from x to the end l of the beam

$$\mathscr{B}(G_2, w) = \mathscr{B}(G_2, w)_{(0,x)} + \mathscr{B}(G_2, w)_{(x,l)}\,. \tag{2.66}$$

Since no work is done at the beam's ends all the work is done by the two moments at the hinge

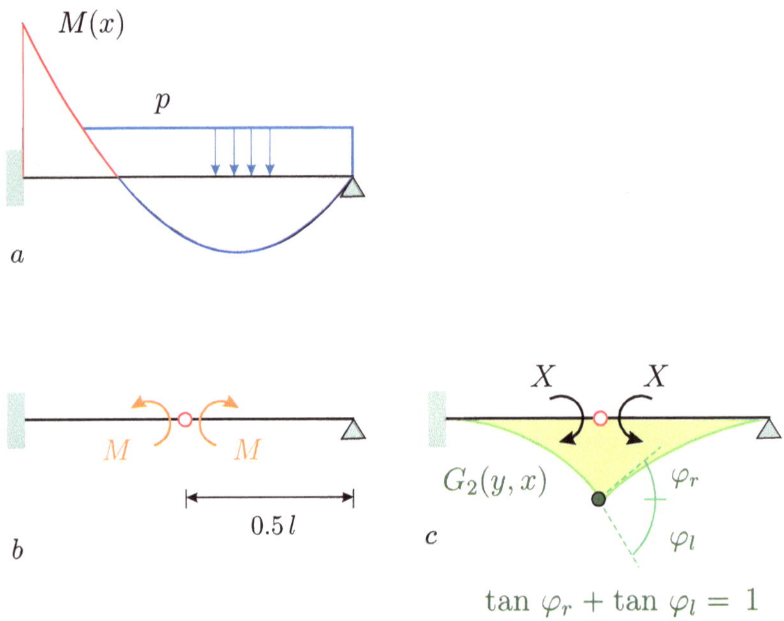

Fig. 2.25 With a hinge added, the internal moments become external moments

$$\underbrace{-M(x)\, w'(x_-)}_{\text{on the left}} + \underbrace{M(x)\, w'(x_+)}_{\text{on the right}} = -M(x) \cdot (\tan\varphi_r + \tan\varphi_l)\,, \tag{2.67}$$

and so

$$\mathscr{B}(G_2, w) = -M(x) \cdot \underbrace{(\tan\varphi_r + \tan\varphi_l)}_{=1} + \int_0^l G_2(y,x)\, p(y)\, dy = 0 \tag{2.68}$$

or

$$M(x) = \int_0^l G_2(y,x)\, p(y)\, dy\,. \tag{2.69}$$

It remains to find the shape $G_2(y, x)$. The unit-spread of the hinge requires the action of two moments $\pm X$. In this situation we first apply two moments $X = \pm 1$ and we scale these moments so that the resulting spread is exactly one.

The work is zero, which is done by the two moments on acting through the deflection of system #1

$$W_{2,1} = -X \cdot w'(x) + X \cdot w'(x) = (-X + X)\, w'(x) = 0\,, \tag{2.70}$$

and therefore

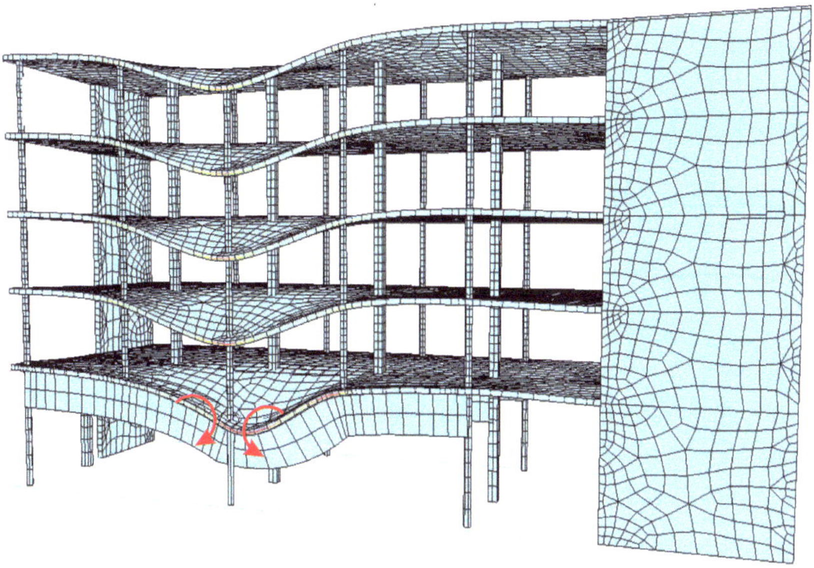

Fig. 2.26 Influence function for the moment in a girder; the work $W_{2,1} = 0$ since in any load case the slope w' at the source point is continuous

$$W_{1,2} = \int_0^l G_2(y,x)\, p(y)\, dy - M \cdot (\tan \varphi_l + \tan \varphi_r) = 0\,, \tag{2.71}$$

must be zero as well since $W_{1,2} = W_{2,1}$.

The result $W_{2,1} = 0$ is no coincidence, it is always zero, see Fig. 2.26. The two (very large) moments which cause the spread have opposite signs but the same size, and since in any load case the rotation $w'(x)$ is continuous at the source point the total work is zero.

When calculating influence functions for forces *Betti's theorem* reduces to

$$W_{1,2} = 0\,. \tag{2.72}$$

2.7 Influence Functions for Support Reactions

A support reaction becomes an external force, when we remove the constraint in the direction of the support, and the influence function is the motion of the released support. If the structure sits on a rigid foundation only the structure can move, and

Fig. 2.27 Influence function for the normal force in a column

the foot of the structure (set free by the hinge) must go the full way of -1 units alone, see Fig. 2.27.

If the soil is elastic, a local analysis must clarify how much the soil contributes to the unit-spread and how much the structure. The force X required to spread the support by 1 m is

$$X = \frac{k_S\, k_B}{k_S + k_B} \qquad k_B = k\text{-soil} \qquad k_S = k\text{-structure}\,. \tag{2.73}$$

It all depends on the stiffness of the pier (or spring), see Fig. 2.28. If the pier is flexible a unit-spread -1 at the foot of the pier is consumed to a large part by the pier itself, and the beam hardly registers the displacement at the foot of the pier, which means the influence function has a very low profile in the beam. Vice versa, if it is a solid, rigid pier nearly the full signal will reach the beam, and so the pier will carry a large part of the moving load.

2.8 Jumps in Internal Forces

The shear forces V_l and V_r on both sides of a support are different and so are the influence functions, see Fig. 2.29. The difference of the two is the influence function for the support reaction $V_r - V_l$.

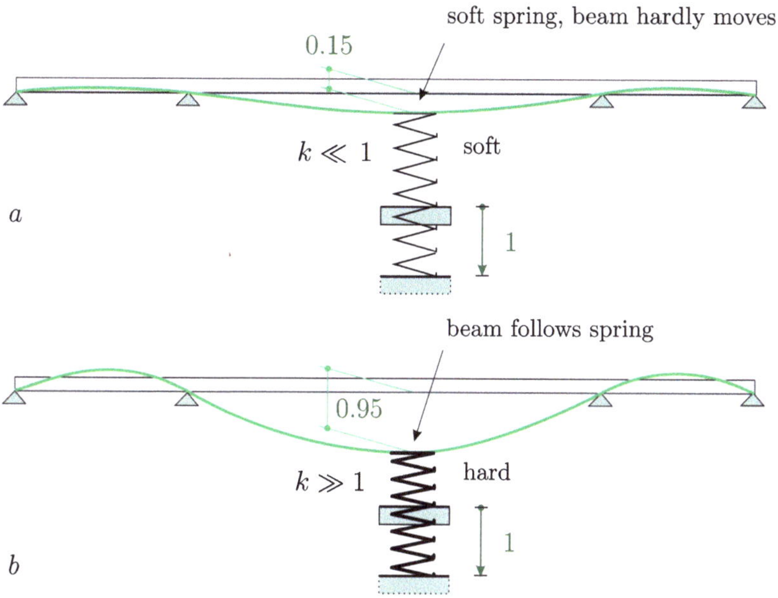

Fig. 2.28 A soft spring dampens much of the motion of the soil, while a stiff spring passes nearly all of it to the beam and so the influence function swings out wide

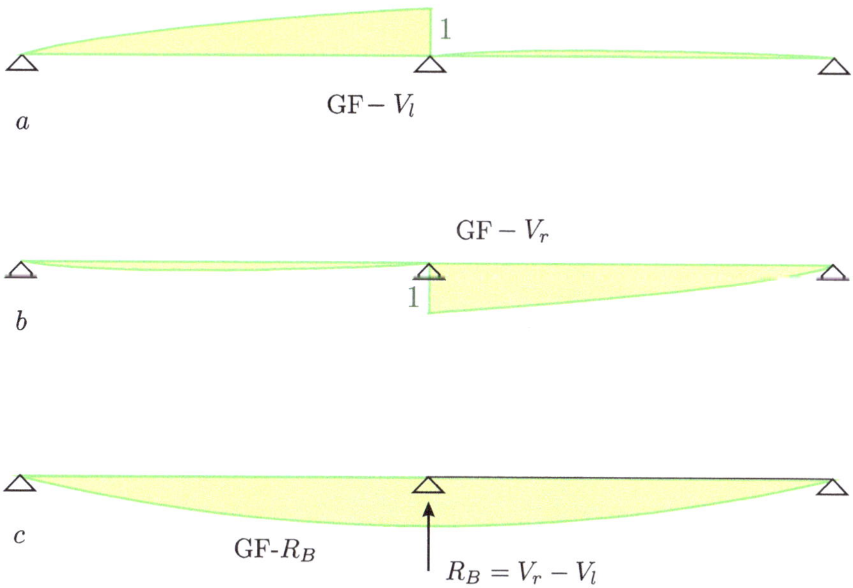

Fig. 2.29 Influence function for the two shear forces, **a** V_l, **b** V_r, and **c** the support reaction $R_B = V_r - V_l$

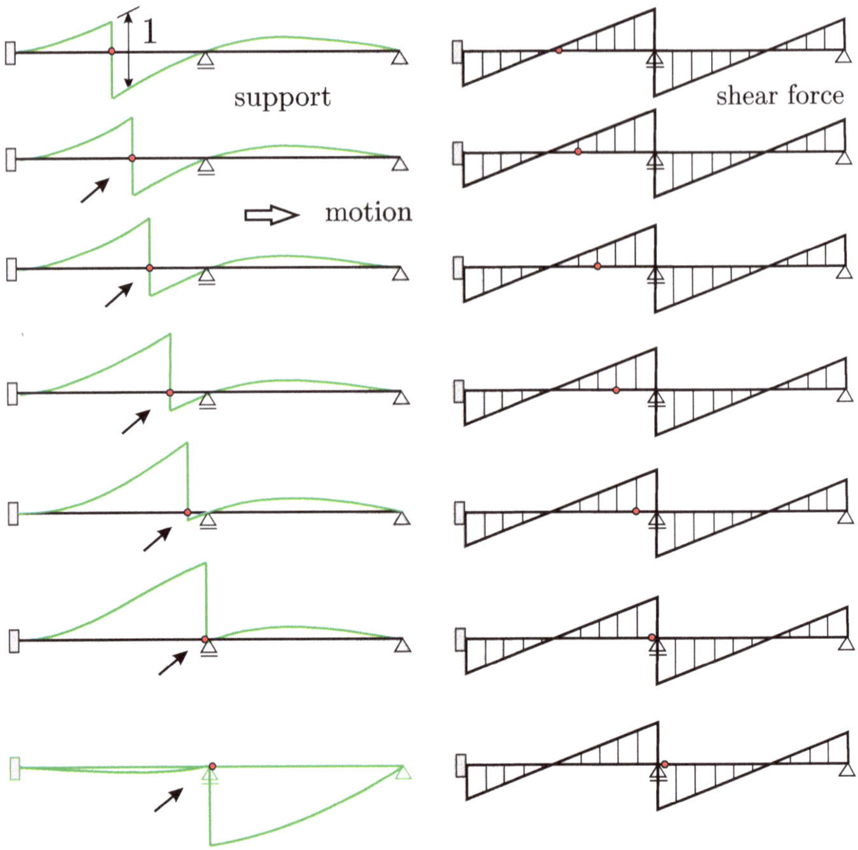

Fig. 2.30 How the influence function for the shear force wanders along the beam, but basically remains the same [2]

The same jump occurs when a travelling point load passes the source point, see Fig. 2.30.

Influence functions for moments are continuous, see Fig. 2.31, but "barely" since the kink indicates that in the next step, $G \to G' = dG/dx$, the kink will become a jump in the derivative. The derivative would be the influence function for $M(x)$ if a moment $\bar{M} = 1$ travels along the beam.

2.9 The Zeros of the Shear Force

A side effect of the equation $EI\, w^{IV} = -V'(x) = p$ is that in a continuous beam **the distance between the zeros** of the shear force is an indicator of the size of the support reactions, see Fig. 2.32.

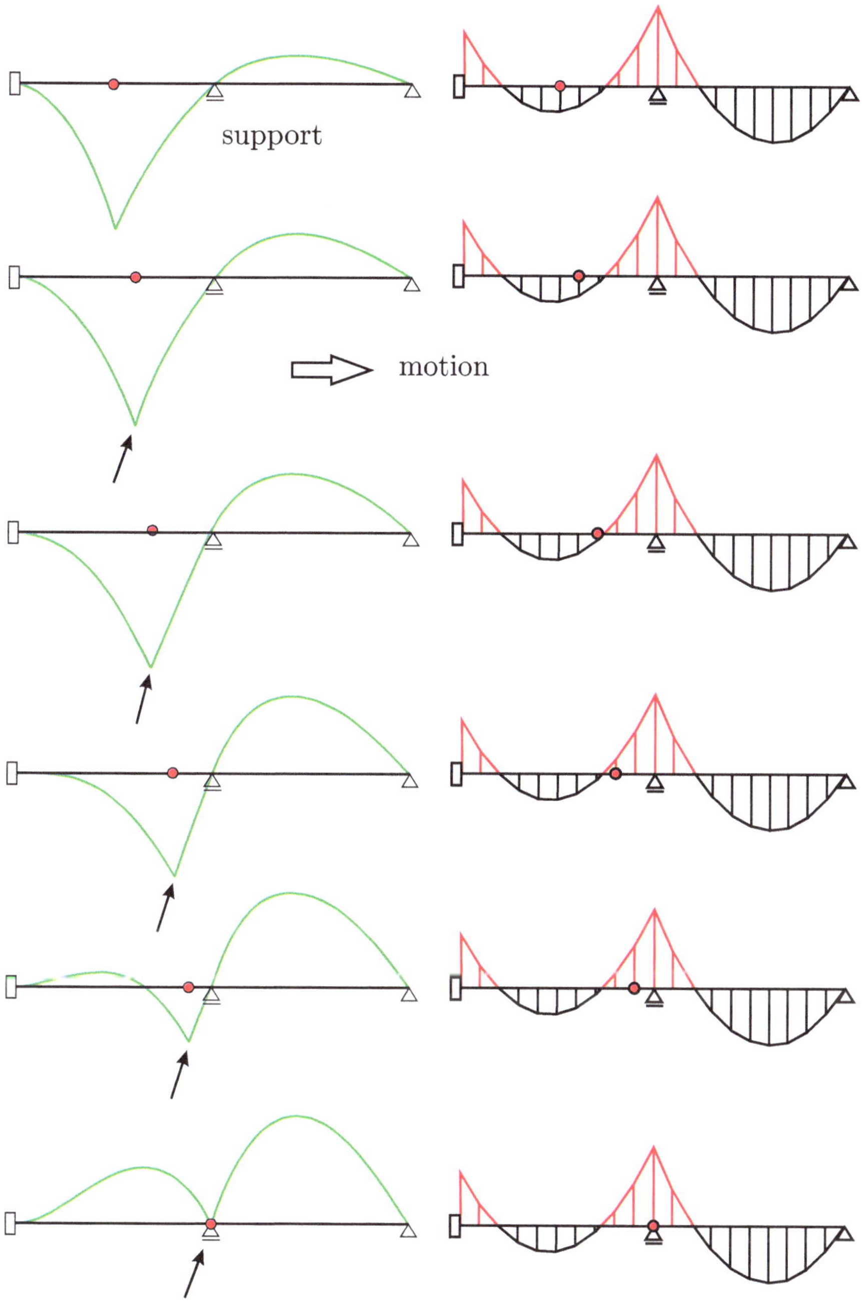

Fig. 2.31 How the influence function for the bending moment wanders along the beam [2]

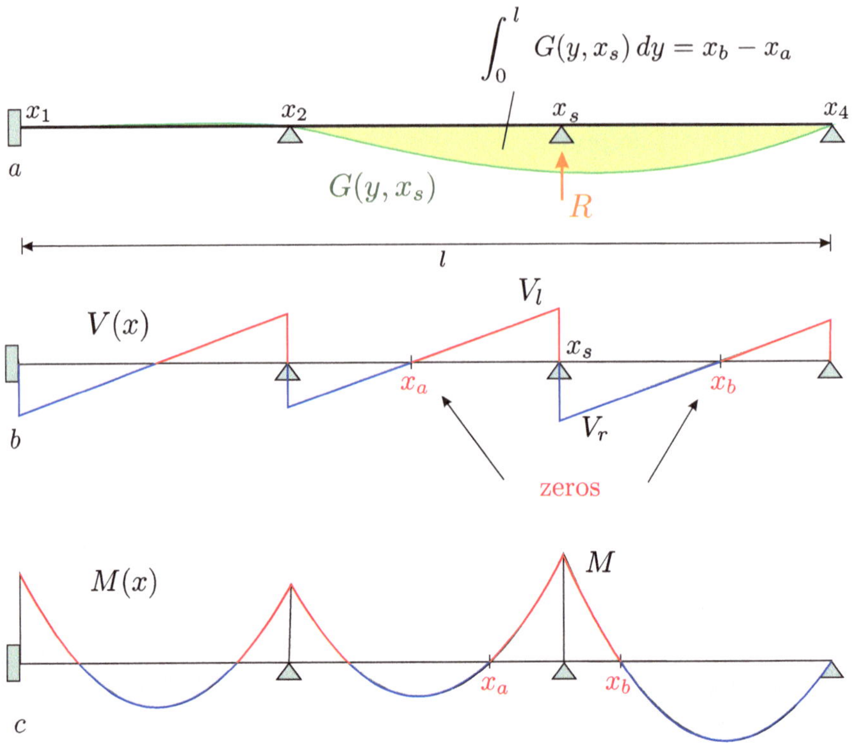

Fig. 2.32 Continuous beam and uniform load, **a** influence function for the support reaction, **b** the zeros of the shear force, **c** moment distribution

Let the two points x_a and x_b be the zeros while the support is located in between at a point x_s. Integrating from x_a up to the support gives

$$\int_{x_a}^{x_s} V'(x)\,dx = V_L(x_s) - V(x_a) = V_L(x_s) - 0 = V_L(x_s)\,, \tag{2.74}$$

and repeating this on the right side of the support gives

$$\int_{x_s}^{x_b} V'(x)\,dx = V(x_b) - V_R(x_s) = 0 - V_R(x_s) = V_R(x_s)\,, \tag{2.75}$$

and so the support reaction is

$$R = V_R(x_s) - V_L(x_s) = \int_{x_s}^{x_b} p(x)\,dx + \int_{x_a}^{x_s} p(x)\,dx = \int_{x_a}^{x_b} p(x)\,dx\,. \tag{2.76}$$

For any constant load p the position of the zeros and the distance between the zeros is the same

$$R = \int_0^l G(y, x_s)\, p\, dy = (x_b - x_a) \cdot p\,. \tag{2.77}$$

2.10 Dirac Deltas

Up to now our approach was rather technical. We have split the beam into two parts, formulated in each part *Betti's theorem* and we have extracted from the zero sum, $0 + 0 = 0$, the influence function.

The Dirac delta in contrast allows a much more direct approach. The Dirac delta is a load which, like a real point force, is zero at all points y not identical with the source point x

$$\delta_0(y - x) = 0 \qquad y \neq x\,, \tag{2.78}$$

and which does the work $w(x) \cdot 1$ on acting through a function $w(x)$

$$\int_0^l \delta_0(y - x)\, w(y)\, dy = w(x) \qquad x \in (0, l)\,. \tag{2.79}$$

Technically, the Dirac delta is a ***distributed load*** [N/m] of infinite height acting over an infinitely short length.

The resulting deflection is the solution of

$$EI \frac{d^4}{dy^4} G_0(y, x) = \delta_0(y - x)\,, \tag{2.80}$$

and with this definition the influence function for $w(x)$ results automatically, so to speak,

$$\begin{aligned}\mathscr{B}(G_0, w) &= \int_0^l \delta_0(y - x)\, w(y)\, dy - \int_0^l G_0(y, x)\, p(y)\, dy \\ &= w(x) - \int_0^l G_0(y, x)\, p(y)\, dy = 0\,.\end{aligned} \tag{2.81}$$

To extend this approach to the slope $w'(x)$, and the section forces $M(x)$ and $V(x)$, we introduce **additional Dirac deltas**

$\delta_0(y - x)$	force $P = 1$	$\delta_1(y - x)$	moment $M = 1$
$\delta_2(y - x)$	kink $\Delta w' = 1$	$\delta_3(y - x)$	dislocation $\Delta w = 1$

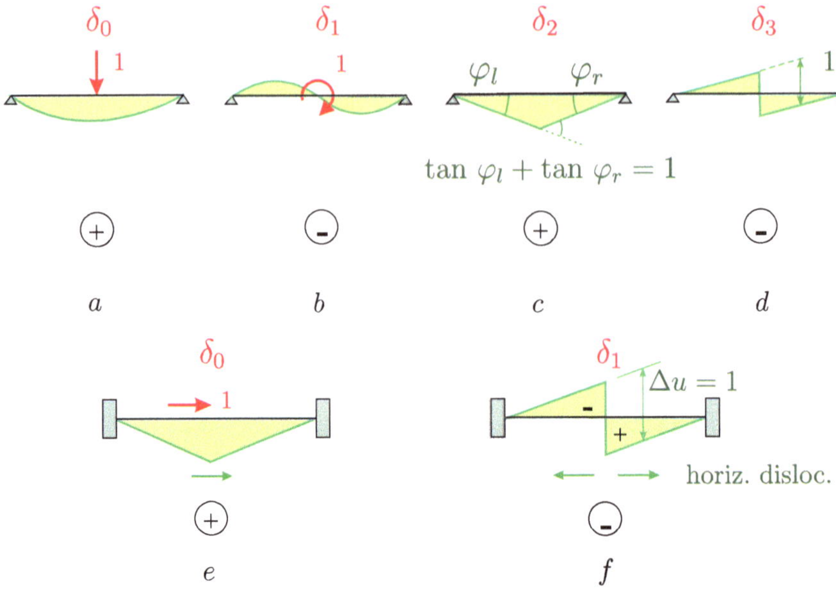

Fig. 2.33 Upper row: influence functions for **a** w, **b** w', **c** M, and **d** V at the midpoint of the beam. The second row are the influence functions for a bar, **e** u, **f** N. The influence functions integrate $+$, or differentiate, $-$, the load

with corresponding properties, see Fig. 2.33,

$$\int_0^l \delta_0(y-x)\, w(y)\, dy = w(x) \qquad \int_0^l \delta_1(y-x)\, w(y)\, dy = w'(x) \tag{2.82a}$$

$$\int_0^l \delta_2(y-x)\, w(y)\, dy = M(x) \qquad \int_0^l \delta_3(y-x)\, w(y)\, dy = V(x)\,. \tag{2.82b}$$

Sure, the Dirac deltas is a very elegant way to circumvent all the cumbersome details of the analytical approach

$$\mathscr{B}(G_0, w) = \lim_{\varepsilon \to 0} \{ \mathscr{B}(G_0^L, w)_{(0,x-\varepsilon)} + \mathscr{B}(G_0^R, w)_{(x+\varepsilon,l)} \} = 0 + 0\,, \tag{2.83}$$

but only this approach provides the necessary details. Later, when the analysis has been done, one can take the shortcut.

And another note: The point-values $w(x)$ etc. do not come from the domain integral, as the Dirac delta wants to make us believe, but they originate on the edge. They are the result of the jump of the internal forces at the interface between the two sides of the beam

$$\underbrace{(V_0^L(x) - V_0^R(x))}_{=1}\, w(x) = 1 \cdot w(x)\,, \tag{2.84}$$

and in 2-D and 3-D problems they are the limits of certain boundary integrals along the perimeter of the shrinking circular hole, which surrounds the source point.

The most important property of Dirac deltas is: They can be integrated, or rather, there are rules for how to interpret an integral such as

$$\int_0^l \delta(y-x)\,\varphi_i(y)\,dy \tag{2.85}$$

since this allows to associate with any Dirac delta an equivalent nodal force

$$f_i = \int_0^l \delta_0(y-x)\,\varphi_i(y)\,dy = \varphi_i(x) \tag{2.86a}$$

$$f_i = \int_0^l \delta_1(y-x)\,\varphi_i(y)\,dy = \varphi_i'(x) \tag{2.86b}$$

$$f_i = \int_0^l \delta_2(y-x)\,\varphi_i(y)\,dy = M(\varphi_i)(x) \tag{2.86c}$$

$$f_i = \int_0^l \delta_3(y-x)\,\varphi_i(y)\,dy = V(\varphi_i)(x)\,. \tag{2.86d}$$

Here, $M(\varphi_i)(x)$ is the moment of the shape function φ_i at the source point x and $V(\varphi_i)(x)$ is the shear force of φ_i at the source point x.

2.11 Dirac Energy

Influence functions express an **energy balance** in the same way as a seesaw or pulley balances the actions (not the forces(!)), see Fig. 2.34. The work of a single force $P = 1$ on a beam on acting through $w(x)$

$$1 \cdot w(x) = \int_0^l G_0(y,x)\,p(y)\,dy\,, \tag{2.87}$$

is the same as the work done by the distributed load p on acting through $G_0(y,x)$, the deflection of the beam due to the single force.

The factor 1 is essential, since without it the dimensions would not match

$$\begin{aligned} \text{force} \cdot \text{displ.} = 1 \cdot u(x) &= \int_0^l G_0(y,x)\,p(y)\,dy \\ &= \text{displ.} \cdot \text{force/length} \cdot \text{length}\,. \end{aligned} \tag{2.88}$$

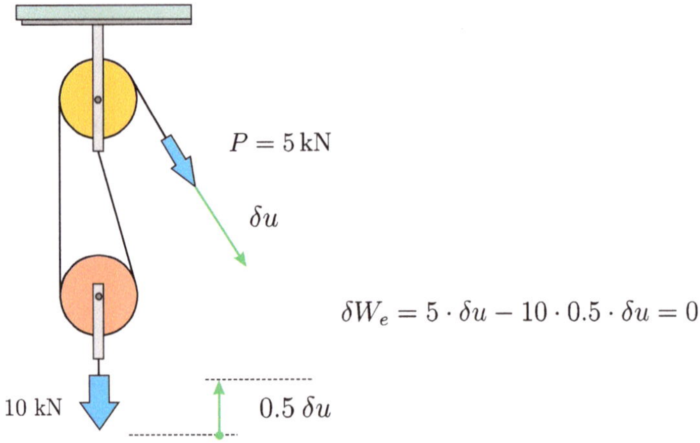

Fig. 2.34 The work done by the two forces is the same

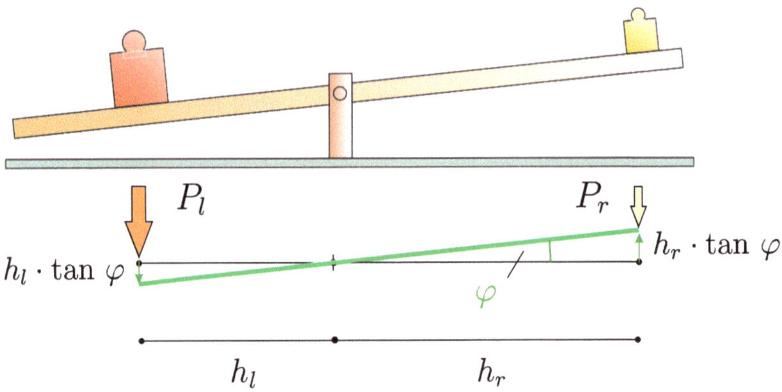

Fig. 2.35 Seesaw

The result of an influence function is an *energy*, an energy quantum, which we call the ***Dirac Energy***.

The Dirac energy is the work done by the load on acting through the influence function.

The simplest example provides a seesaw as in Fig. 2.35, since at each turn φ of the beam the work of the two weights is zero

$$P_l\,u_l - P_r\,u_r = P_l\,\tan\varphi\,h_l - P_r\,\tan\varphi\,h_r = (P_l\,h_l - P_r\,h_r)\,\tan\varphi = 0\,, \quad (2.89)$$

because the two forces balance, $P_l\,h_l = P_r\,h_r$.

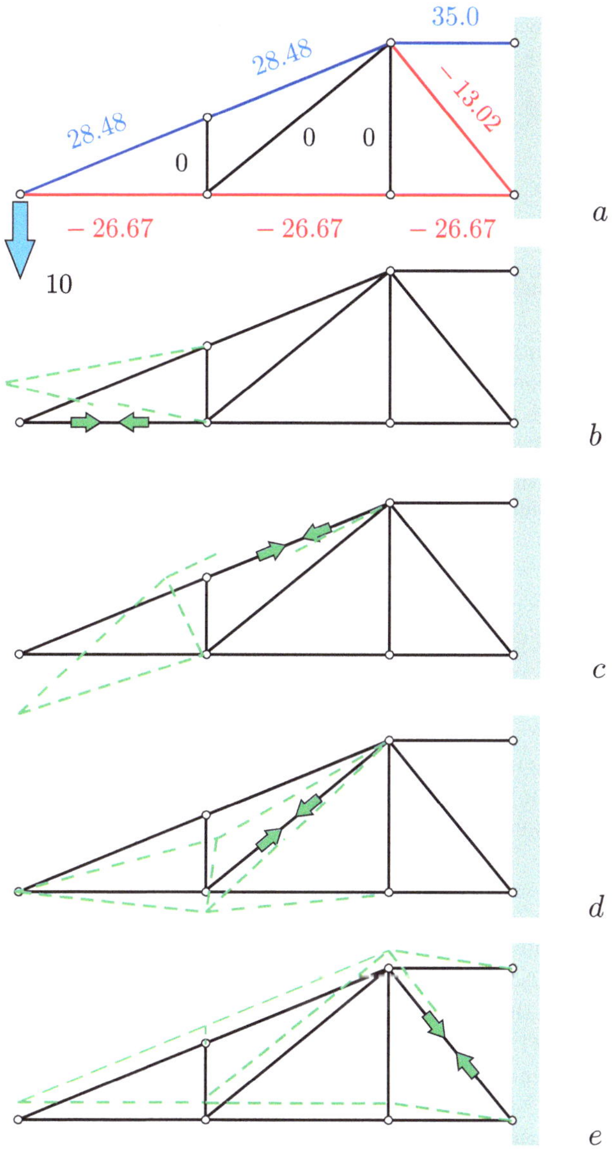

Fig. 2.36 The kinematics of a structure determines the internal forces

In this sense is each influence function a seesaw. To compute the shear force $V(x)$ of a beam at a point x as in Fig. 2.6, we install at x a shear-hinge and we spread the hinge by one unit apart, so that the two shear forces perform the work $-V(x) \cdot 1$

$$-V(x)\, w(x_{-}) - V(x)\, w(x_{+}) = -V(x)\, (w(x_{-}) + w(x_{+})) = -V(x) \cdot 1 \quad (2.90)$$

and the work done by the point load P on acting through w, the response of the beam to this spread, must have the opposite value

$$\underbrace{-V(x)\cdot 1 + P\,w}_{W_{1,2}} = 0\,. \tag{2.91}$$

This is *Betti's theorem*, $W_{1,2} = W_{2,1} = 0$. (The work $W_{2,1}$ is zero, see the following remark).

So, to each internal action $V(x)$, $N(x)$, $M(x)$ etc., belongs a certain mechanism, a "seesaw", see Fig. 2.36, and if we unlock the hinge and compute the work done by the load on acting through the resulting displacement, we learn how large the section force must be to balance the external load.

The problem with finite elements is: We hinder the displacements of a structure, we hobble the structure since the *shape functions* $\varphi_i(x)$ are not flexible enough to represent all possible shapes, and so the load P gets the **wrong signal**. The displacement at the foot of the point load P is w_h

$$-V_h(x)\cdot 1 + P\cdot w_h = 0\,, \tag{2.92}$$

and not the exact value w

$$-V(x)\cdot 1 + P\cdot w = 0\,, \tag{2.93}$$

and so $V_h(x) \neq V(x)$. ***An FE-program gets the Dirac energies wrong***.

We conclude: The kinematics of a mesh, the fineness of details, determines the accuracy of an FE-solution.

mesh = kinematics = precision of influence functions = quality of results

Remark 2.2 Actually, in frame analysis FE-programs get the Dirac energy right, see Fig. 2.38, because the unit-spread of a shear hinge propagates correctly over a frame (disregarding the frame element with the shear hinge itself, see Sect. 3.15 for details) and so $V_h(x) = V(x)$. This can be different when the beams are shaped or other deviations from the beam equation $EI\,w^{IV}(x) = p$ occur.

But in 2-D and 3-D it is neither possible to generate the unit-spread nor does the effect of the (pseudo) unit-spread propagate correctly over the mesh, since FE-solutions (in 2-D and 3-D) are not homogeneous solutions.

We can now also say, what a **good design** is, since the logic of the seesaw

$$V(x) = \frac{P\cdot w}{1} = P\cdot\frac{w}{1} \qquad \leftarrow \qquad \text{make } w \text{ small!} \tag{2.94}$$

signals that in a good design the unit-spread of the shear hinge, the 1 in the denominator, generates a displacement w as small as possible at the foot of the point load P.

Fig. 2.37 The normal force in a truss element depends on how large the displacements at the top are, when the truss element is spread by one unit length

Throw a rock in the pond and watch the waves! The smaller the waves, which reach the load, the better. The **lever of Archimedes** is (quite deliberately) the opposite of a good design. If the short left end is pressed down by one unit, a very large shift w occurs at the right end, and so Archimedes only needs a small force to unhinge the Earth. Conversely, this also means, Archimedes must go long, long ways to lift the globe just one *iota*.

The seesaw logic is universal. When a force $\boldsymbol{P} = \{P_x, P_y, P_z\}^T$ acts at the top of the Eiffel tower, see Fig. 2.37, its effect on the normal force N in a truss element—way down below—depends on how large the displacements g_x, g_y, g_z at the top are, when the truss element is spread by one unit apart

$$1 \cdot N = P_x \cdot g_x + P_y \cdot g_y + P_z \cdot g_z\,. \tag{2.95}$$

Remark 2.3 Equation (2.94) makes clear that influence is a **ratio** $w/1$ of two displacements, and therefore it is of no concern whether the triggering spread is 1 mm, 1 cm or 1 m. It depends only on the ratio of sensed motion to triggering motion.

Figure 2.39 confirms the **truss theory** with which the Eiffel Tower was calculated (+ graphical statics). The nodes do not move when the source point is spread (GF for M) and this means, nodal loads will not generate bending moments in the truss elements.

The near collapse in Fig. 2.40 impressively demonstrates the important role of kinematics in statics. The effects may be too small to stand out at the beginning, but over a long time span they accumulate...

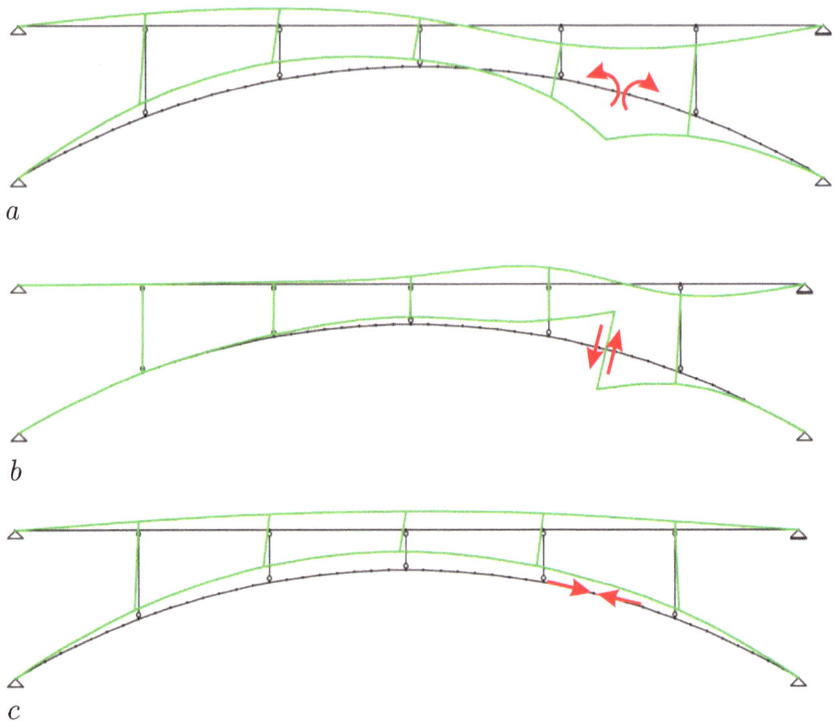

Fig. 2.38 Arch with elevated roadway, influence functions for **a** the moment, **b** the shear force and **c** the normal force at a characteristic point

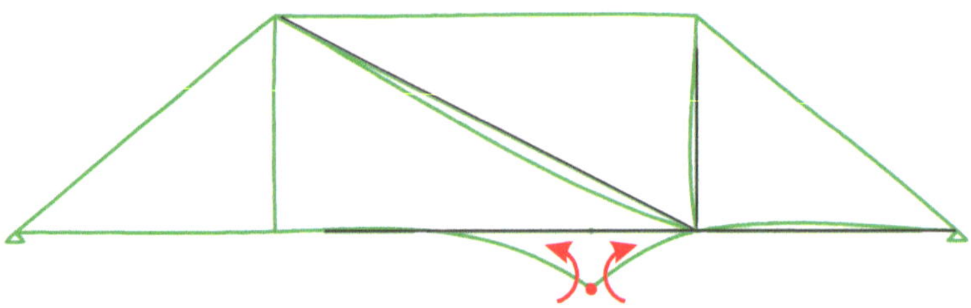

Fig. 2.39 Truss with rigid joints, influence function for a moment. The spread of the hinge generates no displacements at the nodes, and therefore it is admissible to model the truss with pins instead of rigid joints

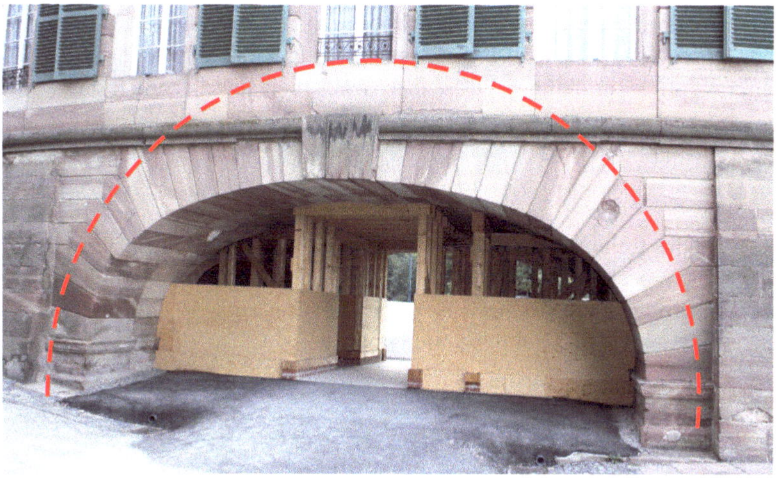

Fig. 2.40 Kinematics rules. After 200 years safety measures were necessary to prevent the collapse of the basket arch. Originally, the arch served as a connecting bridge, but in 1810 the additional weight was added, *Kassel Castle Wilhelmshöhe*

2.12 Point Values in 2-D and 3-D

The formulation of the influence function for the deflection $w(x)$ of a beam is easy, we simply split the beam into two parts, and the discontinuity of the shear force at the interface,

$$[V\,w - M\,w']_0^x + [V\,w - M\,w']_x^l = (V(x_-) - V(x_+))\,w(x) = -1 \cdot w(x) \tag{2.96}$$

makes the single term $w(x)$ jump out

$$1 \cdot w(x) = \int_0^l G_0(y, x)\,p(y)\,dy\,. \tag{2.97}$$

In 2-D and 3-D the situation is different since Green's second identity of the Laplace operator, the simplest 2-D equation,

$$\begin{aligned}\mathscr{B}(w, \hat{w}) = \int_\Omega -\Delta\,w\,\hat{w}\,d\Omega + \int_\Gamma \frac{\partial w}{\partial n}\,\hat{w}\,ds \\ - \int_\Gamma w\,\frac{\partial \hat{w}}{\partial n}\,ds - \int_\Omega w\,(-\Delta \hat{w})\,d\Omega = 0\,,\end{aligned} \tag{2.98}$$

contains no single point value $w(\boldsymbol{x})$.

The **transition to a point** is made possible by a particular property of the deflection surface $G_0(\boldsymbol{y}, \boldsymbol{x})$ of a point load $P = 1$, namely the integral of the shear force $\partial G_0/\partial n$ over circles Γ_{N_ε} of shrinking radius ε approaches the value 1 in the limit

$$\lim_{\varepsilon \to 0} \int_{\Gamma_{N_\varepsilon}} \frac{\partial G_0}{\partial n}\, ds = 1\,. \tag{2.99}$$

So, we start the formulation of *Betti's theorem* on the **punctured domain** $\Omega_\varepsilon = \Omega - N_\varepsilon$, which means we spare a circular neighborhood N_ε of the source point from the domain, and we let the radius $\varepsilon \to 0$ tend to zero. This provides in the limit the influence function for $w(\boldsymbol{x})$ at the source point $\boldsymbol{x}$

$$\lim_{\varepsilon \to 0} \mathscr{B}(G_0, w)_{\Omega_\varepsilon} = \lim_{\varepsilon \to 0} \Big[\int_{\Gamma_{N_\varepsilon}} \frac{\partial G_0}{\partial n}\, w(\boldsymbol{y})\, ds_y - \int_{\Omega_\varepsilon} G_0(\boldsymbol{y}, \boldsymbol{x})\, p(\boldsymbol{y})\, d\Omega_y \Big] = 0\,, \tag{2.100}$$

which means

$$\lim_{\varepsilon \to 0} \mathscr{B}(G_0, w)_{\Omega_\varepsilon} = 1 \cdot w(\boldsymbol{x}) - \int_{\Omega} G_0(\boldsymbol{y}, \boldsymbol{x})\, p(\boldsymbol{y})\, d\Omega_y = 0\,. \tag{2.101}$$

Remark 2.4 $\Omega_\varepsilon = \Omega - N_\varepsilon$ is the domain Ω minus the area N_ε of the punch-out and Γ_{N_ε} is the edge of the punch-out.

In 2-D and 3-D these steps are very technical and not always easy, see [3]. Fortunately, with finite elements things are much simpler, see Chap. 3.

Remark 2.5 The shear force in a rope is $V = H\, w'$, and in a membrane the shear force v_n is the slope of the surface in the direction of the section normal $\boldsymbol{n}$

$$v_n = H\, \frac{\partial w}{\partial n} = H\, \nabla w \bullet \boldsymbol{n} = H\, (w_{,x}\, n_x + w_{,y}\, n_y) \qquad [\text{N/m}]\,. \tag{2.102}$$

2.13 Duality

Betti is duality. Multiply a stiffness matrix $\boldsymbol{K}_{(n \times n)}$ with a vector $\boldsymbol{u}$ and this vector $\boldsymbol{K}\,\boldsymbol{u}$ with a second vector $\delta\boldsymbol{u}$ and you get a number $\delta\boldsymbol{u}^T \boldsymbol{K}\,\boldsymbol{u}$.

Since a real number does not change when it gets transposed, $\pi^T = \pi$, the two sides are the same

$$\delta\boldsymbol{u}^T \boldsymbol{K}\boldsymbol{u} = \boldsymbol{u}^T \boldsymbol{K}\, \delta\boldsymbol{u}\,, \tag{2.103}$$

or

$$\mathscr{B}(\boldsymbol{u}, \delta \boldsymbol{u}) = \delta \boldsymbol{u}^T \boldsymbol{K} \boldsymbol{u} - \boldsymbol{u}^T \boldsymbol{K}\, \delta \boldsymbol{u} = 0\,, \tag{2.104}$$

which is *Betti's theorem* for symmetric, self-adjoint matrices.

To see this equation in action we determine the component u_1 of the displacement vector $\boldsymbol{u}$ of a system $\boldsymbol{K}\boldsymbol{u} = \boldsymbol{f}$.

Since $u_1 = \boldsymbol{u}^T \boldsymbol{e}_1$, where $\boldsymbol{e}_1$ is the unit vector $\boldsymbol{e}_1 = \{1, 0, 0, \ldots, 0\}^T$, we solve the system $\boldsymbol{K}\boldsymbol{g}_1 = \boldsymbol{e}_1$ for $\boldsymbol{g}_1$ and formulate with $\boldsymbol{g}_1$ and the vector $\boldsymbol{u}$ the identity (2.104)

$$\mathscr{B}(\boldsymbol{g}_1, \boldsymbol{u}) = \boldsymbol{g}_1^T \boldsymbol{f} - \boldsymbol{u}^T \boldsymbol{e}_1 = \boldsymbol{g}_1^T \boldsymbol{f} - u_1 = 0\,, \tag{2.105}$$

and so

$$u_1 = \boldsymbol{g}_1^T \boldsymbol{f}\,. \tag{2.106}$$

To each nodal displacement u_i belongs such a vector $\boldsymbol{g}_i$

$$\boldsymbol{K}\, \boldsymbol{g}_i = \boldsymbol{e}_i\,, \tag{2.107}$$

which allows to extract u_i from the right side $\boldsymbol{f}$, see Fig. 2.41,

$$u_i = \boldsymbol{g}_i^T \boldsymbol{f}\,. \tag{2.108}$$

This means: The projection of the vector $\boldsymbol{f}$ onto the n vectors $\boldsymbol{g}_i, i = 1, 2 \ldots n$ is the same as the projection of $\boldsymbol{u}$ onto the unit vectors $\boldsymbol{e}_i, i = 1, 2 \ldots n$

$$u_i = \boldsymbol{g}_i^T \boldsymbol{f} = \boldsymbol{u}^T \boldsymbol{e}_i\,, \tag{2.109}$$

and this is exactly what happens (for all u_i simultaneously), when we multiply the vector $\boldsymbol{f}$ with the inverse matrix $\boldsymbol{K}^{-1}$

$$\boldsymbol{u} = \boldsymbol{K}^{-1} \boldsymbol{f}\,, \tag{2.110}$$

since the rows (= columns) of the symmetric matrix $\boldsymbol{K}^{-1}$ are the vectors $\boldsymbol{g}_i$,

$$\boldsymbol{u} = (\boldsymbol{g}_1^T \boldsymbol{f})\,\boldsymbol{e}_1 + (\boldsymbol{g}_2^T \boldsymbol{f})\boldsymbol{e}_2 + \ldots + (\boldsymbol{g}_n^T \boldsymbol{f})\boldsymbol{e}_n\,. \tag{2.111}$$

When we do the same with a function $u(x)$, say the solution of

$$-EA\, u''(x) = p(x) \qquad u(0) = u(l) = 0\,, \tag{2.112}$$

$\boldsymbol{K}$ has **infinitely many columns**, and the unit vectors become Dirac deltas

$$-EA \frac{d^2}{dy^2}\, G(y, x) = \delta(y - x)\,, \tag{2.113}$$

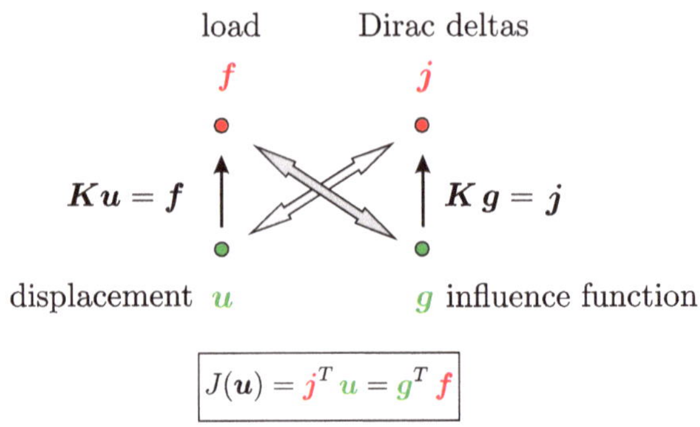

Fig. 2.41 Duality = "crosswise"

but the logic is the same. By projecting the right side p onto the solutions $G(y, x)$ we can determine the solution u at any point x

$$u(x) = \underbrace{\int_0^l G(y, x)\, p(y)\, dy}_{g_i^T f} = \underbrace{\int_0^l \delta(y - x)\, u(y)\, dy}_{u^T e_i} \,. \tag{2.114}$$

2.14 The Adjoint Operator

The adjoint operator L^* of an operator L is the "mirror image" of L

$$(L\,u, v) = \int_0^l L\,u\,v\,dx = \int_0^l u\,L^* v\,dx = (u, L^* v)\,. \tag{2.115}$$

In linear algebra L^* is the **transposed matrix**

$$(\boldsymbol{K}\,\boldsymbol{u}, \boldsymbol{v}) = (\boldsymbol{K}\,\boldsymbol{u})^T \boldsymbol{v} = \boldsymbol{u}^T\, \boldsymbol{K}^T \boldsymbol{v} = (\boldsymbol{u}, \boldsymbol{K}^T \boldsymbol{v})\,, \tag{2.116}$$

and ***adjoint analysis*** is the interplay of L and L^*. Let the vector $\boldsymbol{u}$ be the solution of $\boldsymbol{K}\,\boldsymbol{u} = \boldsymbol{f}$ and assume that, by multiplying $\boldsymbol{u}$ with a vector $\boldsymbol{j}$, we take a reading of $\boldsymbol{u}$, i.e. some value $J(\boldsymbol{u}) = \boldsymbol{u}^T\, \boldsymbol{j}$. If $\boldsymbol{j}$ is the right side of the adjoint equation $\boldsymbol{K}^T \boldsymbol{g} = \boldsymbol{j}$, then holds

$$J(\boldsymbol{u}) = \boldsymbol{u}^T \boldsymbol{j} = \boldsymbol{u}^T \boldsymbol{K}^T \boldsymbol{g} = \boldsymbol{f}^T \boldsymbol{g}\,, \tag{2.117}$$

and this looks familiar: Influence functions and *adjoint analysis* is the same.

2.15 Monopoles and Dipoles

The influence function for the slope w' of a beam is generated by a single moment $M = 1$ or **dipole**

$$M = \lim_{\Delta x \to 0} \frac{1}{\Delta x} \Delta x = 1 , \tag{2.118}$$

a pair of equal-sized but opposite forces $\pm 1/\Delta x$ whose distance Δx shrinks to zero.

A single force alone, in contrast, is a **monopole**. The influence function for a deflection $w(x)$ is the reaction to a monopole.

Influence functions which are generated by monopoles **sum**, they resemble dents or sinks, see Figs. 2.42 and 2.45a. Everything which falls into the sink increases the deflection of the plate.

Dipoles instead generate shear deformations, which are sensitive to imbalances, they **differentiate**, see Figs. 2.42 and 2.45b.

Monopoles integrate and dipoles differentiate.

Each of the four influence functions in Fig. 2.42 is of either type:

- G.F. for deflections and moments *sum*.
- G.F. for rotations, stresses and shear forces *differentiate*

The influence function for the shear force V is generated by a dipole, while the influence function for the bending moment M is generated by two moments $M = \pm 1/\Delta x$, which rotate the beam inwards, and so generate a symmetric deflection with a kink at the source point.[1]

The maximum result is obtained if the load and the influence function are of the same type (***symmetric—symmetric*** or ***antisymmetric—antisymmetric***) and the minimum effect when they are of opposite type, see Fig. 2.43.

The difference between monopoles and dipoles is the reason, why it is easier to approximate displacements and bending moments than stresses and shear forces. It is the difference between **numerical integration** and **numerical differentiation**, see Fig. 2.44.

Remark 2.6 Influence functions for support reactions integrate, though the support reactions are normal forces (stresses) or shear fores, and we therefore would expect the influence functions to differentiate. But if the support sits on a rigid soil, the

[1] To be precise: the correct sequence is: monopole—dipole—quadrupole—octopole, corresponding to the finite differences of w, w', M, V, see Fig. 9.11 page 448, but for our purposes the simple division: monopole—dipole or sum—differentiate will do.

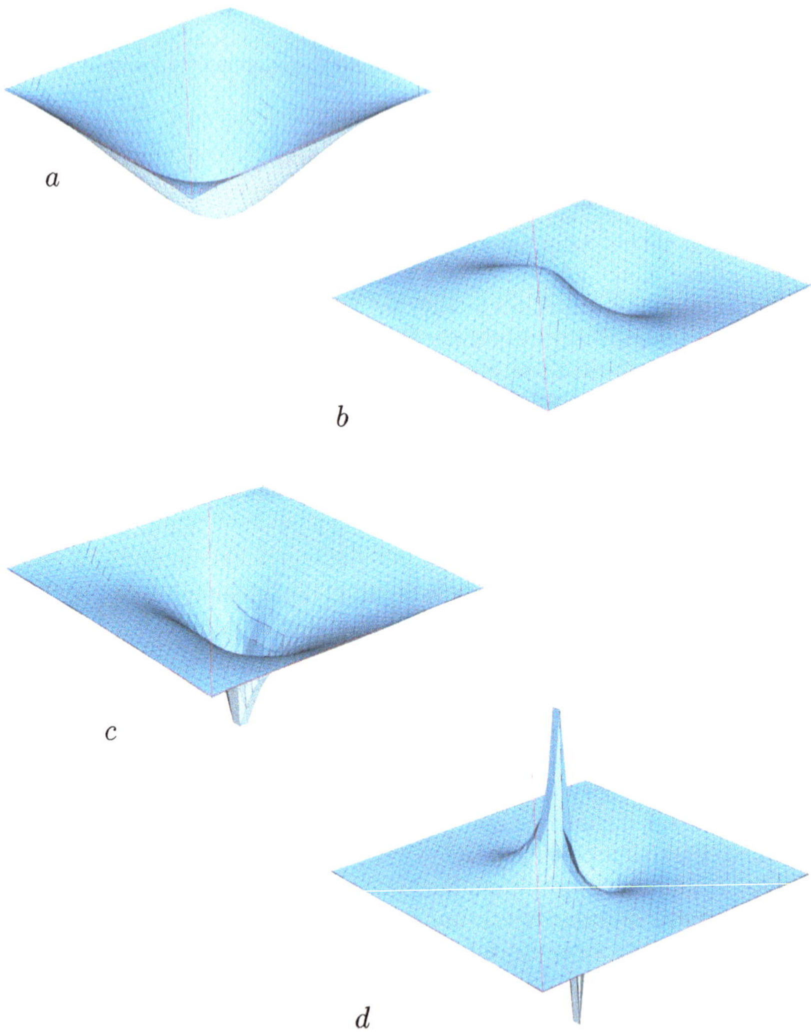

Fig. 2.42 Influence functions are generated by monopoles (on the left) or dipoles (on the right), influence function for **a** deflection, **b** rotation $w_{,x}$, **c** moment m_{xx}, **d** shear force q_x

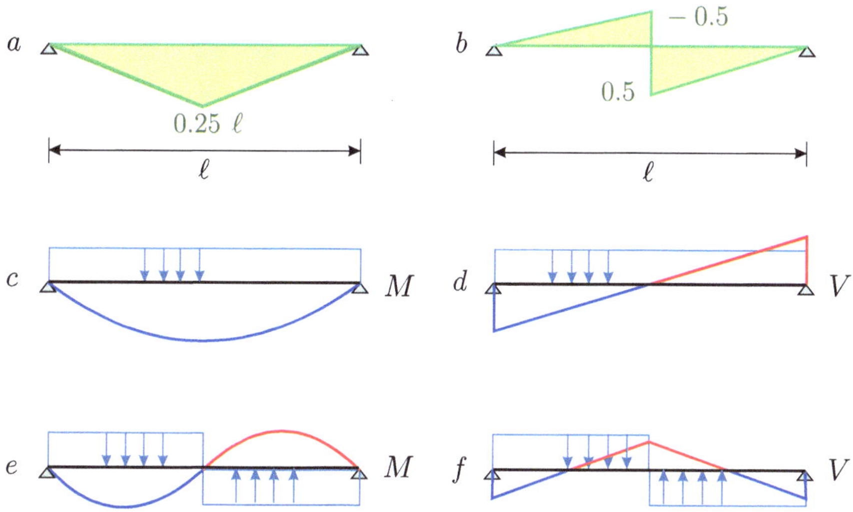

Fig. 2.43 Upper row influence functions for **a** the moment and **b** the shear force at the midpoint, **c** and **d** moments and shear forces under **e** symmetric and **f** antisymmetric load

motion is hindered by the foundation and so the other part must go all the way—alone—to effect a unit dislocation, $[[u]] = 1$, and so the influence function becomes a one-sided integral.

Remark 2.7 Not all influence functions tend to zero. If parts of the structure (after the installation of a hinge) can perform rigid body motions influence functions may blow up, see Fig. 2.46b.

Remark 2.8 The speed with which influence functions decay, depends on the order of the derivative of the target value, which would be 0, 1, 2, 3 in a beam

$$w(x), \quad w'(x), \quad M(x) = -EI\, w''(x), \quad V(x) = -EI\, w'''(x)\,. \qquad (2.119)$$

The lower the order, the more evenly the influence function spreads out, and the more slowly it decays, as the influence function for the deflection $w(\boldsymbol{x})$ of a plate demonstrates, see Fig. 2.45 a. The influence function for the shear force q_x instead is a tightly packed dipole, see Fig. 2.45b, two infinitely large peaks, but the downswing from these peaks is equally steep and fast.

The particular behavior, of course, also depends on the support conditions, see Figs. 2.47 and 2.48, because structures with large overhanging parts (cantilevers) play a special role in this regard. Such parts can swing widely and they can easily blow up any influence function.

Influence functions for forces in statically determinate systems also deserve a remark. Since such systems are kinematic (after the installation of a hinge), deformations can develop **unhindered** because no energy is consumed. Nothing can prevent

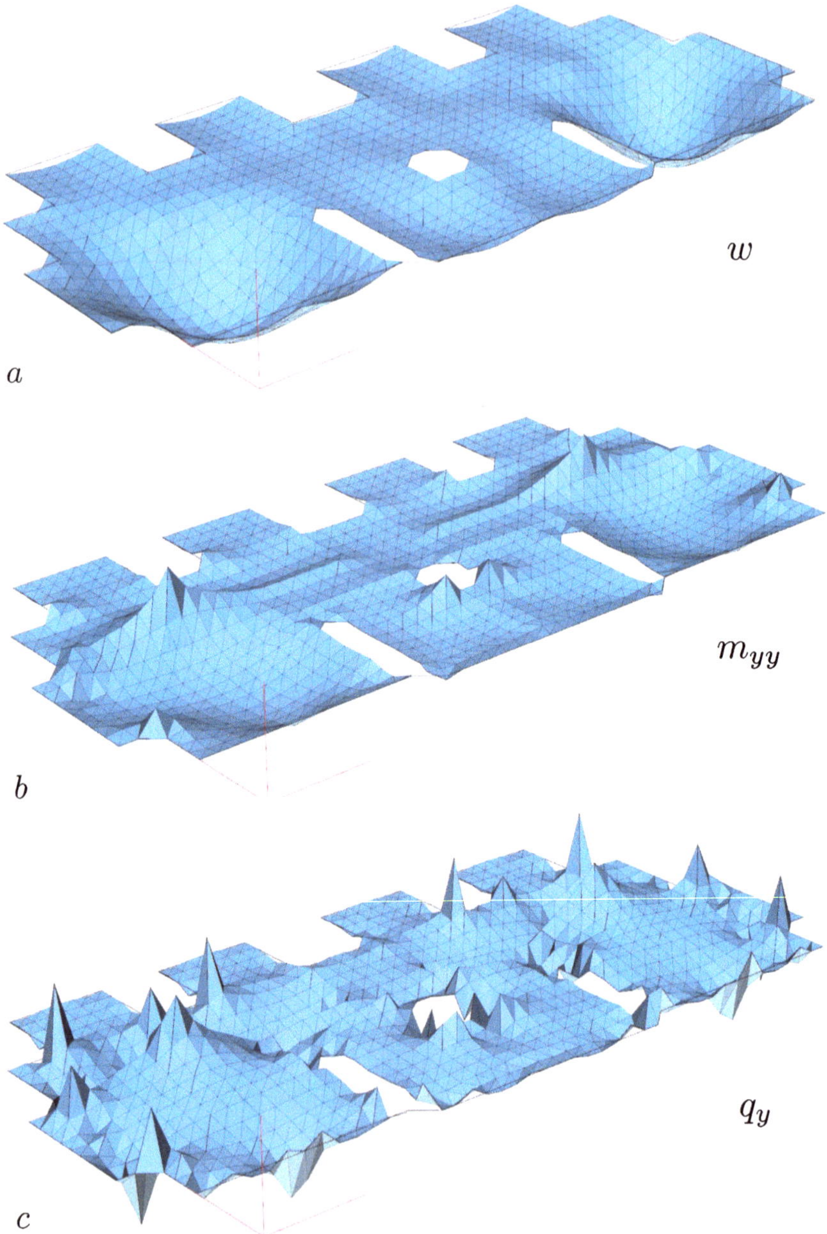

Fig. 2.44 Increasing complexity in a plate, **a** deflection w, **b** moments m_{yy}, **c** shear forces q_y

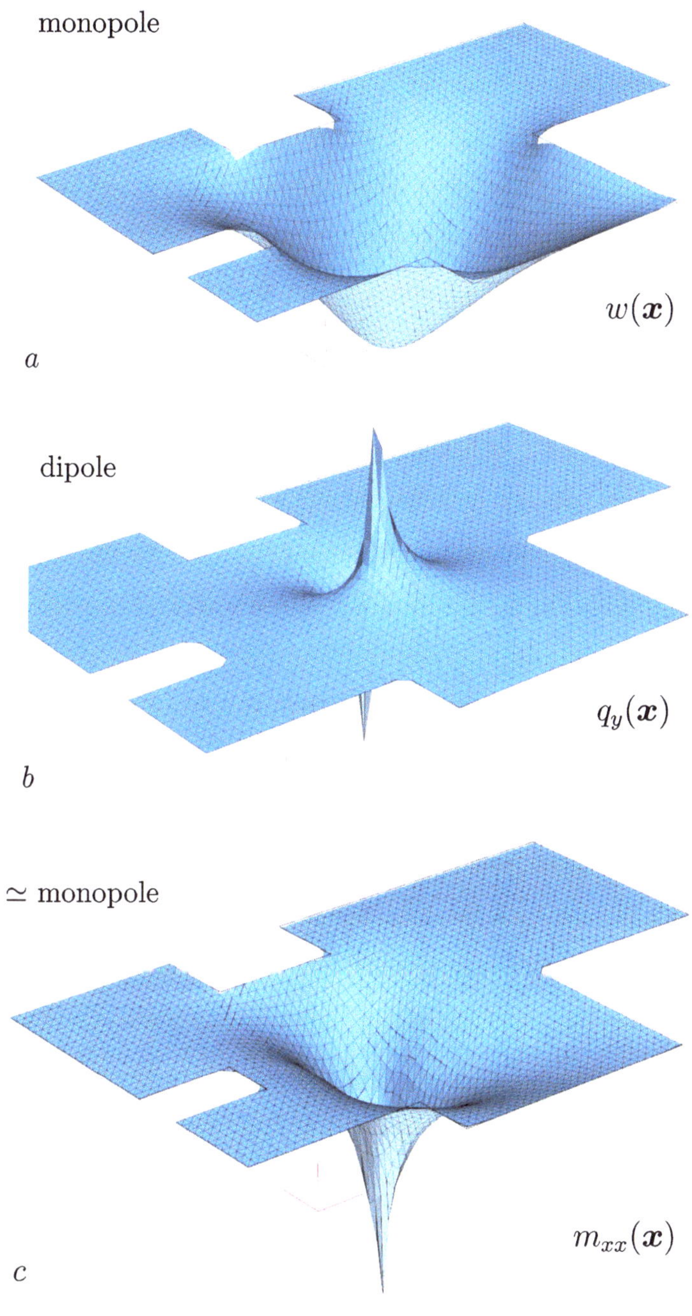

Fig. 2.45 Plate and influence functions, **a** for a deflection ($G_0 = O(r^2 \ln r)$), **b** a shear force ($G_3 = O(r^{-1})$), **c** a moment ($G_2 = O(\ln r)$), see (6.36) page 385 last column of the matrix

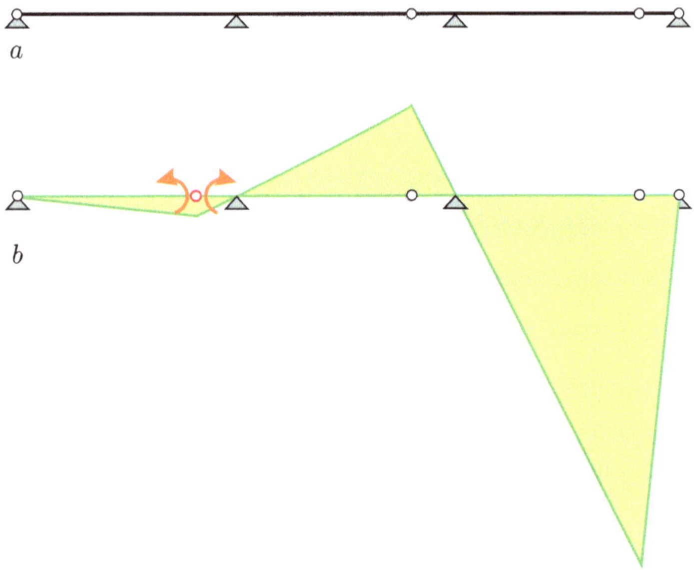

Fig. 2.46 **a** Gerber beam, **b** influence functions for a moment M; not all influence functions decay!

the influence function for the moment in a cantilever beam from running the part to the right of the source point under 45° into the sky, because it costs nothing. This is the reason why kinematic structures collapse so easily, since no energy is needed to trigger the collapse.

Statically indeterminate systems instead dampen the propagation of influence functions, while statically determinate systems lack such a barrier.

2.16 The Leaning Tower of Pisa

The symmetry and antisymmetry, which we observe in influence functions, also plays a prominent role in the leaning tower of Pisa.

The problem is that the soil stiffness under the tower is not uniform, and so the influence function $G_1(y, x)$ for the rotation of the foundation plate is not perfectly antisymmetric, but rather has a bias towards the softer side, and so the tower leans to this side, see Fig. 2.49,

$$w'(x_c) = \int_0^l G_1(y, x_c)\, p(y)\, dy\,. \tag{2.120}$$

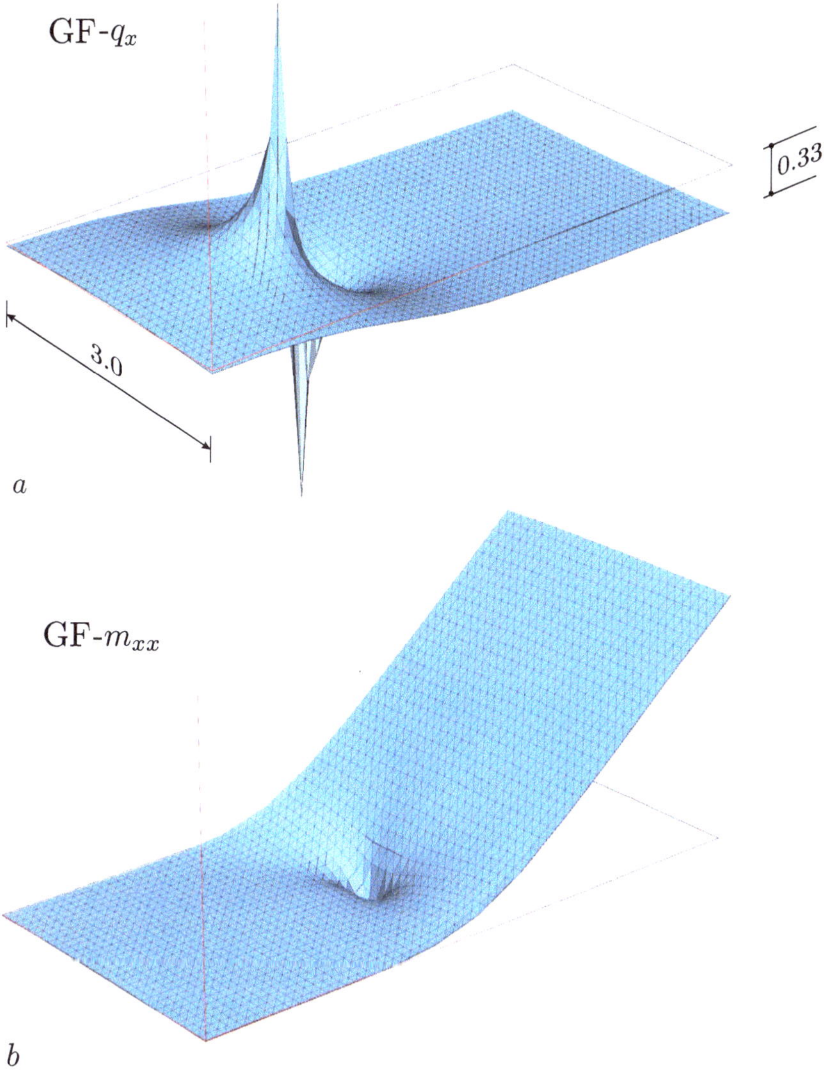

Fig. 2.47 Cantilever plate, **a** influence functions for a shear force q_x, and **b** the moment m_{xx}; it is amazing how with a "numerical" dislocation and a "numerical" kink it is possible to generate a nearly uniform dislocation and 45° rotations. How close can you come to the source point before the singularity, $O(1/r)$ in figure a and $O(\ln r)$ in figure b, dominates the scene?

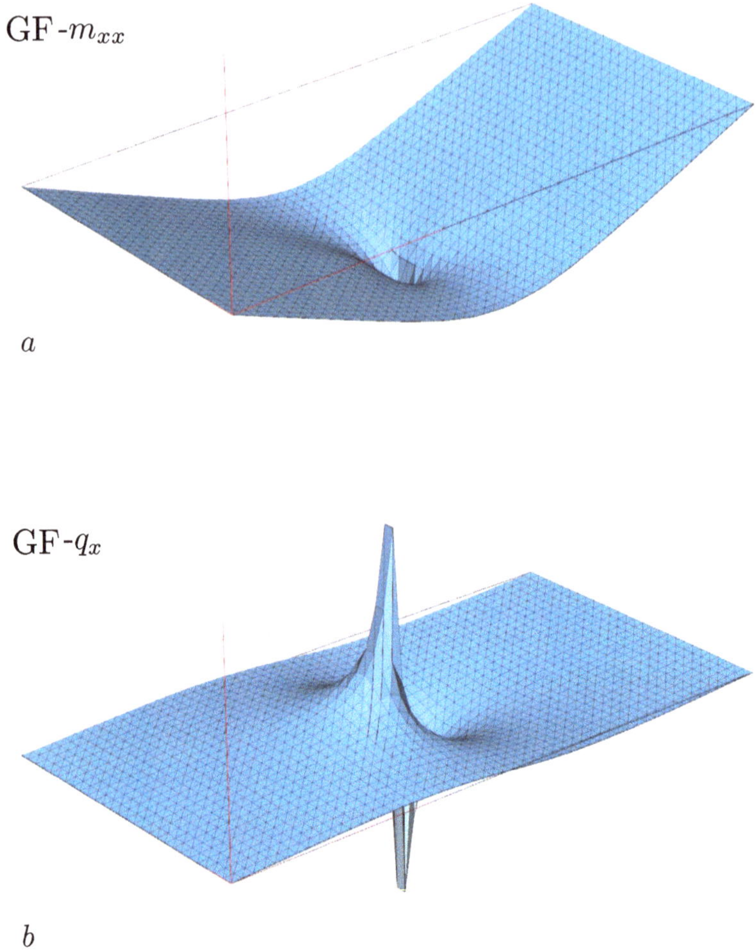

Fig. 2.48 Bridge, **a** influence function for the moment m_{xx}, and **b** for the shear force q_x at the midpoint; the influence function for the integral of q_x across the bridge should be identical with the beam solution

An antisymmetric kernel equals a balance, but if the balance is **misaligned**, even a perfectly symmetric load will rotate the balance. So, a zero rotation requires a perfectly antisymmetric kernel and a perfectly symmetric load.

Exercise: Given a non-perfect antisymmetric kernel, find the load p so that the rotation is zero.

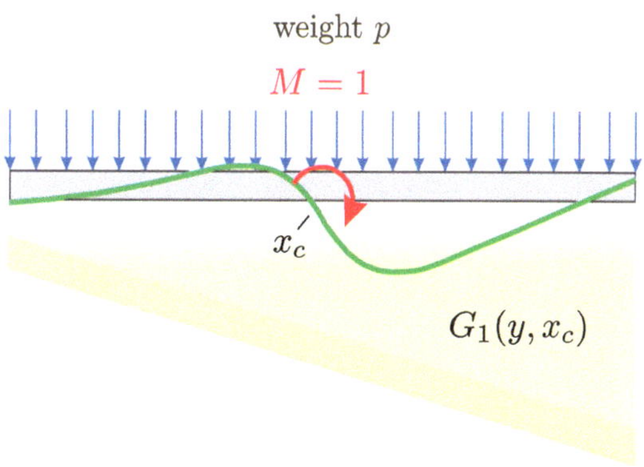

Fig. 2.49 The soil under the leaning tower of Pisa

2.17 Influence Functions for Integral Values

Oscillations near singular points are best smoothed out by **averaging**, by integrating over a short stretch.

Why this helps becomes clear when we look at the influence functions. The influence function for the stress σ_{yy} at a point is a spread of the source point in vertical direction, see Fig. 2.50b, but no mesh can simulate such a spread. If we "stretch" the point, turn it into a short line ℓ, and calculate with the mean of the stresses along this line

$$\sigma_{yy}^{\varnothing} = \frac{1}{\ell} \int_0^{\ell} \sigma_{yy}\, ds\,, \tag{2.121}$$

the influence function is a **simultaneous shift** of all points on the line in vertical direction, and such a displacement is easier to approximate with finite elements than a dislocation at one point. This is the reason why averaging gives better results.

If, as in Fig. 2.50a, the line is a complete cut through the plate, the integral of the stresses

$$N_y = \int_0^{l} \sigma_{yy}\, dx \tag{2.122}$$

is even exact, since this lift lies in $\mathcal{V}_h^+$ (= $\mathcal{V}_h$ + rigid body motions).

The beneficial effect of an integral measure on the bending moment in a plate is also evident in Fig. 2.51.

Influence functions of integral values are based on the same equation $\boldsymbol{K}\boldsymbol{g} = \boldsymbol{j}$ as before. If $J(w)$ is the mean value of the deflection in an element (x_a, x_b),

$$J(w) = \frac{1}{(x_b - x_a)} \int_{x_a}^{x_b} w(x)\, dx\,, \tag{2.123}$$

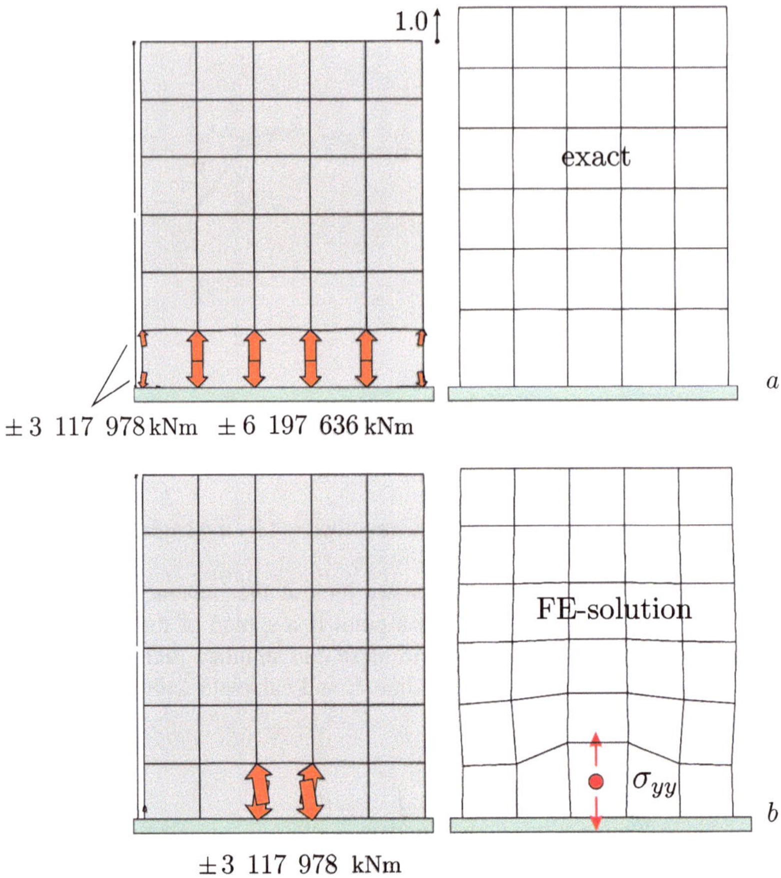

Fig. 2.50 Plate, **a** influence function for N_y (exact above the first row of elements), **b** influence function for σ_{yy} at a point

the j_i are the mean values of the φ_i

$$j_i = \frac{1}{(x_b - x_a)} \int_{x_a}^{x_b} \varphi_i(x)\, dx\,. \tag{2.124}$$

The influence functions for mean values of stresses are particularly simple to generate. The average stress $\sigma_{xx}^{\varnothing}$ in an element Ω_e is the integral

$$\sigma_{xx}^{\varnothing} = \frac{1}{|\Omega_e|} \int_{\Omega_e} \sigma_{xx}\, d\Omega = \frac{E}{|\Omega_e|} \int_{\Omega_e} (\varepsilon_{xx} + \nu\, \varepsilon_{yy})\, d\Omega\,, \tag{2.125}$$

but since $\varepsilon_{xx} = u_{x},_{x}$ and $\varepsilon_{yy} = u_{y},_{y}$, the domain integral can be replaced by a boundary integral over the edge Γ_e of the element

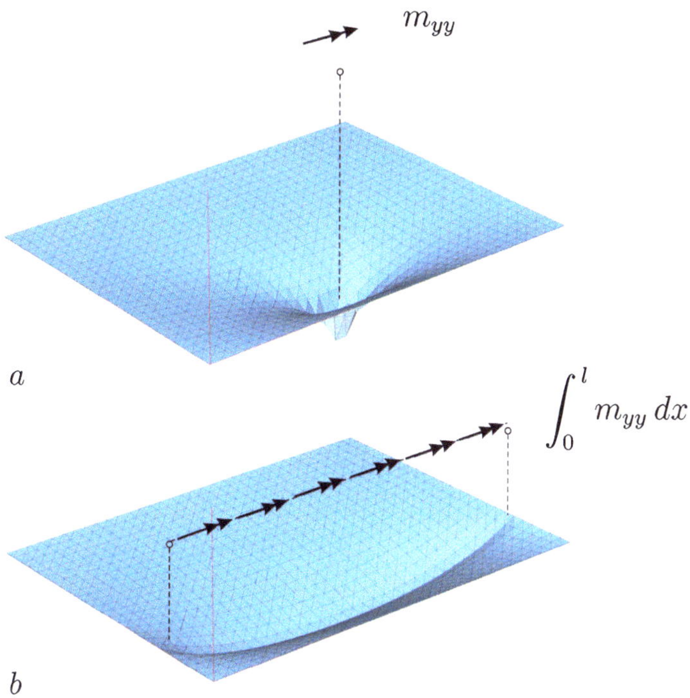

Fig. 2.51 Plate, **a** influence function for m_{yy}, and **b** for the integral of m_{yy}, [4]

$$\sigma_{xx}^{\varnothing} = \frac{E}{|\Omega_e|} \int_{\Omega_e} (\varepsilon_{xx} + \nu\, \varepsilon_{yy})\, d\Omega = \frac{E}{|\Omega_e|} \int_{\Gamma_e} (u_x\, n_x + \nu\, u_y\, n_y)\, ds\,, \tag{2.126}$$

and because the influence function for the displacement u_x or u_y of a boundary point $\boldsymbol{x}$ is generated by a single force $P_x = 1$ or $P_y = 1$ placed at $\boldsymbol{x}$, the influence function for the boundary integral is the displacement field generated by horizontal and vertical line forces, $E/|\Omega_e| \cdot n_x$ and $E/|\Omega_e| \cdot n_y$, along the edge of the element Γ_e, see Fig. 2.52.

An equations says it directly: Let $\boldsymbol{G}_{\varnothing}$ the reaction of the plate to the edge forces $E/|\Omega_e| \cdot n_x$ and $\nu E/|\Omega_e| \cdot n_y$ in Fig. 2.52, which we write as vector $\boldsymbol{t}$, and let $\boldsymbol{u}$ the displacement field of the plate then

$$W_{1,2} = \int_{\Omega} \boldsymbol{G}_{\varnothing} \bullet \boldsymbol{p}\, d\Omega_y = \int_{\Gamma} \boldsymbol{t} \bullet \boldsymbol{u}\, ds_y = W_{2,1}\,, \tag{2.127}$$

and the second integral is identical with (2.126).

The **mean stresses** in a plate held fixed at its edge are therefore zero, since the edge forces which generate the influence function cannot displace the fixed edge.

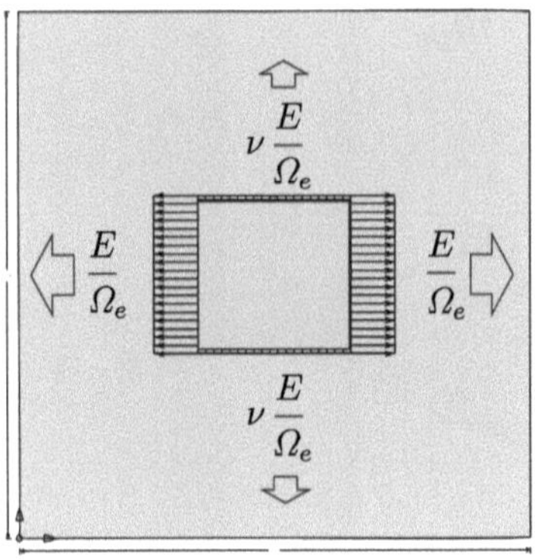

a

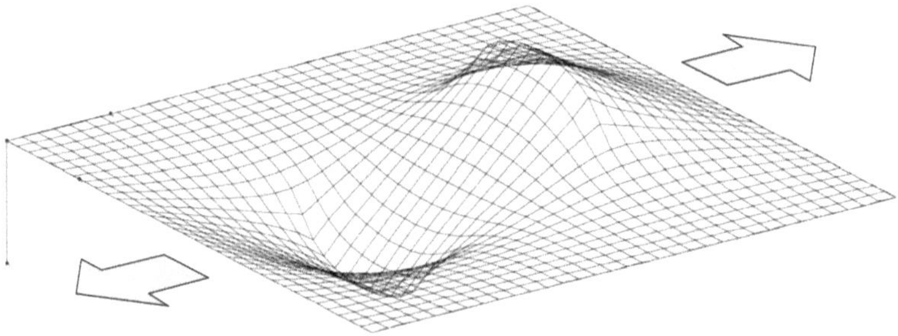

b

Fig. 2.52 Influence function for the average value of σ_{xx} in an element, **a** the "Dirac delta" consists of horizontal forces along the vertical edge and (small, ν-fold) vertical forces along the horizontal edge of Ω_e, **b** the horizontal displacements of the Green's function, here plotted in z-direction

The same is true with slabs: The mean values of the moments of a slab, clamped on all sides, are zero. In the 1-D case we have encountered this phenomenon already in Chap. 1, Eqs. (1.31) and (1.32).

Rule of thumbs: When the edge is fixed the average values of the derivatives are zero.

The stresses in the middle of an element are the most accurate. On the one hand, because the midpoint is the point furthest away from the jumps at the edge, and on the other hand because a dislocation at the midpoint is "relatively" easy to generate. With bilinear elements it is even so, that the influence functions for the stresses at the center and the mean values of the stresses coincide [4]. The stresses at the midpoint of a bilinear element are mean values.

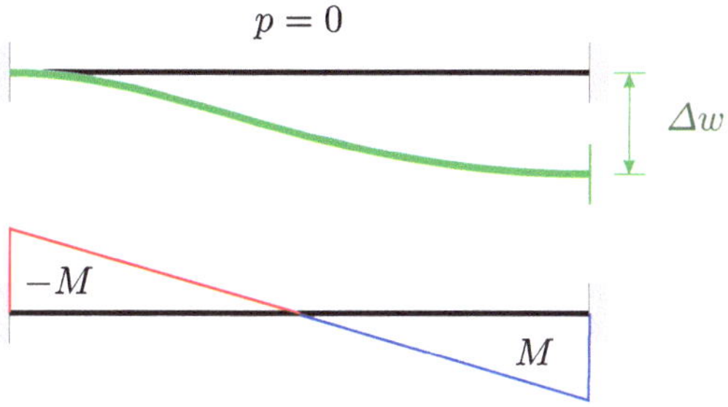

$$\int_0^l M\,dx = -EI(w'(l) - w'(0)) = 0 - 0 = 0$$

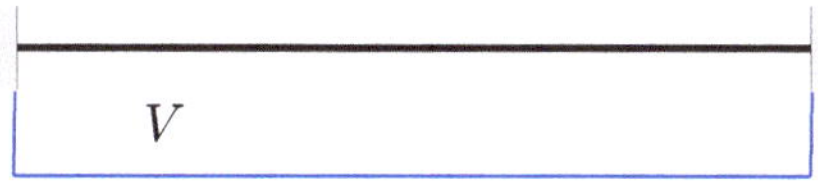

$$\int_0^l V\,dx = M(l) - M(0) = M + M = 2\,M$$

$$\int_0^l p\,dx = -(V(l) - V(0)) = -(V - V) = 0$$

Fig. 2.53 Relations between integrals and end values

Integration by parts is also a statement about the **mean value** of the derivative of a function

$$\frac{1}{l}\int_0^l u'(x)\,dx = \frac{1}{l}(u(l) - u(0))\,. \tag{2.128}$$

This is why the integral of the normal force $N = EA\,u'(x)$ is proportional to the relative displacement of the beam ends

$$\int_0^l N(x)\,dx = EA\,(u(l) - u(0))\,, \tag{2.129}$$

and the integral of $M(x) = -EI\,w''(x)$ is a measure of the relative rotations of the end points, see Fig. 2.53,

$$\int_0^l M(x)\,dx = -EI\,(w'(l) - w'(0))\,. \tag{2.130}$$

The integral of the shear force

$$\int_0^l V(x)\,dx = M(l) - M(0) \tag{2.131}$$

equals the difference between the moments, and the distributed load $p = -V' = EI\,w^{IV}$ is responsible for the difference between the shear forces

$$\int_0^l p\,dx = -(V(l) - V(0))\,. \tag{2.132}$$

2.18 Influence Functions Integrate

Differentiating, $w \to w' \to w'' \to \ldots$ we move forward and integrating we move back. The influence function $G_1(y, x)$ for the normal force integrates the load once

$$N(x) = \int_0^l G_1(y, x)\,p(y)\,dy \qquad ('') \to (')\,, \tag{2.133}$$

and the influence function $G_0(y, x)$ for the displacement $u(x)$ integrates it twice

$$u(x) = \int_0^l G_0(y, x)\,p(y)\,dy \qquad ('') \to (\,)\,. \tag{2.134}$$

Fig. 2.54 illustrates the situation. The deflection w of the cantilever beam is the triple indefinite integral of the shear force $V = -EI\,w'''$

$$w = -\int\int\int V\,dx\,dx\,dx = -\int\int\int P\,dx\,dx\,dx\,, \tag{2.135}$$

and this explains the ℓ^3 in the end deflection

$$w(\ell) = \frac{P\,\ell^3}{3\,EI}\,. \tag{2.136}$$

When a moment $M = -EI\,w''$ is applied we see an ℓ^2

$$w(\ell) = \frac{M\,\ell^2}{2\,EI}\,, \tag{2.137}$$

while a distributed load p produces the deflection

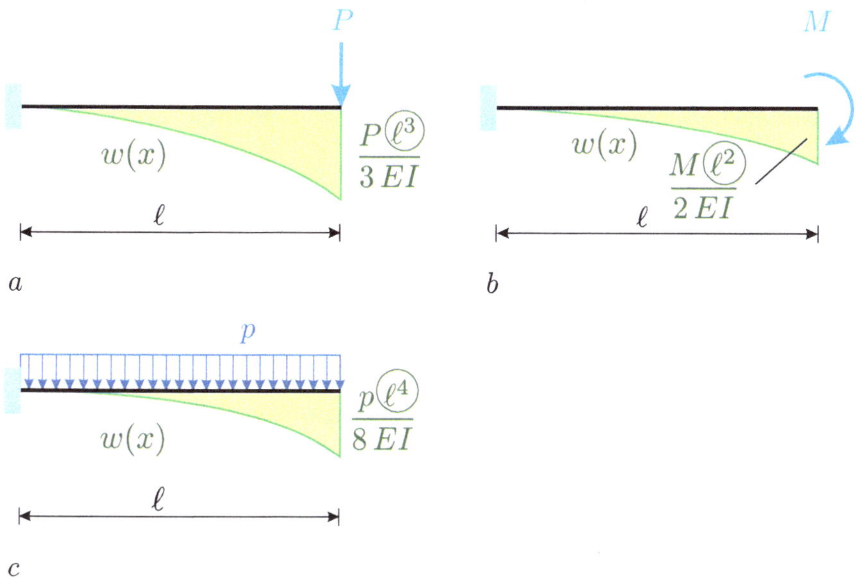

Fig. 2.54 Deflection of a cantilever beam due to **a** a single force—integrate three times, **b** a moment—integrate two times—and **c** distributed load—integrate four times

Fig. 2.55 Grid

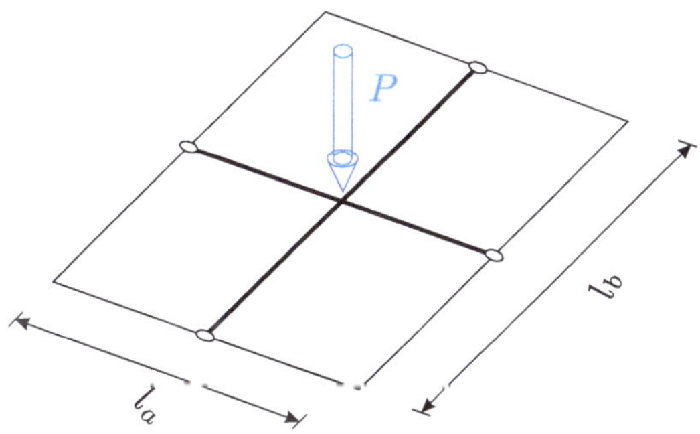

$$w(\ell) = \frac{p\,\ell^4}{8\,EI}\,, \tag{2.138}$$

where the ℓ^4 matches the fourth-order derivative $EI\,w^{IV} = p$.

We find the l^3 of (2.136) also in the formula

$$\frac{P_a}{P_b} = \frac{l_b^3}{l_a^3}\,, \tag{2.139}$$

which regulates how a point load $P = P_a + P_b$ is split between two beams, see Fig. 2.55.

A stiffness matrix, $\boldsymbol{K}\boldsymbol{u} = \boldsymbol{f}$, differentiates, and therefore we find the "inverse" factor EI/l^3 and EA/l up front in the case of a beam and a bar element respectively.

On moving back it is important to know how we arrived at p. Differentiate the function $u(x) = \sin(\pi\, x/l)$ two times and four times respectively

$$-u'' = (\frac{\pi}{l})^2 \sin(\pi\, x/l) \qquad = p(x) \tag{2.140}$$

$$EI\, u^{IV} = (\frac{\pi}{l})^4 \sin(\pi\, x/l) \qquad = \bar{p}(x)\,. \tag{2.141}$$

In the first case it is the response of a rope to a load $p(x)$, and in the second case it is the deflection of a beam which carries a load $\bar{p}(x)$.

To calculate u at the point $x = l/2$ by integrating the right sides $p(x)$ and $\bar{p}(x)$ different influence functions are required, although we ask for the same value, $u(l/2) = \sin(0.5 \cdot \pi)$. We must know which operator created p. Where do the data come from?

2.19 St. Venant's Principle

According to this principle *"the difference between the effects of two different but statically equivalent loads becomes very small at sufficiently large distances from load"* [5].

The reason is simple: Effects spread via influence functions, which are scalar products of the load p and a kernel $G(y, x)$, and the kernel (usually) tends to zero with increasing distance from the source point. If the distance is large enough, a **one-point quadrature** is good enough and the load can be replaced by its resultant. But equivalent loads have the same resultant, and therefore an exchange will have no effect at a large enough distance.

The effects of antisymmetric loads, having zero resultants, in particular subside quickly. And if an antisymmetric load meets a "flat" influence function—no slope—the influence is zero from the start, *symmetric* × *antisymmetric* = *0*.

Antisymmetric loads "differentiate" the influence functions.

This effect plays a significant role in the case of the forces f^+ in Chap. 5.

Fourier

A closely related topic are the effects of oscillating loads, since these have the tendency to cancel each other out, see Fig. 2.56. This is why Fourier series and the

Fig. 2.56 Membrane with movable edge. The shorter the wave length of the induced edge deflection is, the more rapidly it decreases, an effect which is due to the influence functions. Rapidly oscillating line loads on a plate generate probably zero displacements at one meter distance

JPEG method[2] succeed. The higher frequencies in a Fourier series, the short wave oscillations do not get very far.

Decomposing a signal into its frequencies is one thing, but—no less important—is that effects are the **superposition of the signal with influence functions**.

The idea of the **discrete Fourier transform** is to sort a vector by its wave components and to omit the short waves, the high frequencies. The most important vector is the one-vector $\boldsymbol{u} = \{1, 1, 1, \ldots\}^T$, whose integral is simply the area of the influence function. This vector is followed by vectors, which resemble "frozen" waves of increasing frequency. Their effect (= integral) tends to zero with increasing frequency, because they are extinguished when superposed with the influence functions.

The real "trick" of the discrete Fourier transform is the clever basis of wave vectors, "broad brushstrokes", which is much more effective than the "pointillist-like" basis of Euclidean unit vectors $\boldsymbol{e}_i$. If you want to transmit a vector $\boldsymbol{u} = \sum_{i=1}^{64} u_i\, \boldsymbol{e}_i$, you have to send all 64 components u_i, you may not omit one u_i, (it could just be the brightest pixel). Fourier instead can limit himself to the first 10 components without it being too obvious at the other end of the line that something is missing.

2.20 Second-Order Theory

Second-order beam theory is based on

$$EI\, w^{IV}(x) + P\, w''(x) = p(x) \tag{2.142}$$

where P is the compressive force in the beam and $p(x)$ is the distributed load, see Fig. 2.57a. This is a nice fourth-order linear self-adjoint differential equation with constant coefficients, but the problem lies in the coefficient P, which is load case dependent as can be seen in Fig. 2.57. Each P requires a different influence function for $w(x)$ and the section forces.

The more P nears the **buckling load** P_{crit}, the more the influence function for the rotation of the beam's end, Fig. 2.57 b, or for the deflection, Fig. 2.57c, bulge out.

In principle, second-order theory is a **nonlinear problem**, where the longitudinal displacement $u(x)$ and the lateral displacement $w(x)$ are linked in the system

$$-EA\left(u' + \frac{1}{2}\,(w')^2\right)' = p_x \tag{2.143a}$$

$$EI\, w^{IV} - \left(EA(u' + \frac{1}{2}\,(w')^2)\, w'\right)' = p_z\,. \tag{2.143b}$$

[2]In the JPEG method (DCT) a black and white image is divided into blocks of 8×8 pixels and the grayscales in the 64 pixels of a block form a vector, to which we apply a Fourier transform and we leave out the high frequencies.

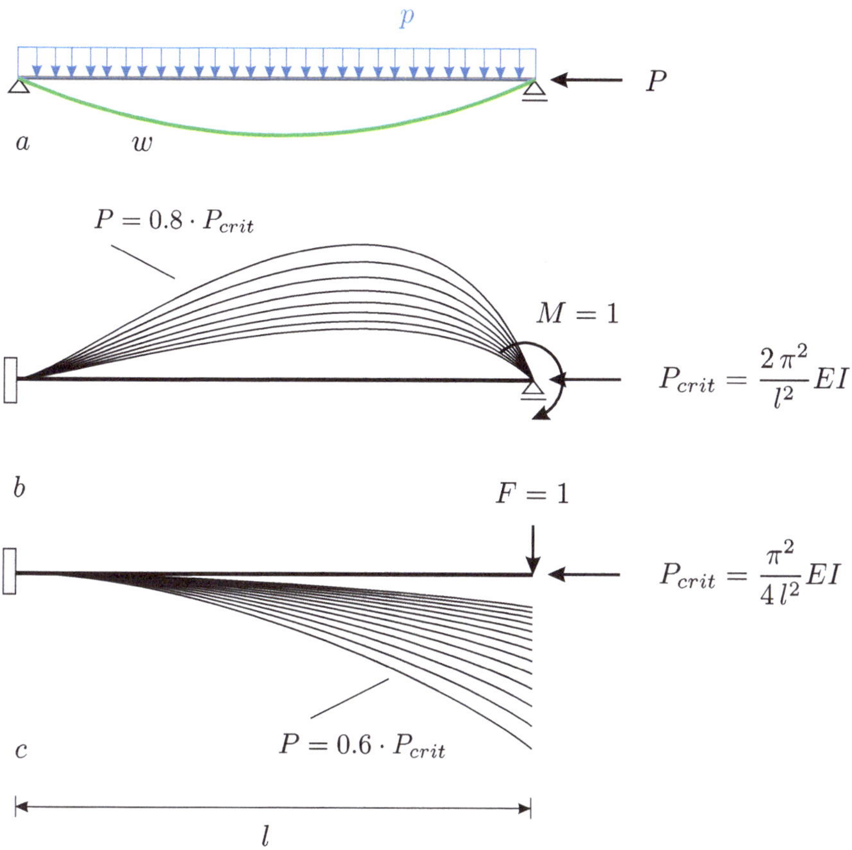

Fig. 2.57 Second-order theory, **a** compressive force P, **b** influence functions for $w'(l)$, and **c** for $w(l)$

Only if the normal force

$$N = EA(u' + \frac{1}{2}\,(w')^2) \tag{2.144}$$

is constant and is known, does this system reduce to (2.142). Note, a negative N is a positive P in (2.142).

References

1. Kurrer K-E (2018) The history of the theory of structures. Wiley Ernst & Sohn
2. Hartmann F (2013) Green's functions and finite elements. Springer, Heidelberg
3. Hartmann F (1989) Introduction to boundary elements. Springer, Heidelberg
4. Hartmann F, Katz C (2010) Structural analysis with finite elements, 2nd edn. Springer, Heidelberg
5. https://en.wikipedia.org/wiki/Saint-Venants-principle. Accessed 11 April 2020

Chapter 3
Finite Elements

Structures are constantly in motion, constantly changing their shape in accordance with the equilibrium conditions, and these tiny corrections are governed by influence functions. As the market woman knows, balance means equal work, $\delta W_e^{left} = \delta W_e^{right}$, means the work of the weights on both sides of the scale is the same. Install a shear hinge in a frame and spread the hinge! You will find that the work of the shear force and the load are the same—the market woman could have told you beforehand.

The structure in the architect's presentation may look pretty, but it does not "live". This requires finite elements. They turn the sketch into a living whole, which can react, deform, and find its balance.

Statics is kinematics and finite elements means life, means motion.

Sure, the FE-structure still more resembles a puppet on a string than a real structure, but even this restricted model has enough life in it, to give an engineer a sufficient idea of the equilibrium position.

Once the model is constructed, the engineer displaces the nodes one by one, $u_i = 1$ (all else are fixed, $u_j = 0$), remembers which forces are necessary for this maneuver, these are the *shape forces* p_i, and combines these loads p_i so, that the FE-load $p_h = \sum_i u_i p_i$ is **shake-equivalent** to the original load p

$$\delta W_e(p, \varphi_i) = \delta W_e(p_h, \varphi_i) \qquad \text{for all } \varphi_i\,. \tag{3.1}$$

This is the system $\boldsymbol{K u} = \boldsymbol{f}$. It is the logic of the market woman.

What we have presented here in anticipation will certainly become clearer in the following. At the beginning, however, we want to introduce finite elements quite classically via the principle of minimum potential energy.

F. Hartmann and P. Jahn, *Statics and Influence Functions*, Springer Series in Solid and Structural Mechanics 13,
https://doi.org/10.1007/978-3-030-55889-5_3

3.1 The Minimum

According to the *principle of minimum potential energy* the deflection of the rope in Fig. 3.1

$$-H\,w''(x) = p(x) \qquad w(0) = w(l) \qquad H = \text{horizontal force in the rope}\,, \quad (3.2)$$

is the function which minimizes the potential energy of the rope

$$\Pi(w) = \frac{1}{2}\int_0^l \frac{V^2}{H}\,dx - \int_0^l p(x)\,w(x)\,dx \qquad (V = H\,w') \qquad (3.3)$$

among all functions, which satisfy the boundary conditions, $w(0) = w(l) = 0$. The set of all these functions we call $\mathcal{V}$ (as in Variety).

Since the set $\mathcal{V}$ is too large to find $w(x)$ by a manual search, we restrict the search to a *finite dimensional* subspace $\mathcal{V}_h \subset \mathcal{V}$ and declare the function w_h, which minimizes the potential energy in this subspace, the best approximation. This is the idea of the **Ritz method**.

We start by subdividing the rope into small elements. An element is a piece of rope on which two linear functions are defined, the so-called element displacements, and by combining the element displacements of neighboring elements, we construct hat functions which represent ***unit displacements of the nodes***, see Fig. 3.1.

The unit displacements of the four internal nodes form the FE-solution

$$w_h(x) = w_1\,\varphi_1(x) + w_2\,\varphi_2(x) + w_3\,\varphi_3(x) + w_4\,\varphi_4(x)\,, \qquad (3.4)$$

and we choose the nodal displacements w_i so that the FE-solution minimizes the potential energy

$$\Pi(w_h) = \frac{1}{2}\int_0^l H\,(w_h')^2\,dx - \int_0^l p\,w_h\,dx \qquad (3.5)$$

on $\mathcal{V}_h \subset \mathcal{V}$, the span of the four hat functions $\varphi_i(x)$.

The function (3.4) wins the competition, if

$$\frac{\partial \Pi}{\partial w_i} = \frac{\partial}{\partial w_i}\{\frac{1}{2}\,\boldsymbol{w}^T\,\boldsymbol{K}\,\boldsymbol{w} - \boldsymbol{f}^T\,\boldsymbol{w}\} = 0 \qquad i = 1, 2, 3, 4\,, \qquad (3.6)$$

if the vector $\boldsymbol{w}$ of the nodal displacements solves the system $\boldsymbol{K}\,\boldsymbol{w} = \boldsymbol{f}$, or

$$\frac{H}{l_e}\begin{bmatrix} 2 & -1 & 0 & 0 \\ -1 & 2 & -1 & 0 \\ 0 & -1 & 2 & -1 \\ 0 & 0 & -1 & 2 \end{bmatrix}\begin{bmatrix} w_1 \\ w_2 \\ w_3 \\ w_4 \end{bmatrix} = \begin{bmatrix} 1 \\ 1 \\ 1 \\ 1 \end{bmatrix}. \qquad (3.7)$$

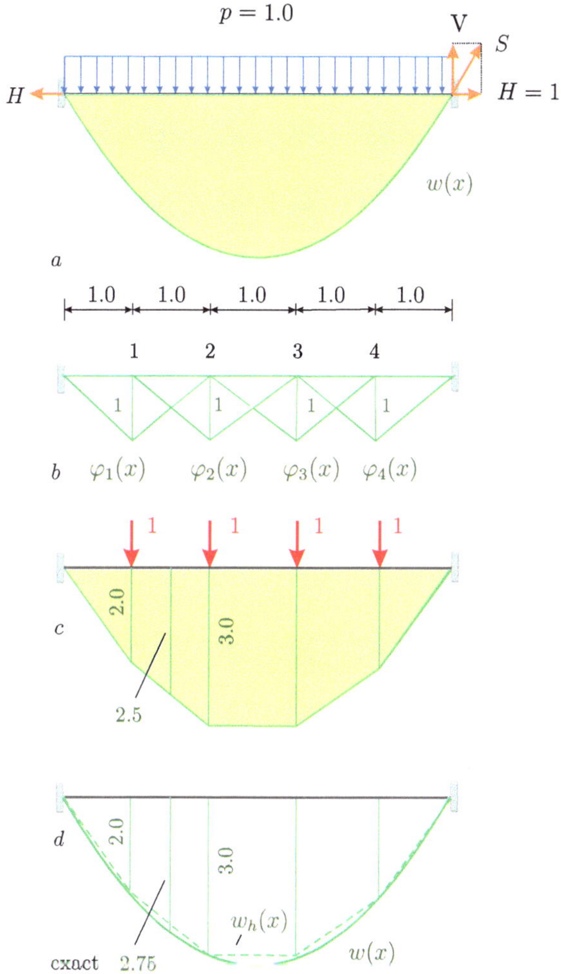

Fig. 3.1 FE-analysis of a rope, **a** system and load, **b** hat functions, **c** FE-solution $w_h(x)$, **d** comparison between $w(x)$ and $w_h(x)$

The elements k_{ij} of the stiffness matrix $\boldsymbol{K}$ are the **strain energy products** between the shape functions

$$k_{ij} = a(\varphi_i, \varphi_j) = \int_0^l H\, \varphi_i'(x)\, \varphi_j'(x)\, dx = \int_0^l \frac{V_i\, V_j}{H}\, dx\,, \tag{3.8}$$

and the **equivalent nodal forces** on the right side are the integrals

$$f_i = \int_0^l p(x)\, \varphi_i(x)\, dx\,. \tag{3.9}$$

If we let $H = 1$ and $l_e = 1$, the system (3.7) has the solution

$$w_1 = w_4 = 2 \qquad w_2 = w_3 = 3\,, \tag{3.10}$$

and so the function

$$w_h(x) = 2 \cdot \varphi_1(x) + 3 \cdot \varphi_2(x) + 3 \cdot \varphi_3(x) + 2 \cdot \varphi_4(x) \tag{3.11}$$

is on $\mathcal{V}_h$ the best approximation.

3.2 Why the Nodal Values of the Rope Are Exact

When we compare the FE-solution with the exact solution

$$w(x) = \frac{1}{2} \cdot (5\,x - x^2)\,, \tag{3.12}$$

we note that the FE-solution agrees with the exact solution at the nodes, $w_i = w(x_i)$. The reason is the following: The FE-program computes the deflection of the rope with the influence function

$$w_h(x) = \int_0^l G_h(y, x)\, p(y)\, dy\,, \tag{3.13}$$

only that it substitutes for the exact kernel $G(y, x)$ an approximate kernel, $G_h(y, x)$, the **best fit** it finds in the trial space $\mathcal{V}_h$. With regard to the nodes the best fit are the original Green's functions since they lie in $\mathcal{V}_h$. This is why the nodal values are exact.

If the Green's function $G(y, x)$ of a point x lies in $\mathcal{V}_h$, the FE-solution is exact at this point, $w(x) = w_h(x)$.

The influence function for the deflection $w(x_1)$ at the first node is the triangle $G(y, x_1)$ in Fig. 3.2b. Since the four shape functions can model this triangle exactly, $G_h(y, x_1) = G(y, x_1)$, the FE-solution agrees with the exact solution at the node x_1

$$w_h(x_1) = \int_0^l G(y, x_1)\, p(y)\, dy = w(x_1)\,. \tag{3.14}$$

This is true in any load case!

We do a test with $p(x) = \sin(\pi x/5)$. The equivalent nodal forces $f_i = (p, \varphi_i)$ are

$$f = \{0.569,\ 0.920,\ 0.920,\ 0.569\}^T\,, \tag{3.15}$$

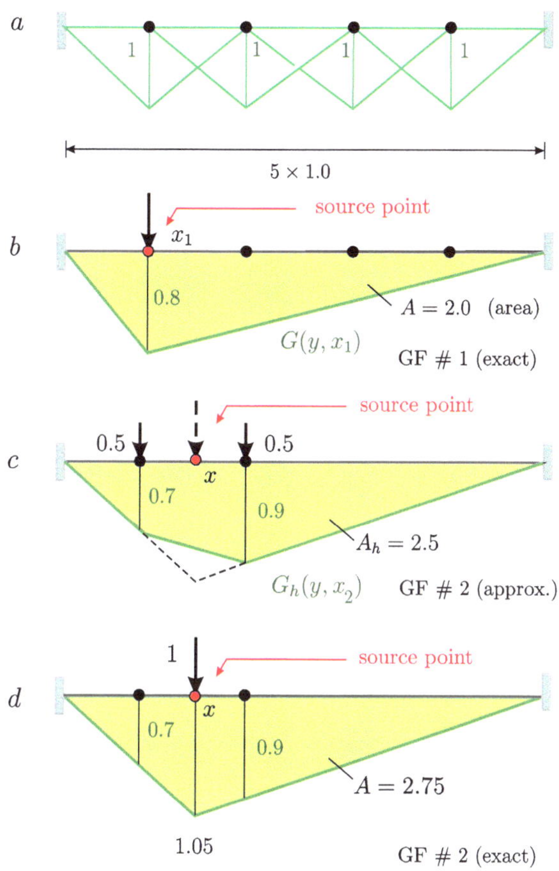

Fig. 3.2 FE-model of a rope, **a** shape functions, **b** influence function (GF) for $w(x_1)$, and **c** for the deflection $w(x)$ at an intermediate point, **d** the exact influence function for $w(x)$

and the solution $\boldsymbol{w} = \{1.489, 2.409, 2.409, 1.489\}^T$ of the system $\boldsymbol{K}\,\boldsymbol{w} = \boldsymbol{f}$ are the nodal values of the solution $w(x) = 25/\pi^2 \cdot \sin(\pi x/5)$.

But if a source point, say the point $x = 1.5$, is not a node, see Fig. 3.2c, the Green's function does not lie in $\mathcal{V}_h$. In this situation the program substitutes for the unit point load at x two half-sized point loads $P = 0.5$ at the neighboring nodes and it calculates with this approximation $G_h(y, x)$

$$w_h(x) = \int_0^l G_h(y, x)\, p(y)\, dy = 2.5 \neq 2.75 = w(x)\,, \tag{3.16}$$

but the result is then of course only an approximate value.

"But doesn't an FE-program calculate the nodal values by solving the system $\boldsymbol{K}\,\boldsymbol{w} = \boldsymbol{f}$ and the values in between by interpolating between the nodes?" Correct, but they are as large ***as if*** they had been calculated with the approximate influence functions. This is the key point.

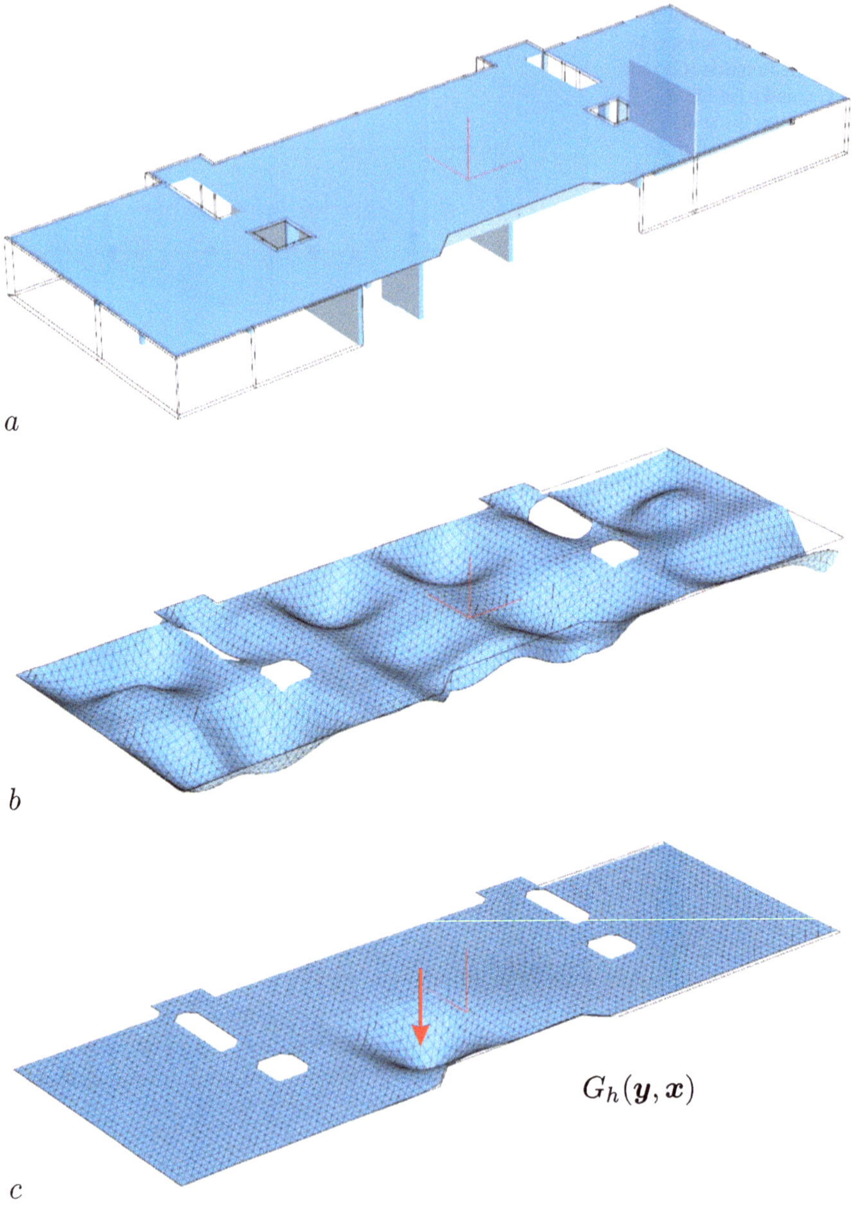

Fig. 3.3 Plate, **a** system, **b** deflection under gravity load, **c** influence function for the deflection w at a node $\boldsymbol{x}$

In 2-D and 3-D this is true as well. The nodal values of the plate in Fig. 3.3 are the result of the superposition of the nodal Green's functions with the load

$$w_h(\boldsymbol{x}_i) = \int_{\Omega} G_h(\boldsymbol{y}, \boldsymbol{x}_i)\, p(\boldsymbol{y})\, d\Omega_{\boldsymbol{y}}\,. \tag{3.17}$$

Of course, the nodal values w_i in the output are the solution of the system $\boldsymbol{K}\boldsymbol{w} = \boldsymbol{f}$, but they are just as large *as if* the FE-program had integrated over the whole plate, the results are the same[1]

$$w_h(\boldsymbol{x}_i) = w_i = \sum_j k_{ij}^{(-1)}\, f_j = \int_{\Omega} G_h(\boldsymbol{y}, \boldsymbol{x}_i)\, p(\boldsymbol{y})\, d\Omega_{\boldsymbol{y}}\,. \tag{3.18}$$

This is the secret, little-known law behind finite elements.

The quality of the influence functions determines the quality of the results.

3.3 Adding the Local Solution

When we calculate the deflection of the rope with an FE-program, we will not see a polygon on the screen, but a well curved parabola, i.e. the exact curve. How does the FE-program do this? It proceeds exactly as we said:

- It splits the rope into small elements.
- It reduces the load to the nodes and calculates the f_i.
- It solves the system $\boldsymbol{K}\boldsymbol{w} = \boldsymbol{f}$.

If it stopped now, we would see a rope polygon on the screen.

But there is one more step to follow. The program calculates in each element the so-called ***local solution*** w_{loc}, this is the deflection if the element is fixed at both ends, and it adds these shapes to the FE-polygon. This is how the curve in Fig. 3.4 was generated.

It is the same technique as in the **slope deflection method**, which first reduces all loads to the nodes, finds the equilibrium position, and then adds the local solutions.

The finite elements do it the same way, since the system of equations actually is

$$\boldsymbol{K}\boldsymbol{w} = \boldsymbol{f} + \boldsymbol{d}\,. \tag{3.19}$$

The f_i are the single forces, which act directly at the nodes, and the d_i are the equivalent nodal forces of the distributed load in between the nodes. In the FE-literature $f_i \equiv f_i + d_i$ usually stands for both parts.

[1] Skeptical readers may integrate the right side of (3.18) by parts.

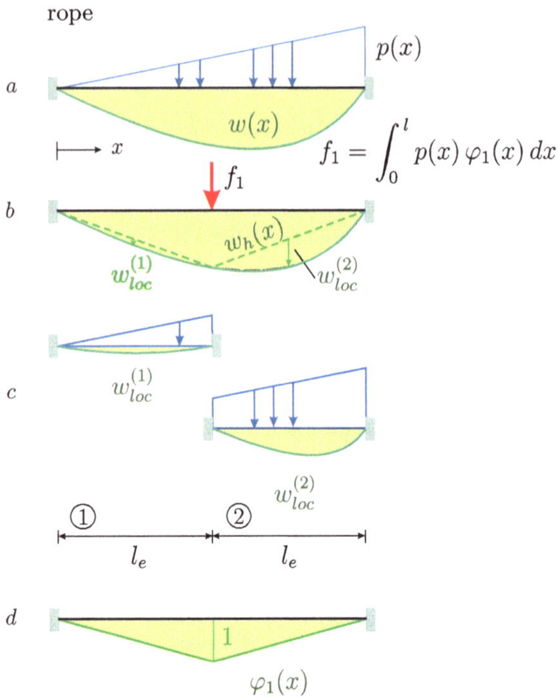

Fig. 3.4 Rope analysis with two elements, **a** load and deflection, **b** FE-solution + local solutions, **c** local solutions, **d** unit deflection of the central node

This way finite elements succeed, despite their **limited kinematics**, i.e. the use of

- piecewise linear functions for the longitudinal displacement
- cubic polynomials for the deflection

in producing the exact deformations, see Fig. 3.5.

Remark 3.1 To be precise this holds only true if EA and EI are constant, because only then are the element shape functions φ_i^e homogeneous solutions of the differential equations. In all other cases the nodal values are not exact.

The equivalence ***finite elements = slope deflection method*** also means: It makes no sense to further divide the individual elements of a frame into short elements. There is no gain in accuracy.

Remark 3.2 The finite element technique is often explained with 1-D problems. We do that too. For the finite elements to remain finite elements, we must agree that all these demonstrations refer to the time ***before*** the local solution is added to the FE-solution.

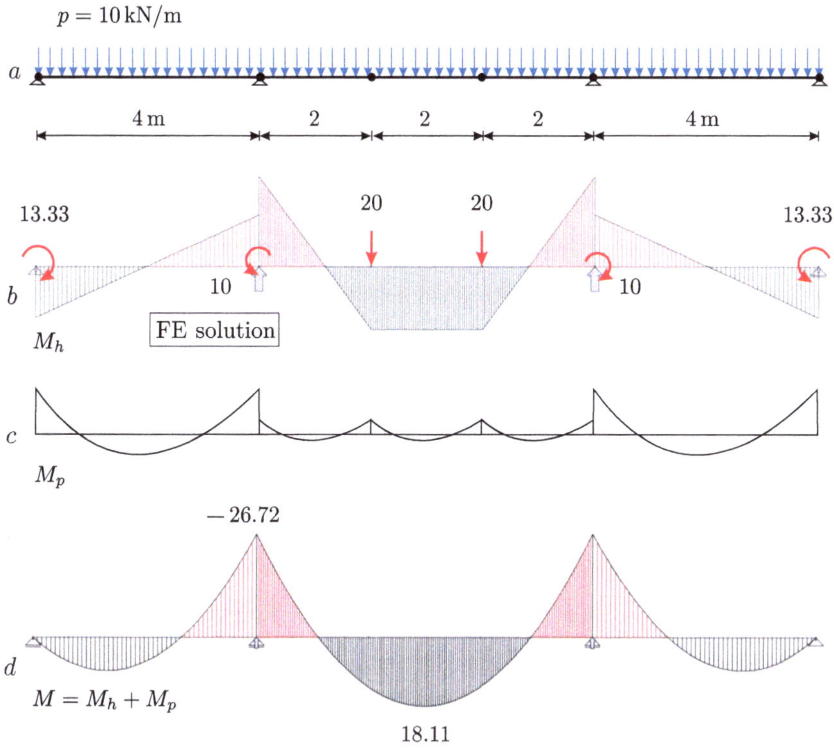

Fig. 3.5 Slope deflection method and finite elements are the same, **a** system, **b** first, the distributed load is reduced to the nodes and equilibrium at the nodes is established, and **c** the local solutions of the single clamped beams **d** are added

3.4 Projection

We arrived at $\boldsymbol{K u} = \boldsymbol{f}$ via the principle of minimum potential energy, however the projection of the exact solution onto the trial space $\mathcal{V}_h$, the ***Galerkin method***, leads to the same equation.

The projection of a vector $\boldsymbol{x} = \{x_1, x_2, x_3\}^T$ onto the $x_1 - x_2$ plane is its **shadow** $\boldsymbol{x}' = c_1 \boldsymbol{e}_1 + c_2 \boldsymbol{e}_2$, see Fig. 3.6, which means the error is orthogonal to $\boldsymbol{e}_1$ and $\boldsymbol{e}_2$

$$(\boldsymbol{x} - \boldsymbol{x}')^T \boldsymbol{e}_i = 0 \qquad i = 1, 2 \qquad \Rightarrow \qquad c_1 = x_1 \,, c_2 = x_2 \,, \tag{3.20}$$

and so the shadow is the vector in the plane with the shortest distance

$$|\boldsymbol{e}| = |\boldsymbol{x} - \boldsymbol{x}'| = \text{minimum} \tag{3.21}$$

to the tip of $\boldsymbol{x}$.

This idea is adopted by the *Galerkin method* for which the best approximation is the function w_h in $\mathcal{V}_h$, whose error $w - w_h$ is orthogonal to all φ_i,

$$a(w - w_h, \varphi_i) = 0 \qquad i = 1, 2, \ldots, n\,. \tag{3.22}$$

Because of Green's first identity

$$\mathcal{G}(w, \varphi_i) = \int_0^l p\,\varphi_i\,dx - a(w, \varphi_i) = 0 \quad \Rightarrow \quad f_i = \int_0^l p\,\varphi_i\,dx = a(w, \varphi_i) \tag{3.23}$$

this is equivalent to

$$a(w_h, \varphi_i) = f_i \qquad i = 1, 2, \ldots, n \tag{3.24}$$

or $\boldsymbol{Ku} = \boldsymbol{f}$. This implies, see (9.50), that w_h has the shortest distance in the energy metric from the exact solution w

$$a(w - w_h, w - w_h) = \text{minimum}\,. \tag{3.25}$$

Looking at Fig. 3.6 we realize that an additional projection will not lead any further. The shadow of the shadow $\boldsymbol{x}'$ is $\boldsymbol{x}'$. ***A projection stops after the first step***. This is the reason why it is not possible to improve the FE-solution on the same mesh, since the difference $p - p_h$ is orthogonal to the shape functions

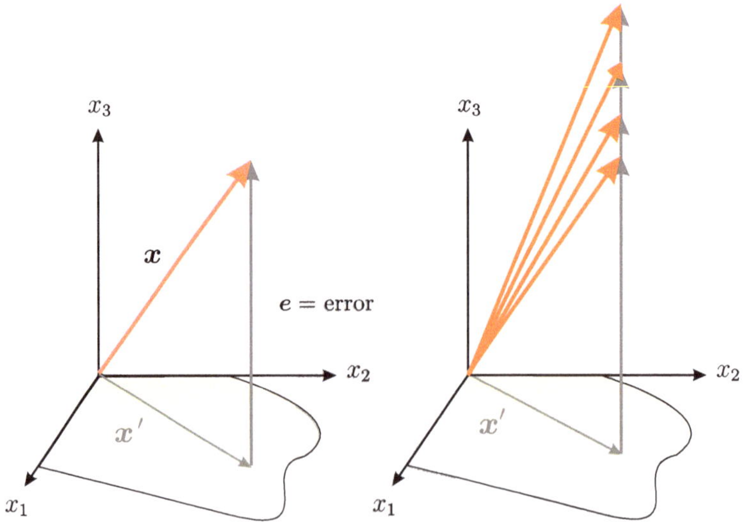

Fig. 3.6 The error vector $\boldsymbol{e}$ is orthogonal to the projection plane and all vectors directly above the vector $\boldsymbol{x}$ have the same shadow

$$f_i = \int_0^l (p - p_h)\,\varphi_i\,dx = 0\,. \tag{3.26}$$

Equation (3.22) is the ***Galerkin-orthogonality***. Because of $\delta W_i = \delta W_e$ it can also be "turned inside out", can be expressed as orthogonality in the virtual external work

$$\text{(inside)} \qquad a(w - w_h, \varphi_i) = \int_0^l (p - p_h)\,\varphi_i\,dx = 0 \qquad \text{(outside)}\,. \tag{3.27}$$

3.5 Equivalent Nodal Forces

The f_i on the right side of the system $\boldsymbol{K u} = \boldsymbol{f}$ are not forces in the true sense, but they are equivalent nodal forces, [Nm], are **energy quanta**[2]

$$f_i = \int_0^l p(x)\,\varphi_i(x)\,dx = [\text{N/m}]\cdot[\text{m}]\cdot[\text{m}] = [\text{Nm}]\,, \tag{3.28}$$

like the entries k_{ij}, which are **energy packets**, for example in a bar,

$$k_{ij} = \int_0^l EA\,\varphi_i'\,\varphi_j'\,dx = [\text{N}]\cdot[\;]\cdot[\;]\cdot[\text{m}] = [\text{Nm}]\,, \tag{3.29}$$

or a beam

$$k_{ij} = \int_0^l EI\,\varphi_i''\,\varphi_j''\,dx = [\text{Nm}^2]\cdot[1/\text{m}]\cdot[1/\text{m}]\cdot[\text{m}] = [\text{Nm}]\,. \tag{3.30}$$

Each derivative means a multiplication with $[\text{m}]^{-1}$.

$$\varphi_i\;[\text{m}] \qquad \varphi_i' = \frac{d\,\varphi_i}{dx} = [\;\;] \qquad \varphi_i'' = \frac{d\,\varphi_i'}{dx} = \frac{1}{[\text{m}]}\,. \tag{3.31}$$

The nodal displacements u_i are (internally) *dimensionless* weights

$$u_h = \textstyle\sum_i u_i\,\varphi_i(x) = [\;\;]\cdot[\text{m}] = [\text{m}] \tag{3.32}$$

In the output they have of course the dimension of a length, as the engineer wants to see it.

---

[2]For an alternative interpretation in beam analysis, see Sect. 9.21, page 449.

Calcoli

The equivalent nodal forces f_i serve bookkeeping purposes—like the calcoli on an *abacus*. The program moves from node to node, displaces the node by one meter in horizontal and vertical direction and makes a note on how much work is done by the load. These notes are the f_i.

A nodal force f_i of 10 Nm means that the loads in the neighborhood of the node contribute a work of 10 Nm on acting through $\boldsymbol{\varphi}_i$.

All what an FE-program does in the end is, it distributes **substitute loads** across the structure in such a way that these loads contribute the same amount of work as the real load, $f_i^h = f_i$.

The system $\boldsymbol{K}\,\boldsymbol{u} = \boldsymbol{f}$ is exactly this match $\boldsymbol{f}_h = \boldsymbol{f}$, since the left side $\boldsymbol{K}\,\boldsymbol{u}$ is the vector $\boldsymbol{f}_h$ of the equivalent nodal forces of the FE-solution.

In the notation of the next section the single f_i^h is the work done by the substitute FE-load

$$\boldsymbol{p}_h = \sum_j u_j\,\boldsymbol{p}_j \tag{3.33}$$

on acting through the nodal displacement $\boldsymbol{\varphi}_i$

$$f_i^h = \sum_j u_j\,\delta W_e(\boldsymbol{p}_j, \boldsymbol{\varphi}_i)\,. \tag{3.34}$$

3.6 Fixed End Forces

Imagine a beam element which is clamped at both ends, and which carries a distributed load p. In such a one-element beam the element shape functions $\varphi_i^e(x)$ are the nodal shape functions, and so the equivalent nodal forces are

$$f_i^e = \int_0^l p(x)\,\varphi_i^e(x)\,dx \qquad actio\,. \tag{3.35}$$

Since the support reactions have the opposite direction

$$f_i^e \times (-1) \qquad reactio\,, \tag{3.36}$$

we conclude that the element shape functions of a bar

$$\begin{aligned} \varphi_1^e(x) &= \frac{1-x}{l_e} \qquad \varphi_1^e(0) = 1\,, \qquad \varphi_1^e(l_e) = 0\,, \\ \varphi_2^e(x) &= \frac{x}{l_e} \qquad \varphi_2^e(0) = 0\,, \qquad \varphi_2^e(l_e) = 1 \end{aligned} \tag{3.37}$$

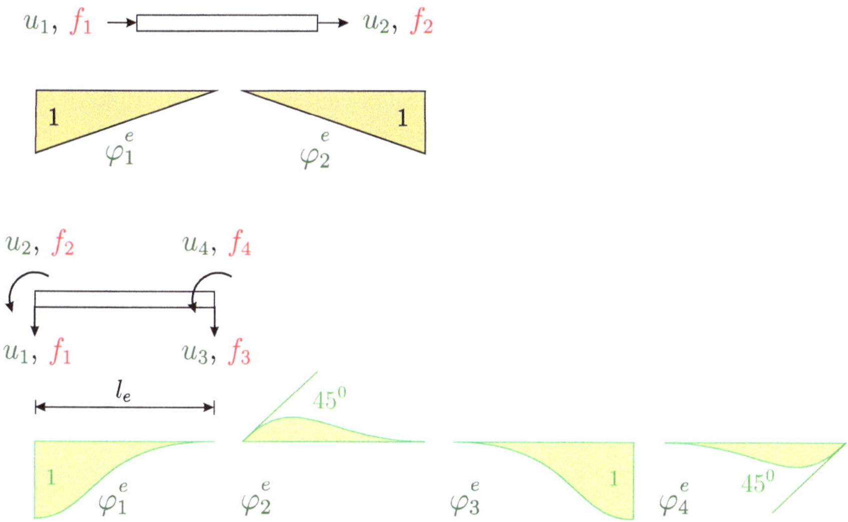

Fig. 3.7 Unit displacements of bar and a beam element

and of a beam, see Fig. 3.7,

$$\varphi_1^e(x) = 1 - \frac{3x^2}{l_e^2} + \frac{2x^3}{l_e^3} \qquad \varphi_3^e(x) = \frac{3x^2}{l_e^2} - \frac{2x^3}{l_e^3}$$

$$\varphi_2^e(x) = -x + \frac{2x^2}{l_e} - \frac{x^3}{l_e^2} \qquad \varphi_4^e(x) = \frac{x^2}{l_e} - \frac{x^3}{l_e^2} \tag{3.38}$$

are also the influence functions for the (negative) support reactions as in *actio*

$$f_i^e = \int_0^{l_e} \underset{\rightarrow}{p(x)}\, \varphi_i^e(x)\, dx \qquad i = 1, 2 \qquad \text{bar} \tag{3.39}$$

$$f_i^e = \int_0^{l_e} \underset{\downarrow}{p(x)}\, \varphi_i^e(x)\, dx \qquad i = 1, 2, 3, 4 \qquad \text{beam} \tag{3.40}$$

and *reactio*, $f_i^e \times (-1)$.

To be precise, this requires the unit displacements φ_i^e to be homogeneous solutions of the governing equations, which means EA and EI must be constant.

In a single frame element, the unit displacements of the nodes are the influence functions for the fixed end forces $\times(-1)$.

3.7 Shape Forces and the FE-load

To push a node of a plate by 1 m in horizontal or vertical direction—while all other nodes are kept fixed—requires a specific set of forces, see Fig. 3.8. We call these forces in analogy to the notion of shape functions, the ***shape forces*** $\boldsymbol{p}_i = \{p_x^{(i)}, p_y^{(i)}\}^T$, which belong to the displacement $u_i = 1$. These are active and restraining forces. The active forces push the node into the direction of the u_i and the restraining forces make that the motion stops at the neighboring nodes. The forces are always balanced, they are equilibrium forces, see Fig. 3.9.

The sum of these *shape-forces*—weighted with the nodal displacements u_i—is the FE-load

$$\boldsymbol{p}_h = \sum_i u_i\, \boldsymbol{p}_i = \sum_i u_i \begin{bmatrix} p_x^{(i)} \\ p_y^{(i)} \end{bmatrix}, \tag{3.41}$$

which produces the shape $\boldsymbol{u}$.

While in 2-D and 3-D problems some effort is necessary to detail the FE-load $\boldsymbol{p}_h$, see Fig. 3.10, in frame analysis (if EA and EI are constant) the FE-load is identical with the nodal forces f_i, which means in 1-D the f_i [Nm] are real forces/moments if you divide by 1 m (not the moments), while in 2-D and 3-D they are "mere" calcoli, forces "as if".

Remark 3.3 Fig. 3.10 illustrates the flexibility (for not to say "ambiguity") of structural analysis with regard to finite elements. What holds the plate fixed at the node, is

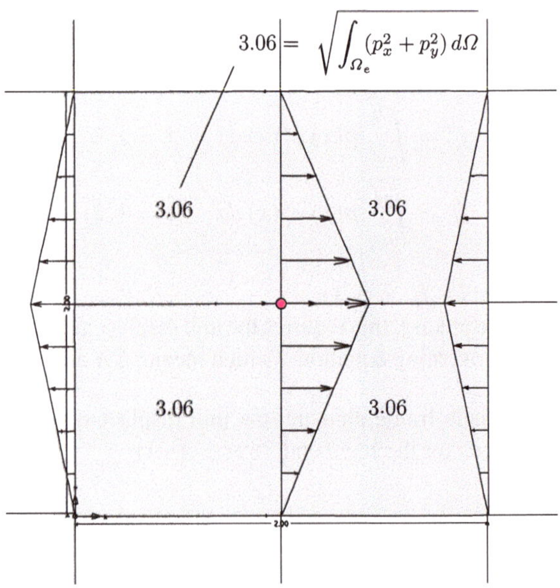

Fig. 3.8 Section of an FE-mesh: The forces, which displace the node by one unit length to the right and simultaneously stop the motion at the neighboring nodes are the *shape forces*. The distributed surface forces, p_x and p_y, are only given as integral values. The other part of the mesh (not shown) is load free, is smooth as a pond

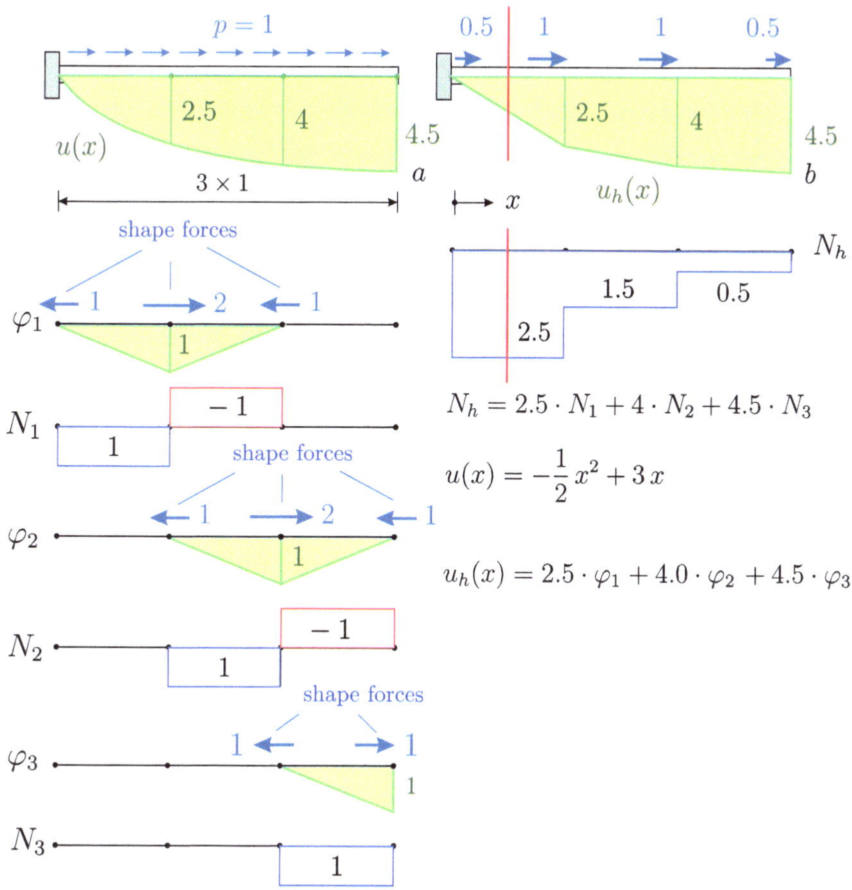

Fig. 3.9 Linear FE-solution of a stretched bar ($EA = 1$) carrying a constant load. Each shape function is plotted together with its set of *shape forces*

not a point force, but the combined action of domain loads and line loads, while in the printout you will only find an equivalent nodal force f_i. This force is the "equivalent sum" so to speak of these distributed forces, and the engineer has no problem (rightly so) to forward this force to the next floor as a real force.

The Role of the u_i

By a proper choice of the nodal displacements u_i the FE-load $\boldsymbol{p}_h$, see (3.41), is made to match the real load in the sense of the ***principle of virtual displacements***, which means the equivalent nodal forces f_i^h agree with the f_i of the applied load

$$f_i^h = \int_\Omega \boldsymbol{p}_h^T \bullet \boldsymbol{\varphi}_i \, d\Omega = \int_\Omega \boldsymbol{p}^T \bullet \boldsymbol{\varphi}_i \, d\Omega = f_i \,. \tag{3.42}$$

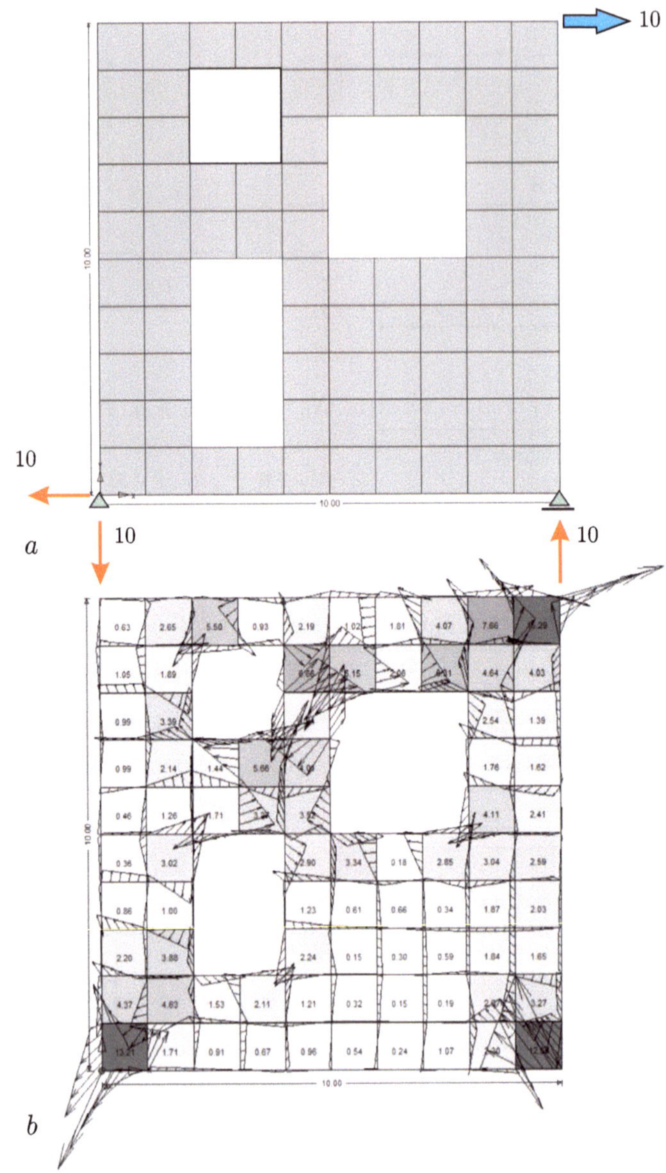

Fig. 3.10 Point load, **a** system, **b** the FE-load $\boldsymbol{p}_h$; the grey shades correspond to the strength of the element loads in the bilinear elements, as in Fig. 3.8

This is guaranteed, if $\boldsymbol{u}$ solves the system $\boldsymbol{K}\boldsymbol{u} = \boldsymbol{f}$, since each row (an internal energy δW_i) is—owing to ***internal = external***—identical to each row of the vector $\boldsymbol{f}_h$ (external work δW_e)

$$\boldsymbol{K}\,\boldsymbol{u} = \boldsymbol{f} \qquad \equiv \qquad \boldsymbol{f}_h = \boldsymbol{f}\,. \tag{3.43}$$

Imagine two plates side-by-side. The first carries the original load $\boldsymbol{p}$ and the second the FE-load $\boldsymbol{p}_h$. When you move from node to node and you displace the node by 1 m in horizontal and vertical direction, you will find that the work is each time the same $f_i^h = f_i$.

The FE-load is "shake-equivalent" to the original load.

Whether a structure carries the original load, or the substitute FE-load cannot be determined by testing the structure with the shape functions. Any such test is not conclusive, since each time the work comes out the same.

In Fig. 3.10 the original load and the FE-load are placed side by side. The single force at the upper right corner gets replaced by a very confusing looking system of domain loads and line loads, which constitute the FE-load $\boldsymbol{p}_h$. Because this load looks so "strange", it is normally not shown by FE-programs, since a user not familiar with the theory would be irritated. (But it is the service load, for which the user designs the structure).

That the FE-results cannot be as poor as Fig. 3.10 seems to indicate, illustrates a direct comparison of the principal stresses in Fig. 3.11. These look okay.

We would like to add a second, **indirect argument**. In the original load case, all $f_i = 0$ are zero, except $f_i = 10$ at the upper right corner, and therefore all $f_{hi} = 0$ must be zero as well, except $f_{hi} = 10$,

$$\int_{\Omega} \boldsymbol{p}_h^T \boldsymbol{\varphi}_i \, d\Omega = f_{hi} = f_i = 0\,. \tag{3.44}$$

The FE-program must juggle vigorously to guarantee this property, and this may explain the "chaos" in Fig. 3.10b. All the forces in this rather confusing looking load case $\boldsymbol{p}_h$ are so balanced that they do zero work when a node gets displaced by one meter in horizontal or vertical direction.

Most of the FE-load $\boldsymbol{p}_h$ is zero in terms of the energy metrics since it generates (as it must(!)) no nodal forces f_{hi}.

Remark 3.4 The *shape forces* $\boldsymbol{p}_i$ of a plate are domain forces and line forces and so the virtual work of these forces is a sum of domain integrals and line integrals. The domain integral in (3.44) is an abbreviating notation for all these integrals.

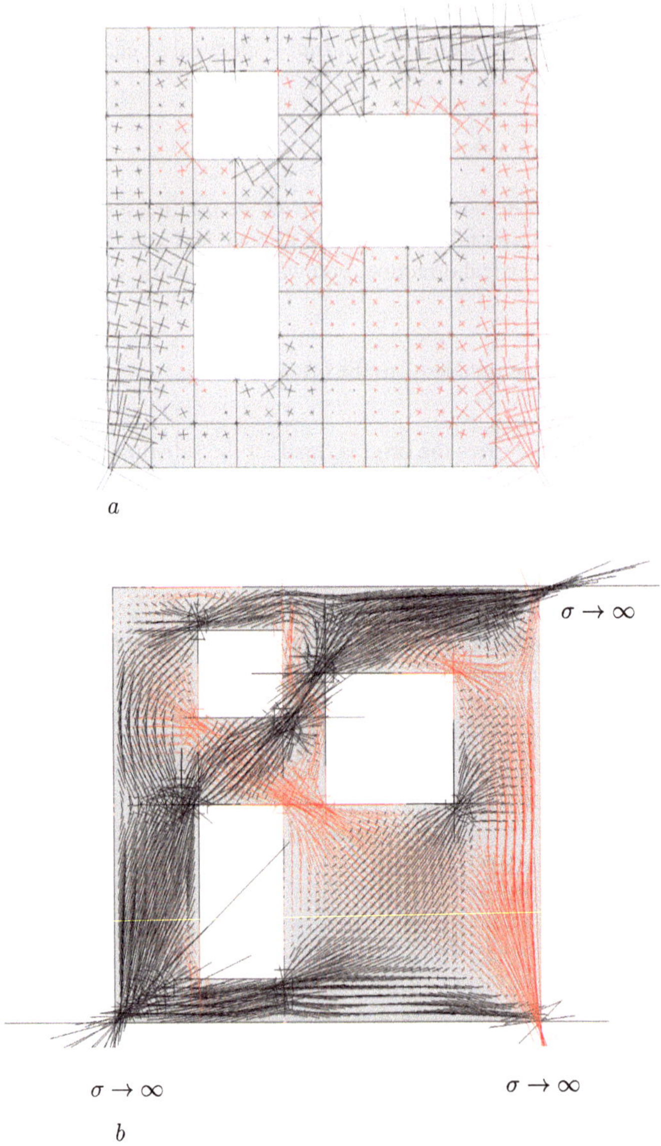

Fig. 3.11 Principal stresses, **a** coarse mesh, and **b** a very refined mesh

3.8 How the Ball Got Rolling

If we read $\boldsymbol{f}_h = \boldsymbol{f}$ as the basic equation of the finite element method, then this agrees with the approach in the **original paper** of *Turner et alteri* (1956) [1]. The authors looked at a *CST element* and formulated a matrix $\boldsymbol{S}$, which links the three (constant) stresses in an element, $\boldsymbol{\sigma} = \{\sigma_{xx}, \sigma_{yy}, \sigma_{xy}\}^T$, to the nodal displacements[3]

$$\boldsymbol{\sigma}_{(3)} = \boldsymbol{S}_{(3\times 6)}\boldsymbol{u}_{(6)} \,. \tag{3.45}$$

The FE-load $\boldsymbol{p}_h$, which generates the shape $\boldsymbol{u}$ consists of edge forces only, since the derivatives of the constant stresses are zero—no volume forces. Next, the authors calculated the six equivalent nodal forces, vector $\boldsymbol{f}_h$, which belong to the load $\boldsymbol{p}_h$ by superposing the element edge forces (tractions) with the 3×2 unit displacements $\boldsymbol{\varphi}_i(\boldsymbol{x})$ of the three nodes. This led to a matrix $\boldsymbol{T}$

$$\boldsymbol{f}_h = \boldsymbol{T}_{(6\times 3)}\boldsymbol{\sigma}_{(3)} \,, \tag{3.46}$$

or with (3.45) to the equation

$$\boldsymbol{f}_h = \boldsymbol{T}_{(6\times 3)}\boldsymbol{S}_{(3\times 6)}\boldsymbol{u}_{(6)} \,, \tag{3.47}$$

and this is exactly the stiffness matrix $\boldsymbol{K} = \boldsymbol{T}\boldsymbol{S}$ of the *CST element*. That an analysis of virtual external work produced the stiffness matrix, whose entries we read as virtual *internal* energies, is based on the ***inside = outside***, which is guaranteed by Green's first identity. If

$$\delta W_e(\boldsymbol{p}_i, \boldsymbol{\varphi}_j) = \int_{\Gamma_e} \boldsymbol{t}(\boldsymbol{\varphi}_i) \cdot \boldsymbol{\varphi}_j \, ds =: f_{ij} \tag{3.48}$$

is the work done by the tractions $\boldsymbol{t}(\boldsymbol{\varphi}_i)$, the edge stresses of the field $\boldsymbol{\varphi}_i$, on acting through $\boldsymbol{\varphi}_j$, then

$$\mathscr{G}(\boldsymbol{\varphi}_i, \boldsymbol{\varphi}_j) = \delta W_e(\boldsymbol{p}_i, \boldsymbol{\varphi}_j) - \delta W_i(\boldsymbol{\varphi}_i, \boldsymbol{\varphi}_j) = f_{ij} - k_{ij} = 0 \,, \tag{3.49}$$

and therewith

$$f_{hi} = \delta W_e(\boldsymbol{p}_h, \boldsymbol{\varphi}_i) = \sum_j f_{ij}\, u_j = \sum_j k_{ij}\, u_j \,. \tag{3.50}$$

The elements k_{ij} of a stiffness matrix $\boldsymbol{K}$ can be read as virtual internal energy as well as virtual external work f_{ij}.

---

[3] We follow here the representation in [2] pp. 882–883.

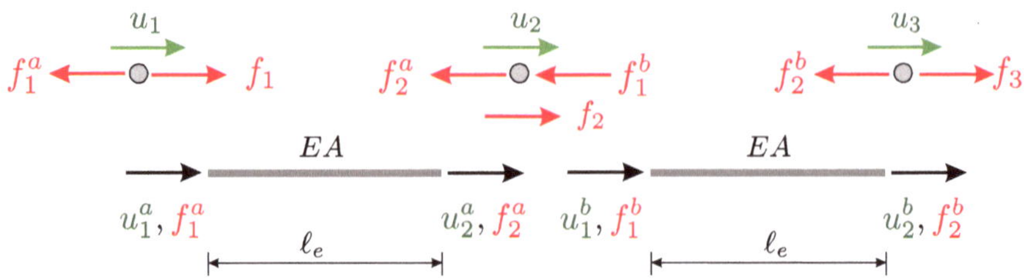

Fig. 3.12 System consisting of two bar elements

The entry $k_{11} = \delta W_i(\varphi_1, \varphi_1)$ of a beam matrix

$$k_{11} = \int_0^l \frac{M_1^2}{EI}\, dx = a(\varphi_1, \varphi_1) = \frac{12\, EI}{l^3} \cdot 1 = \delta W_e(\varphi_1, \varphi_1) = f_{11} \qquad (3.51)$$

equals the work f_{11}, which the force $12\, EI/l^3$—the force, which pushes the node down by one unit length—does on acting through $\varphi_1(0) = 1$.

The "first" stiffness matrix in the history of the FEM was a table of **virtual external work**. Only the equation $\delta W_i = \delta W_e$ leads to the interpretation how we read stiffness matrices $\boldsymbol{K}$ today, $k_{ij} = a(\boldsymbol{\varphi}_i, \boldsymbol{\varphi}_j)$.

What is truly remarkable is, how **simple** the finite elements began—no energy principle, no Galerkin, no higher mathematics—but an old trusted principle, the "shake equivalence"

$$\boxed{\delta W_e(\boldsymbol{p}, \boldsymbol{\varphi}_i) = \delta W_e(\boldsymbol{p}_h, \boldsymbol{\varphi}_i) \qquad \text{the start of the FEM}} \qquad (3.52)$$

got the ball rolling.[4]

3.9 Assembling the Element Matrices

To prepare for the next section on the *equivalent stress transformation*, we shortly repeat, how element matrices are assembled.

At the start the two element matrices of the bar in Fig. 3.12 are placed on the diagonal of a 4×4 matrix $\boldsymbol{K}_{loc}$

[4] Only later, when the mathematicians came on board, it was recognized that the elements could be interpreted as finite functions and that the equivalence $\boldsymbol{f}_h = \boldsymbol{f}$ corresponds to the $\delta\Pi = 0$ of the potential energy.

$$\boldsymbol{f}_{loc} = \begin{bmatrix} f_1^a \\ f_2^a \\ f_1^b \\ f_2^b \end{bmatrix} = \frac{EA}{\ell_e} \begin{bmatrix} 1 & -1 & 0 & 0 \\ -1 & 1 & 0 & 0 \\ 0 & 0 & 1 & -1 \\ 0 & 0 & -1 & 1 \end{bmatrix} \begin{bmatrix} u_1^a \\ u_2^a \\ u_1^b \\ u_2^b \end{bmatrix} = \boldsymbol{K}_{loc} \boldsymbol{u}_{loc} \,, \tag{3.53}$$

and the link between the displacements of the element ends and the nodal displacements u_1, u_2, u_3 can be expressed as

$$\boldsymbol{u}_{loc} = \begin{bmatrix} u_1^a \\ u_2^a \\ u_1^b \\ u_2^b \end{bmatrix} = \begin{bmatrix} 1 & 0 & 0 \\ 0 & 1 & 0 \\ 0 & 1 & 0 \\ 0 & 0 & 1 \end{bmatrix} \begin{bmatrix} u_1 \\ u_2 \\ u_3 \end{bmatrix} = \boldsymbol{A}\boldsymbol{u} \,. \tag{3.54}$$

The element end forces $\boldsymbol{f}_{loc}$ and the nodal forces $\boldsymbol{f}$ do the same work on acting through a virtual displacement $\boldsymbol{\delta u}$ and $\boldsymbol{\delta u}_{loc} = \boldsymbol{A}\, \boldsymbol{\delta u}$, respectively,

$$\boldsymbol{f}_{loc}^T \boldsymbol{\delta u}_{loc} = \boldsymbol{f}^T \boldsymbol{\delta u} \quad \text{or} \quad \boldsymbol{f}_{loc}^T \boldsymbol{A}\, \boldsymbol{\delta u} = \boldsymbol{f}^T \boldsymbol{\delta u} \,, \tag{3.55}$$

and this gives $\boldsymbol{f} = \boldsymbol{A}^T \boldsymbol{f}_{loc}$ or

$$\boldsymbol{f} = \begin{bmatrix} f_1 \\ f_2 \\ f_3 \end{bmatrix} = \begin{bmatrix} 1 & 0 & 0 & 0 \\ 0 & 1 & 1 & 0 \\ 0 & 0 & 0 & 1 \end{bmatrix} \begin{bmatrix} f_1^a \\ f_2^a \\ f_1^b \\ f_2^b \end{bmatrix} = \boldsymbol{A}^T \boldsymbol{f}_{loc} \,, \tag{3.56}$$

which is just the equilibrium condition between the element end forces and the nodal forces f_i.

So, the global stiffness matrix is a product of three matrices

$$\boldsymbol{K} = \boldsymbol{A}^T \boldsymbol{K}_{loc} \boldsymbol{A} = \frac{EA}{\ell_e} \begin{bmatrix} 1 & -1 & 0 \\ -1 & 2 & -1 \\ 0 & -1 & 1 \end{bmatrix} \tag{3.57}$$

3.10 Equivalent Stress Transformation

Coupling identical elements is not a problem. However, to couple beam and plate elements poses a problem since beam ends can rotate, and this degree of freedom is missing on the side of the plate.

The ***equivalent stress transformation*** (EST) by Werkle [3], solves this problem in a very elegant, natural way. In this approach the horse is harnessed from behind, so to speak. Normally, when we formulate a stiffness matrix

$$\boldsymbol{K} = \boldsymbol{A}^T \boldsymbol{K}_{loc} \boldsymbol{A} \tag{3.58}$$

we[5] start with the coupling between the displacements

$$\boldsymbol{u}_{loc} = \boldsymbol{A}\,\boldsymbol{u}_{nodes}\,, \tag{3.59}$$

and the transpose $\boldsymbol{A}^T$ then serves to formulate the equilibrium conditions

$$\boldsymbol{f}_{nodes} = \boldsymbol{A}^T \boldsymbol{f}_{loc}\,. \tag{3.60}$$

Read this as: $\boldsymbol{f}_{loc} = \boldsymbol{K}_{loc}\,\boldsymbol{u}_{loc} = \boldsymbol{K}_{loc}\,\boldsymbol{A}\,\boldsymbol{u}_{node}$ followed by $\boldsymbol{f}_{nodes} = \boldsymbol{A}^T\,\boldsymbol{f}_{loc} = \boldsymbol{A}^T\,\boldsymbol{K}_{loc}\,\boldsymbol{A}\,\boldsymbol{u}_{nodes} = \boldsymbol{K}\,\boldsymbol{u}_{nodes}$.

With the equivalent stress transformation it is the other way round. We first formulate (3.60)—here the static understanding and the skill of the engineer is required—and the transpose $\boldsymbol{A} = (\boldsymbol{A}^T)^T$ is then used in (3.59).

At this point it shows—and this was not so clear before—that there are **two ways** to describe the coupling of the elements via a matrix $\boldsymbol{A}$, the geometric path $\boldsymbol{u}_{node} \rightarrow \boldsymbol{u}_{loc}$ or the static path $\boldsymbol{f}_{loc} \rightarrow \boldsymbol{f}_{node}$. Note, the two approaches need not lead to the same matrix $\boldsymbol{A}$.

Example 3.1 In the following example the technique is used to couple a beam and a plate.

Figure 3.13 shows the situation; three nodes with the degrees of freedom u_i, v_i lie opposite to the beam's end. On the classical, geometric path (3.59) we assume the cross-section of the beam to remain flat, i.e. with $a = d/2$, half the height of the beam, the displacements are

$$u_1 = u_B + a\,\tan\varphi_B, \quad u_2 = u_B, \quad u_3 = u_B - a\,\tan\varphi_B, \quad v_1 = v_2 = v_3 \tag{3.61}$$

and so (3.59) reads

$$\begin{bmatrix} u_1^{(1)} \\ v_1^{(1)} \\ u_2^{(1)} \\ v_2^{(1)} \\ u_2^{(2)} \\ v_2^{(2)} \\ u_3^{(2)} \\ v_3^{(2)} \end{bmatrix} = \begin{bmatrix} 1 & 0 & a \\ 0 & 1 & 0 \\ 1 & 0 & 0 \\ 0 & 1 & 0 \\ 1 & 0 & 0 \\ 0 & 1 & 0 \\ 1 & 0 & -a \\ 0 & 1 & 0 \end{bmatrix} \begin{bmatrix} u_B \\ v_B \\ \tan\varphi_B \end{bmatrix}. \tag{3.62}$$

The linear distribution of the displacements is of course only an approximation. (More) correct in the sense of elasticity theory would be a displacement distribution, where both sides were modeled with plate elements.

[5]The diagonal of $\boldsymbol{K}_{loc}$ contains the element matrices, see (3.53).

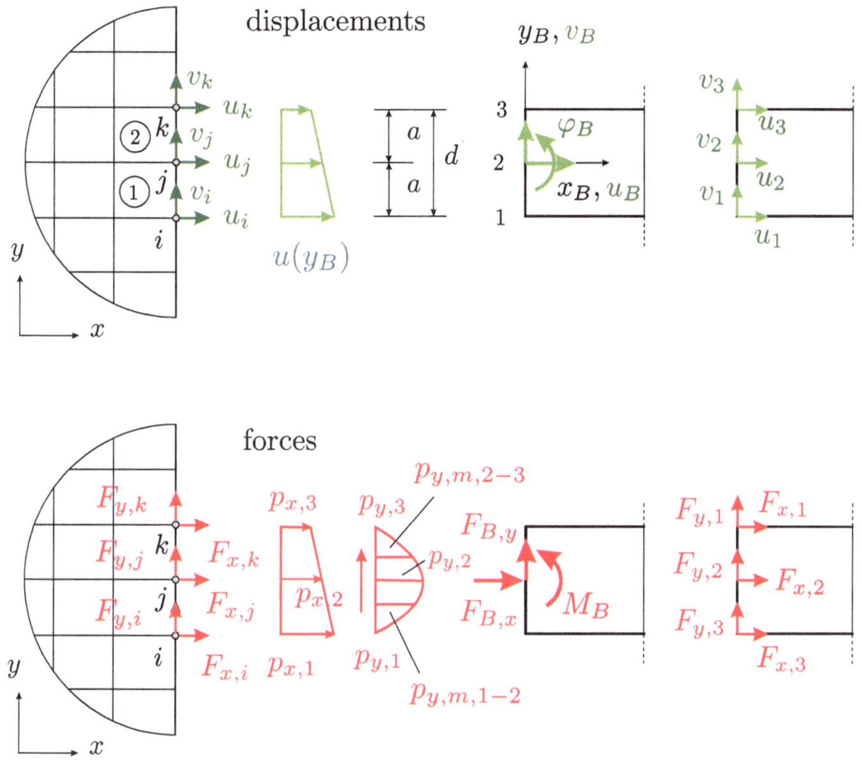

Fig. 3.13 Coupling of a plate and a beam. The actions $F_{B,x}$, $F_{B,y}$, M_B on the side of the beam are transformed into equivalent nodal forces $F_{x,i}$, $F_{y,i}$, (here pictured on the right), so they match the nodal force of the plate on the left [3]

The **static path** on the other hand concentrates on the forces. The section forces F_{Bx}, F_{By} and M_B on the side of the beam, see Fig. 3.13, generate the stresses[6] (rectangular cross-section, t = wall thickness)

$$p_x = \frac{F_{B,x}}{A} - \frac{M_B}{I}\, y_B = \frac{F_{B,x}}{d\,t} - \frac{12\,M_B}{t\,d^3}\, y_B \tag{3.63}$$

$$p_y = \frac{3}{2}\,\frac{1}{d\,t}(1 - 4\,\frac{y_B^2}{d^2})\, F_{B,y}\,, \tag{3.64}$$

which are converted into equivalent nodal forces $\boldsymbol{f}$ on the side of the plate, and therewith establish a relation between the forces on both sides of the interface.

The nodal values p_i of the stresses σ and τ depend on the section forces, the vector $\boldsymbol{f}_B = \{F_{B,x}, F_{B,y}, M_B\}^T$. If we also write the nodal values as a vector $\boldsymbol{p}$, the transition $\boldsymbol{f}_B \to \boldsymbol{p}$ can be formulated with a matrix $\boldsymbol{P}$ (as in polynomials).

[6] signs according to Fig. 3.13.

Given the y-coordinates of points 1, 2 and 3

$$y_{B,1} = -a, \qquad y_{B,2} = 0, \qquad y_{B,3} = a \tag{3.65}$$

the nodal values of the stresses are

$$\begin{bmatrix} p_{x,1} \\ p_{y,1} \\ p_{y,m,1-2} \\ p_{x,2} \\ p_{y,2} \\ p_{y,m,2-3} \\ p_{x,3} \\ p_{y,3} \end{bmatrix} = \frac{1}{16\,t\,a^2} \begin{bmatrix} 8\,a & 0 & 24 \\ 0 & 0 & 0 \\ 0 & 9\,a & 0 \\ 8\,a & 0 & 0 \\ 0 & 12\,a & 0 \\ 0 & 9\,a & 0 \\ 8\,a & 0 & -24 \\ 0 & 0 & 0 \end{bmatrix} \begin{bmatrix} F_{B,x} \\ F_{B,y} \\ M_B \end{bmatrix} \tag{3.66}$$

or

$$\boldsymbol{p} = \boldsymbol{P}\,\boldsymbol{f}_B\,. \tag{3.67}$$

The equivalent nodal forces $\boldsymbol{f}_S$ of the stresses p_x and p_y (=line loads) are the superposition of the stresses with the shape functions. We write this as

$$\boldsymbol{f}_S = \boldsymbol{Q}\,\boldsymbol{p}\,, \tag{3.68}$$

with a matrix $\boldsymbol{Q}$ (as in quadrature).

This relation is first established separately for each element and then the global matrix $\boldsymbol{Q}$ is assembled. On element # 1 we have in axial direction

$$\begin{bmatrix} F^{(1)}_{x,1} \\ F^{(1)}_{x,2} \end{bmatrix} = \frac{a\,t}{6} \begin{bmatrix} 2 & 1 \\ 1 & 2 \end{bmatrix} \begin{bmatrix} p_{x,1} \\ p_{x,2} \end{bmatrix}, \tag{3.69}$$

and in vertical direction

$$\begin{bmatrix} F^{(1)}_{y,1} \\ F^{(1)}_{y,2} \end{bmatrix} = \frac{a\,t}{12} \begin{bmatrix} 3 & 4 & -1 \\ -1 & 4 & 3 \end{bmatrix} \begin{bmatrix} p_{y,1} \\ p_{y,m,1-2} \\ p_{y,2} \end{bmatrix}, \tag{3.70}$$

which gives

$$\begin{bmatrix} F^{(1)}_{x,1} \\ F^{(1)}_{y,1} \\ F^{(1)}_{x,2} \\ F^{(1)}_{y,2} \end{bmatrix} = \frac{a\,t}{12} \begin{bmatrix} 4 & 0 & 0 & 2 & 0 \\ 0 & 3 & 4 & 0 & -1 \\ 2 & 0 & 0 & 4 & 0 \\ 0 & -1 & 4 & 0 & 3 \end{bmatrix} \begin{bmatrix} p_{x,1} \\ p_{y,1} \\ p_{y,m,1-2} \\ p_{x,2} \\ p_{y,2} \end{bmatrix}, \tag{3.71}$$

and correspondingly on element 2

$$\begin{bmatrix} F_{x,2}^{(2)} \\ F_{y,2}^{(2)} \\ F_{x,3}^{(2)} \\ F_{y,3}^{(2)} \end{bmatrix} = \frac{a\,t}{12} \begin{bmatrix} 4 & 0 & 0 & 2 & 0 \\ 0 & 3 & 4 & 0 & -1 \\ 2 & 0 & 0 & 4 & 0 \\ 0 & -1 & 4 & 0 & 3 \end{bmatrix} \begin{bmatrix} p_{x,2} \\ p_{y,2} \\ p_{y,m,2-3} \\ p_{x,3} \\ p_{y,3} \end{bmatrix} . \tag{3.72}$$

Both parts together result in, see Fig. 3.13,

$$\begin{bmatrix} F_{x,1} \\ F_{y,1} \\ F_{x,2} \\ F_{y,2} \\ F_{x,3} \\ F_{y,3} \end{bmatrix} = \frac{a\,t}{12} \begin{bmatrix} 4 & 0 & 0 & 2 & 0 & 0 & 0 & 0 \\ 0 & 3 & 4 & 0 & -1 & 0 & 0 & 0 \\ 2 & 0 & 0 & 8 & 0 & 0 & 2 & 0 \\ 0 & -1 & 4 & 0 & 6 & 4 & 0 & -1 \\ 0 & 0 & 0 & 2 & 0 & 0 & 4 & 0 \\ 0 & 0 & 0 & 0 & -1 & 4 & 0 & 3 \end{bmatrix} \begin{bmatrix} p_{x,1} \\ p_{y,1} \\ p_{y,m,1-2} \\ p_{x,2} \\ p_{y,2} \\ p_{y,m,2-3} \\ p_{x,3} \\ p_{y,3} \end{bmatrix} , \tag{3.73}$$

or

$$\boldsymbol{f}_S = \boldsymbol{Q}\,\boldsymbol{p} , \tag{3.74}$$

where the forces $\boldsymbol{f}_S$ are the forces on the side of the beam, see Fig. 3.13. Continuing with (3.67) we obtain

$$\boldsymbol{f}_S = \boldsymbol{Q}\boldsymbol{P}\boldsymbol{f}_B = \boldsymbol{A}^T \boldsymbol{f}_B , \tag{3.75}$$

and here we see the **transformation matrix**

$$\boldsymbol{A} = \boldsymbol{P}^T \boldsymbol{Q}^T = \begin{bmatrix} 1/4 & 0 & 1/2 & 0 & 1/4 & 0 \\ 0 & 1/8 & 0 & 3/8 & 0 & 1/8 \\ 1/2a & 0 & 0 & 0 & -1/2a & 0 \end{bmatrix} . \tag{3.76}$$

The displacements and forces at the end of the beam

$$\boldsymbol{f}_B = \begin{bmatrix} F_{B,x} \\ F_{B,y} \\ M_B \end{bmatrix} , \quad \boldsymbol{u}_B = \begin{bmatrix} u_B \\ v_B \\ \varphi_B \end{bmatrix} , \quad \boldsymbol{f}_S = \begin{bmatrix} F_{x,1} \\ F_{y,1} \\ F_{x,2} \\ F_{y,2} \\ F_{x,3} \\ F_{y,3} \end{bmatrix} , \quad \boldsymbol{u}_S = \begin{bmatrix} u_1 \\ v_1 \\ u_2 \\ v_2 \\ u_3 \\ v_3 \end{bmatrix} \tag{3.77}$$

transform as

$$\boldsymbol{u}_B = \boldsymbol{A}_{(3\times6)}\boldsymbol{u}_S \qquad \boldsymbol{f}_S = \boldsymbol{A}^T_{(6\times3)}\boldsymbol{f}_B . \tag{3.78}$$

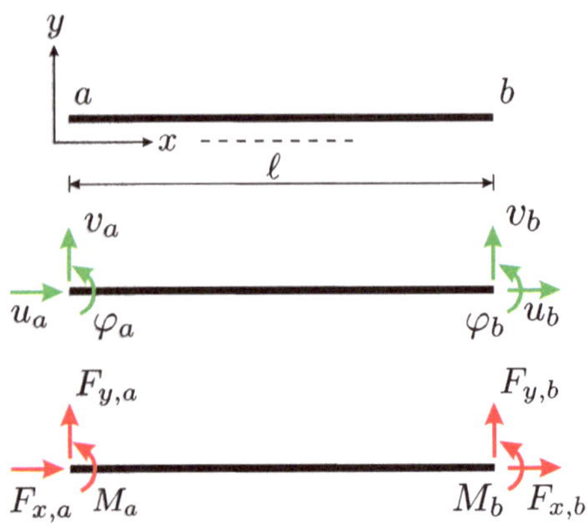

Fig. 3.14 Frame element

At the beam's end, see Fig. 3.14, the relationship between the displacements and forces is

$$\begin{bmatrix} a_0 & 0 & 0 & -a_0 & 0 & 0 \\ 0 & 12\,a_1/\ell^2 & 6\,a_1/\ell & 0 & -12\,a_1/\ell^2 & 6\,a_1/\ell \\ 0 & 6\,a_1/\ell & 4\,a_1 & 0 & -6\,a_1/\ell & 2\,a_1 \\ -a_0 & 0 & 0 & a_0 & 0 & 0 \\ 0 & -12\,a_1/\ell^2 & -6\,a_1/\ell & 0 & 12\,a_1/\ell^2 & -6\,a_1/\ell \\ 0 & 6\,a_1/\ell & 2\,a_1 & 0 & -6\,a_1/\ell & 4\,a_1 \end{bmatrix} \begin{bmatrix} u_a \\ v_a \\ \varphi_a \\ u_b \\ v_b \\ \varphi_b \end{bmatrix} = \begin{bmatrix} F_{x,a} \\ F_{y,a} \\ M_a \\ F_{x,b} \\ F_{y,b} \\ M_b \end{bmatrix} \tag{3.79}$$

with

$$a_0 = \frac{EA}{\ell}, \quad a_1 = \frac{EI}{\ell}, \quad \ell = \text{length of the element.} \tag{3.80}$$

The beam element has two nodes a and b. Accordingly, the above stiffness matrix of the beam is now split into sub-matrices, which refer to node a and b respectively

$$\begin{bmatrix} \boldsymbol{K}_{aa} & \boldsymbol{K}_{ab} \\ \boldsymbol{K}_{ba} & \boldsymbol{K}_{bb} \end{bmatrix} \begin{bmatrix} \boldsymbol{u}_a \\ \boldsymbol{u}_b \end{bmatrix} = \begin{bmatrix} \boldsymbol{f}_a \\ \boldsymbol{f}_b \end{bmatrix}. \tag{3.81}$$

When connecting node a to two plate elements, as in Fig. 3.13, the displacements, and forces at node a are—in the notation of the EST

$$\boldsymbol{u}_a = \boldsymbol{u}_B = \begin{bmatrix} u_B \\ v_B \\ \varphi_B \end{bmatrix} \qquad \boldsymbol{f}_a = \boldsymbol{f}_B = \begin{bmatrix} F_{B,x} \\ F_{B,y} \\ M_B \end{bmatrix}. \tag{3.82}$$

If the plate elements are linear, the quantities transform according to

$$\boldsymbol{u}_B = \boldsymbol{A}_{(3\times 6)}\boldsymbol{u}_S \tag{3.83}$$

with matrix $\boldsymbol{A}$ as in (3.76). If we substitute $\boldsymbol{u}_a = \boldsymbol{u}_B = \boldsymbol{A}\boldsymbol{u}_S$ into (3.81) and multiply the first row from the left with $\boldsymbol{A}^T$

$$\boldsymbol{A}^T_{(6\times 3)}\boldsymbol{f}_a = \boldsymbol{A}^T_{(6\times 3)}\boldsymbol{f}_B = \boldsymbol{f}_S\,, \tag{3.84}$$

the result

$$\begin{bmatrix} \boldsymbol{A}^T\boldsymbol{K}_{aa}\boldsymbol{A} & \boldsymbol{A}^T\boldsymbol{K}_{ab} \\ \boldsymbol{K}_{ba}\boldsymbol{A} & \boldsymbol{K}_{bb} \end{bmatrix} \begin{bmatrix} \boldsymbol{u}_S \\ \boldsymbol{u}_b \end{bmatrix} = \begin{bmatrix} \boldsymbol{f}_S \\ \boldsymbol{f}_b \end{bmatrix} \tag{3.85}$$

is a 9×9 matrix, 6 *dofs* $\boldsymbol{u}_S$ on the side of the plate and 3 *dofs* $\boldsymbol{u}_b$ on the side of the beam, and the $\boldsymbol{f}_S$ are the six nodal forces of the plate.

The stiffness matrix (3.85) is now related at node a to the degrees of freedom of the plate and at node b to the degrees of freedom of the beam. Node b could be transformed as well if it also were connected to a plate on its side.

To avoid any misunderstanding: The static path solely serves to formulate the matrix (3.85). The later assembly of the element matrices does not differ from the standard procedure.

Remark 3.5 The geometric path and the static path are based on different assumptions and lead to different matrices $\boldsymbol{A}$ and $\boldsymbol{A}$. On the geometric path the displacement distribution is specified, and the stresses of the plate elements adapt to these specifications. On the static path, instead, the stress distribution is specified and the displacements (here at points 1, 2 and 3) will adapt accordingly—will probably deviate from the linear distribution of the geometric path.

On the geometric path the connection will be too stiff and on the static path it will be too soft. The geometric path treats the connection like a semi-rigid inclusion (think of the linear displacements on the side of the beam and imagine four or five plate elements lie opposite to the beam, they must all rotate "as one piece"). The FEM however does not handle semi-rigid inclusions well, since they produce stress singularities at the end points of the interface. This is not the case on the static path and therefore it is to be preferred. For further details we refer to [3].

3.11 Calculation of Influence Functions with Finite Elements

We come now to an important subject, the calculation of influence functions with finite elements.

It is the same standard system, $\boldsymbol{K}\boldsymbol{u} = \boldsymbol{f}$, only that we write g_i instead of u_i and instead of f_i we write j_i

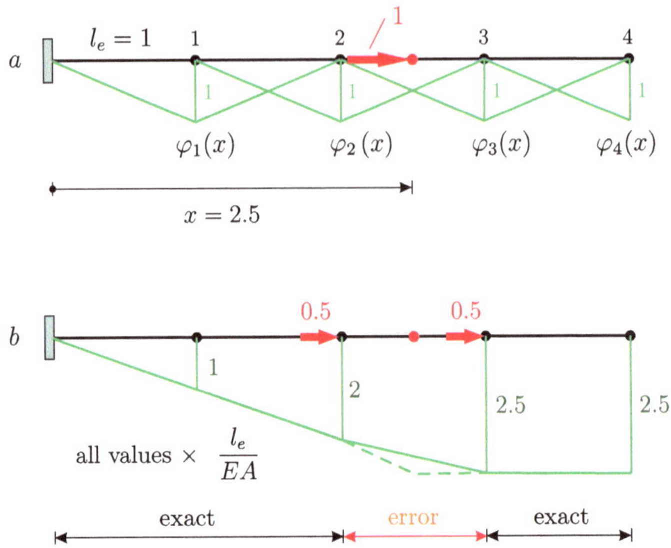

Fig. 3.15 Influence function for a displacement $u(x)$, **a** hat functions and original load, **b** substitute forces and resulting shape

$$\boldsymbol{K}\boldsymbol{g} = \boldsymbol{j}\,. \tag{3.86}$$

This makes it easier to operate with FE-influence functions.

The first function at which we test our skills is the influence function for the longitudinal displacement $u(x)$ of the bar in Fig. 3.15 at the point $x = 2.5$. The equivalent nodal forces are, as we will explain in a moment, the displacements of the shape functions $\varphi_i(x)$ at the point $x = 2.5$ (strictly speaking $\varphi_i(2.5)$ [m] · 1 [N] = displ. × unit force = energy)

$$\varphi_1(2.5) = 0 \qquad \varphi_2(2.5) = 0.5 \qquad \varphi_3(2.5) = 0.5 \qquad \varphi_4(2.5) = 0\,, \tag{3.87}$$

and so $\boldsymbol{K}\boldsymbol{g} = \boldsymbol{j}$ is the system

$$\frac{EA}{l_e}\begin{bmatrix} 2 & -1 & 0 & 0 \\ -1 & 2 & -1 & 0 \\ 0 & -1 & 2 & -1 \\ 0 & 0 & -1 & 2 \end{bmatrix}\begin{bmatrix} g_1 \\ g_2 \\ g_3 \\ g_4 \end{bmatrix} = \begin{bmatrix} 0 \\ 0.5 \\ 0.5 \\ 0 \end{bmatrix}, \tag{3.88}$$

which has the solution,

$$g_1 = 1 \qquad g_2 = 2 \qquad g_3 = 2.5 \qquad g_4 = 2.5 \qquad \times \frac{l_e}{EA}\,, \tag{3.89}$$

and the influence function is therefore

$$G_0^h(y, x = 2.5) = \frac{l_e}{EA}\,[1 \cdot \varphi_1(y) + 2 \cdot \varphi_2(y) + 2.5 \cdot \varphi_3(y) + 2.5 \cdot \varphi_4(y)]\,. \quad (3.90)$$

Since x is the source point, y is the function variable over which we later integrate.

Outside the element with the source point x, the influence function is exact. The error inside the element is corrected automatically when we add the local solution

$$G_0(y, x) = G_0^h(y, x) + \text{local solution} = G_0^h(y, x) + \text{dashed triangle}\,. \quad (3.91)$$

The local solution is the influence function for $u(x)$ if the element is fixed on both sides. This way FE-programs **succeed in generating exact influence functions** in frames—provided EA and EI are constant.

The Key to the Nodal Forces j_i

Why are the equivalent nodal forces j_i ($=f_i$) the values of the shape functions at the source point, why is $j_i = \varphi_i(2.5)$? The answer lies in the definition of the equivalent nodal forces f_i.

A nodal force f_i is an **energy quantum**, it is the work the load $p(x)$ does on acting through the shape function $\varphi_i(x)$

$$f_i = \int_0^l p(x)\,\varphi_i(x)\,dx\,. \quad (3.92)$$

In the case of the influence function the load is a Dirac delta (a line load concentrated at one point)

$$-EA\,\frac{d^2}{dy^2}\,G_0(y, x) = \delta_0(y - x) \qquad [\text{N/m}]\,, \quad (3.93)$$

which resembles a single force at the source point $x = 2.5$. It is, so to speak, the p which belongs to the influence function. (We differentiate with respect to y on the left side since x is the source point).

Integrating we find that—numerically—the equivalent nodal forces ([Nm])

$$j_i = \int_0^l \underbrace{\delta_0(y - x)}_{[\text{N/m}]}\,\underbrace{\varphi_i(y)}_{[\text{m}]}\,\underbrace{dy}_{[\text{m}]} = \varphi_i(x) \qquad x = 2.5 \quad (3.94)$$

are simply the values of the four shape functions φ_i at the source point $x = 2.5$, and this is how the list in (3.87) was generated, see also Fig. 3.16.

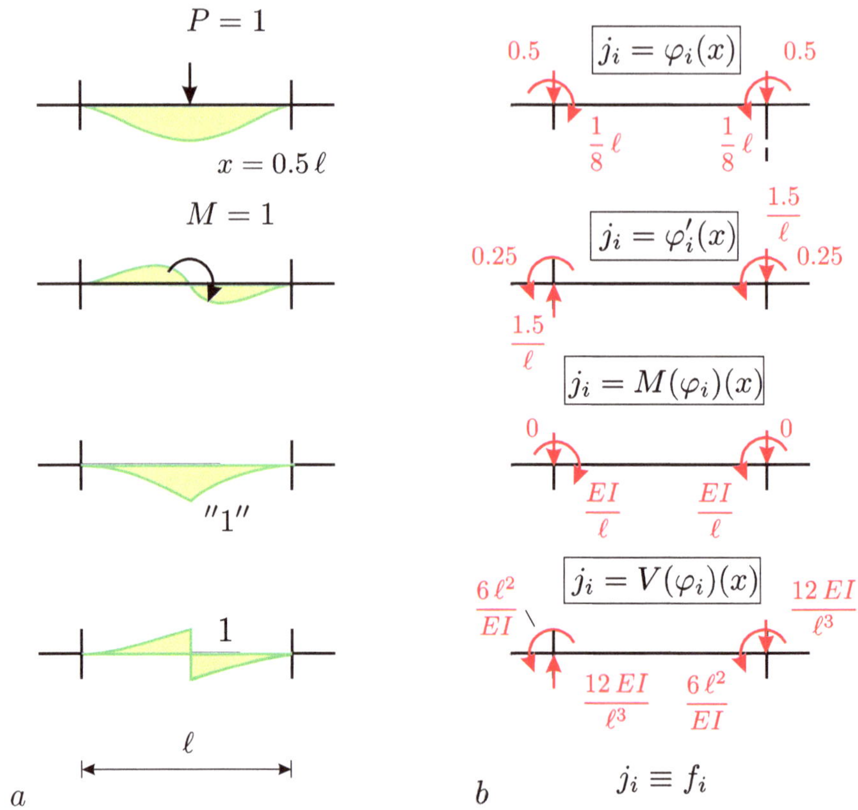

Fig. 3.16 Influence functions for a beam, **a** local solutions, **b** the equivalent nodal forces j_i which produce the FE-influence functions

3.12 Functionals

A functional is a function of functions, like the shear force, the deflection or the moment at a point x of a beam, which are **point functionals**,

$$J(w) = V(x) \qquad J(w) = w(x) \qquad J(w) = M(x)\,. \tag{3.95}$$

Evaluating a functional can be compared to taking a reading, taking a measurement. In linear statics functionals are linear,

$$J(w_1 + w_2) = J(w_1) + J(w_2)\,, \tag{3.96}$$

and all linear functionals, as for example $J_i(\boldsymbol{u}) = u_i$, are **scalar products**

$$J_i(\boldsymbol{u}) = u_i = \boldsymbol{e}_i^T\,\boldsymbol{u}\,. \tag{3.97}$$

The unit vector $\boldsymbol{e}_i$ is the **discrete Dirac delta** and its work on acting through $\boldsymbol{u}$ is u_i. The unit vector is here the *kernel* or *generator* of this functional; the kernel is also the gradient of a linear functional.

This concept can easily be extended to functions. The deflection of a beam at a point x, the functional $J(w) = w(x)$, is the L_2-scalar product of w with a Dirac delta $\delta(y - x)$ (=point load)

$$J(w) = 1 \cdot w(x) = \int_0^l \delta(y - x)\, w(y)\, dy \qquad [\text{Nm}]\,, \tag{3.98}$$

which in its action resembles the action of the unit vector $\boldsymbol{e}_i$.

So, each functional $J(w)$ can be considered the superposition of a certain Dirac delta, (a static or geometric point source) with the function w

$$J(w) = \int_0^l \delta(y - x)\, w(y)\, dy\,. \tag{3.99}$$

Surprisingly, we do not need to know what these different Dirac deltas look like, we just need to know what the result $J(w)$ should be. And since each functional is a scalar product, the result is an ***energy***

$$J(w) = 1 \cdot \text{“anything”} \qquad [\text{Nm}]\,, \tag{3.100}$$

is the product of an unit source with a matching, conjugated term as in

$$J(w) = V(x) = V(x) \cdot 1 \quad = \text{N} \cdot \text{m} \qquad 1 = \text{displacement} \tag{3.101}$$
$$J(w) = w(x) = w(x) \cdot 1 \quad = \text{m} \cdot \text{N} \qquad 1 = \text{force} \tag{3.102}$$
$$J(w) = M(x) = M(x) \cdot 1 \quad = \text{Nm} \cdot [\ ] \quad 1 = \text{kink}\,. \tag{3.103}$$

The kink is a jump in the slope dw/dx = [m]/[m] = [ ]. In the following we will omit the factor 1.

Since we consider the Green’s function the response to a Dirac delta

$$EI\,\frac{d^4}{dy^4}\,G(y, x) = \delta(y - x)\,, \tag{3.104}$$

the equivalent nodal forces j_i $(= f_i)$, which generate the influence function are

$$\boxed{j_i = \int_0^l \delta(y - x)\,\varphi_i(y)\, dy = J(\varphi_i)} \tag{3.105}$$

It could not be any simpler. This means:

(1) The nodal forces j_i, which generate the influence function for the deflection of a beam at a point x, are the deflections of the shape functions φ_i at this point

$$j_i = \varphi_i(x)\,. \tag{3.106}$$

(2) The nodal forces j_i, which generate the influence function for the moment $M(x)$ at a point x

$$j_i = -EI\,\varphi_i''(x) = M(\varphi_i)(x)\,, \tag{3.107}$$

are the moments of the shape functions at this point x—etc.

Theorem 3.1 Nodal forces for influence functions *The influence function for a linear functional $J(u)$ is generated by the nodal forces $j_i = J(\varphi_i)$; they are the values $J(\varphi_i)$ of the shape functions.*

We call the nodal values of the influence functions g_i and the equivalent nodal forces j_i, and so the system $\boldsymbol{Ku} = \boldsymbol{f}$ reads

$$\boldsymbol{Kg} = \boldsymbol{j} \tag{3.108}$$

3.13 Generalized Influence Functions

Now that we have mastered the technique, we can calculate influence functions for *arbitrary* linear functionals, i.e. leave classical statics, which knew or could only calculate influence functions for $w(x)$, $M(x)$ etc., behind.

The functional only must be linear—more is not required. We call these influence functions ***generalized influence functions***, since they exceed the range of classical influence functions.

3.14 Weak and Strong Influence Functions

In Chap. 1 we distinguished between weak and strong influence functions, since it is not possible to calculate forces with weak influence functions.

It may therefore come as a surprise that finite elements (seemingly, but see (9.141)) do not distinguish between weak and strong influence functions, since on $\mathcal{V}_h$ the two formulas

$$J(u_h) = \underbrace{\int_0^l G_h(y,x)\, p(y)\, dy}_{strong} = \underbrace{a(G_h, u_h)}_{weak}\,, \tag{3.109}$$

coincide

$$J(u_h) = \int_0^l G_h(y,x)\, p(y)\, dy = \boldsymbol{g}^T \boldsymbol{f} = \boldsymbol{g}^T \boldsymbol{K} \boldsymbol{u} = a(G_h, u_h)\,. \tag{3.110}$$

In matrix notation the only difference is, how we read the equations

$$J(u_h) = \underbrace{\boldsymbol{g}^T \boldsymbol{f}}_{strong} = \underbrace{\boldsymbol{g}^T \boldsymbol{K} \boldsymbol{u}}_{weak}\,. \tag{3.111}$$

In the weak formulation $\boldsymbol{g}^T \boldsymbol{K}\, \boldsymbol{u}$ we sum over all entries k_{ij}

$$\sum_{i,j} g_i \cdot k_{ij} \cdot u_j = \sum_{i,j} g_i \cdot a(\varphi_i, \varphi_j) \cdot u_j\,, \tag{3.112}$$

which is equivalent to a domain integral (like "Mohr"), while the strong formulation $\boldsymbol{g}^T \boldsymbol{f}$, in contrast, weights the nodal displacements g_i with the nodal forces f_i.

Example 3.2 In Fig. 3.17 we calculate the influence function for the deflection $J(w) = w(x)$ of the rope at the point $x = 1.25$. The nodal forces are the displacements $\varphi_i(x)$ of the shape functions at the source point

$$j_1 = \varphi_1(x) = 0.75 \quad j_2 = \varphi_2(x) = 0.25 \quad j_3 = \varphi_3(x) = 0 \quad j_4 = \varphi_4(x) = 0\,, \tag{3.113}$$

and the system

$$\begin{bmatrix} 2 & -1 & 0 & 0 \\ -1 & 2 & -1 & 0 \\ 0 & -1 & 2 & -1 \\ 0 & 0 & -1 & 2 \end{bmatrix} \begin{bmatrix} g_1 \\ g_2 \\ g_3 \\ g_4 \end{bmatrix} = \begin{bmatrix} 0.75 \\ 0.25 \\ 0 \\ 0 \end{bmatrix} \tag{3.114}$$

has the solution

$$g_1 = 0.75 \qquad g_2 = 0.75 \qquad g_3 = 0.5 \qquad g_4 = 0.25\,, \tag{3.115}$$

and so the Green's function is

$$G(y,x) = 0.75 \cdot \varphi_1(y) + 0.75 \cdot \varphi_2(y) + 0.5 \cdot \varphi_3(y) + 0.25 \cdot \varphi_4(y)\,. \tag{3.116}$$

If the source point is the first node, $x = 1.0$, the nodal forces are

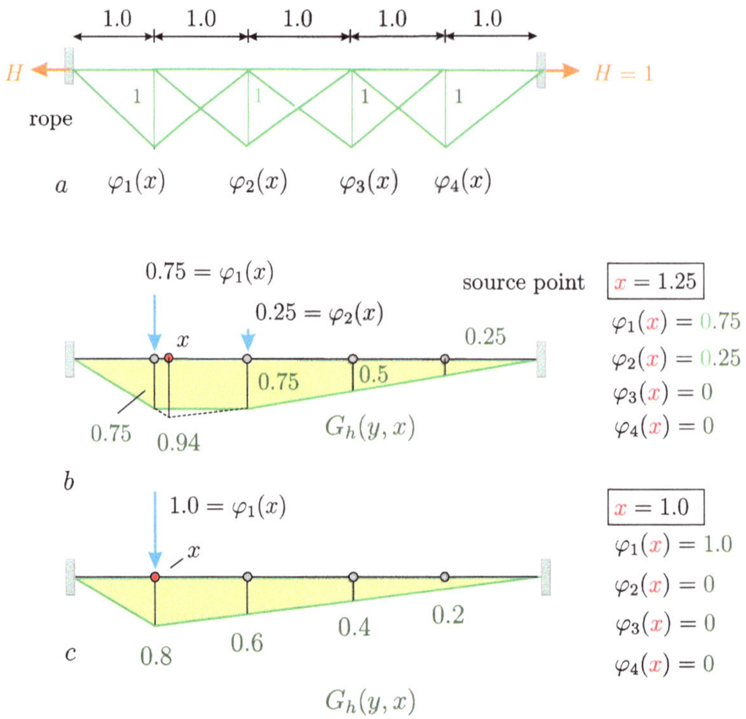

Fig. 3.17 FE-model of a rope, **a** shape functions, **b** FE-influence function for w at $x = 1.25$, **c** FE-influence function for w at the first node; the function is exact, $G_h(y, x) = G(y, x)$

$$j_1 = \varphi_1(x) = 1.0 \quad j_2 = \varphi_2(x) = 0 \quad j_3 = \varphi_3(x) = 0 \quad j_4 = \varphi_4(x) = 0\,, \tag{3.117}$$

and in this case the FE-influence function would be exact, since the solution of $\boldsymbol{K}\boldsymbol{g} = \boldsymbol{j}$, that is the entries

$$g_1 = 0.8 \qquad g_2 = 0.6 \qquad g_3 = 0.4 \qquad g_4 = 0.2\,, \tag{3.118}$$

are the nodal values of the exact influence function, see Fig. 3.17c.

Remark 3.6 If $J(u)$ is a point functional, say $J(u) = u(x)$, the vector $\boldsymbol{j}(x)$ is a function of the source point, and therewith also $\boldsymbol{g}(x) = \boldsymbol{K}^{-1}\boldsymbol{j}(x)$, which means

$$G(y, x) = g_1(x) \cdot \varphi_1(y) + g_2(x) \cdot \varphi_2(y) + g_3(x) \cdot \varphi_3(y) + g_4(x) \cdot \varphi_4(y)\,. \tag{3.119}$$

This factorization looks like an even split of the workload, but the $g_i(x)$ have it easy, since they are called only once, while the $\varphi_i(y)$ carry the influence function from node to node.

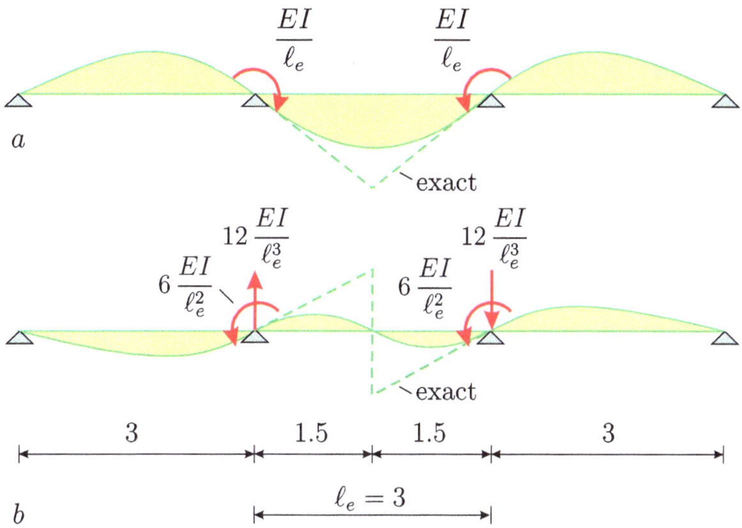

Fig. 3.18 FE-influence function for **a** the bending moment M and **b** the shear force V at $0.5\,\ell_e$. The nodal forces are $j_i = M(\varphi_i)$ and $j_i = V(\varphi_i)$ resp., see (3.123)

Example 3.3 We consider the three span beam in Fig. 3.18 and choose the midpoint, $x = 4.5$, of the second span as source point. The influence function for the bending moment $M = -EI\,w''$ is the response to the equivalent nodal forces, see Fig. 3.19,

$$j_i = -EI\,\varphi_i''(x)\cdot 1 \quad [\text{Nm}]\cdot[\;] \qquad 1 = \text{kink}\;[\;], \tag{3.120}$$

and the influence function for the shear force $V(x) = -EI\,w'''(x)$ is the response to the forces

$$j_i = -EI\,\varphi_i'''(x)\cdot 1 \quad [\text{N}]\cdot[\text{m}] \qquad 1 = \text{disloc.}\;[\text{m}], \tag{3.121}$$

where the φ_i are the element shape functions, ℓ = element length,

$$\varphi_1(x) = 1 - \frac{3x^2}{\ell^2} + \frac{2x^3}{\ell^3} \qquad \varphi_2(x) = -x + \frac{2x^2}{\ell} - \frac{x^3}{\ell^2} \tag{3.122a}$$

$$\varphi_3(x) = \frac{3x^2}{\ell^2} - \frac{2x^3}{\ell^3} \qquad \varphi_4(x) = \frac{x^2}{\ell} - \frac{x^3}{\ell^2}\,. \tag{3.122b}$$

The section forces $M_i = -EI\,\varphi_i''$ and $V_i = -EI\,\varphi_i'''$ are

$$M_1(x) = (\frac{6}{\ell^2} - \frac{12x}{\ell^3})\cdot EI \qquad M_2(x) = (\frac{6x}{\ell^2} - \frac{4}{\ell})\cdot EI \tag{3.123a}$$

$$M_3(x) = (\frac{12x}{\ell^3} - \frac{6}{\ell^2})\cdot EI \qquad M_4(x) = (\frac{6x}{\ell^2} - \frac{2}{\ell})\cdot EI \tag{3.123b}$$

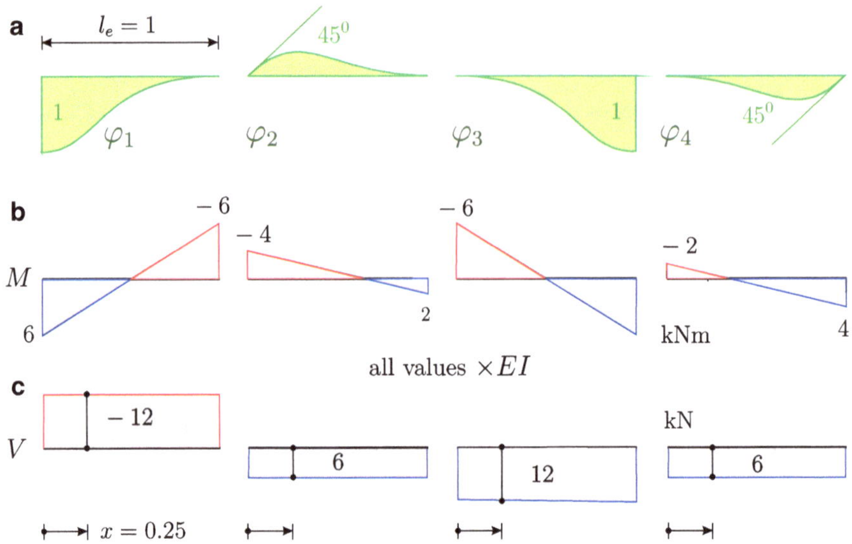

Fig. 3.19 Beam element, **a** the four shape functions φ_i, and **b** the associated bending moments M, and **c** shear forces V, see (3.123)

$$V_1(x) = -\frac{12}{\ell^3} \cdot EI \qquad V_2(x) = \frac{6}{\ell^2} \cdot EI \tag{3.123c}$$

$$V_3(x) = \frac{12}{\ell^3} \cdot EI \qquad V_4(x) = \frac{6}{\ell^2} \cdot EI\,. \tag{3.123d}$$

Only the nodes of the element with the source point x carry nodal forces j_i, since the shape functions φ_i of the nodes lying farther off generate no section forces at the source point x.

The calculated influence functions are **outside** the element with the source point exact, and the computer fixes the error inside the element by adding the local solution (of the GF)—as in the slope deflection method.

Example 3.4 Next, we turn to 2-D problems, and we begin with an **CST element** (*constant strain triangle*), see Fig. 3.20. In such elements the stresses are constant

$$\begin{bmatrix} \sigma_{xx} \\ \sigma_{yy} \\ \sigma_{xy} \end{bmatrix} = \frac{E}{2\,A} \begin{bmatrix} y_{23} & 0 & y_{31} & 0 & y_{12} & 0 \\ 0 & x_{23} & 0 & x_{13} & 0 & x_{21} \\ x_{32} & y_{23} & x_{13} & y_{31} & x_{21} & y_{12} \end{bmatrix} \begin{bmatrix} u_1 \\ v_1 \\ u_2 \\ v_2 \\ u_3 \\ v_3 \end{bmatrix} \tag{3.124}$$

where $x_{ij} = x_i - x_j$ and $y_{ij} = y_i - y_j$.

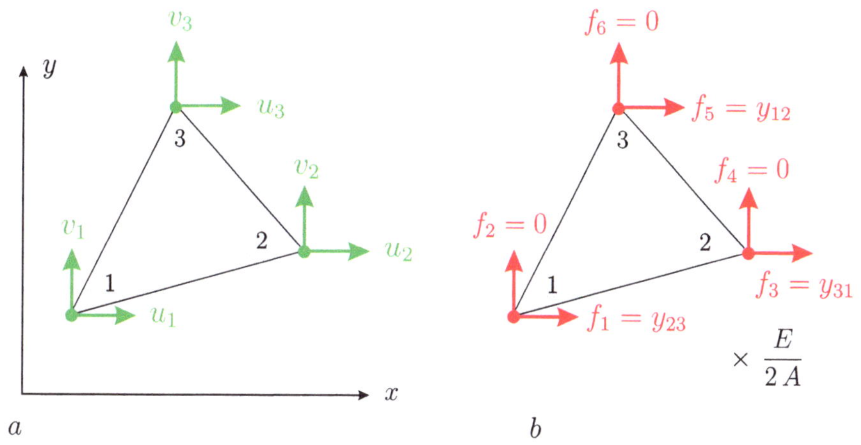

Fig. 3.20 CST element, **a** degrees of freedom, **b** nodal forces which generate the influence function for σ_{xx} at an internal point; the influence function is for each point $\boldsymbol{x}$ the same; A = area of the element

To generate the influence function for the stress σ_{xx} at a point $\boldsymbol{x}$ inside an element, we apply at each node the stresses σ_{xx} of the two unit displacement fields $\boldsymbol{\varphi}_i(\boldsymbol{x})$ of the node as equivalent nodal forces, $j_i = \sigma_{xx}(\boldsymbol{\varphi}_i)(\boldsymbol{x})$, which means

$$j_1 = \frac{E}{2\,A}\,y_{23}\cdot 1 \quad j_3 = \frac{E}{2\,A}\,y_{31}\cdot 1 \quad j_5 = \frac{E}{2\,A}\,y_{12}\cdot 1 \quad j_2 = j_4 = j_6 = 0\,. \tag{3.125}$$

Since the two (horiz. + vert.) displacement fields $\boldsymbol{\varphi}_i(\boldsymbol{x})$ of a node, which is not attached to the element, generate no stresses at the source point, only the three nodes of the element itself will carry nodal forces.

Note that the sum of the j_i is zero

$$j_1 + j_3 + j_5 = \frac{E}{2\,A}(y_2 - y_3 + y_3 - y_1 + y_1 - y_2) = 0\,, \tag{3.126}$$

since the action is a spread of the element by opposite forces j_i.

Example 3.5 An improved element is a bilinear element, which has four nodes and $2\cdot 4$ degrees of freedom, see Fig. 3.21. To each degree of freedom (dof) belongs a unit displacement field $\boldsymbol{\varphi}_i(\boldsymbol{x})$, which represents either a horizontal or vertical displacement of the node

$$\boldsymbol{\varphi}_1(\boldsymbol{x}) = \begin{bmatrix} \psi_1(\boldsymbol{x}) \\ 0 \end{bmatrix} \quad \boldsymbol{\varphi}_2(\boldsymbol{x}) = \begin{bmatrix} 0 \\ \psi_1(\boldsymbol{x}) \end{bmatrix} \quad \boldsymbol{\varphi}_3(\boldsymbol{x}) = \begin{bmatrix} \psi_2(\boldsymbol{x}) \\ 0 \end{bmatrix} \quad \text{etc.} \tag{3.127}$$

The $\psi_i(\boldsymbol{x})$ are the four (scalar-valued) shape functions of the four corner points, see Fig. 3.21,

$$\psi_1(\boldsymbol{x}) = \frac{1}{4\,a\,b}(a-2x)(b-2y) \qquad \psi_2(\boldsymbol{x}) = \frac{1}{4\,a\,b}(a+2x)(b-2y) \quad (3.128)$$

$$\psi_3(\boldsymbol{x}) = \frac{1}{4\,a\,b}(a+2x)(b+2y) \qquad \psi_4(\boldsymbol{x}) = \frac{1}{4\,a\,b}(a-2x)(b+2y)\,. \quad (3.129)$$

So, if a node is pushed in horizontal direction, the vertical displacements are zero and vice versa. Such "monochrome" displacement fields make it easy to control the displacements of the nodes.

Influence Function for u_x

To generate the influence function for the horizontal displacement at a quarter point of an element (length $a = 2$ and height $b = 1$) four horizontal forces j_i are placed at the four corners of the element. These forces correspond to the displacements of the four horizontal displacement fields $\boldsymbol{\varphi}_i(\boldsymbol{x})$, $i = 1, 3, 5, 7$, at the source point $\boldsymbol{x} = (-0.5, -0.25)$ (element coordinates)

$$j_1 = 0.5625 \qquad j_3 = 0.1875 \qquad j_5 = 0.0625 \qquad j_7 = 0.1875\,. \quad (3.130)$$

They produce the shape in Fig. 3.22. Since the vertical displacement fields produce no horizontal displacements, the j_i in vertical direction, $i = 2, 4, 6, 8$, are zero.

Influence Function for σ_{yy}

The influence function for the stress σ_{yy} at a quarter point is generated when the stresses $\sigma_{yy}(\boldsymbol{\varphi}_i)$ of the 4×2 displacement fields $\boldsymbol{\varphi}_i$ at the quarter point are applied as nodal forces j_i.

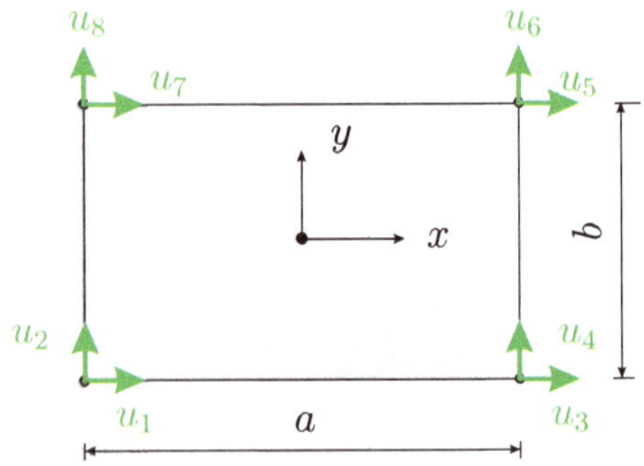

Fig. 3.21 Bilinear element

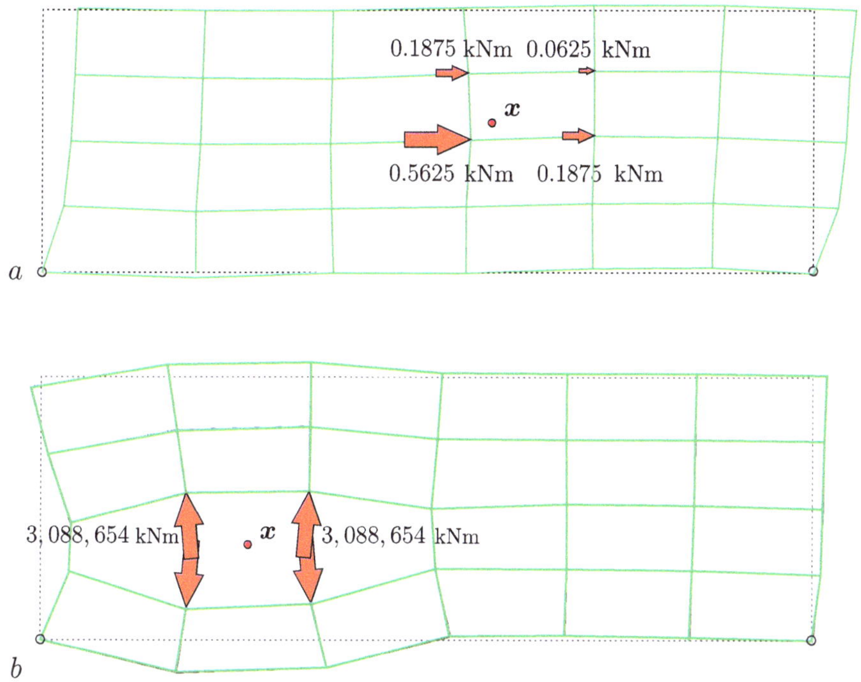

Fig. 3.22 Bilinear elements, **a** influence function for $u_x(\boldsymbol{x})$, and **b** for $\sigma_{yy}(\boldsymbol{x})$

In a bilinear element with length a and height b, as in Fig. 3.21, a displacement $\boldsymbol{u} = \{u_1, u_2, \ldots, u_8\}^T$ of the four corners generates at a point $\boldsymbol{x} = (x, y)$ the stresses

$$\sigma_{xx}(x, y) = \frac{E}{a\,b\,(-1+\nu^2)} \cdot \Big[b\,(u_1 - u_3) + a\,\nu\,(u_2 - u_8) + \\ + x\,\nu\,(-u_2 + u_4 - u_6 + u_8) + y\,(-u_1 + u_3 - u_5 + u_7) \Big] \tag{3.131a}$$

$$\sigma_{yy}(x, y) = \frac{E}{a\,b\,(-1+\nu^2)} \cdot \Big[b\,\nu\,(u_1 - u_3) + a\,(u_2 - u_8) + \\ + x\,(-u_2 + u_4 - u_6 + u_8) + y\,\nu\,(-u_1 + u_3 - u_5 + u_7) \Big] \tag{3.131b}$$

$$\sigma_{xy}(x, y) = \frac{-E}{2\,a\,b\,(1+\nu)} \cdot \Big[b\,(u_2 - u_4) + a\,(u_1 - u_7) + \\ + x\,(-u_1 + u_3 - u_5 + u_7) + y\,(-u_2 + u_4 - u_6 + u_8) \Big]. \tag{3.131c}$$

Since we calculate the influence function for the stress σ_{yy} at the quarter point $\boldsymbol{x} = (x, y)$ we concentrate on the formula for σ_{yy} to calculate the eight nodal forces j_i. The nodal force j_1 is the stress σ_{yy} if we set $u_1 = 1$ and all other $u_i = 0$ in (3.131b)

$$j_1 = \sigma_{yy}(x, y) = \frac{E}{a\,b\,(-1+\nu^2)} \cdot \Big[b\,\nu\,u_1 + y\,\nu\,(-u_1)\Big] = -3.07 \cdot 10^6 \,\text{kNm} \tag{3.132}$$

and in vertical direction ($u_2 = 1$)

$$j_2 = \sigma_{yy}(x, y) = \frac{E}{a\,b\,(-1+\nu^2)} \cdot \Big[a\,(u_2) + x\,(-u_2)\Big] = -3.85 \cdot 10^7 \,\text{kNm}\,. \tag{3.133}$$

The other j_i follow the same pattern. The nodal forces and the shape they produce is plotted in Fig. 3.22b.

Influence Function for N_{xy}

Next, we calculate the influence function for the **integral** of the shear stress

$$N_{xy} = \int_0^l \sigma_{xy}\,dy \tag{3.134}$$

in a vertical section, which passes through a given point $\boldsymbol{x} = (x, y)$. In this situation the equivalent nodal forces are the integrals, see Fig. 3.23,

$$j_i = \int_0^l \sigma_{xy}(\boldsymbol{\varphi}_i)\,dy\,, \tag{3.135}$$

where the $\boldsymbol{\varphi}_i$ are the 4×2 displacement fields of the four nodes of the element through which the section passes. The contribution of the element to this integral is

$$j_i^e = \int_0^b \sigma_{xy}(\boldsymbol{\varphi}_i)\,dy = \frac{-E}{2\,a\,(1+\nu)} \cdot \Big[b\,(u_2 - u_4) + a\,(u_1 - u_7) + \\ + x\,(-u_1 + u_3 - u_5 + u_7) + \frac{b}{2}\,(-u_2 + u_4 - u_6 + u_8)\Big]\,, \tag{3.136}$$

where x is the x-coordinate of the section.

For j_1^e we set $u_1 = 1$ and all other $u_i = 0$. For j_2^e we set $u_2 = 1$ and all other $u_i = 0$, etc. The effective nodal force is then the sum over the element forces of all elements attached to the node.

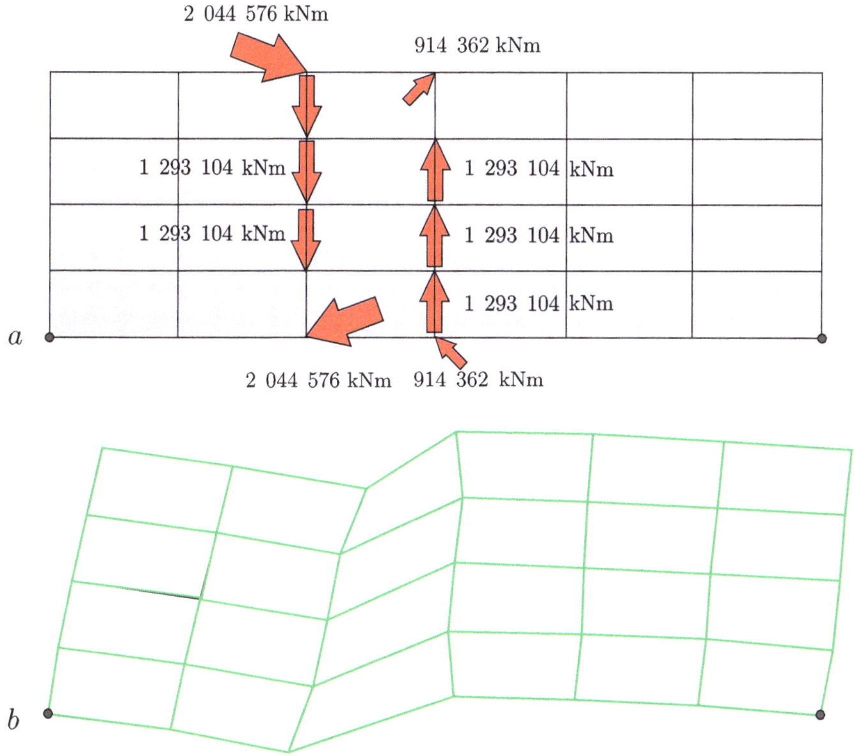

Fig. 3.23 Influence function for the integral of σ_{xy} in a vertical section, **a** equivalent nodal forces, **b** influence function [4]

3.15 The Local Influence Function

In frame elements the local solution (GF) is the influence function of the single clamped element. This solution must be added to the FE-influence function.

The influence function $G_0(y, x)$ for the deflection of a clamped beam, see Fig. 3.24, is the function

$$2G_0(y,x) = \frac{1}{EI}\begin{cases} -\dfrac{(\ell - x)^3}{6}\,\varphi_3^e(y) + \dfrac{(\ell - x)^2}{2}\,\varphi_4^e(y) & y < x \\ -\dfrac{x^3}{6}\,\varphi_1^e(y) - \dfrac{x^2}{2}\,\varphi_2^e(y) & x < y \end{cases} \tag{3.137}$$

where the $\varphi_i^e(y)$ are the local *shape functions*, see (3.38), and the influence function for the longitudinal displacement, see Fig. 3.25,

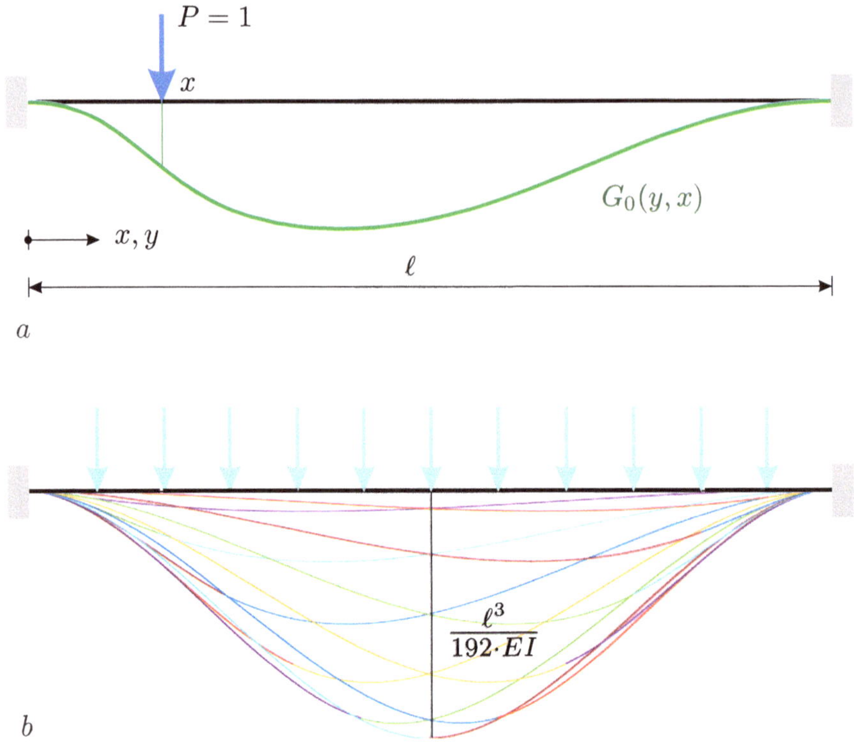

Fig. 3.24 Influence function for the deflection of a clamped beam

$$G_0(y,x) = \frac{1}{EA} \begin{cases} (\ell - x)\,\varphi_2^e(y) & y < x \\ x\,\varphi_1^e(y) & x < y \end{cases} \tag{3.138}$$

is based on the *shape functions* of the bar element.

The formulation of these influence functions is a simple exercise in linear algebra. To compute the local influence function for a shear force V, see Fig. 3.26, we split it into two parts, $w_L(x) + w_R(x)$. The function $w_L(x)$ on the left side is clamped at $x = 0$ and it satisfies the static boundary conditions

$$M(0) = -\frac{6}{3^2} EI \qquad V(0) = -\frac{12}{3^3}\, EI\,, \tag{3.139}$$

and so $w_L(x) = a\,\varphi_3(x) + b\,\varphi_4(x)$ is the natural choice and the coefficients $a = -1$ and $b = 0$ are the solution of the system

$$\begin{bmatrix} M_3(0) & M_4(0) \\ V_3(0) & V_4(0) \end{bmatrix} \begin{bmatrix} a \\ b \end{bmatrix} = EI \begin{bmatrix} 6/3^2 \\ -12/3^3 \end{bmatrix}, \tag{3.140}$$

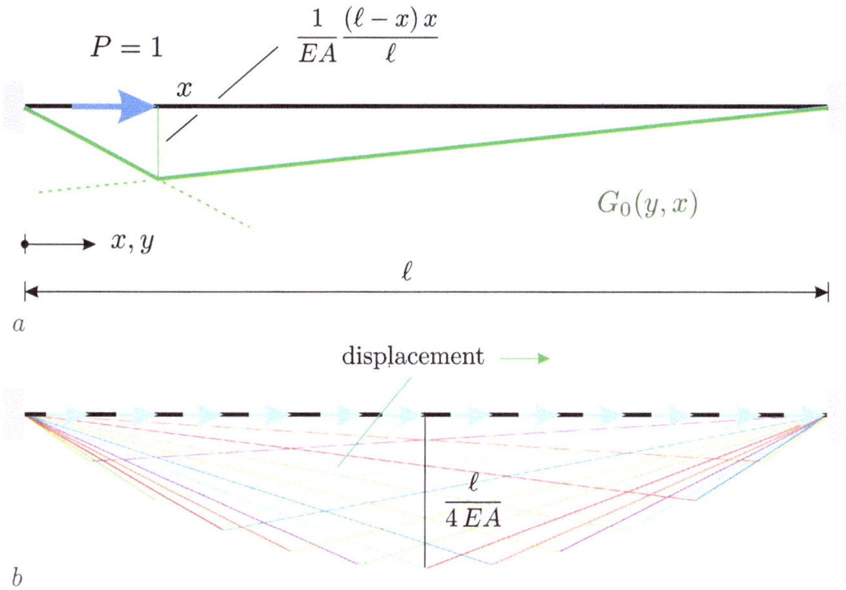

Fig. 3.25 Influence function for the longitudinal displacement of a bar with fixed ends

where M_i and V_i are the bending moments and shear forces respectively of the shape functions $\varphi_i(x)$.

On the right side of the source point we choose the function $w_R(x) = c\,\varphi_1(x) + d\,\varphi_2(x)$ and solve vor $c = 1$ and $d = 0$ by satisfying the static boundary conditions

$$M(l) = -\frac{6}{3^2}EI \qquad V(l) = -\frac{12}{3^3}\,EI\,, \tag{3.141}$$

and so the complete influence function is

$$w(x) = \begin{cases} -\varphi_3(x) & 0 < x < 1.5 \\ \varphi_1(x) & 1.5 < x < 3.0 \end{cases}. \tag{3.142}$$

The local influence function is always an expansion in terms of the 2×2 unit displacements, $\varphi_3(x)$ and $\varphi_4(x)$ on the left and $\varphi_1(x)$ and $\varphi_2(x)$ on the right side of the source point, which must not be the midpoint of the element, see Fig. 3.27b. The influence function for $M(x)$, see Fig. 3.27a, is obtained in the same way.

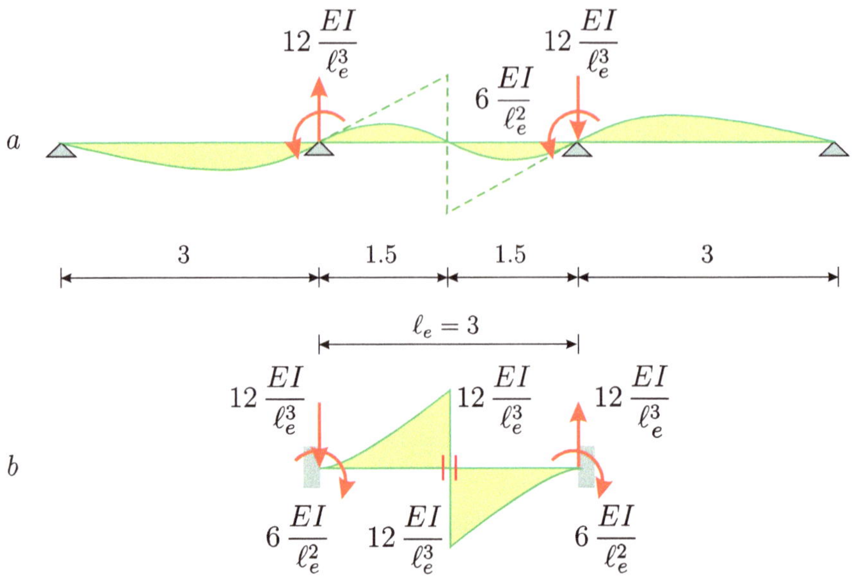

Fig. 3.26 **a** FE-influence function for $V(x)$, and **b** local influence function for $V(x)$

3.16 The Central Equation

FE-influence functions typically are separated into functions of x and y

$$G_h(y, x) = g_1(x)\,\varphi_1(y) + g_2(x)\,\varphi_2(y) + \ldots + g_n(x)\,\varphi_n(y)\,, \tag{3.143}$$

and so the evaluation of an influence function becomes a sum over the nodes

$$u_h(x) = \int_0^l G_h(y, x)\,p(y)\,dy = \sum_{i=1}^{n} g_i(x) \int_0^l p(y)\,\varphi_i(y)\,dy = \sum_{i=1}^{n} g_i(x)\,f_i\,, \tag{3.144}$$

where the nodal displacements g_i of the influence functions act as weights on the equivalent nodal forces f_i.

And as in the discrete case—where $u_i = \boldsymbol{u}^T\,\boldsymbol{e}_i$—we can write $u_h(x)$ also as the scalar product of the Dirac delta and u_h itself

$$u_h(x) = \int_0^l \delta(y - x)\,u_h(y)\,dy\,. \tag{3.145}$$

These two equations, merged into one, constitute the central equation.

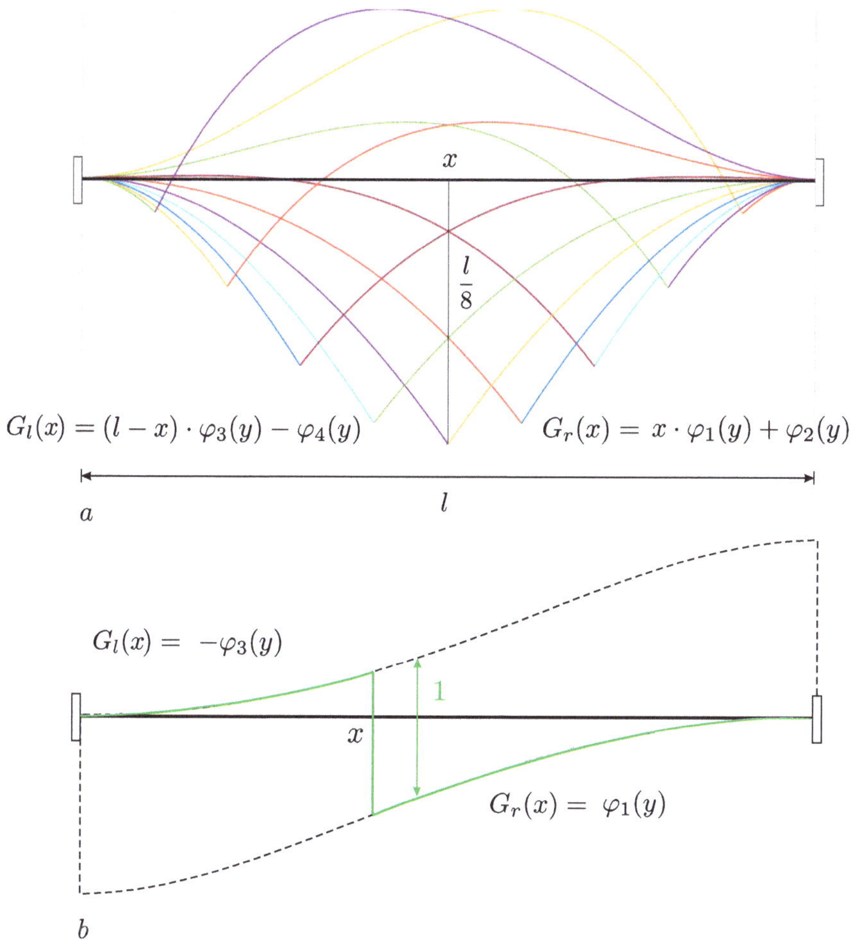

Fig. 3.27 **a** Influence functions for the bending moment in a clamped beam, and **b** influence function for the shear force at a point x

Theorem 3.2 *(The central equation)*

$$\begin{aligned}
u_h(x) &= \int_0^l G_h(y, x)\, p(y)\, dy = \int_0^l \sum_i g_i(x)\, \varphi_i(y)\, p(y)\, dy = \sum_i g_i(x)\, f_i \\
&= \boldsymbol{g}^T \boldsymbol{f} = \boldsymbol{g}^T \boldsymbol{K} \boldsymbol{u} = \boldsymbol{g}^T \boldsymbol{K}^T \boldsymbol{u} = \boldsymbol{j}^T \boldsymbol{u} = \sum_i j_i\, u_i \\
&= \sum_i \varphi_i(x)\, u_i = \int_0^l \sum_i u_i\, \varphi_i(y)\, \delta(y - x)\, dy = \int_0^l u_h(y)\, \delta(y - x)\, dy\,.
\end{aligned} \tag{3.146}$$

Fig. 3.28 An only slightly misplaced boundary node led to singularities in the moments, **a)** plate, **b)** principal moments

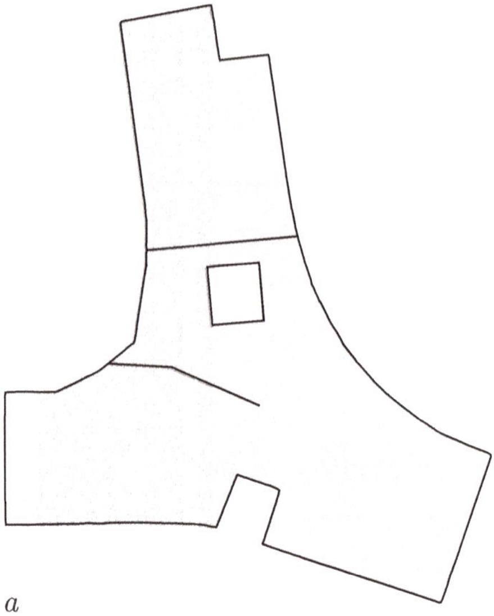

a

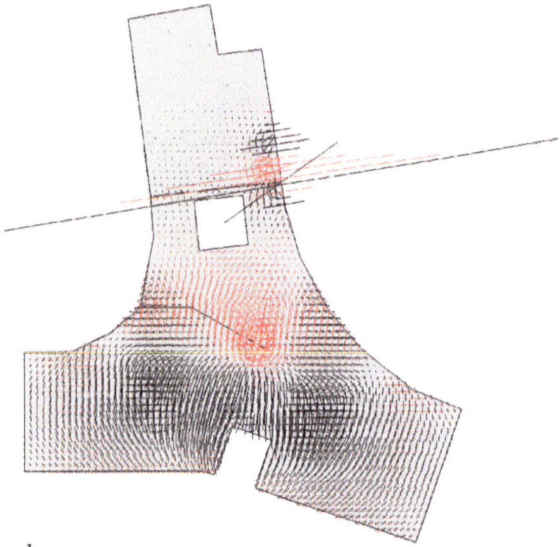

b

The displacement $u_h(x)$ at a point x is the scalar product between the vector $\boldsymbol{g}$ of the influence function and the vector of the equivalent nodal forces $\boldsymbol{f}$ or, vice versa, between the nodal values $\boldsymbol{u}$ of the FE-solution and the equivalent nodal forces $\boldsymbol{j}$ of the influence function

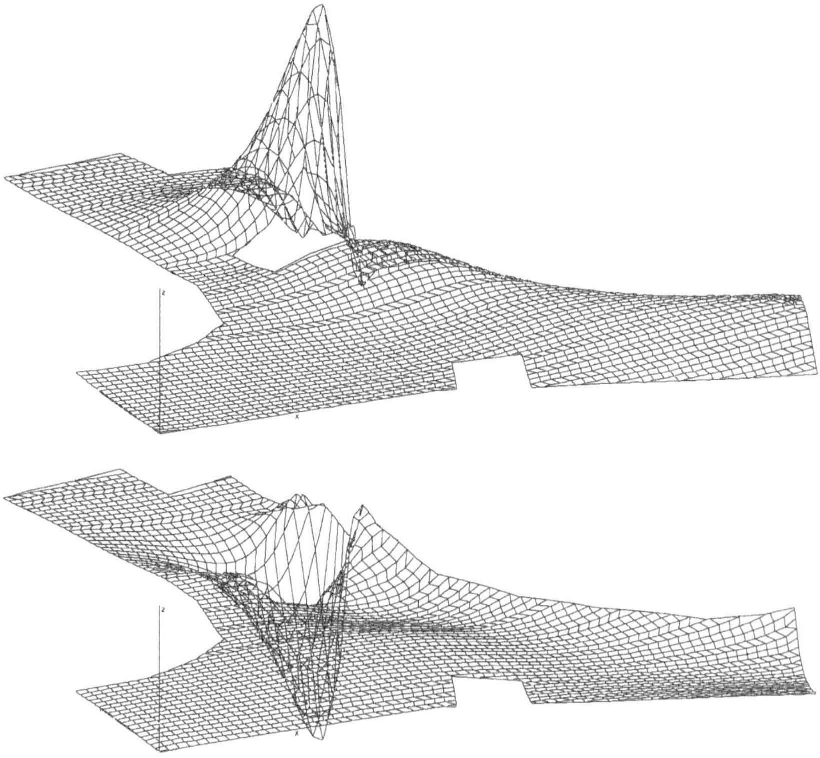

Fig. 3.29 The influence functions for the edge moment m_{nn} at two nearly identical points have an opposite sign

$$u_h(x) = \begin{cases} \displaystyle\int_0^l u_h(y)\,\delta(y-x)\,dy = \sum_i \varphi_i(x)\,u_i = \boldsymbol{j}^T\boldsymbol{u} \\ \displaystyle\int_0^l G_h(y,x)\,p(y)\,dy = \boldsymbol{g}^T\boldsymbol{f}\,. \end{cases} \tag{3.147}$$

This two-way approach

$$J(u) = \begin{cases} \boldsymbol{j}^T\boldsymbol{u} \\ \boldsymbol{g}^T\boldsymbol{f}\,, \end{cases} \tag{3.148}$$

is the shortest representation of a linear functional.

Since the action $J(u) = \boldsymbol{j}^T\boldsymbol{u}$ of the vector $\boldsymbol{j}$ on $\boldsymbol{u}$ resembles the action of the Dirac delta on the function u

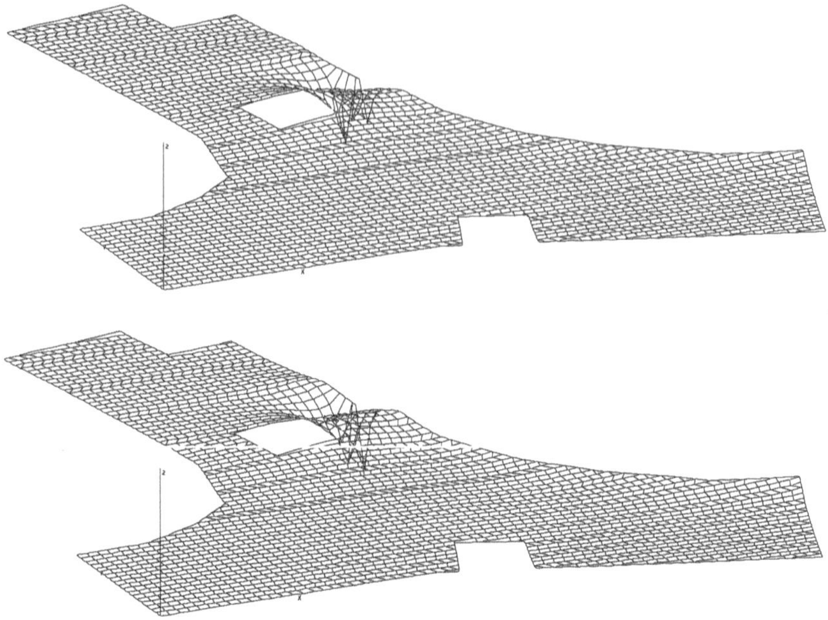

Fig. 3.30 The same influence functions after correcting the position of the node

$$u(x) = \int_0^l \delta(y - x)\, u(y)\, dy\,, \tag{3.149}$$

we call the vector $\boldsymbol{j}$ the **Dirac vector** of the functional $J(u)$. While $\delta(y - x)$ measures directly at the source point, the Dirac vector $\boldsymbol{j}$ resembles a spread of $\delta(y - x)$, since it measures at the nodes y_i near the source point.

The first formula in (3.148)

$$\begin{aligned} J(\boldsymbol{u}) = \boldsymbol{j}^T \boldsymbol{u} &= j_1\, u_1 + j_2\, u_2 + \ldots + j_n\, u_n \\ &= J(\varphi_1)\, u_1 + J(\varphi_2)\, u_2 + \ldots + J(\varphi_n)\, u_n \end{aligned} \tag{3.150}$$

actually, is evident, since it splits $J(\boldsymbol{u})$ into the single contributions $J(\varphi_i)$ weighted with the u_i. In the second formula, $J(\boldsymbol{u}) = \boldsymbol{g}^T \boldsymbol{f}$, the equivalent nodal forces are weighted with the influence coefficients g_i, the nodal values of the FE-influence function. It is the more interesting of the two formulations.

The equation

$$\boxed{J(u) = \boldsymbol{g}^T \boldsymbol{f}} \tag{3.151}$$

is actually the central formula because with finite elements "everything" is a functional. Each derivative

$$\frac{du}{dx} \qquad \frac{\partial u_i}{\partial x_j} \qquad \frac{\partial^2 w}{\partial x\,\partial y} \tag{3.152}$$

is a functional. Each result is a functional and to every linear functional belongs an influence function, a vector $\boldsymbol{g}$, and the scalar product with the load vector $\boldsymbol{f}$ is the value of the functional.

So if you calculate the second derivative of the deflection at a point x, you can differentiate the *shape functions* by hand, or you form the scalar product

$$w''(x) = \boldsymbol{g}^T \boldsymbol{f}\,. \tag{3.153}$$

Of course nobody calculates like this, but the crucial message is, that differentiation, like every evaluation of a linear functional, can be traced back to **linear algebra**, $J(u) = \boldsymbol{g}^T \boldsymbol{f}$. For each measurement on a structure there is a vector $\boldsymbol{g}$ and the *post-processing* is theoretically a repeated application of this formula.

Now think of a **singular corner**. Ten centimeters before and behind the corner we see tensions that could not be more different. But this means that the corresponding vectors $\boldsymbol{g}$ – although the source points are only 20 cm apart – are very different. Singularities in stresses translate into oscillating vectors $\boldsymbol{g}$, into oscillating influence functions, see Figs. 3.28, 3.29 and 3.30.

Note that these vectors $\boldsymbol{g}$ are not load case dependent, but belong to the model. They are created with the model, like all influence functions, only the particular load case makes the differences visible.

3.17 Representation of an FE-solution

We summarize:

Just as the vector $\boldsymbol{u}$ can be written in two ways

$$\boldsymbol{u} = \sum_i f_i\, \boldsymbol{g}_i = \sum_i u_i\, \boldsymbol{e}_i \tag{3.154}$$

also the FE-solution can be written in two ways

$$u(x) = \sum_i f_i\, g_i(x) = \sum_i u_i\, \varphi_i(x) \tag{3.155}$$

The $g_i(x)$ are the influence functions for the nodal displacements u_i, the $\varphi_i(x)$ are the corresponding shape functions and the f_i are the equivalent nodal forces. In the first equation the $\boldsymbol{g}_i$ are the columns of the inverse $\boldsymbol{K}^{-1}$ and the $\boldsymbol{e}_i$ are the unit vectors. Both sets, $\boldsymbol{g}_i$ and $\boldsymbol{e}_i$, form a basis of the $\mathbb{R}^n$ and also the trial space $\mathcal{V}_h$ has two bases, either the g_i or the φ_i, see Sect. 3.22.

3.18 Frame Structures and $J = g^T f$

The result $J(w) = \boldsymbol{g}^T \boldsymbol{f}$ is in general an approximation since the FE-influence function is not exact and the load is not only concentrated at the nodes. But in frame structures we can make it exact by adding the local contribution,

$$J(w) = \boldsymbol{g}^T \boldsymbol{f} + \int_0^{l_e} G_{loc}(y, x)\, p(y)\, dy\,, \tag{3.156}$$

which is the superposition of the local influence function $G_{loc}(y, x)$ (=G.F. of the clamped beam element) with the load.

Of course, we do not integrate, we simply calculate the moment or the shear force—whatever $J(w)$ is—in the clamped element and add this value to $\boldsymbol{g}^T \boldsymbol{f}$.

Equation (3.156) is based on the following:

Reducing the load to the nodes, $p \to f_i$, does not change the nodal displacements u_i of a frame.

This is, by the way, the reason why the slope deflection method works.

3.19 State Vectors and Measurements

In a figurative sense the displacement vector $\boldsymbol{u}$ represents the **state** of a structure, and the evaluation of a functional

$$J(\boldsymbol{u}) = \boldsymbol{g}^T \boldsymbol{f} = \boldsymbol{g}^T \boldsymbol{K} \boldsymbol{u} \tag{3.157}$$

can be considered as taking a measurement of the structure. Since every modification of a structure means a transit from a system $\boldsymbol{K}$ to a system $\boldsymbol{K} \to \boldsymbol{K} + \boldsymbol{\Delta K}$, we are interested in knowing how the measured values change, when we add a matrix $\boldsymbol{\Delta K}$ to $\boldsymbol{K}$.

The original weak formulation

$$\boldsymbol{\delta u}^T \boldsymbol{K} \boldsymbol{u} = \boldsymbol{\delta u}^T \boldsymbol{f} \tag{3.158}$$

and the modified system have the same right side

$$\delta \boldsymbol{u}^T(\boldsymbol{K} + \boldsymbol{\Delta K})\boldsymbol{u}_c = \delta \boldsymbol{u}^T \boldsymbol{f}\,, \tag{3.159}$$

and so with $\boldsymbol{e} = \boldsymbol{u}_c - \boldsymbol{u}$ we have

$$\delta \boldsymbol{u}^T \boldsymbol{K}\boldsymbol{e} = -\delta \boldsymbol{u}^T \boldsymbol{\Delta K}\,\boldsymbol{u}_c\,. \tag{3.160}$$

If we substitute for $\delta \boldsymbol{u}$ the vector $\boldsymbol{g}$ of the influence function, then follows

$$J(\boldsymbol{e}) = -\boldsymbol{g}^T \boldsymbol{\Delta K}\,\boldsymbol{u}_c\,, \tag{3.161}$$

and this is a **local result**, at least as local as the matrix $\boldsymbol{\Delta K}$, since we only have to integrate over the modified element Ω_e (the right side is like Mohr's integral, is a strain energy product).

Imagine, a single element in a frame cracks, $EI \to EI + \Delta EI$, $\Delta EI < 0$. The crack triggers a transition from the original system $\boldsymbol{K}$ to a new system $\boldsymbol{K} + \boldsymbol{\Delta K}$, where $\boldsymbol{\Delta K}$ is a mostly empty array which only contains in some small part the "Delta" of the original element matrix,[7] $\boldsymbol{\Delta K} = \Delta EI/EI \cdot \boldsymbol{K}_e$.

If the frame has $2n$ degrees of freedom u_i, we could theoretically make $2n$ measurements $J_i(\boldsymbol{e}) = u_{ic} - u_i$ of the cracked element with the $2n$ vectors $\boldsymbol{g}_i$

$$J_i(\boldsymbol{e}) = u_{ic} - u_i = -\boldsymbol{g}_i^T \boldsymbol{\Delta K}\,\boldsymbol{u}_c\,, \tag{3.162}$$

and calculate the new nodal displacements

$$u_{ic} = u_i + J_i(\boldsymbol{e})\,. \tag{3.163}$$

In a matrix notation this is

$$\boldsymbol{u}_c = \boldsymbol{u} + \boldsymbol{K}^{(-1)} \boldsymbol{\Delta K}\,\boldsymbol{u}_c\,, \tag{3.164}$$

since the columns of the inverse are just the nodal displacement vectors $\boldsymbol{g}_i$ of the influence functions for the u_i. In Chap. 5 we write this as

$$\boldsymbol{K}\boldsymbol{u}_c = \boldsymbol{K}\boldsymbol{u} + \boldsymbol{\Delta K}\,\boldsymbol{u}_c = \boldsymbol{f} + \boldsymbol{f}^+\,, \tag{3.165}$$

and we will propose various techniques to calculate the vector $\boldsymbol{u}_c$, see Sect. 5.22.

Writing (3.164) as

$$(\boldsymbol{I} + \boldsymbol{K}^{-1}\boldsymbol{\Delta K})\boldsymbol{u}_c = \boldsymbol{u}\,, \tag{3.166}$$

it becomes a coupling condition between the new and the old state.

---

[7] The original element matrix $\boldsymbol{K}_e$ carries the factor EI/l_e^3 up front, and the factor $\Delta EI/EI$ changes this to $\Delta EI/l_e^3$.

Theoretically we could start with a simple stiffness matrix $\boldsymbol{K}_0$, which is enriched step by step with matrices $\boldsymbol{\Delta K}_i$

$$\boldsymbol{K}_i = \boldsymbol{K}_o + \boldsymbol{\Delta K}_1 + \boldsymbol{\Delta K}_2 + \cdots + \boldsymbol{\Delta K}_i\,, \tag{3.167}$$

and so, we would produce a chain of solutions $\boldsymbol{u}_i$, where each new vector $\boldsymbol{u}_{i+1}$ is entangled with the previous vector

$$(\boldsymbol{I} + \boldsymbol{K}_i^{-1}\boldsymbol{\Delta K}_{i+1})\boldsymbol{u}_{i+1} = \boldsymbol{u}_i\,. \tag{3.168}$$

3.20 Maxwell's Theorem

Actually, *Maxwell's theorem* belongs in Chap. 2, but we first had to introduce the concept of a functional and equate influence functions with functionals.

Maxwell's theorem is the equation

$$w_1(x_2) = w_2(x_1)\,. \tag{3.169}$$

The deflection a force $P = 1$ at a point x_1 produces at a remote point x_2 is as large as the deflection a force $P = 1$ at the point x_2 produces at the point x_1, see Fig. 4.1.

We can generalize this to any two linear functionals and their kernels G_i

$$\boxed{J_1(G_2) = J_2(G_1)} \tag{3.170}$$

since *Maxwell's theorem* is not limited to displacements. Equation (3.170) is *Betti's theorem* to the point.

To see this duality applied, let us consider a square plate as in Fig. 3.31, where the focus is on two points $\boldsymbol{x}_a$ and $\boldsymbol{x}_b$. At the first point $\boldsymbol{x}_a$ we measure the **horizontal displacement** of a given displacement field $\boldsymbol{u}$

$$J_1(\boldsymbol{u}) = u_x(\boldsymbol{x}_a)\,, \tag{3.171}$$

and at the point $\boldsymbol{x}_b$ we measure the **stress** σ_{xx}

$$J_2(\boldsymbol{u}) = \sigma_{xx}(\boldsymbol{u})(\boldsymbol{x}_b) \tag{3.172}$$

of this field.

The influence function G_1 of the functional J_1 is generated by a single horizontal force $P = 1$, and the influence function $\boldsymbol{G}_2$ of J_2 is generated by a horizontal dislocation at $\boldsymbol{x}_b$. According to *Maxwell's theorem* we have

$$J_1(\boldsymbol{G}_2) = J_2(\boldsymbol{G}_1) \tag{3.173}$$

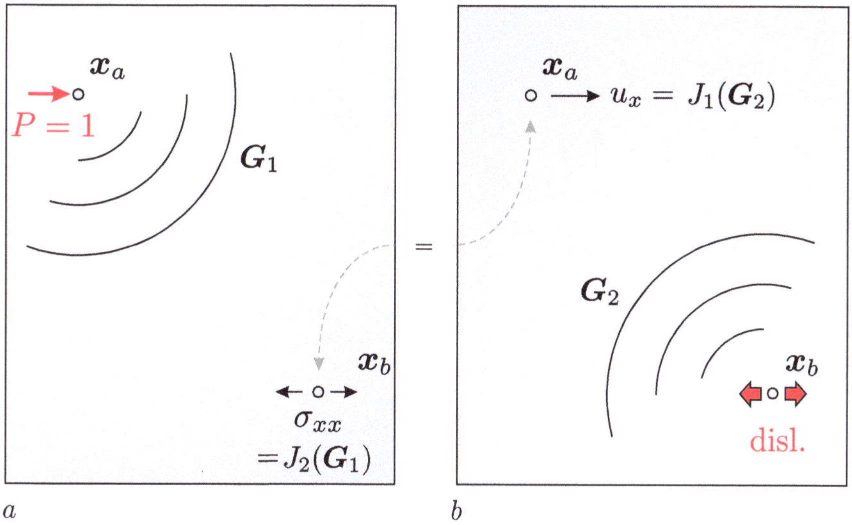

Fig. 3.31 Two influence functions, **a** a single force generates the influence function $\boldsymbol{G}_1$ for $u_x(\boldsymbol{x}_a)$, and **b** a dislocation generates the influence function $\boldsymbol{G}_2$ for the stress σ_{xx} at the point $\boldsymbol{x}_b$; the two kernels are adjoint, $J_2(\boldsymbol{G}_1) = J_1(\boldsymbol{G}_2)$

which means: *The displacement at the point* $\boldsymbol{x}_a$ *due to the dislocation at point* $\boldsymbol{x}_b$ *is the same as the stress at* $\boldsymbol{x}_b$ *due to the single force at* $\boldsymbol{x}_a$.

Example 3.6 In this example we apply *Maxwell's theorem* to the continuous beam in Fig. 3.32. The first functional is the **moment** at the point x_c

$$J_1(w) = M(x_c)\,, \tag{3.174}$$

and the second functional

$$J_2(w) = R_B\,, \tag{3.175}$$

is the **support reaction** at the support B.

To G_1 (=influence function for J_1) belongs the support reaction $J_2(G_1) = -8\,112\,\text{kN} \times 1\,\text{m}$ and to the influence function G_2 (=influence function for R_B) belongs a moment $J_1(G_2) = -8\,112$ kNm, and both values are the same[8]

$$J_1(G_2) = -8\,112 = -8\,112 = J_2(G_1)\,, \tag{3.176}$$

since ***Betti extended***, see Chap. 4, guarantees, that *Maxwell's theorem* also applies to the FE-approximations G_1^h and G_2^h

[8]The result of an influence function is an energy.

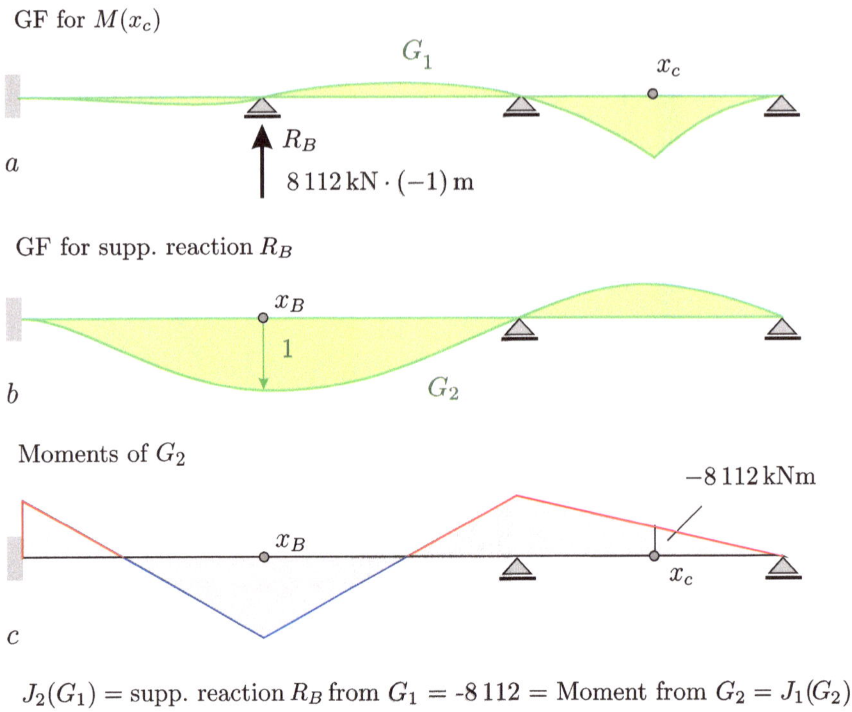

Fig. 3.32 The two influence functions are adjoint, $J_2(G_1) = J_1(G_2)$

$$J_1(G_2^h) = J_2(G_1^h)\,. \tag{3.177}$$

Remark 3.7 The classical Maxwell, $\delta_{12} = \delta_{21}$, see Fig. 4.1, is a simple consequence of the **symmetry of the strain energy product**

$$\delta_{12} = a(w_1, w_2) = a(w_2, w_1) = \delta_{21}\,. \tag{3.178}$$

Settlement of a Support

Maxwell's theorem also answers the question of how to evaluate influence functions when a support settles, as for example the central support of the wall plate in Fig. 3.33. The spread of the source point generates a vertical support reaction R_y of 119 259 kN in the support and so the stress is

$$\sigma_{yy} = -119\,259 \cdot \Delta_y = \text{supp. react. of GF} \times \text{supp. settlement}\,, \tag{3.179}$$

where Δ_y (positive in y direction) is the settlement of the support.

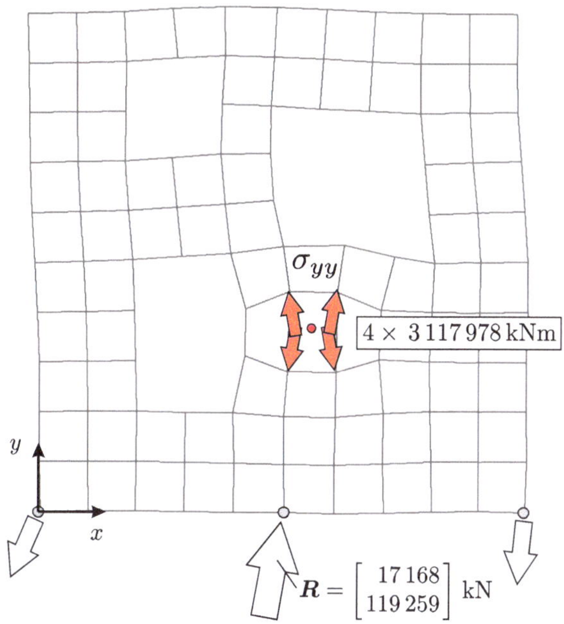

Fig. 3.33 FE-influence function for the stress σ_{yy} at the midpoint of an element, equivalent nodal forces and support reactions due to the dislocation at the source point

3.21 The Inverse Stiffness Matrix

The FE-influence function for the displacement $u(x)$ at a node x_k is

$$G_h(y, x_k) = \sum_i g_i(x_k)\, \varphi_i(y)\,, \tag{3.180}$$

where the vector $\boldsymbol{g} = \{g_1, g_2, \ldots, g_n\}^T$ is the solution of the $n \times n$ system

$$\boldsymbol{K}\, \boldsymbol{g} = \boldsymbol{e}_k \qquad \text{(unit vector } \boldsymbol{e}_k)\,. \tag{3.181}$$

So, the n **columns** $\boldsymbol{g}_k$ of the inverse stiffness matrix $\boldsymbol{K}^{-1}$

$$\boldsymbol{g}_k = \boldsymbol{K}^{-1} \boldsymbol{e}_k \tag{3.182}$$

are the n **influence functions** $G_h(y, x_k)$ of the nodal displacements $u(x_k)$

$$G_h(y, x_k) = \sum_i g_{ki}\, \varphi_i(y) = \boldsymbol{g}_k^T \boldsymbol{\varphi}(y)\,, \tag{3.183}$$

where $\boldsymbol{\varphi}(y) = \{\varphi_1(y), \varphi_2(y), \ldots, \varphi_n(y)\}^T$.

This explains why the inverse of a **tri-diagonal difference matrix** (think of a rope $(\ldots 0\ \ -1\ \ 2\ \ -1\ 0\ \ \ldots) \cdot H/l_e$) is a fully populated matrix. The compensating motion when a single point load $P = 1$ is placed on the rope affects **all nodes**, $\boldsymbol{u} = \boldsymbol{K}^{-1}\boldsymbol{f}$.

A stiffness matrix $\boldsymbol{K}$ "differentiates" and its inverse $\boldsymbol{K}^{-1}$ "integrates". The inverse is *always* a fully populated matrix and it is symmetric (because of Maxwell).

Remark 3.8 Strictly speaking are the columns $\boldsymbol{g}_i$ the nodal values of the nodal influence functions $G(y, x_i)$, not the functions themselves, but it will simplify matters, if we allow this little slip.

Example 3.7 The rope in Fig. 3.34 a is pre-stressed by a force H and divided into five linear elements. Its stiffness matrix $\boldsymbol{K}$

$$\boldsymbol{K} = \frac{H}{l_e}\begin{bmatrix} 2 & -1 & 0 & 0 \\ -1 & 2 & -1 & 0 \\ 0 & -1 & 2 & -1 \\ 0 & 0 & -1 & 2 \end{bmatrix} \tag{3.184}$$

is tri-diagonal while the inverse

$$\boldsymbol{K}^{-1} = \frac{l_e}{5\,H}\begin{bmatrix} 4 & 3 & 2 & 1 \\ 3 & 6 & 4 & 2 \\ 2 & 4 & 6 & 3 \\ 1 & 2 & 3 & 4 \end{bmatrix} \tag{3.185}$$

is fully populated. Its columns $\boldsymbol{g}_k$, see Fig. 3.34b–f, are the displacements of the nodes if a single force $P = 1$ acts at the nodes x_k.

The **row sums** $r_i = \sum_j k_{ij}^{(-1)}$ of the inverse constitute a **sensitivity profile** of the structure, since each sum r_i is the displacement, if all $f_j = 1$ (or choose other selection criteria, say horizontal vs. vertical loads, etc.)

$$w_h(x_i) = \sum_j k_{ij}^{(-1)} f_j = \sum_j k_{ij}^{(-1)} = r_i\,. \tag{3.186}$$

The nodes with values max r_i, max deflection, are here the nodes 2 and 3.

Example 3.8 The stiffness matrix of the beam in Fig. 3.35, with $EI = 1$ and $l_e = 1$, is

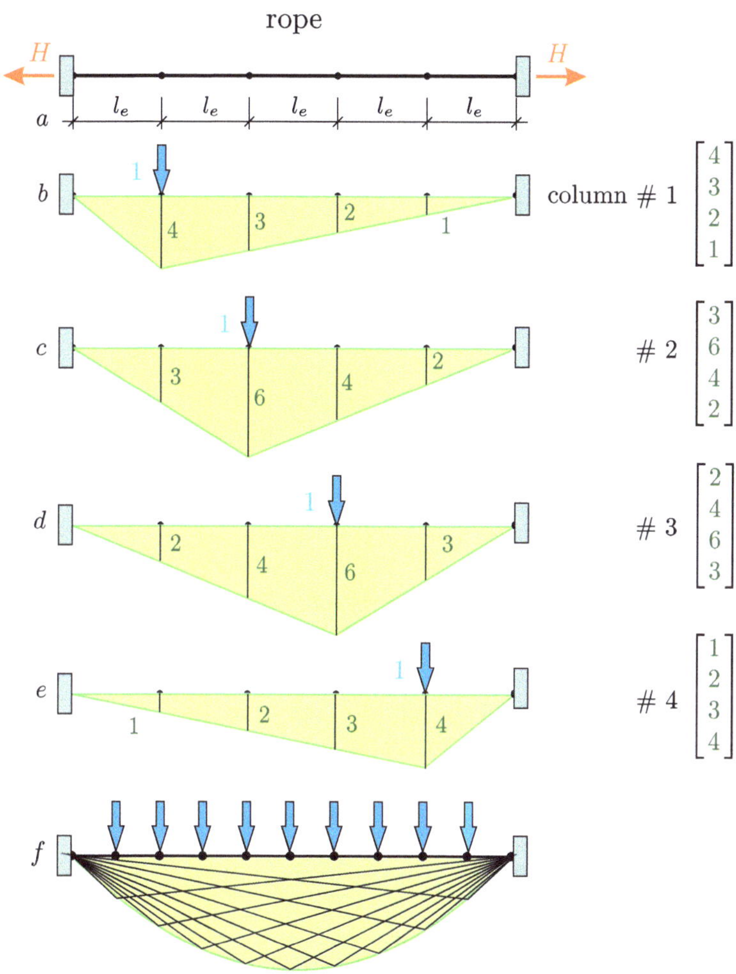

Fig. 3.34 Rope, **a** $n = 5$ elements, **b–e** the deflections are the columns of the inverse stiffness matrix (all values times $l_e/(5\,H)$), **f** if n increases, the columns of $\boldsymbol{K}^{-1}$ more and more look alike, $\det(\boldsymbol{K}^{-1}) \to 0$, that is $\boldsymbol{K}^{-1}$ becomes singular

$$\boldsymbol{K} = \begin{bmatrix} 4 & 6 & 2 & 0 & 0 & 0 \\ 6 & 24 & 0 & -12 & -6 & 0 \\ 2 & 0 & 8 & 6 & 2 & 0 \\ 0 & -12 & 6 & 24 & 0 & -6 \\ 0 & -6 & 2 & 0 & 8 & 2 \\ 0 & 0 & 0 & -6 & 2 & 4 \end{bmatrix} \tag{3.187}$$

and the **columns of the inverse**

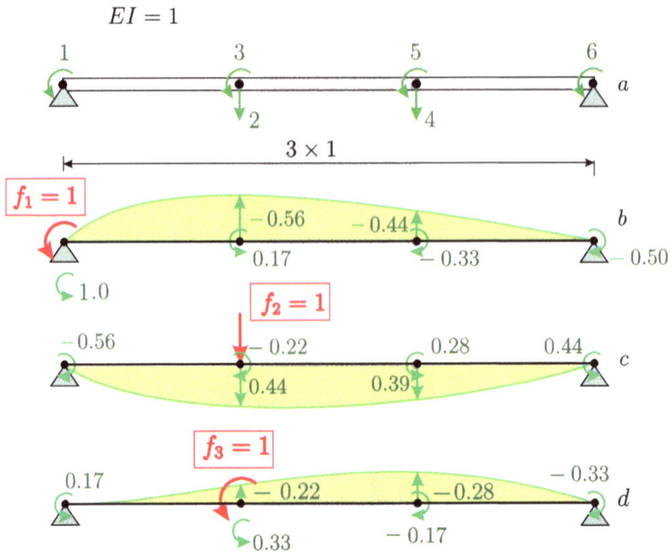

Fig. 3.35 Beam, **a** three elements, **b** deflection due to $f_1 = 1$ (column 1 of $\boldsymbol{K}^{-1}$), **c** due to $f_2 = 1$ (column 2 of $\boldsymbol{K}^{-1}$), **d** due to $f_3 = 1$ (column 3)

$$\boldsymbol{K}^{-1} = \begin{bmatrix} 1.00 & -0.56 & 0.17 & -0.44 & -0.33 & -0.50 \\ -0.56 & 0.44 & -0.22 & 0.39 & 0.28 & 0.44 \\ 0.17 & -0.22 & 0.33 & -0.28 & -0.17 & -0.33 \\ -0.44 & 0.39 & -0.28 & 0.44 & 0.22 & 0.56 \\ -0.33 & 0.28 & -0.17 & 0.22 & 0.33 & 0.17 \\ -0.50 & 0.44 & -0.33 & 0.56 & 0.17 & 1.00 \end{bmatrix} \tag{3.188}$$

are the nodal displacements in the six unit load cases $\boldsymbol{f} = \boldsymbol{e}_i$.

This is also true for complete frames, see Fig. 3.36, where each vector $\boldsymbol{g}_i$ is fully populated because each node of the frame feels the effect of a single force $f_i = 1$—even if it is far away.

If a single force $f_i = 1$ pushes a node in the direction u_i, the associated strain energy is the entry g_{ii} of $\boldsymbol{K}^{-1}$, since

$$a(\boldsymbol{g}_i, \boldsymbol{g}_i) = \boldsymbol{g}_i^T \boldsymbol{K} \boldsymbol{g}_i = \boldsymbol{g}_i^T \boldsymbol{K} \boldsymbol{K}^{-1} \boldsymbol{e}_i = g_{ii} \,. \tag{3.189}$$

The larger g_{ii} is, the more flexible the structure is in the direction of the displacement u_i.

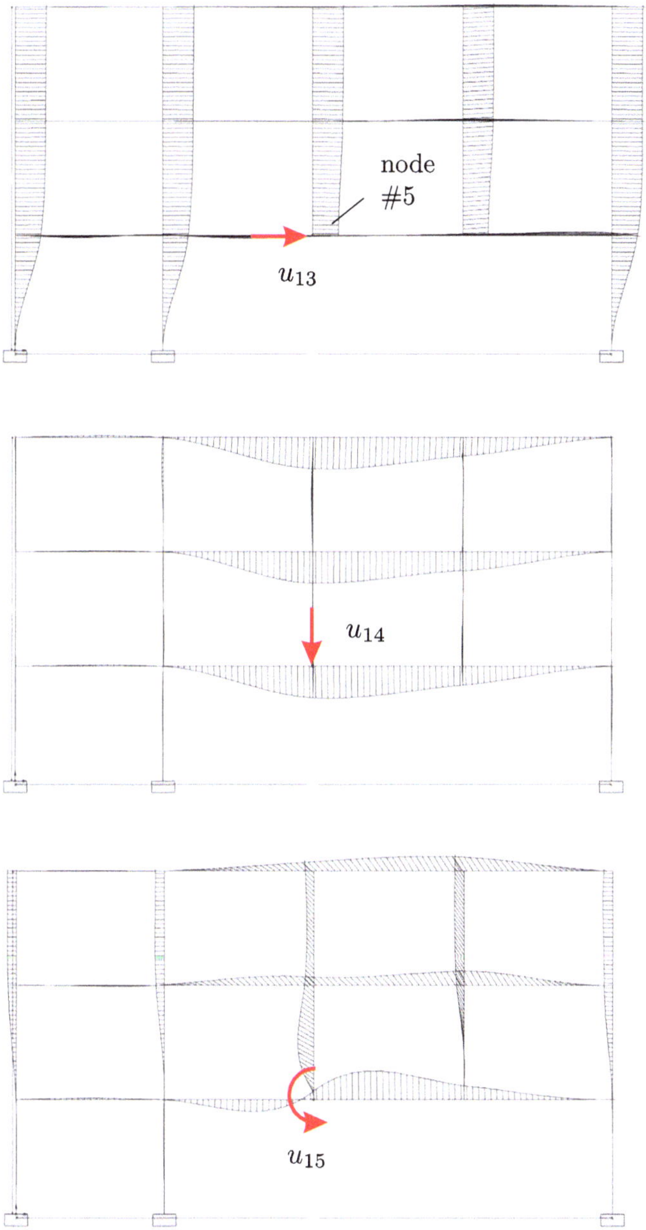

Fig. 3.36 The nodal displacements of these three influence function form columns #13, #14, #15, of the inverse $\boldsymbol{K}^{-1}$

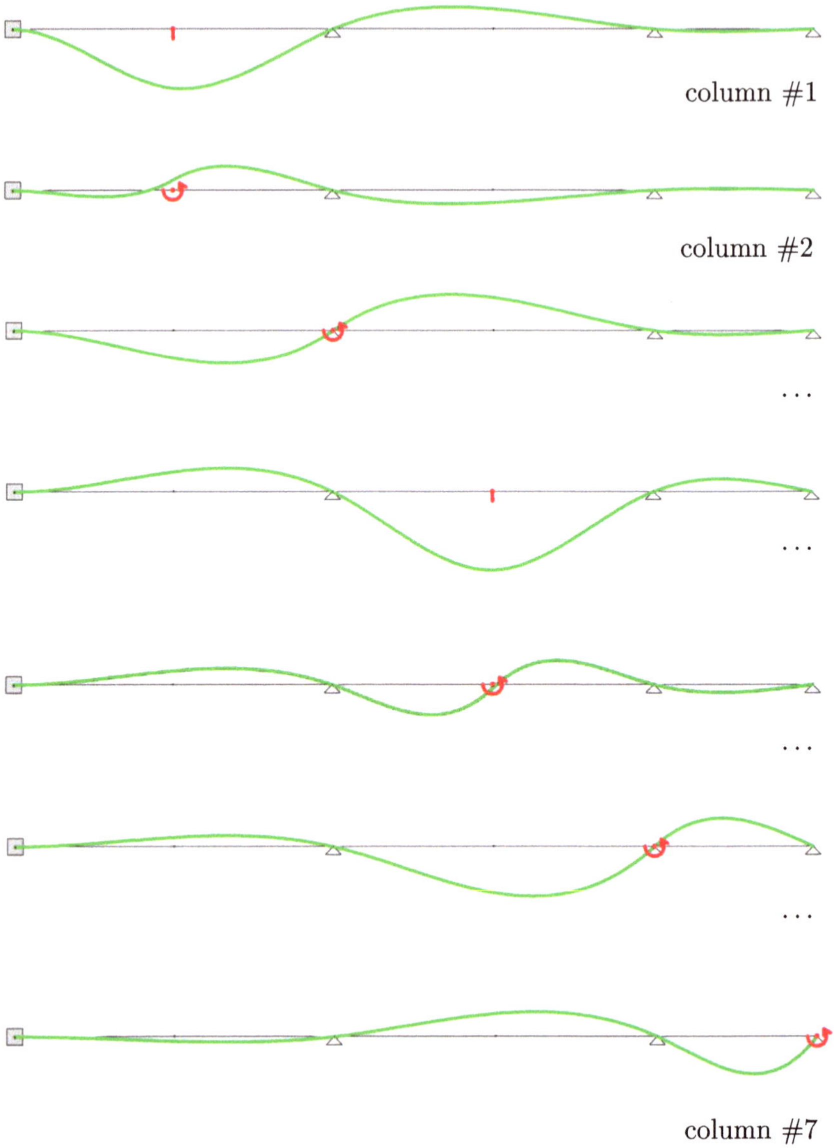

Fig. 3.37 The column-functions of the beam. Each curve corresponds to a column of the inverse $\boldsymbol{K}^{-1}$. These functions are also the influence functions of the nodal displacements u_i

3.22 The Trial Space $\mathcal{V}_h$ Has Two Bases

All displacements, all shapes which an FE-model can take, are expansions with respect to the shape functions $\varphi_i(x)$

$$u_h(x) = \sum_i u_i\, \varphi_i(x)\,, \tag{3.190}$$

and the vector $\boldsymbol{u}$ is the **address** of the shape with respect to this basis. But if we look carefully, we see that there exists a second basis in $\mathcal{V}_h$

$$g_j(x) = \sum_i g_{ij}\, \varphi_i(x)\,, \tag{3.191}$$

since

$$\boxed{u_h(x) = \sum_i u_i\, \varphi_i(x) = \sum_{i,j} g_{ij}\, f_j\, \varphi_i(x) = \sum_j g_j(x)\, f_j} \tag{3.192}$$

The function $g_j(x)$ is the shape when the u_i are the entries g_{ij} in column j, which means $u_i = g_{ij}$; it is the living image of column j of $\boldsymbol{K}^{-1}$, see Fig. 3.37. The address with respect to this basis are the nodal forces $\boldsymbol{f}$.

While the first basis, the $\varphi_i(x)$, consists of a string of ***shortwave*** "elementary events" at the nodes, the other basis is ***longwave***, since the $g_i(x)$, as the influence functions of the nodal displacements u_i, extend their reach over the whole structure.

The $g_i(x)$ represent the engineering basis of $\mathcal{V}_h$, the response of the structure to single point loads at the nodes, $\boldsymbol{f} = \boldsymbol{e}_i$.

Note, any other influence function

$$G_h(y, x) = \sum_i g_i(y)\, j_i(x) \qquad j_i(x) = J(\varphi_i)(x) \tag{3.193}$$

is a "remix" in terms of this basis $g_i(x)$.

3.23 General Form of an FE-Influence Function

Theorem 3.3 (General Form of an FE-Influence Function) *Let $\boldsymbol{K}$ be the stiffness matrix of the structure.*
(i) The FE-influence function for $u_h(x)$, the value of the FE-solution at a point x, is

$$G_h(y, x) = \boldsymbol{\varphi}(y)^T \boldsymbol{K}^{-1} \boldsymbol{\varphi}(x) \tag{3.194}$$

where

$$\boldsymbol{\varphi}(x) = \{\varphi_1(x), \varphi_2(x), \ldots, \varphi_n(x)\}^T \tag{3.195}$$

are the values of the shape functions at the point x, and $\boldsymbol{\varphi}(y)$ is the same list, except that x has been replaced by y (the integration variable).
(ii) The influence function for a linear point functional J is

$$G_h(y, x) = \boldsymbol{\varphi}(y)^T \boldsymbol{K}^{-1} \boldsymbol{j}(x) \tag{3.196}$$

where

$$\boldsymbol{j}(x) = \{J(\varphi_1), J(\varphi_2), J(\varphi_3), \ldots, J(\varphi_n)\}^T . \tag{3.197}$$

The deflection of a rope is the **envelope** of infinitely many influence functions, each of which is the influence of an infinitesimal part $p(y)\,dy$ of the load on the deflection $w(x)$, see Fig. 3.38,

$$w(x) = \int_0^l dw = \int_0^l G(y, x)\,dp = \int_0^l G(y, x)\,p(y)\,dy . \tag{3.198}$$

In contrast to this infinite sum the FE-solution $w_h(x) = \boldsymbol{w}^T \boldsymbol{\varphi}(x)$ is a sum of the n influence functions of the n nodes weighted with the equivalent nodal forces f_i

$$w_h(x) = f_1\, G_h(x_1, x) + f_2\, G_h(x_2, x) + \cdots + f_n\, G_h(x_n, x) , \tag{3.199}$$

since the vector $\boldsymbol{w}$ of the nodal values is

$$\begin{aligned} \boldsymbol{w} &= \boldsymbol{K}^{-1}\boldsymbol{f} = \boldsymbol{K}^{-1}(f_1\boldsymbol{e}_1 + f_2\boldsymbol{e}_2 + \cdots + f_n\boldsymbol{e}_n) \\ &= f_1\boldsymbol{g}_1 + f_2\boldsymbol{g}_2 + \cdots + f_n\boldsymbol{g}_n = \sum_{k=1}^{n} f_k \cdot \left[\boldsymbol{g}_k\right], \end{aligned} \tag{3.200}$$

and the column $\boldsymbol{g}_k$ of $\boldsymbol{K}^{-1}$ corresponds to $G_h(x_k, x)$.

The **cantilever beam** in Fig. 3.39 may serve as an illustration. The stiffness matrix $\boldsymbol{K}$ and its inverse $\boldsymbol{K}^{-1}$ are

$$\boldsymbol{K} = \begin{bmatrix} 12 & 6 \\ 6 & 4 \end{bmatrix} \quad \boldsymbol{K}^{-1} = \begin{bmatrix} 0.333 & -0.500 \\ -0.500 & 1.000 \end{bmatrix}, \tag{3.201}$$

and the FE-solution $\boldsymbol{w}$ is a linear combination of the columns of $\boldsymbol{K}^{-1}$

$$\boldsymbol{w} = f_1\,\boldsymbol{g}_1 + f_2\,\boldsymbol{g}_2 = 0.5\begin{bmatrix} 0.333 \\ -0.500 \end{bmatrix} + \frac{1}{12}\begin{bmatrix} -0.500 \\ 1.000 \end{bmatrix} = \begin{bmatrix} 0.125 \\ -0.166 \end{bmatrix} . \tag{3.202}$$

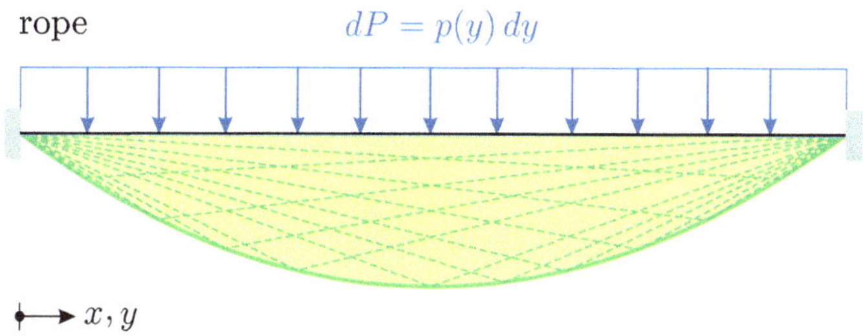

Fig. 3.38 The deflection curve under the distributed load p is the envelope of the rope polygons of the point forces dP

What applies to the nodal values w_i, applies also to the displacement at the points in between since

$$w_h(x_P) = \int_0^l G_h(y, x_P)\, p(y)\, dy = \sum_k g_k(x_P) \int_0^l \varphi_k(y)\, p(y)\, dy$$
$$= \sum_{k=1}^n g_k(x_P)\, f_k = \boldsymbol{g}(x_P)^T\, \boldsymbol{f}\,, \tag{3.203}$$

where $\boldsymbol{g}(x_P)$ is the solution to $\boldsymbol{K}\, \boldsymbol{g}(x_P) = \boldsymbol{j}(x_P)$ and the entries of the vector $\boldsymbol{j}$ are the nodal values of the shape functions $\varphi_k(x_P)$ at x_P. If we use linear elements and x_P lies between two nodes x_i and x_{i+1}, then

$$\boldsymbol{K}\, \boldsymbol{g}(x_P) = \boldsymbol{j}(x_P) = \varphi_i(x_P)\, \boldsymbol{e}_i + \varphi_{i+1}(x_P)\, \boldsymbol{e}_{i+1} \tag{3.204}$$

and so, the vector $\boldsymbol{g}(x_P)$ is a linear combination of the two columns $\boldsymbol{g}_i$ and $\boldsymbol{g}_{i+1}$ of the inverse $\boldsymbol{K}^{-1}$

$$\boldsymbol{g}(x_P) = \varphi_i(x_P)\, \boldsymbol{g}_i + \varphi_{i+1}(x_P)\, \boldsymbol{g}_{i+1}\,. \tag{3.205}$$

3.24 The Dominance of the Columns $\boldsymbol{g}_i$ of the Inverse

We can formulate the above results even more succinctly by focusing on the nodal vectors. The entries in the vector $\boldsymbol{g}$ of an influence function

$$G(y, x) = \boldsymbol{g}^T(x)\, \boldsymbol{\varphi}(y) = \sum_i g_i(x)\, \varphi_i(y) \tag{3.206}$$

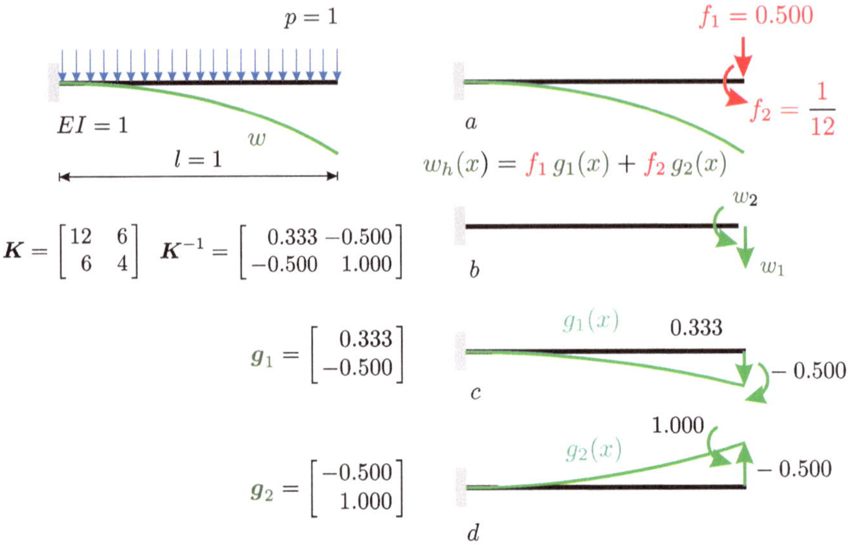

Fig. 3.39 Cantilever beam, **a** equivalent nodal forces, **b** positive nodal displacements, **c** influence function for w_1, and **d** for w_2

are the weights with respect to the shape functions, and the vector $\boldsymbol{g}$ is the solution of the system $\boldsymbol{K}\,\boldsymbol{g} = \boldsymbol{j}$, which means $\boldsymbol{g}$ is an expansion in terms of the columns $\boldsymbol{g}_i$ of the inverse

$$\boldsymbol{g}(x) = \sum_i \boldsymbol{g}_i\, j_i(x)\,. \tag{3.207}$$

And this holds true for every (!) influence function. It is like a game of sticks.

All influence functions—whether for displacements, stresses or else—are expansions of the same n columns $\boldsymbol{g}_i$, the same n nodal influence functions, only the weights differ.

3.25 Nature Makes No Jumps, but Finite Elements Do

On crossing the line between two elements the stresses jump. However, this means the influence functions jump. How does this happen?

The reason why can be seen in Fig. 3.40. The equivalent nodal forces, which generate the influence function for σ_{yy} at the upper point $\boldsymbol{x}_1$, are the stresses σ_{yy} of the nodal displacements $\boldsymbol{\varphi}_i$ at this point. Since only the shape functions of the element itself generate stresses at $\boldsymbol{x}_1$, only the four nodes of the element carry forces.

If the point moves to the next element, $\boldsymbol{x}_1 \rightarrow \boldsymbol{x}_2$, these nodal forces disappear and reappear at the four nodes of the neighboring element. This sudden jump in the load carrying nodes is the reason why the stresses jump: ***The influence functions jump***.

Displacements do not jump, since the shape functions are continuous (else they would be non-conforming). The influence functions for the displacement at two points $\boldsymbol{x}_1$ and $\boldsymbol{x}_2$, immediate neighbors but lying on either side of an interelement boundary, are the same since the load carrying nodes and the loads are the same.

According to Fig. 3.41 displacements are best evaluated at the nodes and stresses at the center of the elements. Averaging the stresses at a node is like doubling the element size.

3.26 The Path from the Source Point to the Load

A **correct communication** between the source point and the load requires that the stiffness along **all possible paths** from the source point to the load is modeled correctly, see Figs. 3.42 and 3.43.

This communication becomes more and more "shaky", the further the source point and the load lie apart, since with increasing distance more and more elements must be traversed, and so the modeling errors will accumulate. Fortunately, however, the

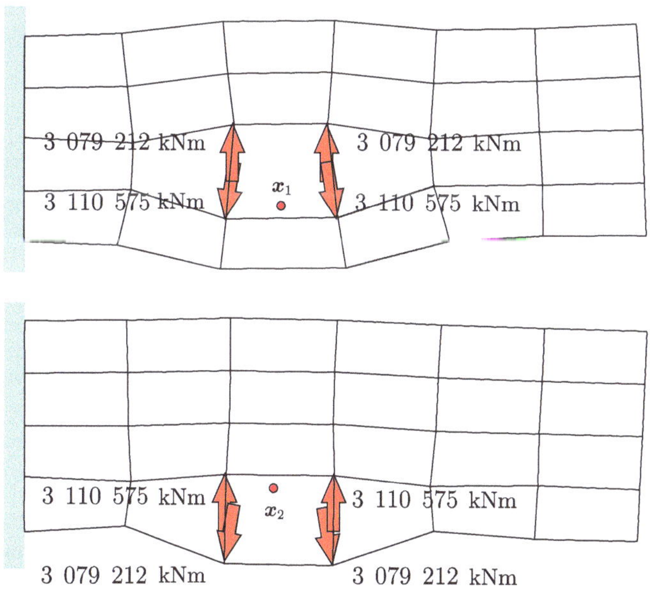

Fig. 3.40 FE-influence functions for σ_{yy} at two neighboring points

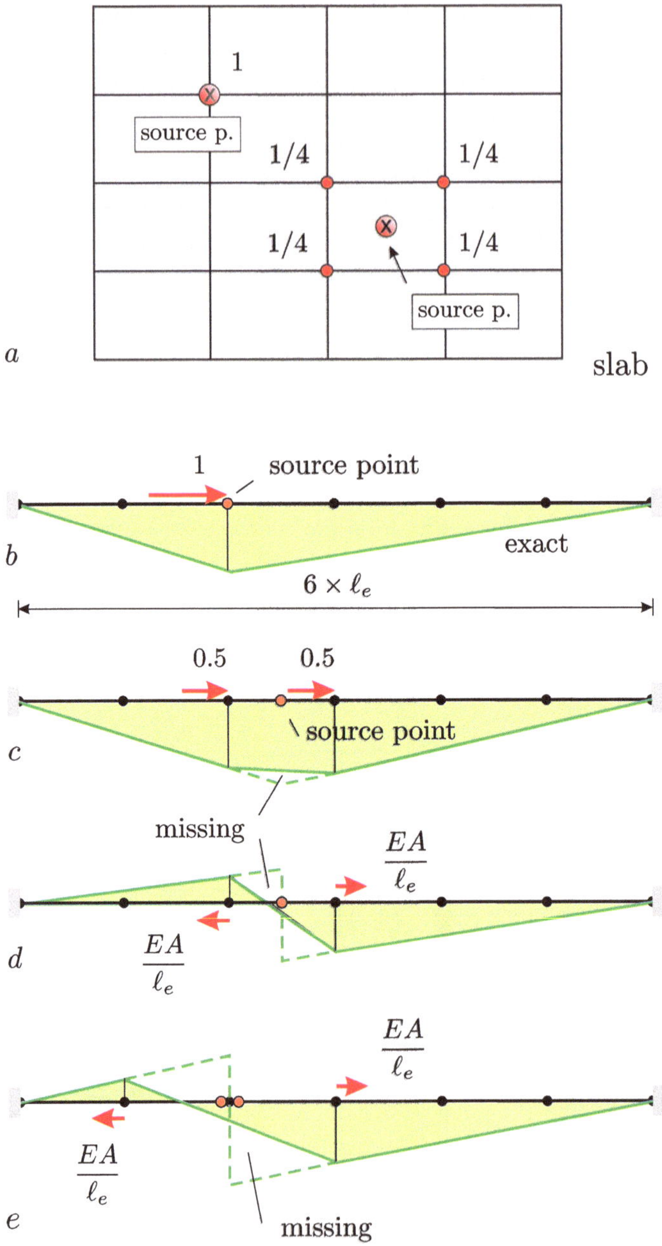

Fig. 3.41 Influence functions in a plate and a bar, **a** the nodal forces which generate the influence function for the deflection at a node and at the center of an element, **b** influence function for a nodal displacement, **c** for a displacement at midpoint, **d** for the normal force at midpoint, and **e** for the average value of the normal force at a node, sum of left and right

influence coefficients usually decrease with increasing distance and therefore also the effects of possible errors.

Remark 3.9 The situation is similar to that in quantum mechanics where the path integral, the influence function is the "sum" of all possible paths the wave functions can take. Just change the elastic properties in a small region and the sum, the influence function $G(y, x)$, will change.

3.27 The Columns of K and K^{-1}

It will be interesting to compare the columns of the two matrices $\boldsymbol{K}$ and $\boldsymbol{K}^{-1}$.

The Columns of K

The columns of a stiffness matrix can be read in two ways

$$\boldsymbol{K} = \underbrace{[\boldsymbol{k}_1, \boldsymbol{k}_2, \ldots, \boldsymbol{k}_n]}_{\delta W_i} = \underbrace{[\boldsymbol{f}_1, \boldsymbol{f}_2, \ldots, \boldsymbol{f}_n]}_{\delta W_e} . \tag{3.208}$$

According to the second interpretation, see Fig. 3.44a,

$$\boldsymbol{K}\boldsymbol{u} = u_1 \cdot \boldsymbol{f}_1 + u_2 \cdot \boldsymbol{f}_2 + \cdots + u_n \cdot \boldsymbol{f}_n = \boldsymbol{f}_h , \tag{3.209}$$

the columns $\boldsymbol{f}_i$ are the nodal forces, which produce the shape $\boldsymbol{u} = \boldsymbol{e}_i$, since $\boldsymbol{K}\boldsymbol{e}_i = \boldsymbol{f}_i$.

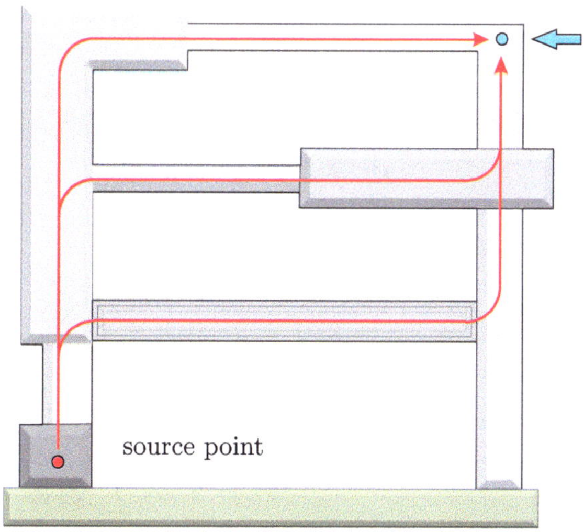

Fig. 3.42 On the path from the source point to the single force the stiffness must be modeled correctly to establish a correct communication between the two points

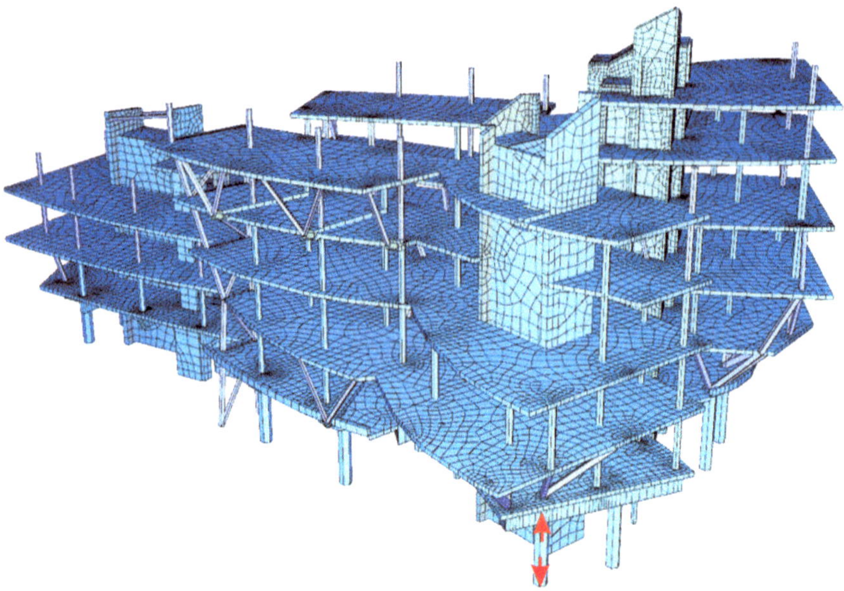

Fig. 3.43 A spread of the pier generates the influence function for the pier reaction. The correct propagation depends on the correct modeling of the stiffness of the structural elements

With regard to the first interpretation we note: Since $\mathcal{G}(\varphi_i, \varphi_j) = \delta W_e - \delta W_i = 0$, the entry f_{ij} in the vector $\boldsymbol{f}_i$ equals the strain energy product, $f_{ij} = a(\varphi_i, \varphi_j) = k_{ij}$, and so we can identify the column $\boldsymbol{f}_i$ also with a vector $\boldsymbol{k}_i$, whose entries are the strain energy products of the shape function φ_i and the n shape functions φ_j, $j = 1, 2, \ldots n$.

The Columns of K^{-1}

The matrix $\boldsymbol{K}^{-1} = [\boldsymbol{g}_1, \boldsymbol{g}_2, \ldots, \boldsymbol{g}_n]$ maps the vector $\boldsymbol{f}$ of the equivalent nodal forces onto the vector $\boldsymbol{u}$

$$\boldsymbol{K}^{-1}\boldsymbol{f} = f_1 \cdot \boldsymbol{g}_1 + f_2 \cdot \boldsymbol{g}_2 + \cdots + f_n \cdot \boldsymbol{g}_n = \boldsymbol{u}\,, \tag{3.210}$$

or componentwise

$$u_i = \boldsymbol{g}_i^T \boldsymbol{f} = \sum_j g_{ij} f_j = \int_0^l \sum_j g_{ij}\, \varphi_j(y)\, p\, dy\,, \tag{3.211}$$

which proves that

$$G_h(y, x_i) = \sum_j g_{ij}\, \varphi_i(y) \tag{3.212}$$

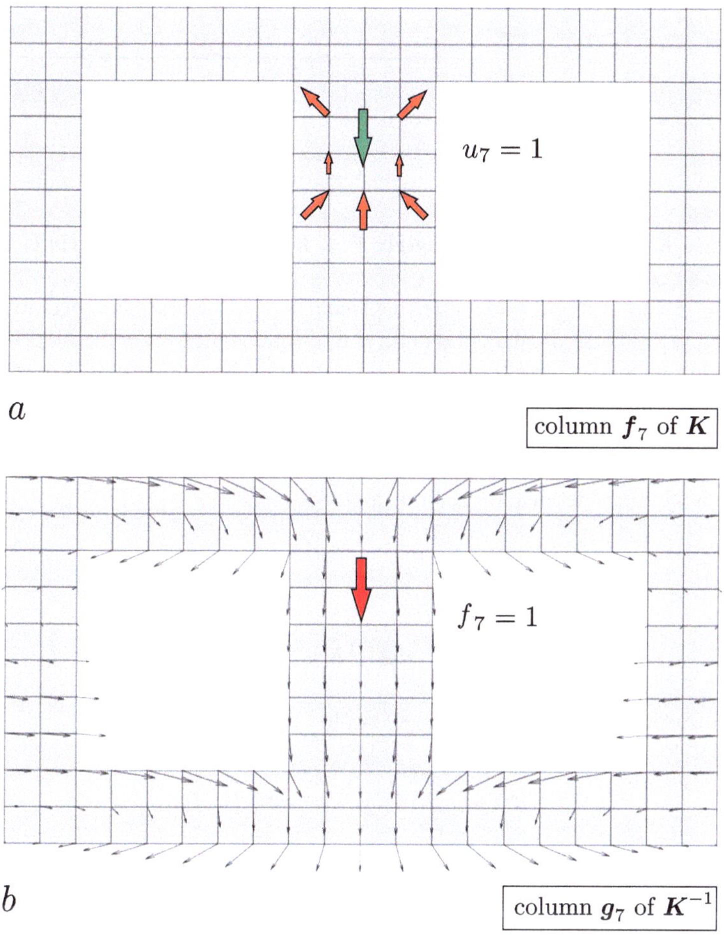

Fig. 3.44 $K\,u = f$ and $K^{-1}\,f = u$. Two columns of K and K^{-1} illustrate the difference, **a** the forces behind $u_7 = 1$ and all else is fixed, $u_i = 0$, are the entries in column f_7 of K, and **b** the displacements u_i due to $f_7 = 1$ and $f_i = 0$ otherwise, are the entries of column g_7 of $F = K^{-1}$

is the influence function for u_i

$$u_i = \int_0^l G_h(y, x_i)\,p(y)\,dy\,. \tag{3.213}$$

Note,

$$e_i = K^{-1} f_i \qquad g_i = K^{-1} e_i \qquad K\,K\,g_i = f_i\,. \tag{3.214}$$

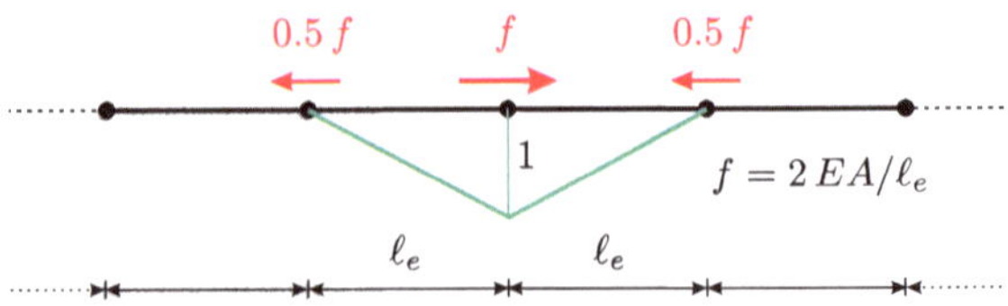

Fig. 3.45 The transition to a Dirac delta with $\ell_e \to 0$

The inverse $\boldsymbol{K}^{-1}$ is the **flexibility matrix** $\boldsymbol{F} = \boldsymbol{K}^{-1}$, but **Green's matrix** $\boldsymbol{G} = \boldsymbol{K}^{-1}$ would be an equally suitable name.

In Fig. 3.44a a single node gets displaced, $u_7 = 1$, while all other nodes are kept fixed, $u_j = 0$. To displace the node and to immediately stop the motion at the neighboring nodes, requires forces, which we call the shape forces p_7. The entries in column $\boldsymbol{f}_7$ of $\boldsymbol{K}$ are the equivalent nodal forces of this shape $\boldsymbol{u} = \boldsymbol{e}_7$. It is a "local" event, because *"the neighborhood remains small even if the world is big"* (G. Strang).

The opposite is when a single force, say $f_7 = 1$, acts at a node, see Fig. 3.44b since then all nodes get displaced—it is a global event.

The Identity $K^{-1}K$

Since $\boldsymbol{u} = \boldsymbol{K}^{-1}\boldsymbol{f}$ and $\boldsymbol{f} = \boldsymbol{K}\boldsymbol{u}$ it is

$$\boldsymbol{u} = \boldsymbol{K}^{-1}\boldsymbol{K}\,\boldsymbol{u}\,, \tag{3.215}$$

which corresponds in the limit, $1/\ell_e \to \infty$, to

$$u(x) = \int_0^l \delta_0(y - x)\,u(y)\,dy\,. \tag{3.216}$$

Fig. 3.45 shows the situation. The smaller the element length l_e gets, the more the action of the three nodal forces corresponds to the action of a Dirac delta.

Stiffness and Flexibility

The equation

$$\frac{1}{g_{ii}} \stackrel{?}{=} k_{ii} \tag{3.217}$$

is intriguing, but the diagonals of $\boldsymbol{K}$ and $\boldsymbol{K}^{-1}$ are of course not the respective reciprocal values.

The entries on the diagonal of the inverse $\boldsymbol{K}^{-1}$ are the flexibilities $g_i \equiv g_{ii}$ in the direction of u_i, and the reciprocal value $1/g_{ii}$ is the stiffness k_i, if we want to call it that, in the direction of u_i. So, one can invert the terms as one would in a spring, only $1/g_{ii} \neq k_{ii}$.

The entry k_{ii} on the diagonal of $\boldsymbol{K}$ is the force $k_{ii} \cdot 1 = f_i$ required for $u_i = 1$ while all other $u_j = 0$ are fixed. The entry k_{ii} does not know anything about the wider neighborhood, does not know whether three or four stories are to follow, while this information is embedded into g_{ii}. The k_{ij} are "local" quantities, which depend only on the neighbourhood, could be found in any other matrix as well, while the g_{ij} are **system responses**.

3.28 The Inverse as an Analysis Tool

The inverse $\boldsymbol{K}^{-1}$ offers simple possibilities to analyze a structure.

Since the elements g_{ij} of the inverse are the nodal values of the influence functions

$$u_i = \int_0^l G_h(y, x_i)\, p(y)\, dy = \sum_{j=1}^{n} \int_0^l g_{ij}\, \varphi_j(y)\, p(y)\, dy = \sum_{j=1}^{n} g_{ij}\, f_j = \boldsymbol{g}_i^T \boldsymbol{f}\,, \tag{3.218}$$

they can steer us to the nodes, which should carry a force, if we aim at producing the maximum possible displacement at a node x_i. Assuming the f_i are all the same, say $f_i = 1$, the maximum displacement at the node is the sum of the values $g_{ij} > 0$ in row i

$$\max u_i = \sum_{j=1}^{n} g_{ij} \qquad g_{ij} > 0\,. \tag{3.219}$$

The question, whether a load f_j on the 3rd floor has an influence on a nodal displacement u_i on the 1st floor is directly answered by the size of the element g_{ij} of the inverse.

One could also generate charts, how far the forces f_j can stray from a node, for to have still some influence on a displacement u_i. These would be all the nodes with $|g_{ij}| > \varepsilon$, i.e. larger than a certain bound ε.

This can be done with any quantity, any functional $J(u)$. One just calculates the vector

$$\boldsymbol{g} = \boldsymbol{K}^{-1} \boldsymbol{j} \qquad j_i = J(\varphi_i)(x) \tag{3.220}$$

and sorts the components g_i into positive and negative values

$$g_i \gtrless 0\,. \tag{3.221}$$

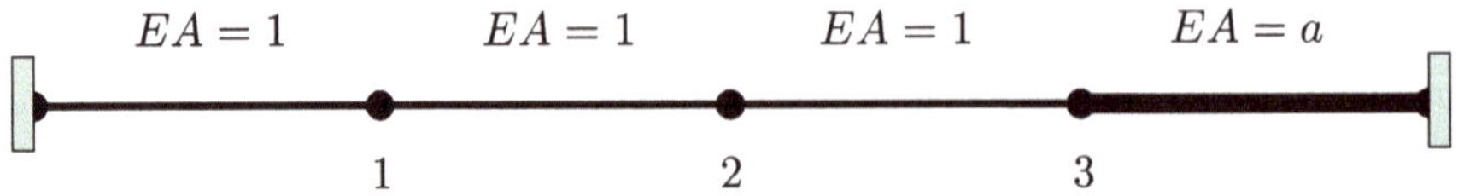

Fig. 3.46 Bar modeled with four elements, $l_e = 1$

3.29 Local Changes and the Inverse

Local modifications of an element affect the whole inverse $\boldsymbol{K}^{-1}$, since we divide by the determinant when we solve the system $\boldsymbol{K}\boldsymbol{u} = \boldsymbol{f}$.

Unlike the first three elements, $EA = 1$, the fourth element of the bar in Fig. 3.46 has the stiffness $EA = a$, and so a appears only once in

$$\boldsymbol{K} = \begin{bmatrix} 2 & -1 & 0 \\ -1 & 2 & -1 \\ 0 & -1 & 1+a \end{bmatrix}, \tag{3.222}$$

but it shows up in each element of the inverse $\boldsymbol{K}^{-1}$

$$\boldsymbol{K}^{-1} = \begin{bmatrix} \frac{2a+1}{3a+1} & \frac{a+1}{3a+1} & \frac{1}{3a+1} \\ \frac{a+1}{3a+1} & \frac{2a+2}{3a+1} & \frac{2}{3a+1} \\ \frac{1}{3a+1} & \frac{2}{3a+1} & \frac{3}{3a+1} \end{bmatrix}. \tag{3.223}$$

A *local* change affects all influence functions.

However, we will see in Chap. 5 that there is a way how to assess the effects of stiffness changes by looking only at the affected element—at least approximately.

3.30 Mohr and the Flexibility Matrix K^{-1}

Let $\boldsymbol{K}$ be the stiffness matrix of a truss. The stiffness k_i in the direction of a displacement u_i is

$$k_i = \frac{1}{u_i}, \tag{3.224}$$

if u_i is the displacement due to a single force $f_i = 1$. Since the displacement vector in this load case is $\boldsymbol{u} = \boldsymbol{K}^{-1}\boldsymbol{e}_i$ the displacement

$$u_i = \boldsymbol{e}_i^T \boldsymbol{u} = \boldsymbol{e}_i^T \boldsymbol{K}^{-1} \boldsymbol{e}_i = g_{ii} \tag{3.225}$$

is the entry g_{ii} on the diagonal of the flexibility matrix $\boldsymbol{F} = \boldsymbol{K}^{-1}$.

But we could calculate u_i also with **Mohr's integral**

$$u_i = g_{ii} = \sum_e \frac{N_e^2 \, l_e}{E A_e}\,, \tag{3.226}$$

where N_e is the normal force in the bar e in the load case $\boldsymbol{f} = \boldsymbol{e}_i$.

In a frame the same equation would read

$$u_i = g_{ii} = \sum_e \int_0^{l_e} \left(\frac{N_e^2}{E A_e} + \frac{M_e^2}{E I_e}\right) dx\,, \tag{3.227}$$

where N_e and M_e have the corresponding meaning.

In modern notation we could write this as

$$g_{ii} = a(\boldsymbol{u}^{(i)}, \boldsymbol{u}^{(i)})\,, \tag{3.228}$$

where $\boldsymbol{u}^{(i)}$ is the response to $\boldsymbol{f} = \boldsymbol{e}_i$. The generalization to the off-diagonal terms is obvious

$$g_{ij} = \boldsymbol{e}_j^T \boldsymbol{K}^{-1} \boldsymbol{e}_i = \sum_e \frac{N_e^{(i)} \, N_e^{(j)} \, l_e}{E A_e} = a(\boldsymbol{u}^{(i)}, \boldsymbol{u}^{(j)})\,. \tag{3.229}$$

3.31 Non-uniform Plates

The **flow of the stresses** in a plate mirrors the stiffness of the single zones. Solid zones attract forces while the stresses tend to bypass lesser zones.

If, as in Fig. 3.47b, the center element is twice as stiff as the rest of the plate, the forces $\boldsymbol{f}$ needed to spread the element (GF σ_{yy}) are approximately also twice as large, but the real reason behind the increase in u_y is: The element better can resist the pressure of the neighboring elements, which want to push it back to its original shape when it is spread to make room for the dislocation. The larger value u_y means a larger part of the load will flow through the center element. Vice versa, if the material is less solid, the neighboring zones will hinder the propagation of the dislocation and so only a small part of the load will flow through the soft element.

How stiff zones in a plate attract loads can be seen in Fig. 3.48, where the uniform load is mostly carried by the more solid zones of the plate, $h = 40$ cm, versus $h = 20$ cm for the inclusions.

In the limit, if the inclusions have zero stiffness, $E = 0$, we see results as in Fig. 3.49, where all stresses are forced to flow around the holes.

Drop panels produce similar effects, since they cause the bending moments to migrate to the column capitals. Note, the jump in the plate thickness makes that the bending moments m_{yy} in a horizontal cross section—perpendicular to the discontinuity—jump while the moments m_{xx} are not affected, see Fig. 3.50. In a vertical cross section through the center it would just be the opposite.

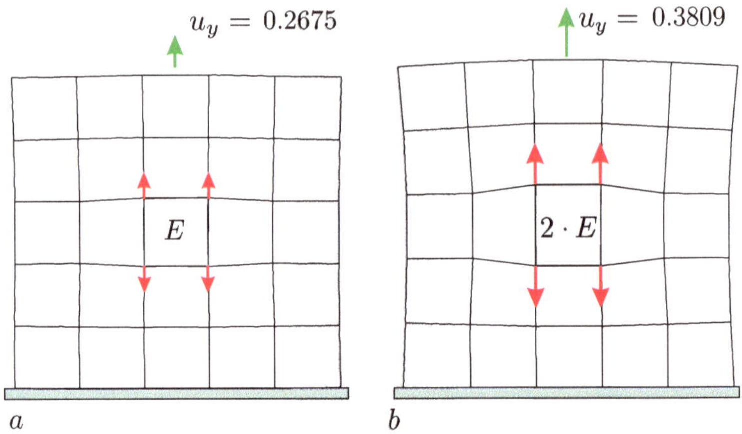

Fig. 3.47 Influence function for σ_{yy}, **a** original E, and **b** $2 \cdot E$

3.32 Sensitivity Plots

In FE-analysis any linear functional $J(u_h) = \boldsymbol{g}^T \boldsymbol{f}$ is the scalar product of the nodal values $\boldsymbol{g}$ of the influence function and the vector $\boldsymbol{f}$ of the equivalent nodal forces. This scalar product can be read as a sum over the N nodes of the mesh

$$J(u_h) = \sum_{i=1}^{N} \boldsymbol{g}_i^T \boldsymbol{f}_i \qquad i = \text{nodes}\,, \tag{3.230}$$

where the vectors $\boldsymbol{g}_i$ and $\boldsymbol{f}_i$ are the parts of the larger vectors $\boldsymbol{g}$ and $\boldsymbol{f}$, which refer to node i

$$\boldsymbol{g} = \{\underbrace{g_1, g_2}_{\boldsymbol{g}_1}, \underbrace{g_3, g_4}_{\boldsymbol{g}_2}, \ldots, g_{2N}\}^T \qquad \text{2-D}\,. \tag{3.231}$$

So, if $\boldsymbol{f}_i$ is orthogonal to $\boldsymbol{g}_i$ at a node, the contribution of this node to $J(u_h)$ is zero, and therefore the plot of the vectors $\boldsymbol{g}_i$ constitutes a **sensitivity plot** of the functional $J(u_h)$, see Figs. 3.51, 3.52 and 3.53. Nodal forces $\boldsymbol{f}_i$ which point in the

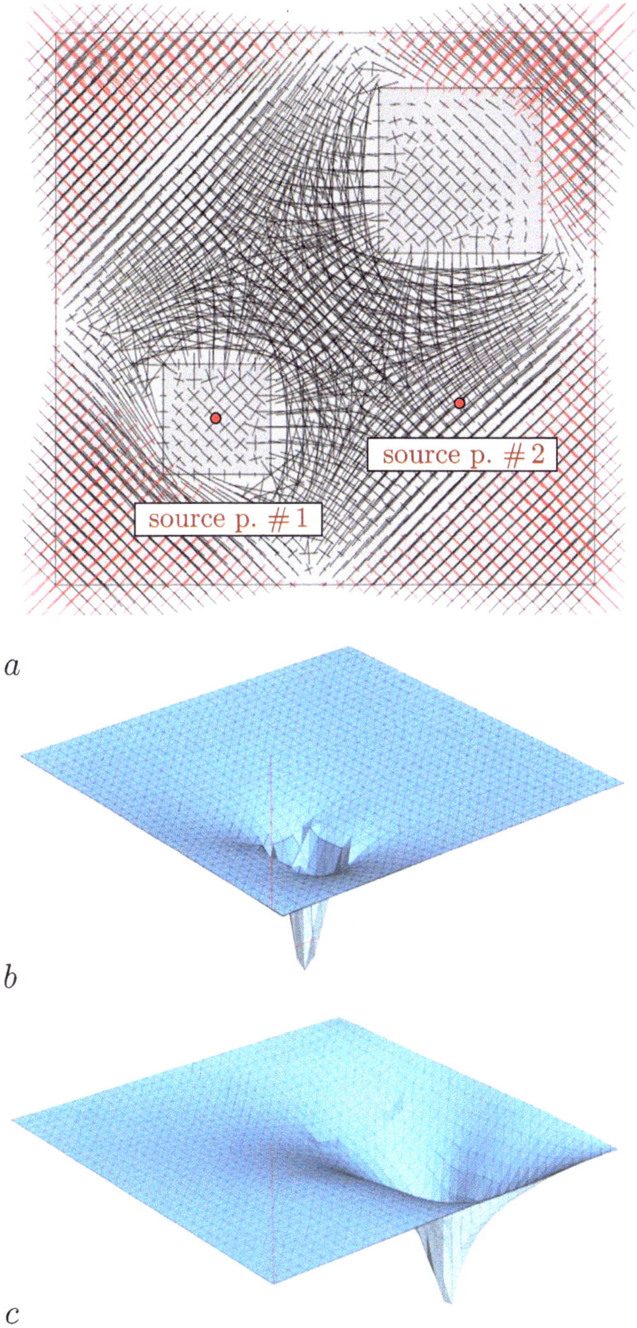

Fig. 3.48 Plate and gravity load, in two small zones the plate thickness is 20 cm instead of 40 cm, **a** principal moments, **b** influence function for m_{xx} at source point #1, and **c** at source point #2

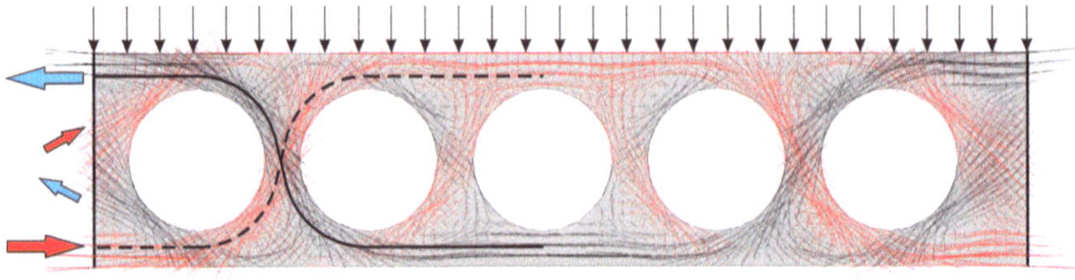

Fig. 3.49 Principal stresses in a cellular beam with fixed vertical edges

same direction as the $\boldsymbol{g}_i$ contribute the most to $J(u_h)$.

The vector $\boldsymbol{g} = \nabla J$ is the gradient $(\partial J/\partial f_i)$ of the functional $J(u_h) = \boldsymbol{g}^T \boldsymbol{f}$

In Fig. 3.54, a plot of the influence function for the stress σ_{yy} at the crack tip of the plate, we note two quiet zones, where the influence of the nodal forces on $J(u_h)$ is practically zero. We call these points ***Lagrange points***. In astronomy the Lagrange points are the points, at which the gravitational forces of the Sun and the Moon balance, because they pull with opposite forces at a satellite parked at this point.[9] Such Lagrange points can be found in almost all of these plots and also in frames, see Fig. 3.55.

3.33 Support Reactions

The support reactions are the nodal forces f_i at the fixed nodes. An FE-program calculates the f_i when it has solved the system $\boldsymbol{K}\,\boldsymbol{u} = \boldsymbol{f}$ as follows:

- It first expands the vector $\boldsymbol{u}$ by adding the previously deleted $u_i = 0$ at the supports, $\boldsymbol{u} \rightarrow \boldsymbol{u}_G$,
- and multiplies the non-reduced global stiffness matrix $\boldsymbol{K}_G$ with the full vector $\boldsymbol{u}_G$.
- The entries f_i in the vector $\boldsymbol{f}_G = \boldsymbol{K}_G \boldsymbol{u}_G$ belonging to the fixed degrees of freedom are the nodal forces *without* the part of the load which was reduced directly to the supports. These forces, which we call R_d, see Fig. 3.56, must be added to the nodal forces

$$f_i(complete) = f_i + R_d = R_{FE} + R_d\,. \tag{3.232}$$

[9] The relative position of the Lagrange points—there are 5 such points—to the Sun and the Moon does not change when S and M rotate about their common center of gravity (which lies in the sun).

Fig. 3.50 Plate with drop panel, **a** moments m_{xx}, and **b** m_{yy}

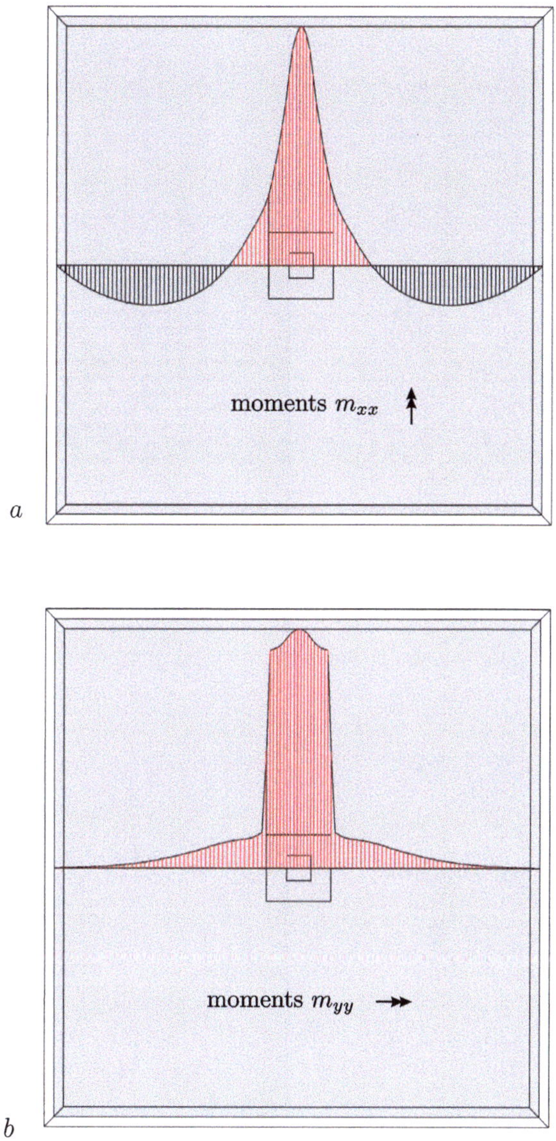

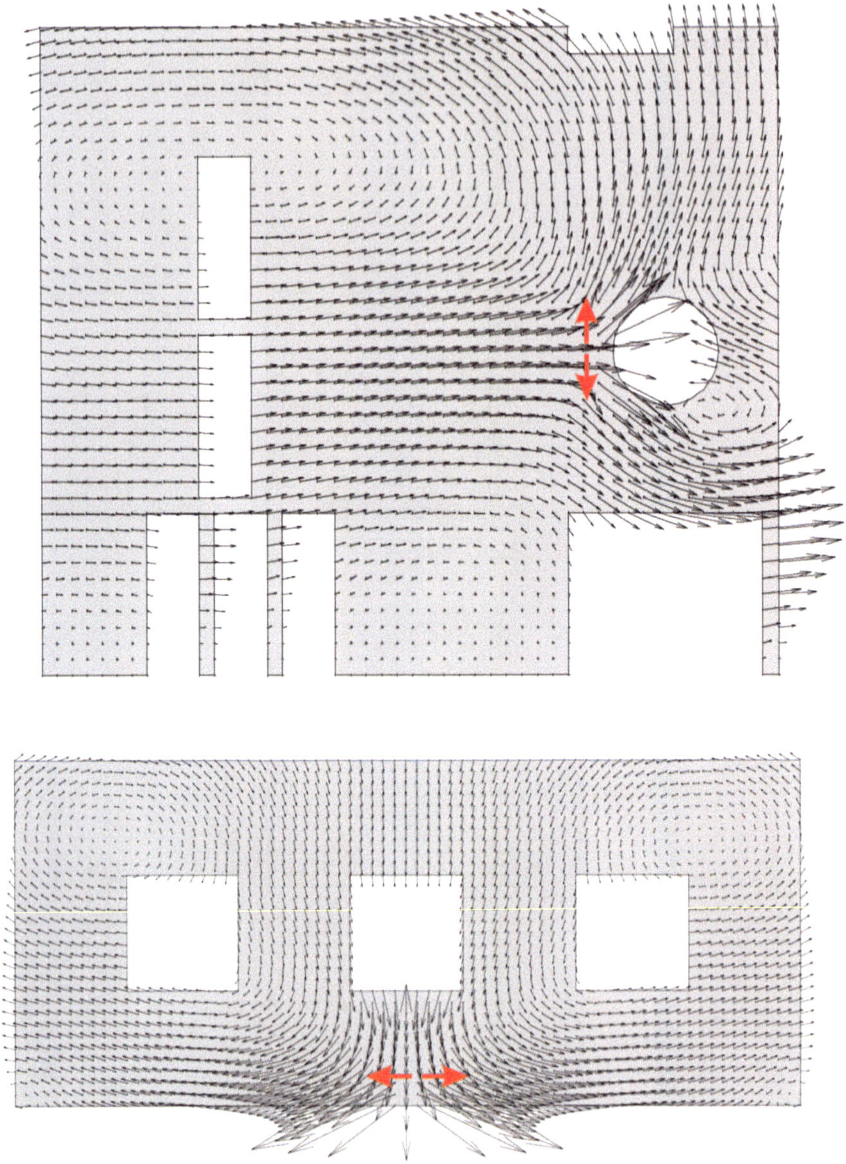

Fig. 3.51 Influence functions for stresses σ_{yy} and σ_{xx} in two wall plates. The arrows correspond to the displacements $\boldsymbol{g}_i$ at the nodes $\boldsymbol{x}_i$ resulting from the dislocation at the source point. Nodal forces $\boldsymbol{f}_i$ which point in the direction of the $\boldsymbol{g}_i$ have maximum effect, while forces which are orthogonal to the $\boldsymbol{g}_i$ have zero influence

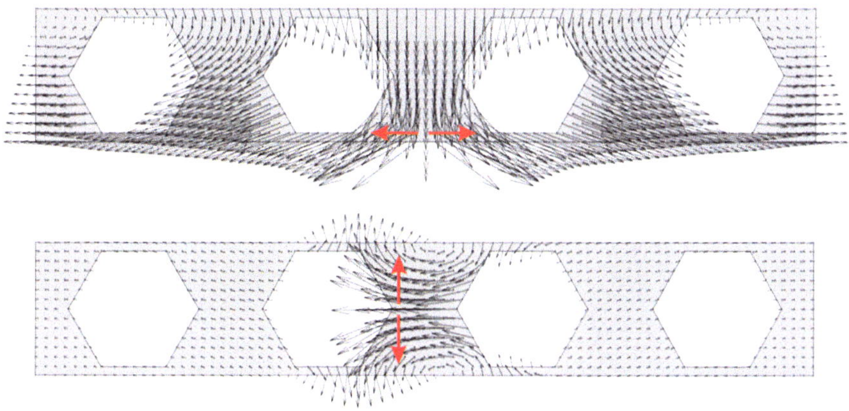

Fig. 3.52 Wall plate, plot of the nodal vectors $\boldsymbol{g}_i$ of the functional $J(u_h) = \sigma_{xx}$ (upper figure) and $J(u_h) = \sigma_{yy}$ (lower figure)

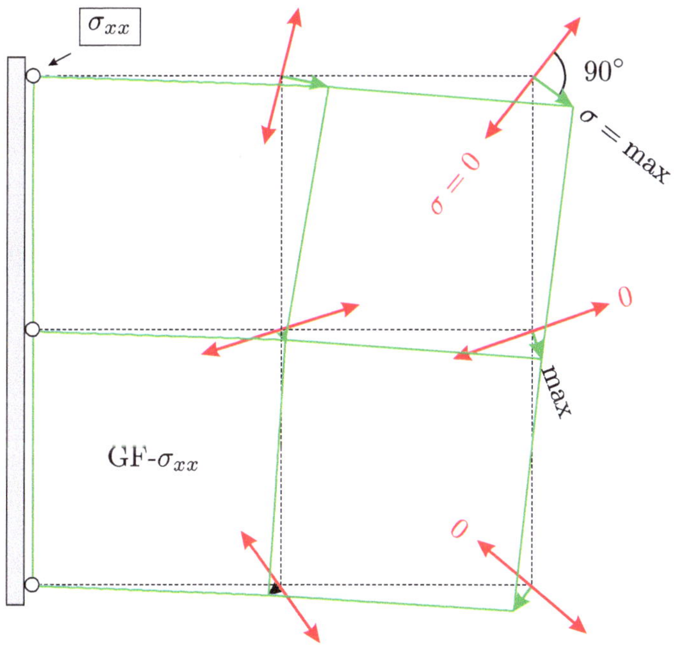

Fig. 3.53 Four bilinear elements, influence function for the stress σ_{xx} at the uppermost node at the fixed edge. Nodal forces which are orthogonal to the nodal displacements of the influence function generate no stress σ_{xx} at the node

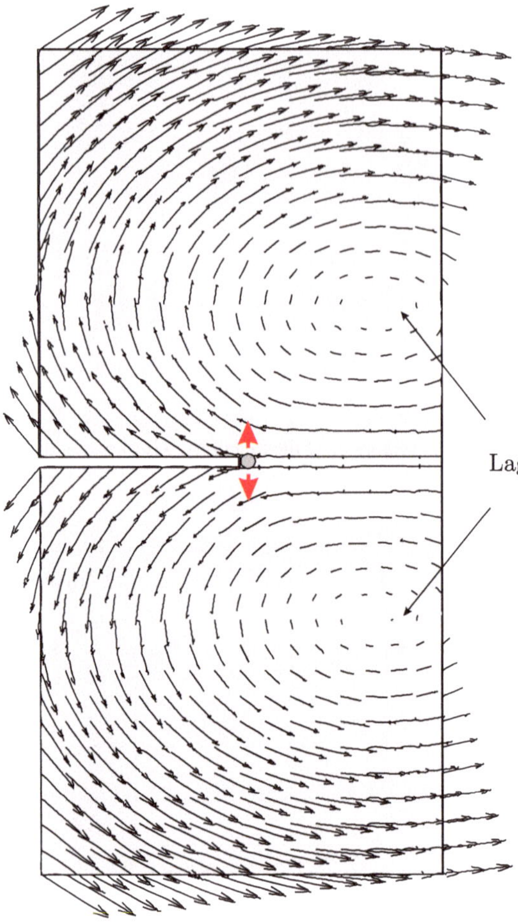

Fig. 3.54 Plot of the nodal vectors $\boldsymbol{g}_i$ of the functional $J(u_h) = \sigma_{yy}$, the vertical stress at the crack tip. At the *Lagrange points* the influence of the nodal forces $\boldsymbol{f}_i$ on σ_{yy} is practically zero

- At elastic supports the last maneuver is not necessary, since at such supports $R_{FE} = u \cdot k$, the displacement u of the support times its stiffness k, is the complete support reaction.

The sum of the f_i $(complete)$ at the supports must match the applied load, since the shape functions form a ***partition of unity***, which means $\sum_i \varphi_i(x) = 1$ at each point x. If this is true, the rigid body motions of the released structure lie in $\mathcal{V}_h^+$.

The space $\mathcal{V}_h^+$ is the **span of the shape functions** φ_i. By eliminating the φ_i, which belong to fixed degrees of freedom, we construct the subspace $\mathcal{V}_h \subset \mathcal{V}_h^+$, which contains the FE-solution; $\mathcal{V}_h^+ = \mathcal{V}_h$ + rigid body motions.

Locally, given an arbitrary patch Ω_p of the mesh, equilibrium cannot be guaranteed between the section forces along the edge of the patch and the load p, since the rigid

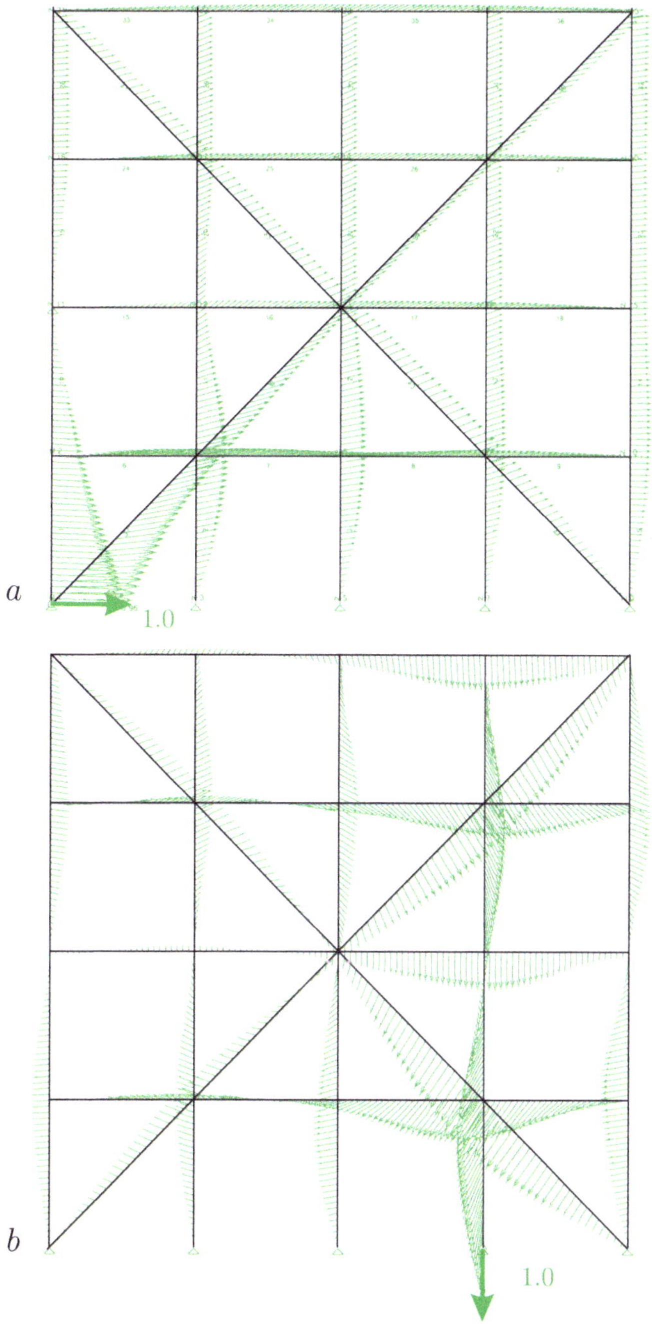

Fig. 3.55 Braced frame, **a** influence function for the horizontal support reaction on the first node, and **b** the vertical support reaction at the second to last node

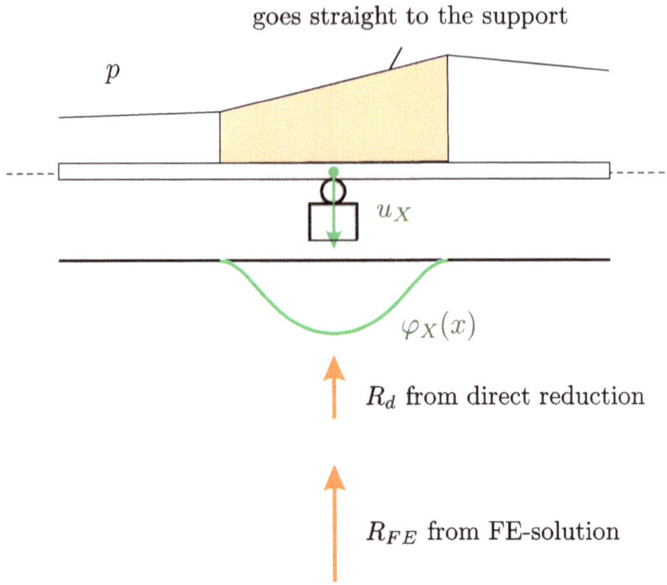

Fig. 3.56 Rigid support, the support reaction is the sum of the FE-support reaction R_{FE} plus the portion of the load p, which was reduced directly to the support

body motions u_0^p of the local patches Ω_p do not lie in $\mathcal{V}_h$, and so the program cannot test for $(p - p_h, u_0^p) = 0$. The u_0^p have the value 1 inside the patch and 0 outside, and such discontinuous "lifts" cannot be modeled with conforming shape functions.

3.34 If a Support Settles

If the end support of the beam in Fig. 3.57 settles by one unit, $u_5 = \Delta w = 1$, the FE-program places column $\boldsymbol{f}_5$ of $\boldsymbol{K}$ on the right side, deletes row 5 and solves $\boldsymbol{K}\boldsymbol{u} = -\Delta w \cdot \boldsymbol{f}_5$ for $\boldsymbol{u}$, which results in

$$w(x) = \sum_{\substack{i \neq 5 \\ i=1}}^{n} u_i\, \varphi_i(x) + \Delta w \cdot \varphi_5(x) =: w_V(x) + \Delta w \cdot \varphi_5(x)\,. \tag{3.233}$$

This can also be read as an application of the principle of virtual displacements, since the deflection w of the beam is **orthogonal** to the virtual displacements of the beam, $\delta W_i(w, \delta w) = 0$, see Eq. (1.122). If we let δw one of the φ_i—except for φ_5—it holds

$$\delta W_i(w, \varphi_i) = a(w_V + \Delta w \cdot \varphi_5, \varphi_i) = a(w_V, \varphi_i) + \Delta w \cdot a(\varphi_5, \varphi_i) = 0\,, \tag{3.234}$$

or

$$a(w_V, \varphi_i) = -\Delta w \cdot a(\varphi_5, \varphi_i) \qquad i = 1, 2, \ldots, n \quad i \neq 5\,, \tag{3.235}$$

and this is exactly the system $\boldsymbol{K}\boldsymbol{u} = -\Delta w\, \boldsymbol{f}_5$, since w_V is an expansion in terms of the shape functions φ_i (with a gap at φ_5).

Example 3.9 The system $\boldsymbol{K}\boldsymbol{u} = \boldsymbol{f}$ of the beam in Fig. 3.57 consists of the three equations

$$\begin{bmatrix} 8 & 6 & 2 \\ 6 & 12 & 6 \\ 2 & 6 & 4 \end{bmatrix} \begin{bmatrix} u_3 \\ u_5 \\ u_6 \end{bmatrix} = \begin{bmatrix} f_3 \\ f_5 \\ f_6 \end{bmatrix}. \tag{3.236}$$

If the end support settles by 1 m we place the column belonging to u_5 on the right side

$$\begin{bmatrix} 8 & 2 \\ 6 & 6 \\ 2 & 4 \end{bmatrix} \begin{bmatrix} u_3 \\ u_6 \end{bmatrix} = \begin{bmatrix} -6 \\ -12 \\ -6 \end{bmatrix} \cdot u_5\ (= 1.0)\,, \tag{3.237}$$

delete row # 2

$$\begin{bmatrix} 8 & 2 \\ 2 & 4 \end{bmatrix} \begin{bmatrix} u_3 \\ u_6 \end{bmatrix} = - \begin{bmatrix} 6 \\ 6 \end{bmatrix} \tag{3.238}$$

and solve for $u_3 = -0.4286$, $u_6 = -1.2857$ and so, since $u_5 = 1.0$, the deflection of the beam is

$$w(x) = -0.4286 \cdot \varphi_3(x) + 1.0 \cdot \varphi_5(x) - 1.2857 \cdot \varphi_6(x)\,. \tag{3.239}$$

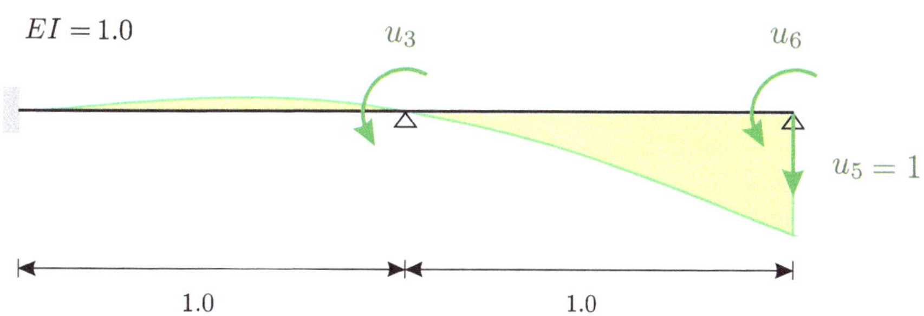

Fig. 3.57 Support settlement

3.35 Influence Function for a Rigid Support

The classical approach is to remove the support and to push the beam down by one unit length.

But we could approach it also from the finite element point of view, since a support reaction R is a **functional**

$$R = J(w)\,, \tag{3.240}$$

and the nodal forces j_i which generate the influence function $\boldsymbol{g} = \boldsymbol{K}^{-1}\boldsymbol{j}$ are the support reactions of the shape functions

$$j_i = R(\varphi_i)\,, \tag{3.241}$$

and the single j_i is just the **strain energy product** $\times(-1)$ between φ_i and the function φ_l, the unit displacement of the support

$$a(\varphi_i, \varphi_l) = \int_0^l EI\, \varphi_i''\, \varphi_l''\, dx = -R(\varphi_i) \cdot 1\,. \tag{3.242}$$

This result is based on Green's first identity, since the last equation is identical with $\delta W_i = \delta W_e$

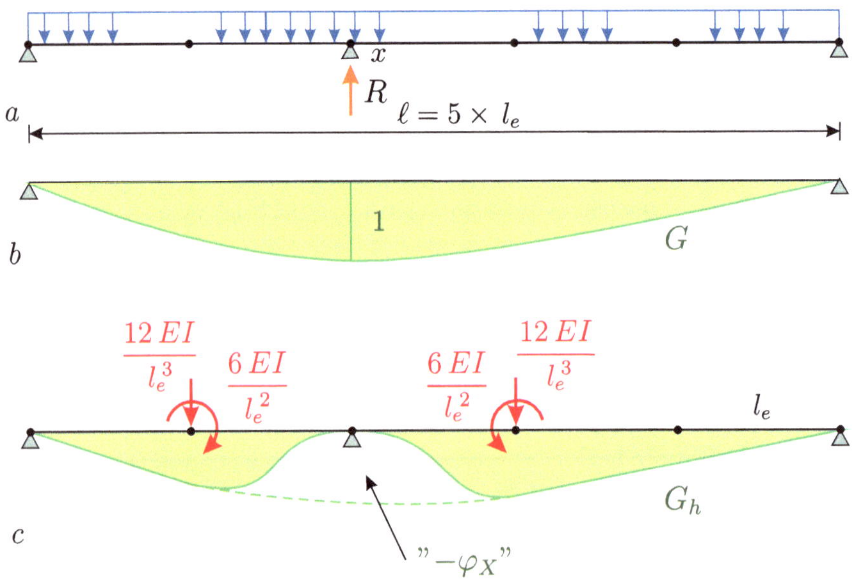

Fig. 3.58 Continuous beam, **a** support reaction R, **b** influence function G for R, **c** FE-influence function G_h; the nodal forces j_i, which generate G_h, are the support reactions R of the shape functions

$$\begin{aligned}\mathscr{G}(\varphi_i, \varphi_l) &= \mathscr{G}(\varphi_i, \varphi_l)_{(0,x)} + \mathscr{G}(\varphi_i, \varphi_l)_{(x,l)} \\ &= \underbrace{V(\varphi_i)(x_-) - V(\varphi_i)(x_+)}_{-R(\varphi_i)} - \int_0^l EI\, \varphi_i''\, \varphi_l''\, dx = 0\,. \end{aligned} \tag{3.243}$$

To explain: The φ_i are homogeneous solutions of the beam equation, and φ_l is zero at all nodes except the support, where $\varphi_l = 1$, while $V(\varphi_i)$ jumps in agreement with the support reaction, and so the virtual external work simplifies to the expression $-R(\varphi_i) \cdot 1$.

The support reaction R of a shape function $\varphi_i(x)$ is equal to the strain energy product $a(\varphi_i, \varphi_l) \times (-1)$ between $\varphi_i(x)$ and $\varphi_l(x)$.

So, the vector $\boldsymbol{j}$ is identical to column $\boldsymbol{f}_l$ of $\boldsymbol{K}$ (where l is the index of the support), and since $\boldsymbol{g} = \boldsymbol{K}^{-1}\boldsymbol{j}$ it follows that

$$G_h(y, x) = \sum_i g_i\, \varphi_i(y) = \boldsymbol{\varphi}(y)^T \boldsymbol{g} = -\boldsymbol{\varphi}(y)^T \boldsymbol{K}^{-1} \boldsymbol{f}_l\,. \tag{3.244}$$

Now it seems that something is missing, see Fig. 3.58c, namely the piece φ_l directly under the support. But no, this must be, because $\mathcal{V}_h$ does not contain the shape φ_l, since the node is fixed. But then why do finite elements obtain the correct support reaction? The answer lies in how FE-programs proceed.

Given a distributed load $p(x)$, the program places at each node the corresponding equivalent nodal force and at the support the force

$$f_l = \int_0^l p(x)\, \varphi_l(x)\, dx\,. \tag{3.245}$$

But this force directly flows off, and the influence function in Fig. 3.58 is exactly the influence function of the support reaction in these **"amputated load cases"**.

The nodal force f_l placed on the support will trigger an opposite support reaction $R_d = -f_l$, and at the end the FE-program adds to the support reaction R_{FE} of the "amputated" load case this support reaction R_d, and so everything fits together

$$R = R_{FE} + R_d\,. \tag{3.246}$$

This is the same result as if you would simply add the missing part φ_l to the FE-influence function G_h

$$G = G_h + \varphi_l\,, \tag{3.247}$$

and superpose the load with this influence function

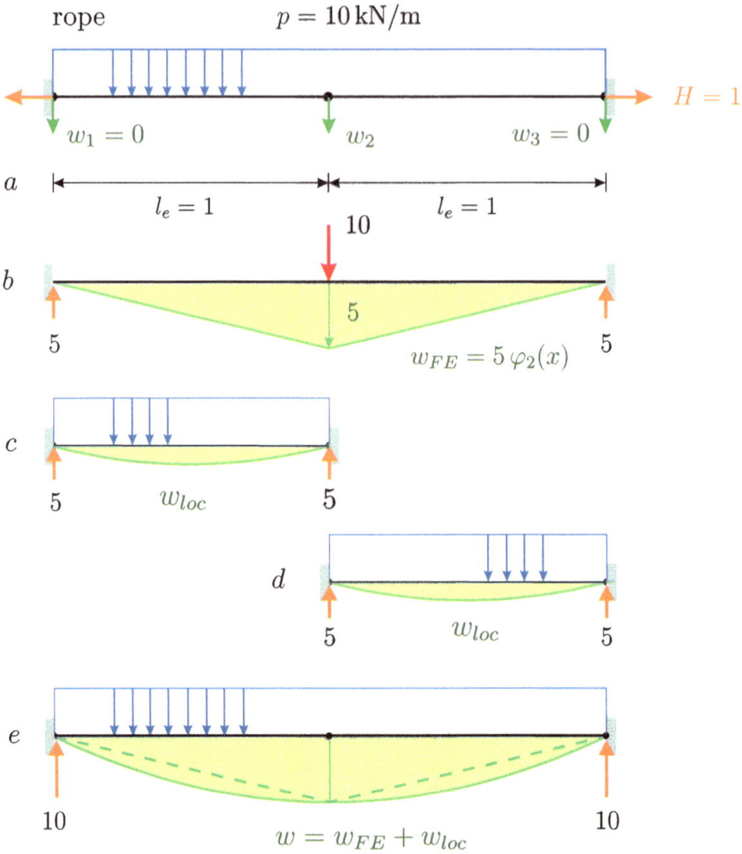

Fig. 3.59 Rope, **a** system, **b** FE-solution **c** - **d** local solutions, **e** exact solution

$$R = R_{FE} + R_d = \int_0^l G(y,x)\,p(y)\,dy\,. \tag{3.248}$$

In frame analysis the addition $R = R_{FE} + R_d$ **happens automatically**, when the local solutions are added to the FE-solution.

The rope in Fig. 3.59 is subject to a constant load $p = 10\,\text{kN/m}$. A subdivision into two linear elements[10]

$$\begin{bmatrix} 2 & -1 & 0 \\ -1 & 2 & -1 \\ 0 & -1 & 2 \end{bmatrix} \begin{bmatrix} 0 \\ 5 \\ 0 \end{bmatrix} = \begin{bmatrix} f_1 \\ 10 \\ f_3 \end{bmatrix} \tag{3.249}$$

[10]This is the non-reduced global stiffness matrix $\boldsymbol{K}_G$.

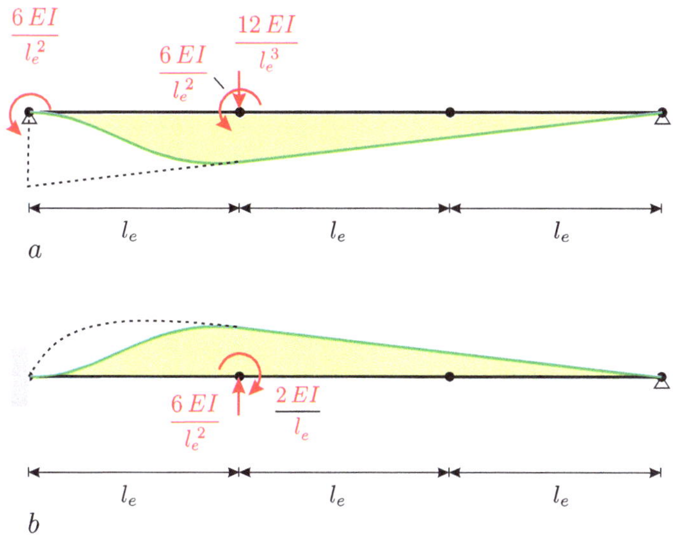

Fig. 3.60 Three-element model of a beam, **a** FE-influence function for the shear force next to the left support. The f_i are the shear forces of the shape functions in the first element, **b** FE-influence function for the fixed-end moment

produces the result $w_2 = 5$ and the support reactions $f_1 = f_3 = -5$. When the local solutions are added, see Fig. 3.59c, d, the shape $w = w_{FE} + w_{loc}$ becomes a well-rounded curve and the support reactions fit exactly.

3.36 Shear Forces

The effect that equivalent nodal forces placed on a rigid support are "lost", also affects the shear forces at an intermediate or end support of a plate. The shear stresses are not as large as they should be, since the part of the load, which is carried by the fixed node, never flows through the cross section marked for the design of the shear reinforcement, see Fig. 3.58.

This observation can be made more precise by comparing the FE-influence function for the shear force with the exact influence function. In a beam, as in Fig. 3.60, the FE-influence function does not exhibit the typical jump but instead it glides in a smooth curve to zero. The curved triangle (superposed with the load) is the part, which the FE-program misses.

The fixed end moments are subject to the same logic, see Fig. 3.60b, and at a flexible support we essentially observe the same effect, see Fig. 3.61.

In 1-D problems (frames) this error is automatically corrected when we add the local solutions. It is only in 2-D and 3-D that cross-sections eventually do not carry the full load.

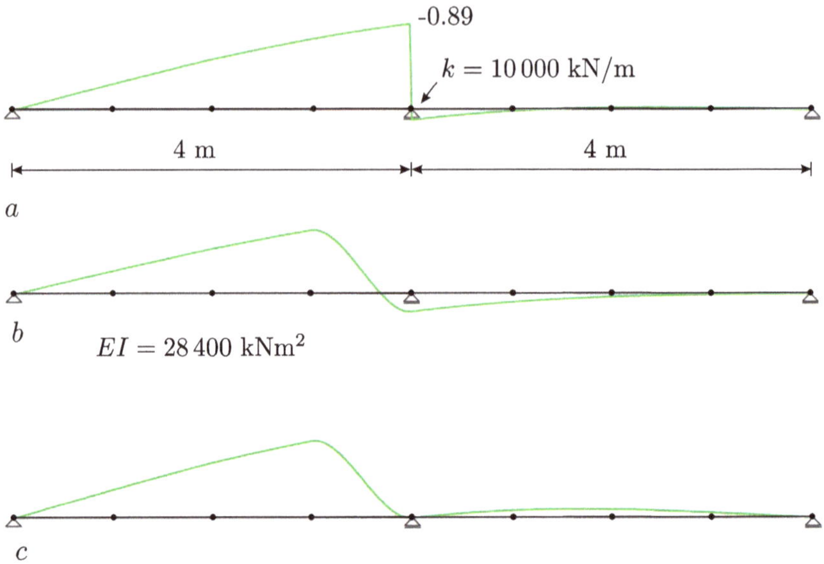

Fig. 3.61 Influence function for the shear force V_l, **a** elastic support, **b** FE-influence function, **c** if support is rigid

Remark 3.10 Above, we have implicitly assumed that the design of the **shear reinforcement** is based on the shear forces q_n, but actually with a Kirchhoff plate it should be the Kirchhoff shear $v_n = q_n + d/ds\, m_{nt}$, since this is the quantity, which is conjugated to w (and therewith to the lift $w = 1$, the motion which controls the equilibrium). That it is done differently in practice can perhaps be justified by the fact that the twisting moment m_{nt} (together with m_{nn} and m_{tt}) determines the principal moments, which then regulate the bending reinforcement.

It is different in Reissner-Mindlin plates, where the shear force $\boldsymbol{q} \bullet \boldsymbol{n} = q_x\, n_x + q_y\, n_y$ perpendicular to the edge is really the equilibrium force, see (7.56), since

$$\mathscr{G}(\boldsymbol{\varphi}, w; \mathbf{0}, 1) = \int_\Omega p \cdot 1\, d\Omega + \int_\Gamma \boldsymbol{q} \bullet \boldsymbol{n} \cdot 1\, ds = 0\,. \qquad (3.250)$$

3.37 Influence Function for an Elastic Support

To generate the influence function for a pier reaction R, we place a **force** $P = 1$ on top of the pier and multiply the response $G_0(y, x)$ of the structure (=influence function for the compression of the support) with the stiffness k of the support, see Fig. 3.62a,

$$R = k \int_0^l G_0(y, l)\, p(y)\, dy\,. \tag{3.251}$$

In a **torsional spring**, see Fig. 3.62b, it is the same. The response $G_1(y, x)$ of the beam to a moment $M = 1$ times the stiffness k_φ of the spring, is the influence function for the moment in the spring

$$M = k_\varphi \int_0^l G_1(y, l)\, p(y)\, dy\,. \tag{3.252}$$

A single point load does not require a very refined mesh (if the focus is only on the displacements), and therefore pier reactions are quite accurate even if the mesh is relatively simple.

Instead of placing a unit load on the pier one can also pull the foot of the pier down by one meter and see how much of this displacement makes it to the top of the pier; this is how it was done in Fig. 3.63.

3.38 Elasticity Theory and Point Supports

In elasticity theory a point support equals an infinitely thin needle, which offers no resistance to the highly stressed and plastified material, and so the plate simply slips

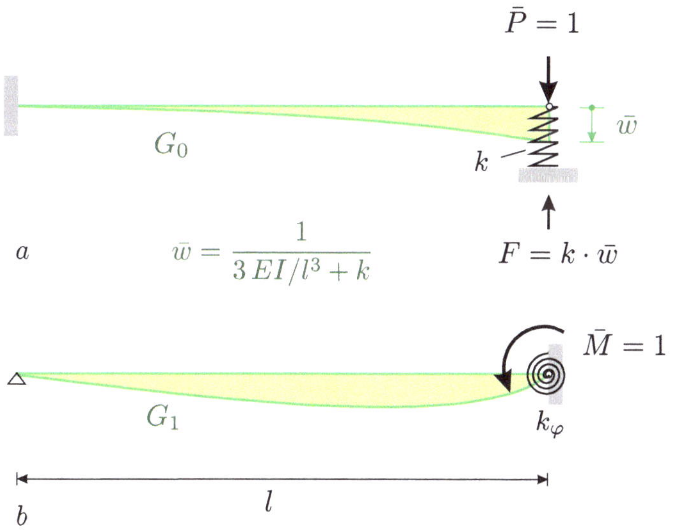

Fig. 3.62 Elastic supports, **a** influence function G_0 for the support reaction, and **b** influence function G_1 for the moment in the torsional spring

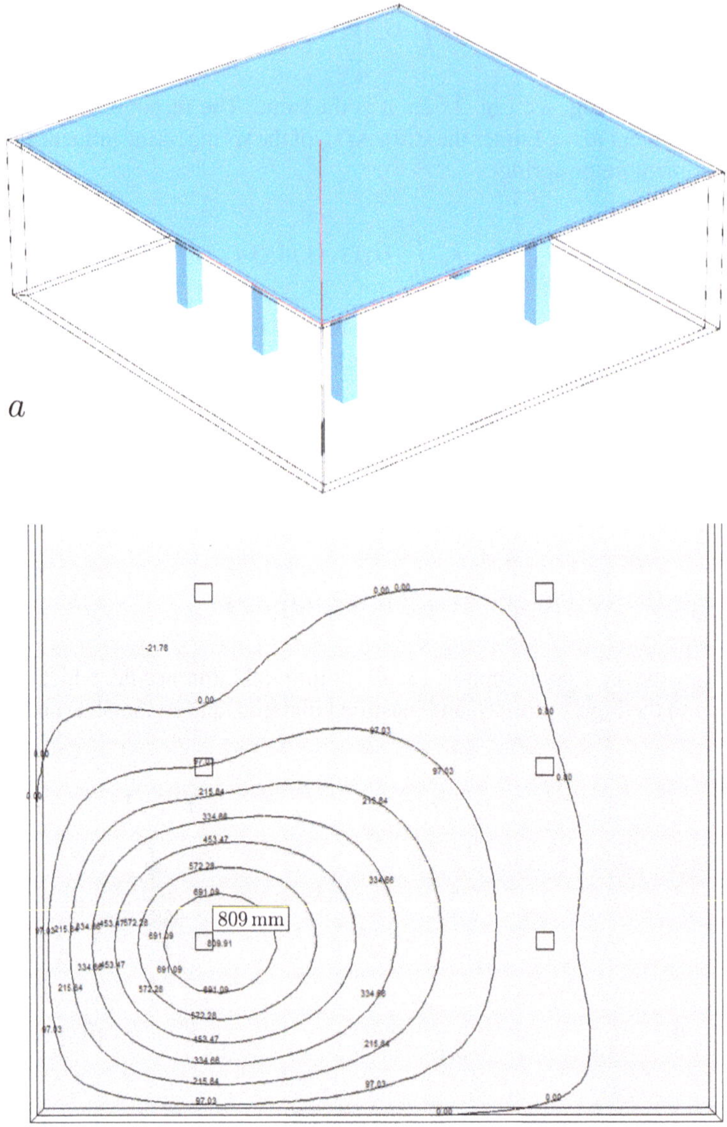

Fig. 3.63 Hinged plate on piers, **a** system, **b** influence function for a pier. The foot of the pier is pulled down 1 000 mm. About 80 % $\doteq$ 800 mm/1000 mm of this displacement reach the top of the pier and so 80% of a force placed on top of the pier are carried by the pier and 20% by the plate. If the pier would be rigid, $EA = \infty$, it would be 100%

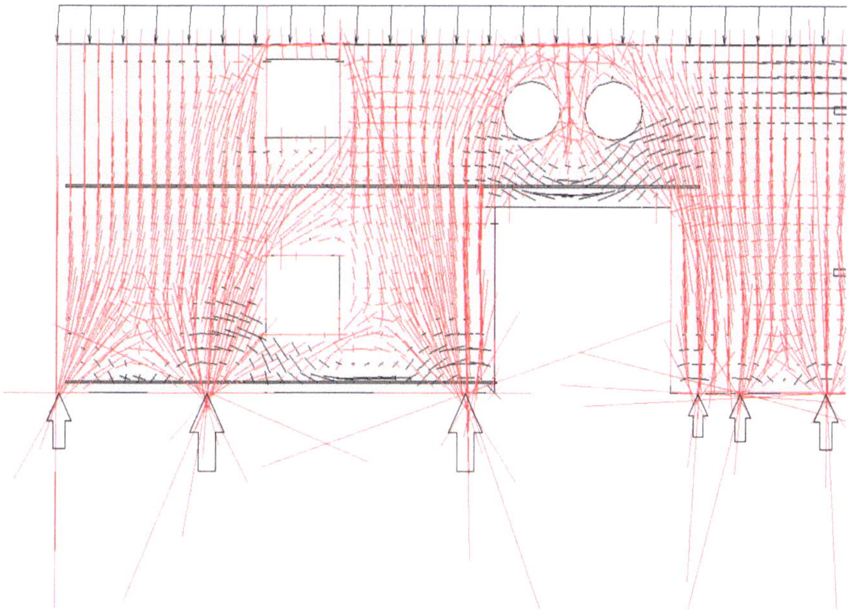

Fig. 3.64 Wall plate on point supports (cutout)

by, $f_i = 0$. But finite elements have no problems with point supports—at least not with the support reactions f_i.

The reason is simple: The nodal forces f_i in the printout are not genuine point forces, but work-equivalent substitutes for the domain and line loads which keep the plate fixed.

The plate in Fig. 3.10, which sits on two point supports and carries a point load is—in mathematical terms—an "ill-posed problem", because of the singularities. But with finite elements we are yet some distance away from the danger, that the material yields, and so it is only natural to treat the f_i like real forces.

3.39 Point Supports Are Hot Spots

When we keep a node fixed, the plate gets grounded: It is as if you touch with one hand a wire carrying a high voltage and with the other hand the ground. The rapid increase in the displacements from the fixed node out to the free nodes, $u = O(h^\alpha)$, $\alpha < 1$, see Fig. 6.1 a, produces large stresses in the elements, which are attached to the node, see Fig. 3.64, and the smaller the elements become, the larger the displacement gradient will be

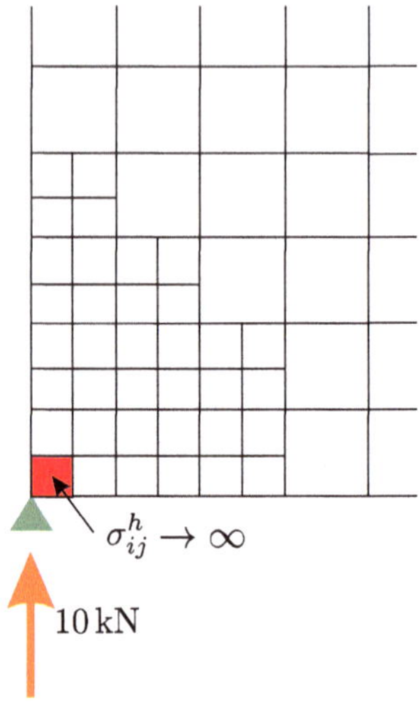

Fig. 3.65 The full support reaction must pass through this element

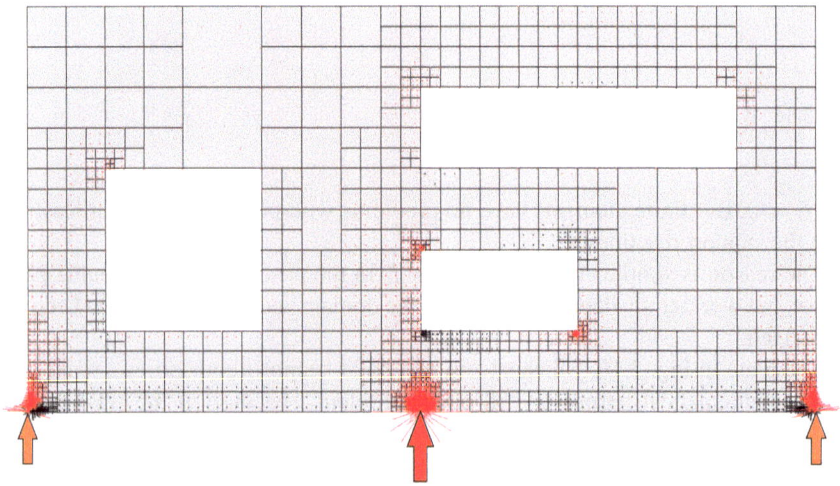

Fig. 3.66 Gravity load—the adaptive refinement near the point support leads in the limit, $h \to 0$, to infinitely sharp concentrated forces and the material would yield under the pressure

$$\sigma = E \cdot \varepsilon = E \cdot \lim_{h \to 0} \frac{u(h) - u(0)}{h} = \infty \,. \tag{3.253}$$

Why the stresses become infinitely large—why they must—will be understood when we study an FE-solution near such a support.

Let us assume the support reaction has a value of $f_i = 10$ kNm (an equivalent nodal force), which means if we push the support up by one meter, the work done by the support reaction will be $\delta W_e = 10$ kNm, see Fig. 3.65.

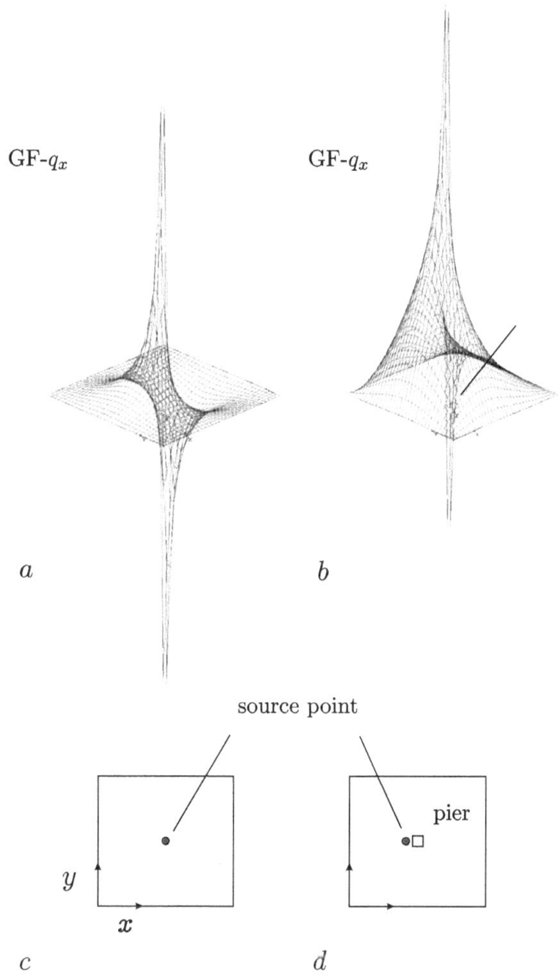

Fig. 3.67 Influence function for the shear force q_x, **a** the spread can form freely, **b** the rigid pier hinders the spread, the nodal forces further off have an advantage and they so manage to push up the plate—also in regions which lie farther off

The displacement of the node is only sensed by the element Ω_e to which the node belongs, and so the element must react to the displacement by generating an equal amount of virtual internal energy δW_i

$$\delta W_e = 1 \cdot f_i = \int_{\Omega_e} \sigma_{kj}\, \delta\varepsilon_{kj}\, d\Omega = \delta W_i\,, \tag{3.254}$$

where the strains $\delta\varepsilon_{kj}$ are the strains produced by the displacement $\delta u_i = 1$ of the node. Note, the other elements sense nothing since all other nodes are kept fixed.

This last element has to muster all the energy necessary, to balance the external work δW_e of the node!

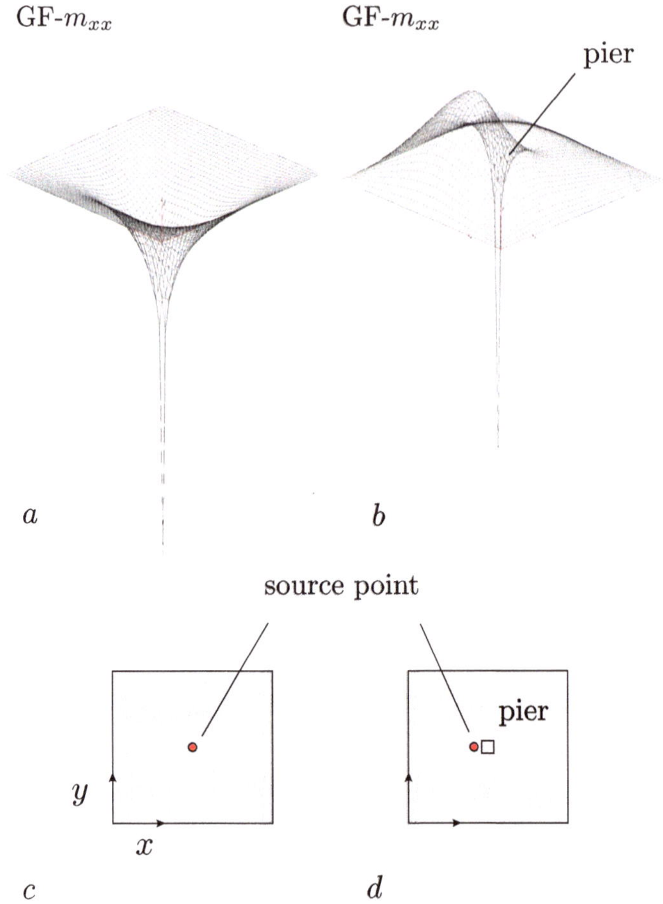

Fig. 3.68 The same plate as in the previous figure; the influence function for the bending moment m_{xx} exhibits a similar behavior

Hence if the element shrinks, because we want to have accurate results, the stresses within the element must increase, since less and less expanse must supply the energy δW_i to balance the external energy $\delta W_e = f_i \cdot 1$, and so we have no chance to produce any reasonable results near the node, see Fig. 3.66.

3.40 The Amputated Dipole

On switching from 1-D to 2-D jumps become poles, turn into singularities like $1/r$, since to extract from a 2-D function via *Betti's theorem* a point value, one needs a point source of **infinite strength**, since only a $1/r$ singularity can counterbalance

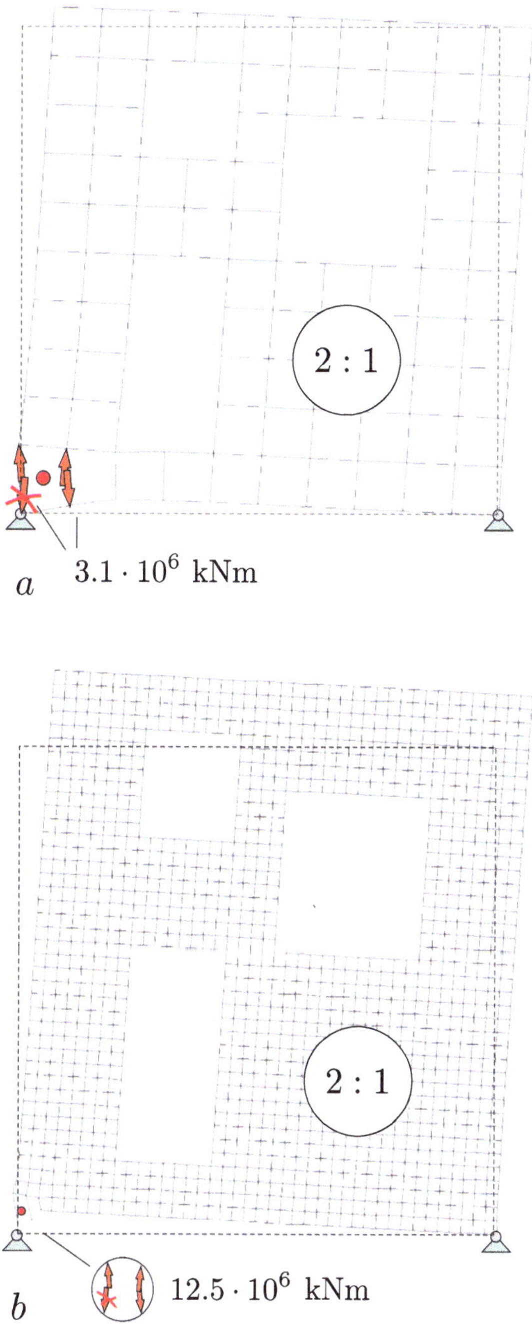

Fig. 3.69 Influence function for σ_{yy}, **a** the size of the nodal forces which generate the unit-spread depends on the mesh width h, **b** the smaller the elements are the larger the nodal forces will be and therewith the displacements. The ratio between the forces is 2 : 1, since the fixed support neutralizes one force ("amputated dipole")

the shrinking radius r of the circle, which surrounds the source point. But these point sources need a certain run-up to develop their strength, and this can become a problem at non-smooth points.

The influence function for the shear force q_x in a Kirchhoff plate, see Fig. 3.67, is generated by a dipole.[11] If the dislocation can form unhindered, the effects of the up- and down-displacements will balance in the distance. But if the source point comes—on one side (!)—close to a point support, the balance gets skewed[12] and this disturbance lets the deflection "explode" also in the distance, see Fig. 3.67b. Practically the same effect can be observed in Fig. 3.68.

The nature of this problem is better understood, when we look at the influence functions for the stresses σ_{yy} in a plate, since these too are generated by such dislocations, see Fig. 3.69.

In the FE-model the joint effort of four nodal forces[13] generates (approximately) the influence function for σ_{yy} at the center of an element.

If the source point lies on an element with a fixed node, the ratio between the upward and downward forces is 2:1, which means ***two*** nodal forces push the element up, but only ***one*** nodal force pushes it down, because the nodal force at the support effects nothing. The net effect of this imbalance is that the upper edge of the plate is pushed "skywards", towards the point ∞. If, instead, the source point lies in any of the neighboring elements, see Fig. 3.70, where no node is fixed, all ***four = two + two*** forces act simultaneously and so they keep a balance, the shear deformation is finite outside of the immediate neighborhood of the source point.

Note, when you add the force at the fixed node to the count then it is four forces again, but it is not 2+2 as in the case of a true dipole. Numerically the two forces up and the one force down at the free nodes are the same, but not the force at the fixed node.

Where the singularity comes from, is best seen when we imagine the element to be a small disk with radius R, since polar coordinates simplify the mathematics. The strains caused by the virtual displacement of the node behave as

$$\delta\varepsilon_{ij} = O(\frac{1}{r}) \tag{3.255}$$

and therefore we have

$$\int_{\Omega_e} \sigma_{ij}\, \delta\varepsilon_{ij}\, d\Omega \sim \int_0^{2\pi} \int_0^R \sigma_{ij}\, \frac{1}{r}\, r\, dr\, d\varphi = \int_0^{2\pi} \sigma_{ij}\, \frac{1}{2}\, R\, d\varphi\,, \tag{3.256}$$

which means σ_{ij} must shrink like $1/r$ for a force f_i to appear in the limit, $R \to 0$,

---

[11] Actually it is an octopole, similar to Fig. 9.11, which generates the dislocation but for our purposes we may consider it similar to a dipole, see page 448.

[12] As in the case of the leaning tower of Pisa: Soft soil in itself is not dangerous but soft on one side, and stiff on the other produces problems, see Sect. 2.16.

[13] like Doctor Dolittle's "pushmi-pullyu".

$$\lim_{R \to 0} \int_{\Omega_e} \sigma_{ij} \, \delta\varepsilon_{ij} \, d\Omega = f_i \,. \tag{3.257}$$

Remark 3.11 **Line supports** in 3-D have the same problems as **point supports** in 2-D, since the high stresses in a line support (=infinitely thin wire) will make the material flow. Luckily, our FE-models are a far distance away from such subtleties of the theory of elasticity.

3.41 A Dipole at the Edge

Dipoles are also amputated at edges. The influence function for the transverse force $q_x = H\,w_{,x}$ in a membrane equals a dipole since it is a dislocation (in vertical direction) of the source point, see Fig. 3.71.

If the source point lies inside the membrane as in Fig. 3.71, the two peaks + and − can form unhindered, and integration will make them disappear. However, if the source point nears the 270° corner, see Fig. 3.72, the influence function has only one peak and the singularity gets through.

But what about a smooth edge, where the influence function for the support reaction $H\,w_{,x}$ (at a vertical edge) also is an amputated dipole, but the superposition with the load p exhibits no singularity?

We think the form of the edge and the support conditions are crucial. At a smooth point on the boundary the angle is 180° and the fixed edge, see Fig. 3.73, hinders the formation of a supercritical singularity, while at the 270° corner the singularity becomes supercritical (infinite volume), even though the membrane is fixed on both sides of the corner point.

Perhaps we can explain it by counting on our fingers: At a smooth edge, two nodes are free to move, and two nodes are fixed, see Fig. 3.74a. This ratio, 2:2, seems to tie the effect of the two+two nodal forces $\pm f_i \to \infty$ back, the integral is bounded, while at the 270° corner the ratio is 3:1, see Fig. 3.74b, and the three f_i at the free nodes have free reign, their "imbalance" lets the field explode

$$v_x = \int_{\Omega} G_1^h(\mathbf{y}, \mathbf{x})\, p(\mathbf{y})\, d\Omega_{\mathbf{y}} = \begin{cases} < \infty & \text{at edge} \\ \infty & \text{at corner}\,. \end{cases} \tag{3.258}$$

3.42 Single Force at a Node

The FE-program translates a single force P directly into an equivalent nodal force $f_i = P \cdot 1$ Nm. Initially such a force only signals the presence of loads in the vicinity of the node, which perform the work $P \cdot 1$ Nm, when the node is displaced by one unit, see Fig. 3.75. But if the f_i at the neighboring nodes are zero, it must be really a lone single force.

Fig. 3.70 Influence function for σ_{yy} in two elements, which can move freely and so the actions of the up and down forces balance, the displacements in the distance remain bounded, also in the limit, $h \to 0$, ("true dipole")

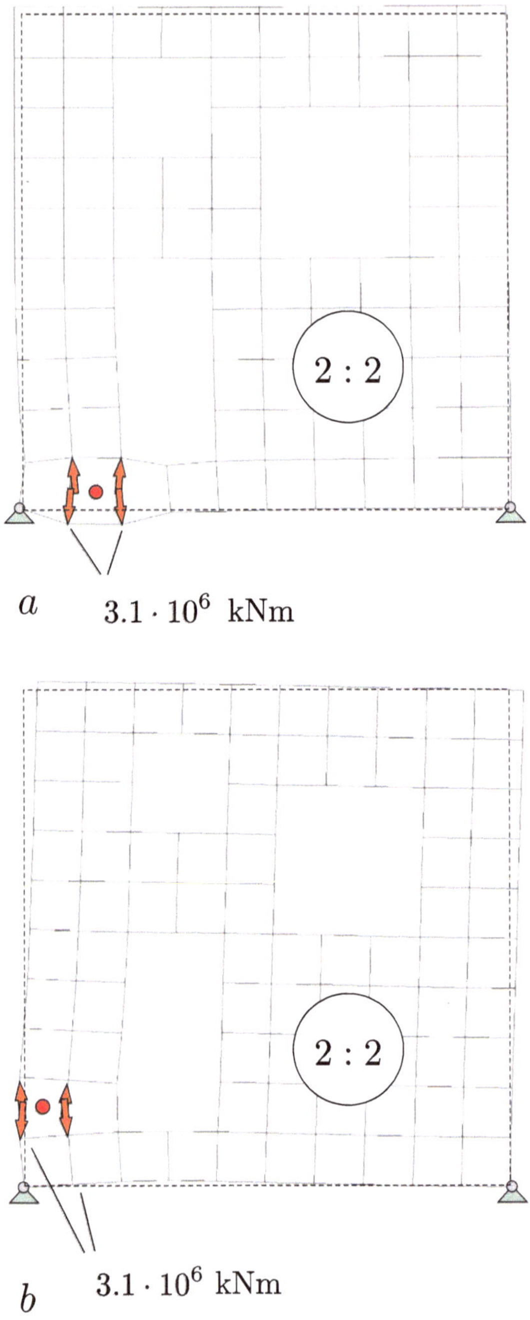

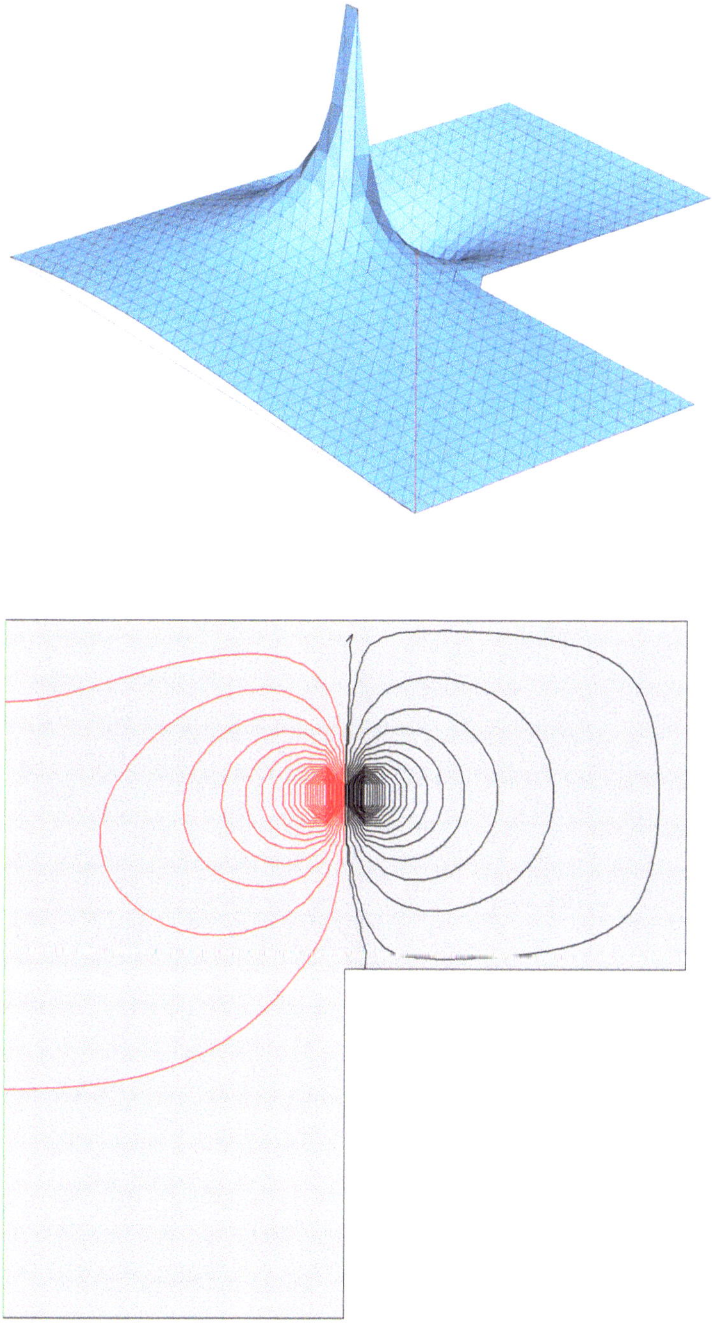

Fig. 3.71 Influence surface of the shear force $v_x = w,_x$ in a membrane (with a free edge on the left), 3-D plot and contour plot of the function

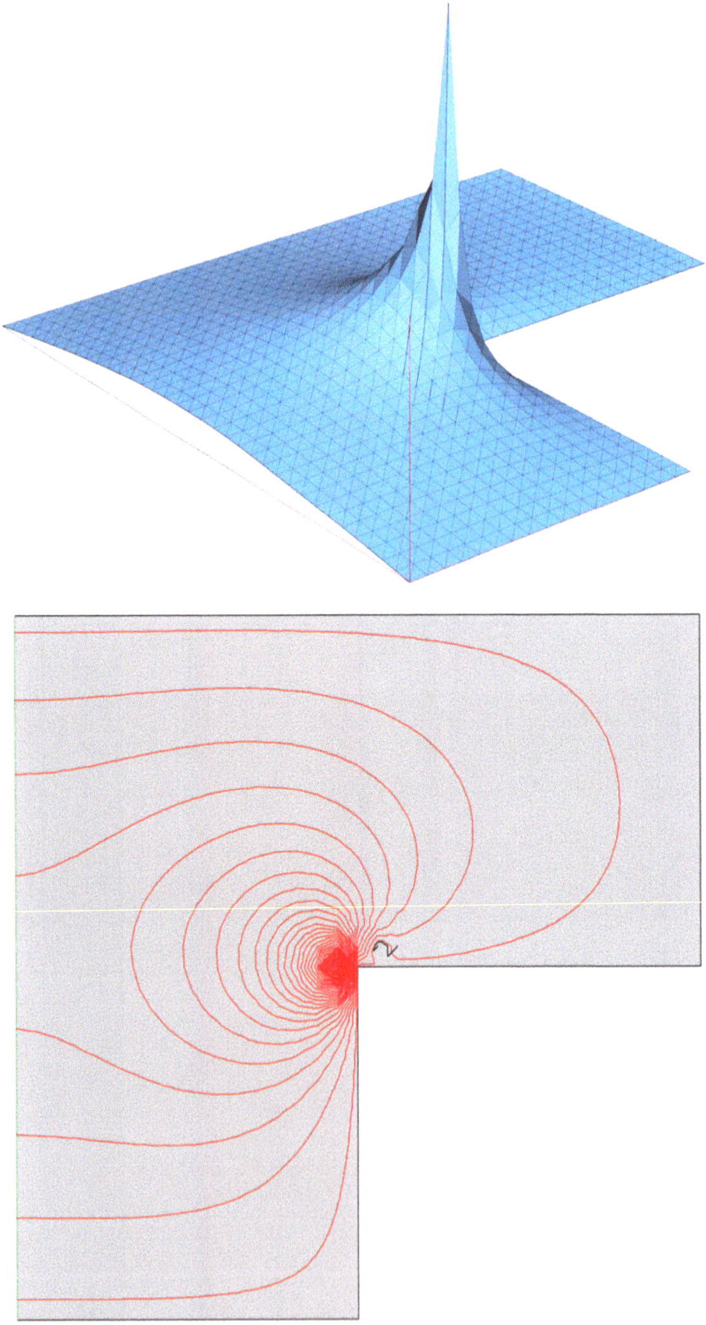

Fig. 3.72 The same influence function when the source point lies at a corner, and so the singularity dominates since the opposite peak is missing

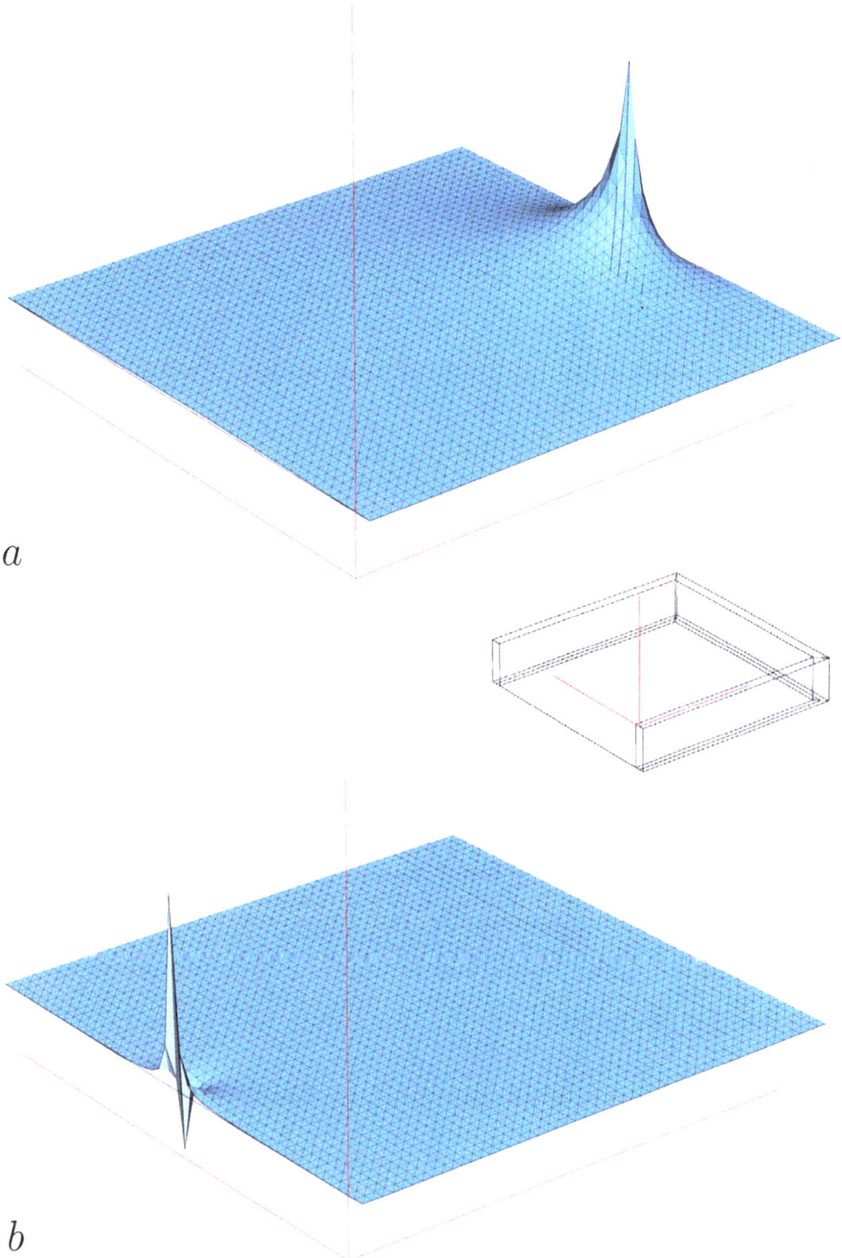

Fig. 3.73 Membrane, 5 m × 5 m, and influence function for $w_{,x}$, **a** near a fixed edge, and **b** near a free edge, near = 5 cm from the edge

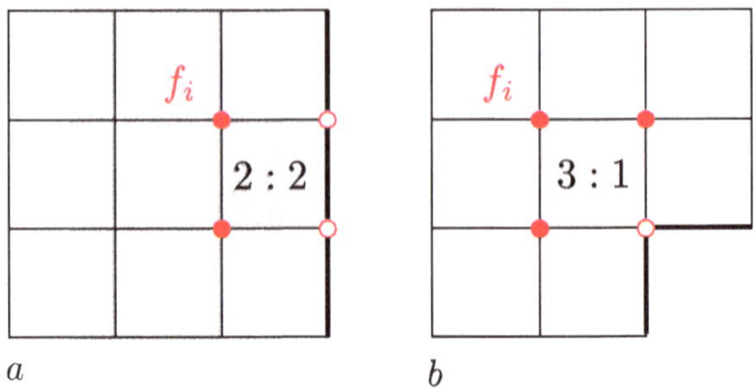

Fig. 3.74 Single element and nodal forces $f_i = O(1/h)$ of the Green's function, near **a** a smooth edge, and **b** a corner point

When the elements shrink, this is a signal for the program to let the stresses tend to infinity, since otherwise the balance

$$\int_\Omega \sigma_{ij}\, \delta\varepsilon_{ij}\, d\Omega = f_i \qquad (3.259)$$

cannot be maintained. The strains $\delta\varepsilon_{ij}$ come from the unit displacement of the node in the direction of the force (all other nodes are kept fixed) and because only the elements attached to the node will contribute to (3.259) ever larger and larger stresses σ_{ij} and strains $\delta\varepsilon_{ij}$ acting on less and less expanse must balance the work of the nodal force. That *both* tend to ∞ has been proved above. This too is a situation where singular stresses develop, only here it is a protective measure on the side of the FE-program to cope with the exigencies of a point load.

This has the peculiar side effect that the size h of the elements directly determines **the singularity** under a single force. Surface forces or line forces instead, i.e. everything, which has an *extension*, are not susceptible to this effect, since the f_i become smaller if h becomes smaller. But here f_i retains its size and so $h \to 0$ triggers the reaction $\sigma \to \infty$.

3.43 Predeformations

We want to briefly mention how predeformations in frames can be handled with the FEM.

"In the presence of predeformation, the equilibrium position corresponds to the total deformed *system, where total deformation = predeformation + load deformation"*. We write this as

$$w(x) = s(x) + w_L(x) \qquad s \text{ as in "sinuous line"} \qquad (3.260)$$

The deflection must satisfy the differential equation

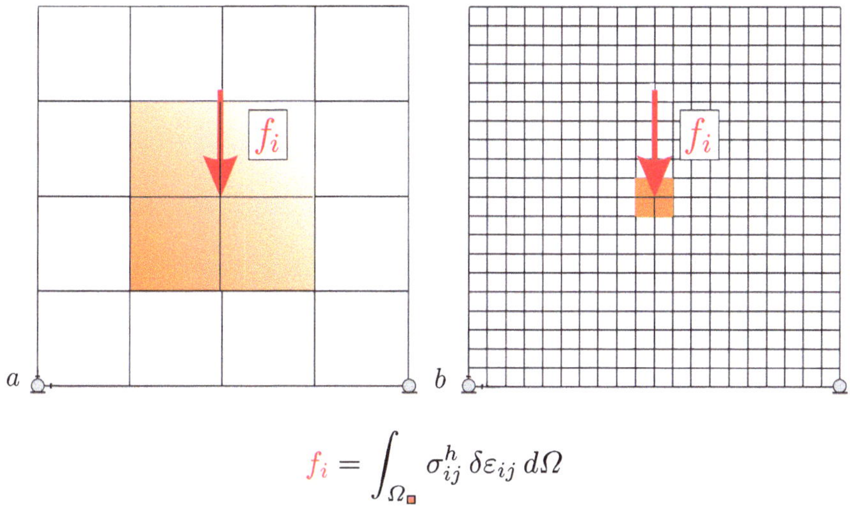

Fig. 3.75 The more the four elements shrink, the more the stresses must grow to produce the same equivalent nodal force f_i from the four elements

$$EI\, w^{IV}(x) + P\, w'''(x) = p(x) \qquad P = |P| \text{ as compressive force} \tag{3.261}$$

or

$$EI\, w_L^{IV}(x) + P\, w_L''(x) = p(x) - (EI\, s^{IV}(x) + P\, s''(x))\,. \tag{3.262}$$

The predeformation $s(x)$ gets replaced by the shape

$$s_I(x) = \sum_j s_j\, \varphi_j(x)\,, \tag{3.263}$$

where the s_j are the nodal values of $s(x)$ (Hermite interpolation). Because the shape functions are homogeneous solutions, $EI\, \varphi_i^{IV} = 0$, this reduces in each element to

$$EI\, w_L^{IV}(x) + P\, w_L''(x) = p(x) - P\, s_I''(x)\,. \tag{3.264}$$

To the equivalent nodal forces of the load must be added the equivalent nodal forces of the function $s_I(x)$. On a single element holds (integration by parts)

$$\mathscr{G}_e(s_I, \varphi_i) = \int_0^{l_e} -P\, s_I''\, \varphi_i\, dx + [P\, s_I'\, \varphi_i]_0^{l_e} - P \int_0^{l_e} s_I'\, \varphi_i'\, dx = 0\,, \tag{3.265}$$

and since the function s_I has continuous first derivatives at the nodes and because also the $\varphi_i(x)$ are continuous, the brackets are zero[14]

$$\sum_e \mathcal{G}_e(s_I, \varphi_i) = \int_0^l -P\, s_I''\, \varphi_i\, dx - P \int_0^l s_I'\, \varphi_i'\, dx = 0\,, \tag{3.266}$$

and with (3.263) this is just the result

$$f_{si} - \sum_j P \int_0^l \varphi_i'\, \varphi_j'\, dx\, s_j = 0\,. \tag{3.267}$$

If we let $\boldsymbol{K}_{geo}$ the ***geometrical stiffness matrix*** with entries (φ_i', φ_j') the vector of the equivalent nodal forces of the predeformations is

$$\boldsymbol{f}_s = P \cdot \boldsymbol{K}_{geo}\, \boldsymbol{s}\,. \tag{3.268}$$

The system to solve is $\boldsymbol{K}_{II}\boldsymbol{u} = \boldsymbol{f} + \boldsymbol{f}_s$, where $\boldsymbol{K}_{II}$ is the second-order stiffness matrix which normally gets replaced by the matrix $\boldsymbol{K}$ of the first-order theory and the geometric stiffness matrix $\boldsymbol{K}_{geo}$

$$\boldsymbol{K}_{II} \simeq \boldsymbol{K} + \boldsymbol{K}_{geo}\,. \tag{3.269}$$

3.44 The Limits of FE-Influence Functions

When we set up the system $\boldsymbol{K}\,\boldsymbol{g} = \boldsymbol{j}$ to calculate the influence function for the second derivative $u''(x)$ in a bar (linear elements) we soon realize that all j_i are zero

$$j_i = \varphi_i''(x) = 0\,, \tag{3.270}$$

and so, the FE-influence function must be zero

$$G_2^h(y, x) = 0\,. \tag{3.271}$$

It is always like this. FE-Influence functions can only be formulated for derivatives up to the order of the shape functions,

$$\text{bar} \quad \varphi_i(x) = \text{linear} \quad \rightarrow \quad \max u'(x) \tag{3.272}$$

$$\text{beam} \quad \varphi_i(x) = \text{cubic} \quad \rightarrow \quad \max u'''(x)\,. \tag{3.273}$$

This agrees with the ***h*- *permutation rule***, see page 264,

[14]at free ends remains a part $P\, s_I' \cdot \varphi_i$ which is to be added to the nodal force f_{si}.

$$J_h(u) = J(u_h)\,. \tag{3.274}$$

Let $J(u) = u''(x)$ and let u_h be a linear FE-solution. Since its second derivative is zero, $J(u_h) = 0$, the "adjoint" formulation $J_h(u) = (G_2^h, p) = 0$ must be zero as well—and this must be true for all functions u, whatever their right side $p = -EAu''$, and this is only possible, if $G_2^h(y, x)$ is identically zero.

To put this in a wider context: Imagine a series of linear elements (1-D). The j_i (of the influence functions) for the displacements $u(x)$ at the points $0 < x < l_e$ of an element vary linearly, $j_i = \varphi_i(x)$. The j_i for the derivatives $u'(x)$ are at each point x the same, $j_i = \varphi_i'(x) = \pm 1/l_e$, and the j_i for $u''(x)$ are zero, $j_i = \varphi_i'' = 0$—*linear*, *constant*, *zero*.

3.45 Checking Results

The simpler the influence function is, the more likely it is that an independent check will confirm the FE-result.

Support Reactions

The influence function for the sum of the vertical forces, *load* + *support reactions*, is identical to a unit shift $\delta w = \pm 1$ of the released structure (free body diagram). There is no simpler influence function than this lift. Only the sliding motion to the left or to the right, $\delta u = \pm 1$, which controls the sum of the horizontal forces, is also this simple. The next more complicated displacement would be a rotation of the released structure about a fixed point to check the sum of the moments about this point.

All these three influence functions lie in $\mathcal{V}_h^+$, since the shape functions form a partition of unity, no matter how coarse the mesh is, and so there should be no surprise here.

Normal Forces in Columns and Walls

Next comes the check of the support reactions in columns, piers, and walls. This too should not offer any surprises, since the influence functions are simple dents, which are easy to generate, even on coarse meshes, and therefore no large discrepancy should be found. Deviations rather indicate that loads have been forgotten.

Nodal Displacements

Of similar "simplicity" are the influence functions for the nodal displacements, which in turn means that an agreement with independent results does not mean much. In the early days of finite elements, the deflection at the center of a plate was compared with a series solution. At that time, the result was appreciated, but the evidential value of such a result is low—it is the most undemanding test imaginable, because the influence function is a simple dent.

Section Forces

If the check of the support reactions was successful, the section forces should be controlled next at “smooth” points, which means preferably in the middle of the spans, because the influence functions for moments at such points are relatively simple. Regarding vertical loads, gravity load and traffic loads, the moments are primarily determined by the load in the immediate vicinity, because the influence functions quickly subside. This is especially true for the transverse forces because the influence functions of these section forces live practically only in the vicinity of the source point.

With regard to wind loads, which means horizontal loads, though the situation is quite different, since these loads make their presence felt practically at all floors and in all columns.

Moments at Intermediate Supports

The moments at intermediate supports are usually uncritical, although the first discrepancies in the size of the moments will be observed since they end in peaks. Building codes allow to round off these moments, but does it make sense to round off an approximate value?

Really difficult are of course the moments at piers since these have steep flanks, and everything depends on the modeling of the support. In chapter 3.10 we have given hints on how to tackle the problem with the *equivalent stress transformation*.

Shear Forces

Shear forces in a plate are the least reliable quantities since the accuracy of section forces decreases with the order of the derivative. One can only argue “backwards”, from the supports back into the plate, because the support reactions should be relatively accurate and with this information one can examine the distribution of the transverse forces in front of the supports.

In any case it is pointless to argue about single values, the transverse forces are too erratic. It is also to ask, and not only with regard to shear forces, what “is real”—comes directly from the computer—and what is a result of clever post-processing, since often the “raw” data are replaced by smoothed values.

Singularities

At each corner of a plate you will find singularities, you only have to look (want to look) closely, see Figs. 6.19 and 6.20. Engineers rightly keep a healthy distance from these points because a constructive reinforcement is often sufficient to cover the stress peaks. The cracks, which appear when the reinforcement is missing, conversely, prove that the stresses really exist.

But there are also singularities, which are more serious, such as the singularity in the support reactions of a skew bridge, see Fig. 3.76.

3.46 Transient Problems

In transient problems like

$$
\begin{aligned}
&- EAu''(x,t) + m\,\ddot{u}(x,t) = p(x,t) \\
&u(0,t) = 0\,, N(l,t) = 0\,, u(x,0) = a(x)\,, \dot{u}(x,0) = b(x)
\end{aligned}
\tag{3.275}
$$

we let

$$
u_h(x,t) = \sum_i u_i(t)\,\varphi_i(x)\,,
\tag{3.276}
$$

which leads to the semi-discrete problem

$$
\boldsymbol{M}\ddot{\boldsymbol{u}}(t) + \boldsymbol{K}\,\boldsymbol{u}(t) = \boldsymbol{f}(t)
\tag{3.277}
$$

and therewith to finite differences in time. It is this mixture of techniques, which hinders the transfer of adjoint methods to dynamic problems.

The integral theorems, which belong to initial value problems, are substantially different from the static case. *Betti's theorem* gets replaced by **Graffi's theorem** [6], and the L_2-scalar product becomes a **convolution**. Not the least problem is that initial value problems must be artificially converted to "boundary value problems" to apply Green's functions, see Sect. 9.14. Apart from ***Duhamel's integral*** dual formulations are rarely applied in dynamics.

3.47 The Intelligence of Functions

Each C^1 function over an interval $(0, l)$ has the **integral representation**

$$
u(x) = u(0) + \int_0^x u'(y)\cdot 1\,dy = \int_\Gamma \ldots + \int_\Omega \ldots
\tag{3.278}
$$

—this is just integration by parts—and every C^2 function over a domain Ω has the **integral representation**, see Sect. 9.17,

$$
u(\boldsymbol{x}) = \int_\Gamma [g(\boldsymbol{y},\boldsymbol{x})\,\frac{\partial u(\boldsymbol{y})}{\partial n} - \frac{\partial g(\boldsymbol{y},\boldsymbol{x})}{\partial n}\,u(\boldsymbol{y})]\,ds_y + \int_\Omega g(\boldsymbol{y},\boldsymbol{x})\,(-\Delta u(\boldsymbol{y}))\,d\Omega_y\,.
\tag{3.279}
$$

The kernel $g(\boldsymbol{y},\boldsymbol{x}) = -1/(2\,\pi)\ln|\boldsymbol{x}-\boldsymbol{y}|$ is the fundamental solution of the Poisson equation, $-\Delta g = \delta(\boldsymbol{y}-\boldsymbol{x})$. (The kernel in the first equation is the *Heaviside-function* $g = 1$ up to the point x).

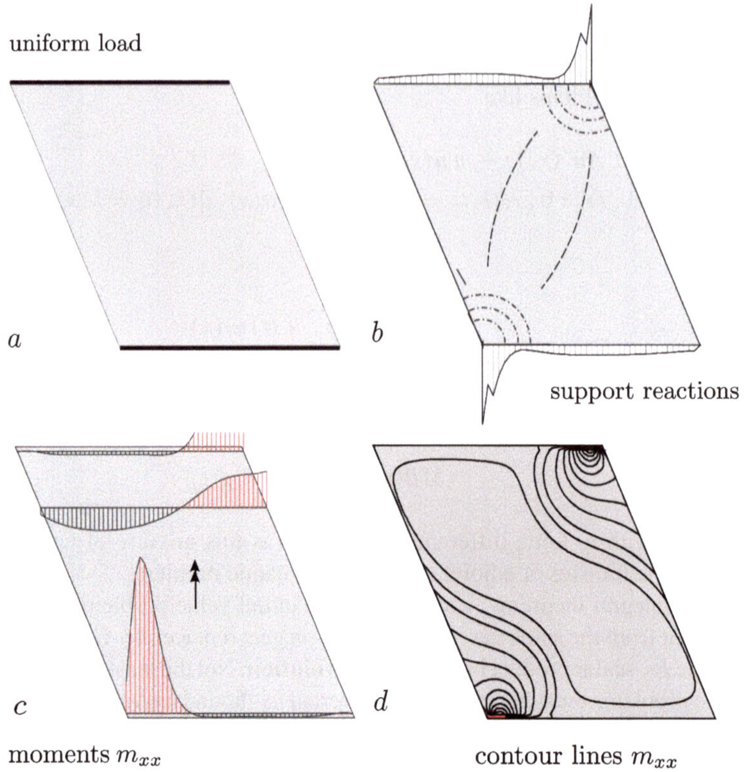

Fig. 3.76 At the obtuse-angled corners the support reactions and the bending moments m_{xx} are singular. In the sixties plaster models were used to determine the influence functions for such plates [5]

This means, a C^2 function in a domain Ω is uniquely determined by its boundary value u, its "trace", and the slope $\partial u/\partial n$ on the boundary Γ, and the "load" in the domain Ω, the sum of the second derivatives $\Delta u = u_{,11} + u_{,22}$, see Fig. 3.77.

We consider (3.279) to be the key to calculus: ***domain, boundary*** and ***function*** form a unit and the finite elements are the logical implementation of this idea

$$\text{Each function is an element} = \text{domain} + \text{edge} + \text{displacement}\,. \tag{3.280}$$

By associating a function with a differential equation, reading the function as a displacement, the function becomes an element. This is how the civil engineer *Clough* came to the concept of a finite element. Mathematicians or electrical engineers do not think in *shapes*, they could not have invented finite elements. They lack—forgive us the prejudice—the intuitive understanding of this context which characterizes a civil or mechanical engineer.

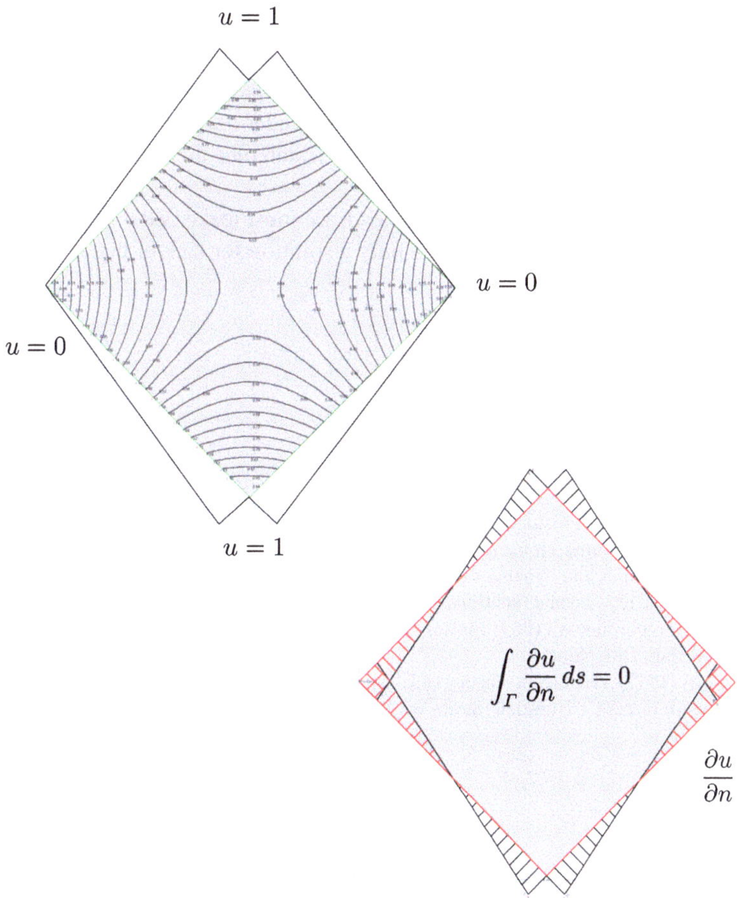

Fig. 3.77 Shape of a membrane under linear edge displacements. The solution, a saddle surface $-\Delta u = 0$, automatically adjusts itself so, that the integral of the normal derivative is zero—this is the "intelligence" built into the solution

It was not about replacing $\sin(x)$ and $\cos(x)$ with short hat functions—other authors also had this idea—but the element concept, which grows naturally out of (3.279), is the actual idea behind finite elements.

The accountant, who connects the dots on a paper with a ruler, does not see like the engineer, who recognizes a rope polygon in the straight lines, that nodal forces f_i act at the corners, and that the sum over all kinks $f_i = \tan\varphi_r(x_i) - \tan\varphi_l(x_i)$ is zero, since the equilibrium condition dictates this. And because of his static understanding, the engineer can convert the nodal forces f_i into line loads and try to put a smoother curve through the dots on the paper. Statics solves interpolation problems.

You gain a completely different perspective of plots, if you look at the curves and surfaces as solutions to differential equations, because you automatically have the whole apparatus, we are tempted to say, the whole intelligence at your disposal, which is built into a differential equation.[15]

The functions $\sin(k\,x)$ and $\cos(k\,x)$ are eigensolutions of the rope equation $-u'' = p$ and so, each Fourier series is a model of a rope under oscillating loads. Or think of the *splines*, the wooden rulers of shipbuilding, which owe their good nature directly to the beam equation $EI\,w^{IV}(x) = 0$.

It seems to us that if we want to get ahead, we must make better use of **the intelligence built into functions**, $h \to 0$ is only a solution for lack of better ideas.

References

1. Turner MJ, Clough RW, Martin HC, Topp LJ (1956) Stiffness and deflection analysis of complex structures. J Aeronautic Sci 23(9):805–823
2. Kurrer K-E (2018) The history of the theory of structures. Wiley Ernst & Sohn
3. Werkle H (2020) Finite elements in structural analysis - theoretical concepts and modeling procedures in statics and dynamics of structures. Springer, Cham
4. Hartmann F (2013) Green's functions and finite elements. Springer, Heidelberg
5. Rüsch H, Hergenröder A (1969) Einflussfelder der Momente schiefwinkliger Platten, 3rd edn. Werner-Verlag, Düsseldorf
6. Gurtin ME (1972) The Linear Theory of Elasticity, Handbuch der Physik Band VIa/2 Festkörpermechanik II, ed. C. Truesdell, Series ed. S. Flügge, Springer, Heidelberg

[15]What is contained in a differential equation can of course be learned—the reader might have guessed it—from Green's first's identity.

Chapter 4
Betti Extended

In the previous chapter we repeatedly made use of the fact that the FE-solution $u_h(x)$ is the superposition of the **approximate influence function** $G_h(y, x)$ and the load $p(y)$

$$u_h(x) = \int_0^l G_h(y, x)\, p(y)\, dy\,. \tag{4.1}$$

This result is based on a theorem which we call *Betti extended* .

Theorem 4.1 (*Betti extended*) *One may replace the exact solutions u_1 and u_2 in Betti's theorem*

$$W_{1,2} = \int_0^l p_1 \underset{\uparrow}{u_2}\, dx = \int_0^l p_2 \underset{\uparrow}{u_1}\, dx = W_{2,1} \tag{4.2}$$

with the FE-approximations u_{1h} and u_{2h}

$$W_{1,2}^h = \int_0^l p_1 \underset{\uparrow}{u_{2h}}\, dx = \int_0^l p_2 \underset{\uparrow}{u_{1h}}\, dx = W_{2,1}^h\,. \tag{4.3}$$

The claim is not that $W_{1,2}^h$ is the same as $W_{1,2}$, but rather: If $W_{1,2} = W_{2,1}$ is true, then $W_{1,2}^h = W_{2,1}^h$ is true as well; in short,

$$(p_1, u_2) = (p_2, u_1) \qquad \Rightarrow \qquad (p_1, u_{2h}) = (p_2, u_{1h})\,. \tag{4.4}$$

The best way to understand *Betti extended* is to focus on ***Maxwell's theorem***, see Fig. 4.1, which is just a particular application of *Betti's theorem*

$$P_1 \cdot w_2(x_1) = P_2 \cdot w_1(x_2)\,. \tag{4.5}$$

F. Hartmann and P. Jahn, *Statics and Influence Functions*, Springer Series in Solid and Structural Mechanics 13,
https://doi.org/10.1007/978-3-030-55889-5_4

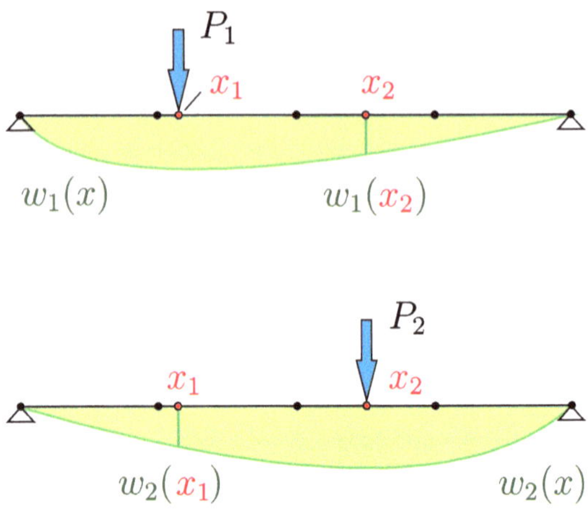

Fig. 4.1 *Maxwell's theorem*

If the two curves $w_1(x)$ and $w_2(x)$ are approximated with finite elements, the deflections at the two points x_1 and x_2 are not exact

$$w_{1h}(x_2) \neq w_1(x_2) \qquad w_{2h}(x_1) \neq w_2(x_1)\,, \tag{4.6}$$

but *Betti extended* guarantees that *Maxwell's theorem* holds true also in this situation

$$P_1 \cdot w_{2h}(x_1) = P_2 \cdot w_{1h}(x_2)\,. \tag{4.7}$$

This feature extends *Maxwell's theorem* to FE-solutions, set $P_1 = P_2 = 1$, which is not necessarily self-evident. That it had to be true for the nodal values, is a consequence of the symmetry of the stiffness matrices. ***Betti extended*** guarantees this also for all points in between.

4.1 Proof

The proof of ***Betti extended*** is based on the two equations

$$\int_\Omega p_{1h}\, u_{2h}\, d\Omega = \int_\Omega p_1\, u_{2h}\, d\Omega \tag{4.8a}$$

$$\int_\Omega p_{2h}\, u_{1h}\, d\Omega = \int_\Omega p_2\, u_{1h}\, d\Omega\,, \tag{4.8b}$$

and ***Betti's theorem*** itself

$$\mathscr{B}(u_{1h}, u_{2h}) = \int_\Omega p_{1h}\, u_{2h}\, d\Omega - \int_\Omega p_{2h}\, u_{1h}\, d\Omega = 0\,. \tag{4.9}$$

This gives

$$\int_\Omega p_1\, u_{2h}\, d\Omega = \int_\Omega p_{1h}\, u_{2h}\, d\Omega = \int_\Omega p_{2h}\, u_{1h}\, d\Omega = \int_\Omega p_2\, u_{1h}\, d\Omega\,, \tag{4.10}$$

or

$$\int_\Omega p_1\, u_{2h}\, d\Omega = \int_\Omega p_2\, u_{1h}\, d\Omega\,, \tag{4.11}$$

which is *Betti extended*.

At Eq. (4.8a) we arrive as follows: According to the *Galerkin-orthogonality* we have

$$\delta W_i = a(u_1 - u_{1h}, \varphi_i) = 0\,, \tag{4.12}$$

or, if we write it as external instead of internal virtual work, $\delta W_i = \delta W_e$,

$$\int_\Omega (p_1 - p_{1h})\, \varphi_i\, d\Omega = 0 \quad i = 1, 2, \ldots n \quad \Rightarrow \quad \int_\Omega (p_1 - p_{1h})\, u_{2h}\, d\Omega = 0\,, \tag{4.13}$$

since u_{2h} is a linear combination of the φ_i. In the same way we arrive at the second equation.

With *Betti extended* the proof of the central Eq. (4.1) is easy, since in the influence function for $u(x)$

$$W_{1,2} = 1 \cdot u(x) = \int_0^l \delta(y - x)\, u(y)\, dy = \int_0^l G(y, x)\, p(y)\, dy = W_{2,1}\,, \tag{4.14}$$

we may replace u and G with the FE-solutions u_h and G_h

$$W^h_{1,2} = \int_0^l \delta(y - x)\, \underset{\uparrow}{u_h}(y)\, dy = \int_0^l \underset{\uparrow}{G_h}(y, x)\, p(y)\, dy = W^h_{2,1}\,, \tag{4.15}$$

and so

$$u_h(x) = \int_0^l G_h(y, x)\, p(y)\, dy\,. \tag{4.16}$$

This **switch**, $u \to u_h$ and $G \to G_h$, can be applied to all linear functionals

$$J(u) = \int_0^l \delta(y - y)\, u(y)\, dy = \int_0^l G(y, x)\, p(y)\, dy\,, \tag{4.17}$$

resulting in

$$J(u_h) = \int_0^l \delta(y - y)\, u_h(y)\, dy = \int_0^l G_h(y, x)\, p(y)\, dy\,. \tag{4.18}$$

4.2 At Which Points Is the FE-Solution Exact?

With the help of *Betti extended* we can now also specify when and where FE-results are exact.

We study this question with a prestressed rope. The influence function $G(y, x)$ for the deflection $u(x)$ of the rope, see Fig. 4.2, at the point $x = 1.5$ is the response of the rope to a single force $P = 1$, a **Dirac delta** $\delta(y - x)$.

Since the FE-program cannot generate this shape, it instead places two half as large single forces at the two neighboring nodes. This is—in our notation—the load case $\delta_h(y, x)$ and the corresponding deflection $G_h(y, x)$ is the approximate influence function.

So, there are **two Dirac deltas**, the exact and the approximate

$$\delta(y - x) \quad \downarrow \qquad \delta_h(y - x) \quad \frac{1}{2} \downarrow + \frac{1}{2} \downarrow\,, \tag{4.19}$$

and also, **two influence functions**

$$G(y, x) \quad \text{(one peak)} \qquad G_h(y, x) \quad \text{(two peaks)}\,. \tag{4.20}$$

With finite elements we search for an approximate solution in the space $\mathcal{V}_h$, i.e. all the rope polygons, which can be generated with the three nodal shape functions $\varphi_i(x)$. Note, the dual of $\mathcal{V}_h$ is the space $\mathcal{V}_h^*$ of all load cases (nodal forces f_1, f_2, f_3), which create the rope polygons in $\mathcal{V}_h$.

Now, if a function u_h lies in $\mathcal{V}_h$ (is a rope polygon), the approximate Dirac delta (2 half-sized single forces) is as good as the exact Dirac delta (one single force)

$$u_h(x) = \int_0^l \delta(y - x)\, u_h(y)\, dy = \int_0^l \delta_h(y - x)\, u_h(y)\, dy\,. \tag{4.21}$$

In concrete terms this means

$$1 \cdot u_h(x) = \frac{1}{2} \cdot u_h(x_1) + \frac{1}{2} \cdot u_h(x_2)\,, \tag{4.22}$$

and this makes sense since the height of a straight line in between two nodes is just the average of the nodal values.

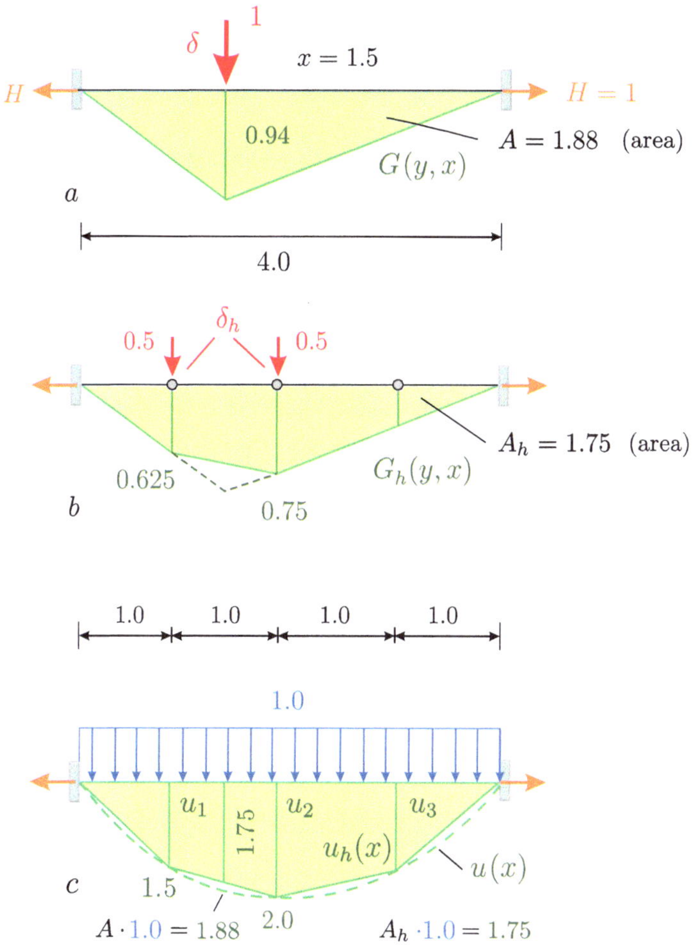

Fig. 4.2 Influence function for the deflection at the point $x = 1.5$, **a** exact influence function, **b** approximation, **c** FE-solution under uniform load $p = 1$

And because the **FE-load case** p_h lies in $\mathcal{V}_h^*$, i.e. consists of three nodal forces, the approximate influence function $G_h(y, x)$ is as good as the exact influence function

$$u_h(x) = \int_0^l G(y, x)\, p_h(y)\, dy = \int_0^l G_h(y, x)\, p_h(y)\, dy\,. \tag{4.23}$$

This too, is easy to understand. Since the load p_h consists of nodal forces f_i, the influence function is a sum over the nodes

$$u_h(x) = \int_0^l G_h(y, x)\, p_h(y)\, dy = \sum_{i=1}^{3} G_h(y_i, x)\, f_i\,. \tag{4.24}$$

But each of the three nodal influence functions is exact, $G_h(y_i, x) = G(y_i, x)$, and this explains why the sum in (4.24) is $u_h(x)$ at each point x.

On $\mathcal{V}_h$ and $\mathcal{V}_h^*$ the results obtained with $\delta_h(y, x)$ and $G_h(y, x)$ respectively are exact.

And this is true for **any** function in $\mathcal{V}_h$, since (4.21) applies to all $u_h \in \mathcal{V}_h$. And since p_h lies in $\mathcal{V}_h^*$ one can calculate any value $u_h(x)$ of the rope polygon with the approximate influence function $G_h(y, x)$. This is the essence of (4.24).

But that is not the end of it. The approximate Dirac delta, the two "half-sized" point loads at the neighboring nodes, constitute themselves a functional

$$J_h(u) = \int_0^l \delta_h(y - x)\, u(y)\, dy = \frac{1}{2}\,(u(x_1) + u(x_2))\,, \tag{4.25}$$

which can be applied to any function—not just the rope polygons in $\mathcal{V}_h$. Applied to $u(x) = \sin \pi\, x/4$ the result is

$$J_h(u) = \frac{1}{2}\,(\sin \frac{1.0\,\pi}{4} + \sin \frac{2.0\,\pi}{4}) = 0.85\,, \tag{4.26}$$

while $J(u) = \sin(1.5\,\pi/4) = 0.92$. So, there is a difference between J and J_h.

However, regarding the exact solution $u(x)$ and its FE-approximation $u_h(x)$, the following ***h-permutation rule*** applies

$$\boxed{J_h(u) = J(u_h)} \tag{4.27}$$

which can easily be verified, since

$$J_h(u) = \frac{1}{2}\,(u(1.0) + u(2.0)) = \frac{1}{2}\,(1.5 + 2.0) = 1.75 \tag{4.28}$$

$$J(u_h) = u_h(1.5) = 1.75\,. \tag{4.29}$$

The functional $J_h(u)$ measures u at the two points $x = 1.0$ and $x = 2.0$, while the functional $J(u_h)$ measures u_h only at the source point $x = 1.5$. But both measurements produce the same result!

The *h-permutation rule* is based on the fact that an FE-solution can be written in six separate ways

$$\begin{aligned} u_h(x) &= \int_0^l G(y,x)\, p_h(y)\, dy = \int_0^l G_h(y,x)\, p(y)\, dy \\ &= \int_0^l G_h(y,x)\, p_h(y)\, dy \\ &= \int_0^l \delta(y,x)\, u_h(y)\, dy = \int_0^l \delta_h(y,x)\, u_h(y)\, dy \\ &= \int_0^l \delta_h(y,x)\, u(y)\, dy\,, \end{aligned} \tag{4.30}$$

and if we also count the three formulas

$$u_h(x) = a(G, u_h) = a(G_h, u_h) = a(G_h, u)\,, \tag{4.31}$$

which are **variants of Mohr's integral**, then there are even nine.

The first two equations

$$J(u_h) = \int_0^l G(y,x)\, p_h(y)\, dy = \int_0^l G_h(y,x)\, p(y)\, dy = J_h(u) \tag{4.32}$$

formulate the *h-permutation rule*, which is of course applicable to any functional $J(u_h)$, not only the displacements $u(x)$.

Whether we superpose the exact influence function G with the FE-load p_h, or the approximate influence function G_h with the original load p, makes no difference—the result is the same.

Figure 4.3 illustrates this with a pier placed under a plate, which carries a heavy-duty truck. The exact influence for the pier reaction is plotted in Fig. 4.3a and the approximate function in Fig. 4.3b. The wheel loads of the truck represent the load p, and the block load is a graphical substitute for the FE-load p_h. We have only one formula for the exact pier reaction

$$R = \int_\Omega G(\mathbf{y}, \mathbf{x})\, p(\mathbf{y})\, d\Omega_y\,, \tag{4.33}$$

but **three ways** to calculate the approximate pier reaction R_h

$$R_h = \int_\Omega G(\mathbf{y}, \mathbf{x})\, p_h(\mathbf{y})\, d\Omega_y = \int_\Omega G_h(\mathbf{y}, \mathbf{x})\, p_h(\mathbf{y})\, d\Omega_y = \int_\Omega G_h(\mathbf{y}, \mathbf{x})\, p(\mathbf{y})\, d\Omega_y\,. \tag{4.34}$$

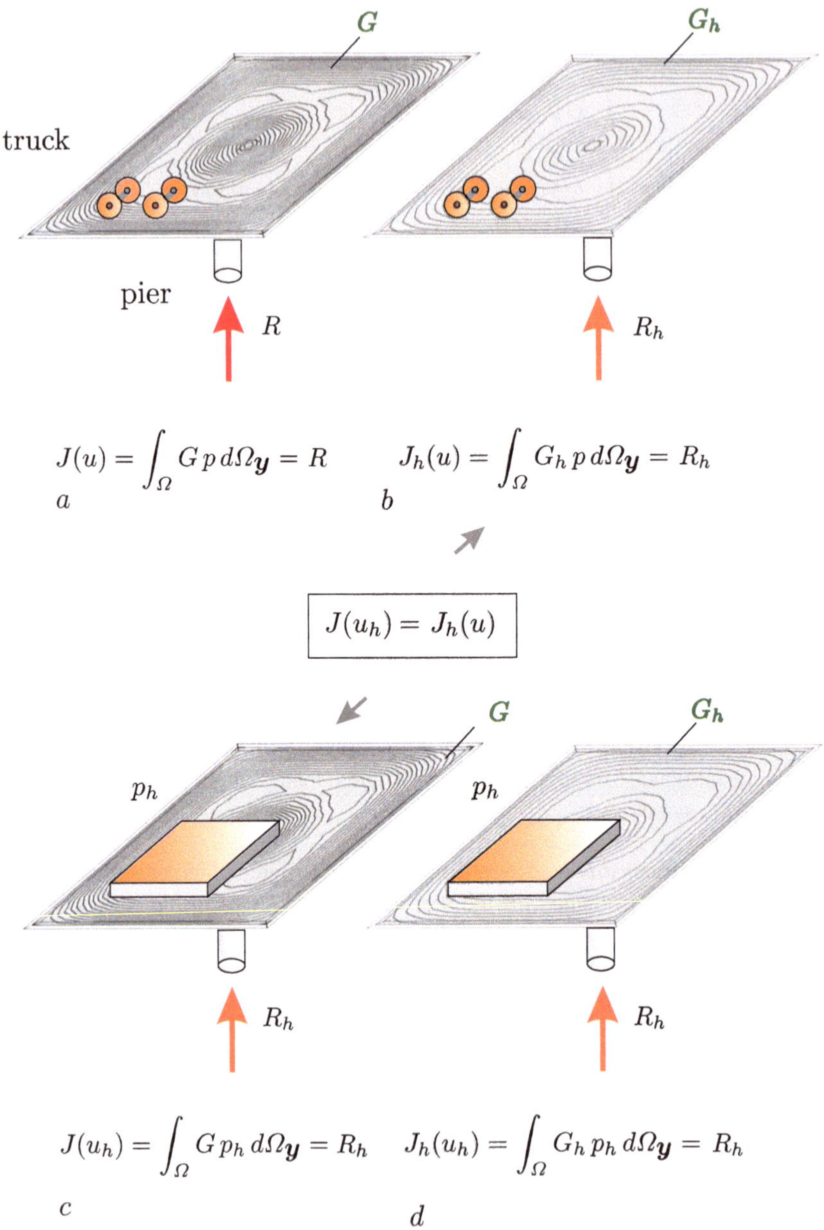

Fig. 4.3 Hinged plate with central pier. The four wheels of the truck, LC p, placed on the influence surface G_h gives the pier reaction R_h of the FE-solution. The same result is obtained, if the FE-load p_h (here pictured as a block load) is placed on the exact influence surface. And it is as well $R_h = (G_h, p) = (G_h, p_h)$, see Figure b and d

4.3 Exact Values

We can now also state, when the FE-solution is exact at a point.

Theorem 4.2 (*Exact values*)

Sufficient conditions

1. *If the influence function G of a functional J lies in $\mathcal{V}_h$, its FE-approximation G_h is identical with G and therefore*

$$J(u_h) = J_h(u) = J(u)\,, \tag{4.35}$$

or

$$J(u_h) = (G, p_h) = (G_h, p) = (G, p) = J(u)\,. \tag{4.36}$$

2. *If the exact solution lies in $\mathcal{V}_h$, $u = u_h$, the error in any influence function is orthogonal to the right side p of the solution*

$$J(u) - J(u_h) = \int_{\Omega} (G(\boldsymbol{y}, \boldsymbol{x})) - G_h(\boldsymbol{y}, \boldsymbol{x}))\, p(\boldsymbol{y})\, d\Omega_{\boldsymbol{y}} = 0\,. \tag{4.37}$$

Necessary condition

1. *If a value is exact, $J(u_h) = J(u)$, the error in the influence function must be orthogonal to the right side p*

$$J(u) - J(u_h) = \int_{\Omega} (G(\boldsymbol{y}, \boldsymbol{x})) - G_h(\boldsymbol{y}, \boldsymbol{x}))\, p(\boldsymbol{y})\, d\Omega_{\boldsymbol{y}} = 0\,. \tag{4.38}$$

4.4 One-Dimensional Problems

Since all influence functions are piecewise homogeneous solutions of the governing differential equation, exact nodal values, $u_h(x_i) = u(x_i)$, require the trial space $\mathcal{V}_h$ to contain these solutions.

In 1-D problems such as $-EA\,u'' = p_x$ and $EI\,w^{IV} = p_z$ this is true, since the homogeneous solutions

$$u_h(x) = c_1 + c_2\,x \tag{4.39a}$$

$$w_h(x) = c_1 + c_2\,x + c_3\,x^2 + c_4\,x^3 \tag{4.39b}$$

lie in $\mathcal{V}_h$, see Fig. 4.2, but if a bar must work against some friction (c)

$$-EA\,u''(x) + c\,u(x) = p_x\,, \tag{4.40}$$

the homogeneous solution is

$$u(x) = c_1\,e^{\alpha x} + c_2\,e^{-\alpha x} \qquad \alpha = \sqrt{\frac{c}{EA}}\,. \tag{4.41}$$

And to quote further examples: To a beam on an elastic foundation (c)

$$EI\,w^{IV} + c\,w(x) = p_z\,, \tag{4.42}$$

belongs the homogeneous solution

$$w(x) = e^{\beta x}(c_1 \cos\beta x + c_2 \sin\beta x) + e^{-\beta x}(c_3 \cos\beta x + c_4 \sin\beta x) \tag{4.43a}$$

$$\beta = \sqrt[4]{\frac{c}{EI}} \tag{4.43b}$$

and in a tensile chord bridge

$$EI\,w^{IV}(x) - H\,w''(x) = p(x) \qquad H = \text{prestress} \tag{4.44}$$

the homogeneous solution is

$$w_h(x) = c_1\,e^{\alpha x} + c_2\,e^{-\alpha x} + c_3 + c_4\,x \qquad \alpha = \sqrt{\frac{H}{EI}}\,. \tag{4.45}$$

Since the standard trial spaces $\mathcal{V}_h$ do not contain such "exotic" functions, the FE-solutions does not interpolate the exact solution at the nodes.

One also misses the exact value (in general) at the mid-nodes of quadratic shape functions, see Fig. 4.4, since the ***bubble function*** of the mid-node is **too smooth**. The influence function for the deflection has a jump in the first derivative—corresponding to the action of a point force—but the derivative of a bubble function is continuous and so the FE-solution of

$$-u''(x) = x \qquad u(0) = u(l) = 0\,, \tag{4.46}$$

does not agree with the exact solution at the mid-nodes, since the influence functions for the mid-nodes do not lie in $\mathcal{V}_h$. This problem would get unnoticed, if $u(x)$ would lie in $\mathcal{V}_h$, but here this is not the case.

An FE-program must therefore **find a balance** between the regularity required by the strain energy in order to be able to calculate the contributions $k_{ij} = a(\varphi_i, \varphi_j)$ of the stiffness matrix and the "non-regularity" needed to generate a jump in the first derivative.

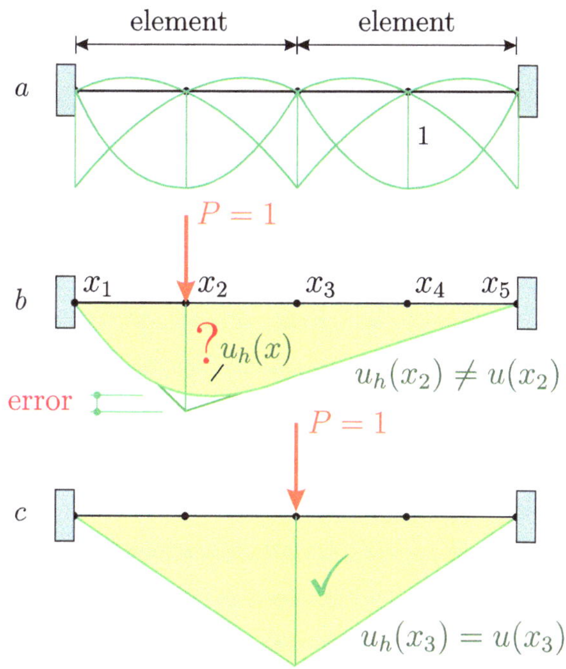

Fig. 4.4 Quadratic elements do not interpolate the exact solution at the midpoints, because the *bubble function* of a midpoint is too smooth, **a** shape functions, **b** point load at a midpoint and FE–solution, **c** point load at an end point and the FE-solution which is exact

4.5 Isogeometric Analysis

The *isogeometric analysis* uses the ***splines*** (*B-splines* and *nurbs*) of the CAD-model also as shape functions for the FE-solution [1, 2].

Here, we are only interested in one aspect of the method, namely that the shape functions no longer interpolate the FE-solution at the nodes, the familiar $u_i(x_j) = \delta_{ij}$ (Kronecker delta) is lost. This raises the question: Do our results still apply? Yes, everything is still correct. *Betti extended* applies and it remains true that the quality of the FE-results depends on the quality of the influence functions.

Trigonometric functions face the same problem and therefore we focus here on the deflection of a long rope, $l = 10$ m,

$$-w''(x) = 10 \qquad 0 \leq x \leq l \qquad w(0) = w(l) = 0\,, \tag{4.47}$$

whose shape we approximate with four sine-functions, see Fig. 4.5a,

$$\varphi_i(x) = \sin(\frac{i\,\pi\,x}{l}) \qquad i = 1, 2, 3, 4\,. \tag{4.48}$$

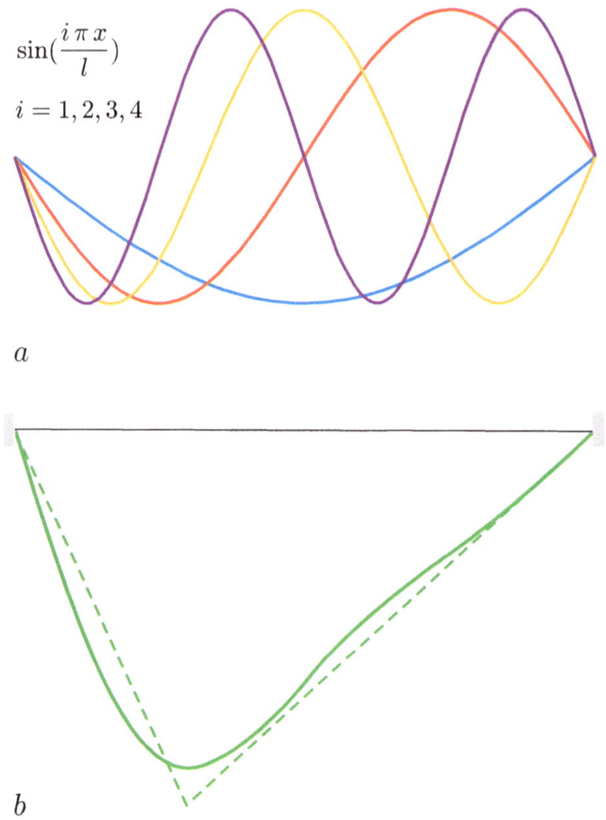

Fig. 4.5 **a** *Sine*-functions, **b** approximate and exact influence function for the rope deflection at the quarter point $x = 2.5$

The stiffness matrix

$$k_{ij} = \int_0^l \varphi_i' \, \varphi_j' \, dx = \begin{cases} \pi^2 \, i^2/20 & i = j \\ 0 & i \neq j \end{cases} \tag{4.49}$$

is a diagonal matrix and the f_i are the integrals of the $\varphi_i(x)$ times $10\,(= p)$. As expected, the solution $w_h(x)$ does not interpolate the exact solution (at which points?), but ***Betti extended*** still is valid

$$w_h(x) = \int_0^l G_h(y, x) \, p(y) \, dy \,, \tag{4.50}$$

if $G_h(y, x) = \sum_i g_i(x) \, \varphi_i(x)$ is the expansion of the influence function with respect to the sine-functions, see Fig. 4.5; it is $\boldsymbol{g} = \boldsymbol{K}^{-1} \boldsymbol{j}$ where $j_i = \varphi_i(x)$.

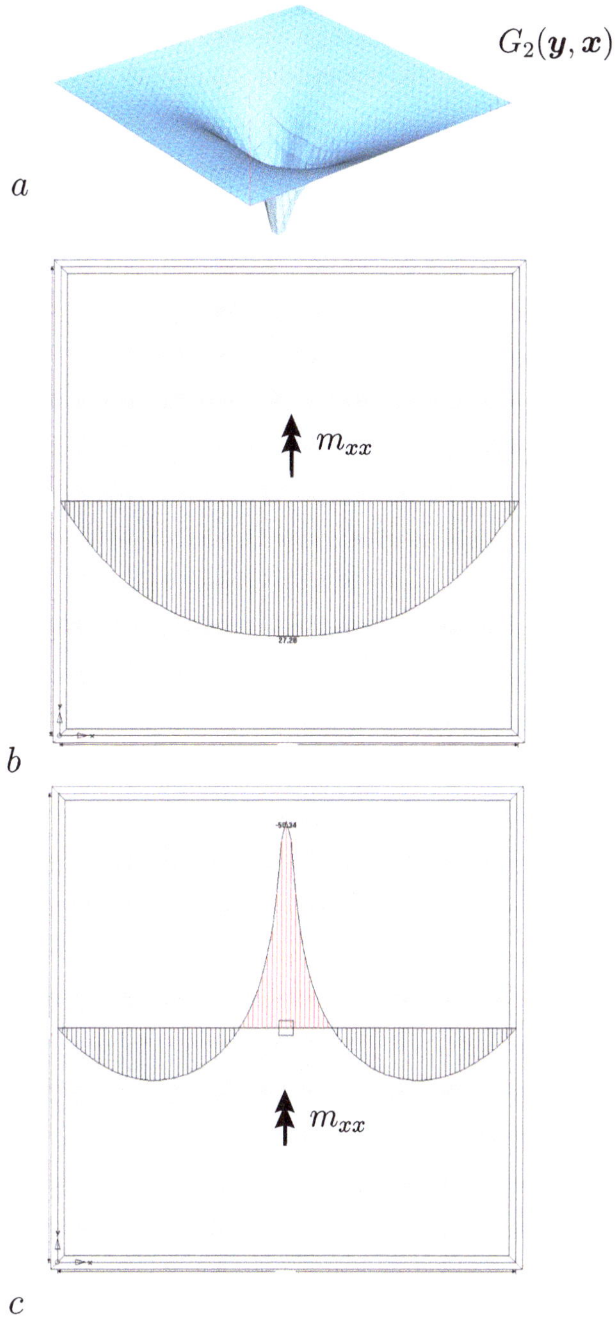

Fig. 4.6 Plate and gravity load, **a** influence function for m_{xx} at the center, **b** moment m_{xx} in a horizontal section, **c** moment distribution with pier

At the point $x = 2.5$, for example, the integral is exactly the value $w_h(x) = 94.6$ of the FE-solution, but the exact value is $w(x) = 93.75$. Beauty has its price. Smooth is not always better when it comes to approximating influence functions. Linear elements can do it exactly.

Remark 4.1 The four sine functions do not form a ***partition of unity*** of the rope, and so the sum of the two FE-support reactions $2 \cdot 45.032 \neq 100$ is not the applied load, and also the equivalent nodal forces

$$f_1 + f_2 + f_3 + f_4 = \frac{200}{\pi} + 0 + \frac{200}{3\pi} + 0 = 84.8826\,, \tag{4.51}$$

do not add up to $100 = p \cdot l$, since this would require the translation $w = 1$ to lie in $\mathcal{V}_h^+$, which is not the case.

4.6 Planar Problems

When the solution is exact at a point $\boldsymbol{x}$, this is normally a **coincidence** and a signal that at this particular point the error $G(\boldsymbol{y}, \boldsymbol{x}) - G_h(\boldsymbol{y}, \boldsymbol{x})$ in the influence function is orthogonal to the load p,

$$u(\boldsymbol{x}) - u_h(\boldsymbol{x}) = \int_\Omega (G(\boldsymbol{y}, \boldsymbol{x})) - G_h(\boldsymbol{y}, \boldsymbol{x}))\, p(\boldsymbol{y})\, d\Omega_{\boldsymbol{y}} = 0\,. \tag{4.52}$$

No other explanation is possible, because the exact influence functions of 2-D and 3-D problems do not lie in $\mathcal{V}_h$, since the shape functions are not homogeneous solutions of the governing equations.

But we also expect the spread of the load

$$\textit{uniform load} \qquad \textit{line load} \qquad \textit{point load} \tag{4.53}$$

to have an influence on the size of the error. The more evenly the load is distributed, the more likely it is for local errors in the influence functions to cancel each other out.

Figure 4.6 is a plot of the influence function for the moment m_{xx} at the midpoint of a quadratic plate. Given a uniformly distributed load p the moment

$$m_{xx}(\boldsymbol{x}) = \int_\Omega G_2(\boldsymbol{y}, \boldsymbol{x})\, p\, d\Omega_{\boldsymbol{y}} = p \cdot \int_\Omega G_2(\boldsymbol{y}, \boldsymbol{x})\, d\Omega_{\boldsymbol{y}} = p \cdot Vol(\boldsymbol{x}) \tag{4.54}$$

at any point $\boldsymbol{x}$ is proportional to the volume $Vol(\boldsymbol{x})$ of the influence function G_2, and apparently the FE-program can approximate the various values $Vol(\boldsymbol{x})$—we traverse

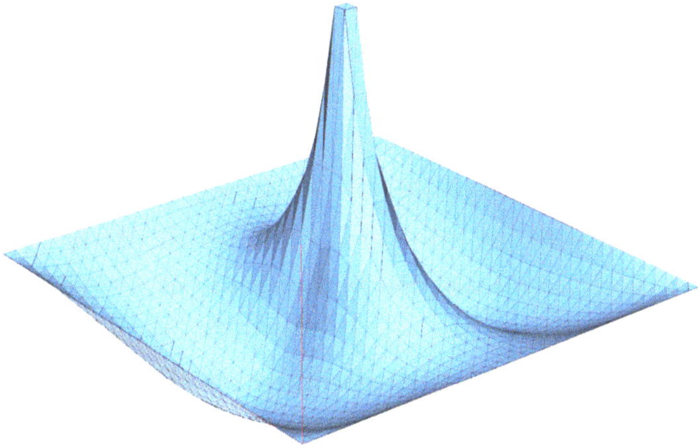

Fig. 4.7 3-D plot of the bending moment m_{xx} in a plate with a central pier

the plate from left to right—relatively well, since the resulting moment distribution in Fig. 4.6b is a smooth curve.

When we place the plate on a point support, the support reaction R gets added to the influence function ($|\Omega_P|$ is the cross section of the pier)

$$m_{xx}(\boldsymbol{x}) = \int_{\Omega} G_2(\boldsymbol{y}, \boldsymbol{x})\, p(\boldsymbol{y})\, d\Omega_{\boldsymbol{y}} + \frac{R}{|\Omega_P|} \int_{\Omega_P} G_2(\boldsymbol{y}, \boldsymbol{x})\, d\Omega_{\boldsymbol{y}}\,, \tag{4.55}$$

and due to the steep flanks in the influence function this second term will be difficult to evaluate when the source point $\boldsymbol{x}$ comes close to the support, see Fig. 4.7.

Strictly speaking does the program calculate with an equivalent load $p_h(\boldsymbol{y})$, which contains the support reaction "equivalently"

$$m^h_{xx}(\boldsymbol{x}) = \int_{\Omega} G^h_2(\boldsymbol{y}, \boldsymbol{x})\, p_h(\boldsymbol{y})\, d\Omega_{\boldsymbol{y}}\,, \tag{4.56}$$

but also this formula will be affected by the sharp rise of G_2 near the support.

4.7 Point Supports

There is a remarkable difference between plates and slabs regarding point supports. Support reactions are relatively accurate (even exact, if the support of a structure is statically determinate), but this does not help to set a limit to the nearly singular stresses in the vicinity of a point support. In the case of a plate, however, we can replace the support reaction by an equivalent uniform stress distribution.

But this is not applicable to the bending moments in a slab, since the moments and the support reaction are in no way related, the **offsets** e_x and e_y of the support reaction remain unknown.

4.8 If the Solution Lies in $\mathcal{V}_h$

If the solution lies in the trial space $\mathcal{V}_h$ of a bilinear mesh, because bilinear elements are good enough to model the deformations of a plate, then $\boldsymbol{p}_h = \boldsymbol{p}$, i.e. then the FE-load case is identical with the original load case.

But one question is still open. An FE-program calculates the stresses with **approximate influence functions**

$$\sigma_{xx}^h(\boldsymbol{x}) = \int_\Omega \boldsymbol{G}_h(\boldsymbol{y}, \boldsymbol{x}) \cdot \boldsymbol{p}(\boldsymbol{y})\, d\Omega_{\boldsymbol{y}}\,, \tag{4.57}$$

and so we expect the FE-stresses to be only approximations, but why are they exact?

The reason is that the error in the influence functions is orthogonal to the load if the solution lies in $\mathcal{V}_h$,

$$\sigma_{xx}(\boldsymbol{x}) - \sigma_{xx}^h(\boldsymbol{x}) = \int_\Omega \underbrace{(\boldsymbol{G}(\boldsymbol{y}, \boldsymbol{x}) - \boldsymbol{G}_h(\boldsymbol{y}, \boldsymbol{x}))}_{error} \cdot \boldsymbol{p}(\boldsymbol{y})\, d\Omega_{\boldsymbol{y}} = 0\,, \tag{4.58}$$

since any such $\boldsymbol{p}$ "neutralizes" the error in the influence functions.

The error in an approximate influence function is orthogonal to all load cases p, which can be solved exactly on $\mathcal{V}_h$.

The plate in Fig. 4.8 gets stretched by horizontal forces at its right edge, and for this simple stress state bilinear elements are good enough, and so the exact solution lies in $\mathcal{V}_h$.

But the influence function for the horizontal stress σ_{xx}, no matter which point $\boldsymbol{x}$ we choose, does not lie in $\mathcal{V}_h$, since these influence functions are produced by dislocations, and such gaps cannot be modeled with bilinear elements.[1]

The only possible solution is: The **integral** of the (horizontal) edge displacement of the FE-influence function agrees with the exact value, otherwise σ_{xx}^h would not be σ_{xx}

$$\sigma_{xx}^h(\boldsymbol{x}) = \int_\Gamma \boldsymbol{G}_h(\boldsymbol{y}, \boldsymbol{x}) \cdot \boldsymbol{t}(\boldsymbol{y})\, ds_{\boldsymbol{y}} = \sigma_{xx}(\boldsymbol{x})\,. \tag{4.59}$$

Here Γ is the right edge which gets pulled and $\boldsymbol{t} = \{t, 0\}^T$ is the edge load.

[1] Only in 1-D is a gap a gap. In 2-D one sees no gap, one must circle the source point once and then one will experience a displacement shift in the direction of the dislocation.

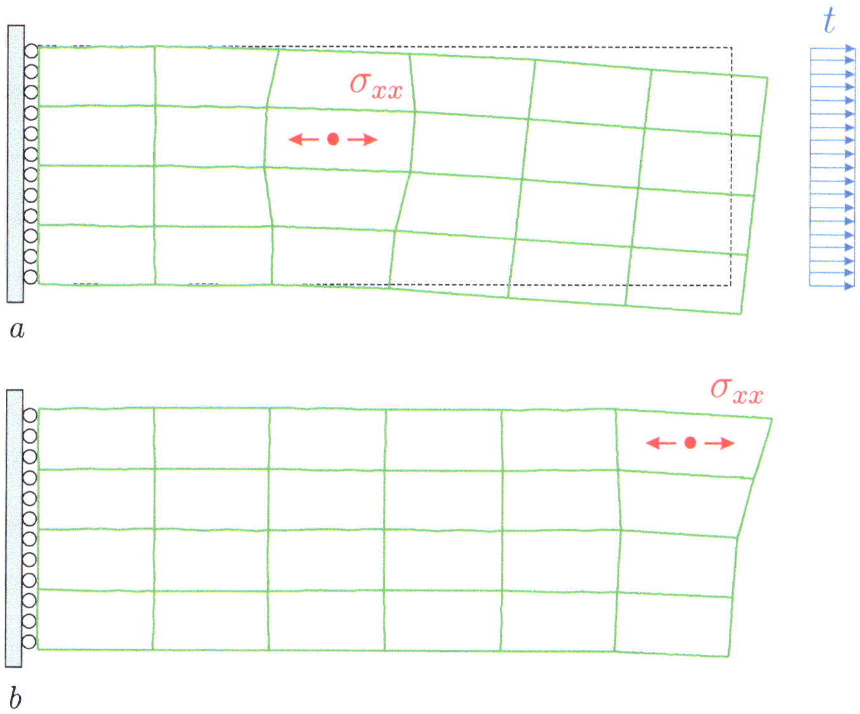

Fig. 4.8 FE-influence functions for σ_{xx}. These are crude approximations, but the integral value along the right edge is exact, is the exact σ_{xx}

Remark 4.2 The equation

$$\int_\Omega (\boldsymbol{G}(\boldsymbol{y},\boldsymbol{x}) - \boldsymbol{G}_h(\boldsymbol{y},\boldsymbol{x})) \bullet \boldsymbol{p}_h(\boldsymbol{y})\, d\Omega_y = 0 \tag{4.60}$$

is the **counterpart** to the ***Galerkin-orthogonality***, which states that the difference between the FE-load $\boldsymbol{p}_h$ and the exact load $\boldsymbol{p}$ is orthogonal to all shape functions $\boldsymbol{\varphi}_i$

$$\int_\Omega (\boldsymbol{p}(\boldsymbol{x}) - \boldsymbol{p}_h(\boldsymbol{x})) \bullet \boldsymbol{\varphi}_i(\boldsymbol{x})\, d\Omega = 0 \qquad \text{Galerkin}\,. \tag{4.61}$$

Galerkin tests with the shape functions, while the test with the influence functions (4.60) are the load cases $\boldsymbol{p}_h$, which can be solved exactly on $\mathcal{V}_h$. The error in the influence function must be orthogonal to each $\boldsymbol{p}_h(\boldsymbol{\varphi}_i) \in \mathcal{V}_h^*$

$$\int_\Omega (\boldsymbol{G}(\boldsymbol{y},\boldsymbol{x}) - \boldsymbol{G}_h(\boldsymbol{y},\boldsymbol{x})) \bullet \boldsymbol{p}_h(\boldsymbol{\varphi}_i)(\boldsymbol{y})\, d\Omega_y = 0\,, \tag{4.62}$$

where $\boldsymbol{p}_h(\boldsymbol{\varphi}_i)$ are the forces, which produce the shape $\boldsymbol{\varphi}_i$.

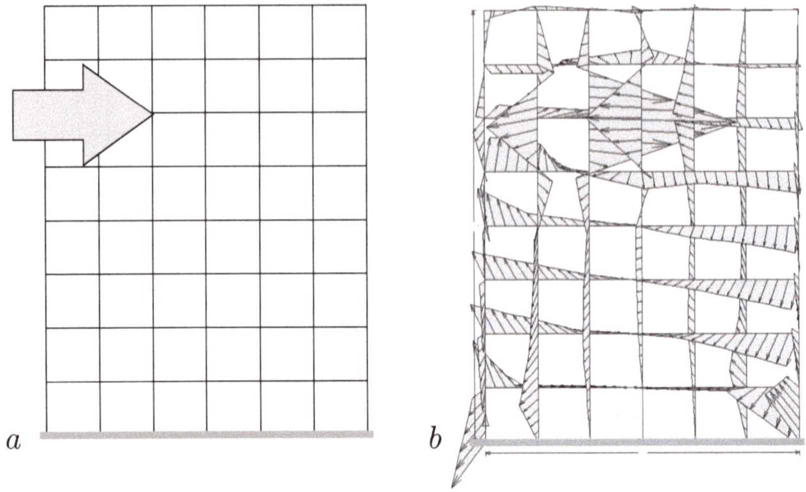

Fig. 4.9 Plate with exact Dirac delta $\delta(\boldsymbol{y}-\boldsymbol{x})$ and approximate Dirac delta $\delta_h(\boldsymbol{y},\boldsymbol{x})$, on $\mathcal{V}_h$ both provide the same value, $u_h(\boldsymbol{x}) = (u_h, \delta) = (u_h, \delta_h)$

The *Galerkin-orthogonality* holds also true for point loads, see Fig. 4.9, and therefore there is no difference on $\mathcal{V}_h$ between the exact and the approximate Dirac delta, they produce the same result

$$\varphi_{xi}(\boldsymbol{x}) = \int_\Omega \boldsymbol{\delta}(\boldsymbol{y}-\boldsymbol{x}) \cdot \boldsymbol{\varphi}_i(\boldsymbol{y})\, d\Omega_{\boldsymbol{y}} = \int_\Omega \boldsymbol{\delta}_h(\boldsymbol{y},\boldsymbol{x}) \cdot \boldsymbol{\varphi}_i(\boldsymbol{y})\, d\Omega_{\boldsymbol{y}}\,. \tag{4.63}$$

Here, the exact Dirac delta $\boldsymbol{\delta}(\boldsymbol{y}-\boldsymbol{x})$ is a unit point force in x-direction, $\boldsymbol{\delta}_h(\boldsymbol{y},\boldsymbol{x})$ is the FE-approximation and $\varphi_{xi}(\boldsymbol{x})$ is the horizontal displacement of the field $\boldsymbol{\varphi}_i(\boldsymbol{x}) \in \mathcal{V}_h$ at the source point $\boldsymbol{x}$.

4.9 Patch Test

In a ***patch test*** we construct a solution, which is simple enough to lie in $\mathcal{V}_h$ and we check whether we can recover the exact stresses. In the previous section we have explained why this test makes sense: The error in the influence functions is orthogonal to the load if the solution u lies in $\mathcal{V}_h$, i.e. if $p_h = p$, since, see (4.68),

$$\int_\Omega (G(\boldsymbol{y},\boldsymbol{x}) - G_h(\boldsymbol{y},\boldsymbol{x}))\, p(\boldsymbol{y})\, d\Omega_{\boldsymbol{y}} = \int_\Omega G(\boldsymbol{y},\boldsymbol{x})\, (p(\boldsymbol{y}) - p_h(\boldsymbol{y})) d\Omega_{\boldsymbol{y}} = 0\,. \tag{4.64}$$

Tests, which are bound to fail, are point forces in 2-D and 3-D. They overtax any mesh. No $\mathcal{V}_h$ contains the exact solution.

4.10 Adaptive Refinement

The error of an FE-solution cannot be determined directly, because neither the exact displacements nor the exact stresses are known. We can only measure the deviation between the original load $\boldsymbol{p}$ and the FE-load $\boldsymbol{p}_h$ and refine the mesh, see Fig. 4.10, where these differences are large.

Eventually it is even easier, see Fig. 4.11: One pays attention only to the **"kinks"** (the jumps in the stresses) and refines the mesh at these points.[2]

Since, as so often, only the algebra matters, we write the equations in scalar form and we study the bending surface $u(\boldsymbol{x})$ of a membrane under a pressure p.

Given the exact and the approximate solution

$$u(\boldsymbol{x}) = \int_\Omega G(\boldsymbol{y}, \boldsymbol{x}) p(\boldsymbol{y})\, d\Omega_{\boldsymbol{y}} \qquad u_h(\boldsymbol{x}) = \int_\Omega G_h(\boldsymbol{y}, \boldsymbol{x}) p(\boldsymbol{y})\, d\Omega_{\boldsymbol{y}}\,, \tag{4.65}$$

the error in the FE-solution is

$$u(\boldsymbol{x}) - u_h(\boldsymbol{x}) = \int_\Omega (G(\boldsymbol{y}, \boldsymbol{x}) - G_h(\boldsymbol{y}, \boldsymbol{x}))\, p(\boldsymbol{y})\, d\Omega_{\boldsymbol{y}}\,. \tag{4.66}$$

But the error can also be attributed to the difference between p and p_h

$$u(\boldsymbol{x}) - u_h(\boldsymbol{x}) = \int_\Omega G(\boldsymbol{y}, \boldsymbol{x})(p(\boldsymbol{y}) - p_h(\boldsymbol{y}))\, d\Omega_{\boldsymbol{y}}\,, \tag{4.67}$$

and on combining these two equations we obtain

$$u(\boldsymbol{x}) - u_h(\boldsymbol{x}) = \int_\Omega G(\boldsymbol{y}, \boldsymbol{x})(p(\boldsymbol{y}) - p_h(\boldsymbol{y}))\, d\Omega_{\boldsymbol{y}} \tag{4.68a}$$

$$= \int_\Omega (G(\boldsymbol{y}, \boldsymbol{x}) - G_h(\boldsymbol{y}, \boldsymbol{x}))\, p(\boldsymbol{y})\, d\Omega_{\boldsymbol{y}} \tag{4.68b}$$

$$= \int_\Omega (G(\boldsymbol{y}, \boldsymbol{x}) - G_h(\boldsymbol{y}, \boldsymbol{x}))\, (p(\boldsymbol{y}) - p_h(\boldsymbol{y}))\, d\Omega_{\boldsymbol{y}}\,. \tag{4.68c}$$

The step from the second equation to the third equation is made possible by the *Galerkin orthogonality*

[2]For even order shape functions, $p = 2, 4, \ldots$, the residual error r dominates while for odd order, $p = 1, 3, \ldots$, the kinks dominate (*"dichotomy"*), see [3], p. 424.

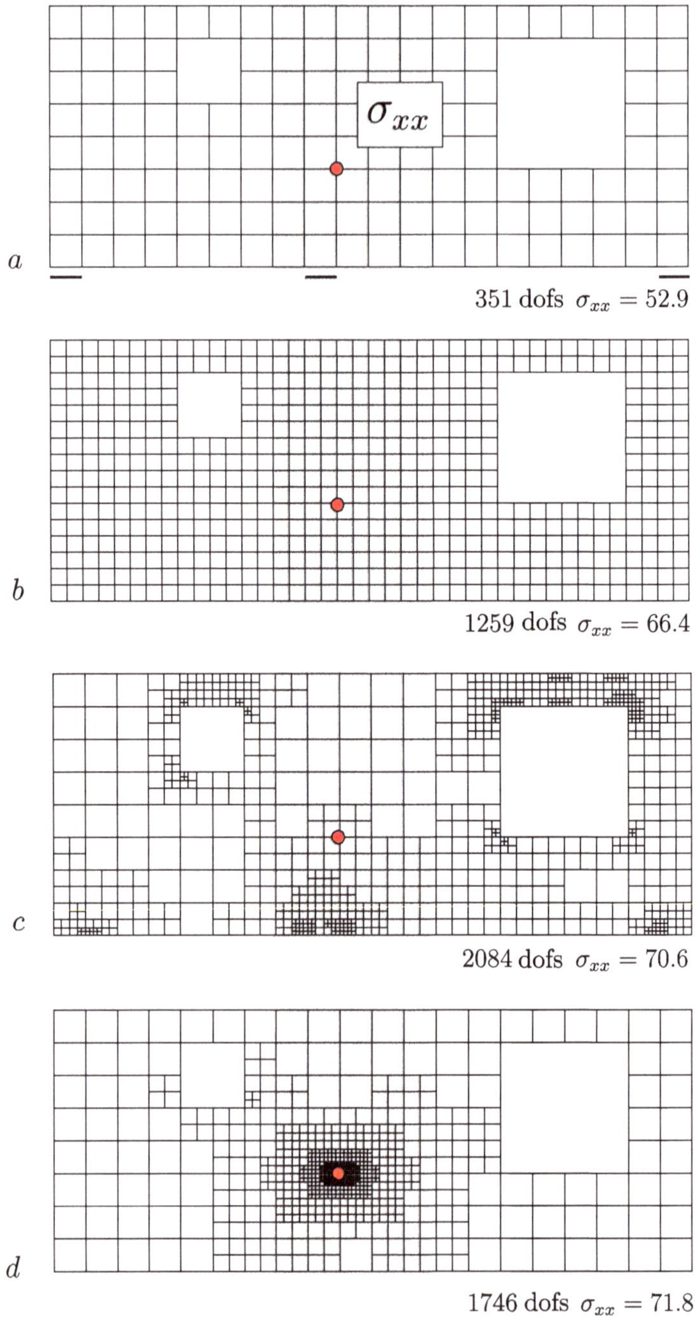

Fig. 4.10 Bilinear elements, gravity load and focus on σ_{xx} at one point, **a** initial mesh, **b** element length split in half, **c** adaptive refinement, **d** goal oriented adaptive refinement

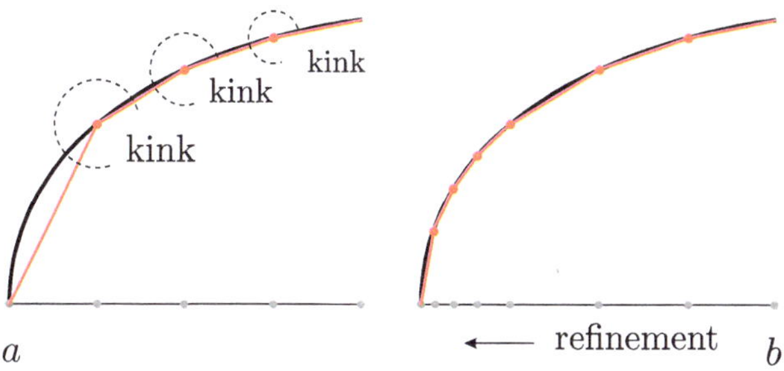

Fig. 4.11 To refine a mesh you do not need to know the exact solution, kinks in the FE-solution indicate the critical spots

$$\int_{\Omega} (G(\boldsymbol{y}, \boldsymbol{x}) - G_h(\boldsymbol{y}, \boldsymbol{x}))\, p_h(\boldsymbol{y})\, d\Omega_y = 0\,, \tag{4.69}$$

since it means, we only added a zero.

Classical adaptive refinement is based on the first Eq. (4.68a), because the mesh is refined where the deviations between p and the FE-load p_h are large.

The alternative would be to reduce the error $G - G_h$ in the influence function as suggested by (4.68b). This raises the question: How do we estimate the distance $G - G_h$, since the exact G is unknown? We solve this problem as follows: First, we replace the influence function (4.68b) with its weak form

$$u(\boldsymbol{x}) - u_h(\boldsymbol{x}) = \int_{\Omega} (G(\boldsymbol{y}, \boldsymbol{x}) - G_h(\boldsymbol{y}, \boldsymbol{x}))\, p(\boldsymbol{y})\, d\Omega_y = a(G - G_h, u) \tag{4.70}$$

and we write this strain energy product as external work[3]

$$u(\boldsymbol{x}) - u_h(\boldsymbol{x}) - a(G - G_h, u) = \int_{\Omega} (\delta(\boldsymbol{y} - \boldsymbol{x}) - \delta_h(\boldsymbol{y} - \boldsymbol{x}))\, u(\boldsymbol{y})\, d\Omega_y\,, \tag{4.71}$$

and the deviation $\delta - \delta_h$ we can measure, see Fig. 4.9. Not directly, because δ is only a fleeting symbol, but we can plot the forces δ_h and refine the mesh, so they will largely disappear, become a small dot on the screen.

The third variant is the ***goal-oriented adaptive refinement***, where the error $G - G_h$ in the influence function and the error $p - p_h$ in the load are made smaller simultaneously, see Fig. 4.10. This strategy is based on (4.68c). If we write this equation in weak form

$$u(\boldsymbol{x}) - u_h(\boldsymbol{x}) = a(\underbrace{G(\boldsymbol{y}, \boldsymbol{x}) - G_h(\boldsymbol{y}, \boldsymbol{x})}_{f}, \underbrace{u - u_h}_{g}) =: a(f, g)\,, \tag{4.72}$$

[3]This is *Betti's theorem*, (4.70) = $W_{12} = W_{21}$ = (4.71).

we can apply ***Schwarz' inequality***,

$$|u(\boldsymbol{x}) - u_h(\boldsymbol{x})| = |a(f, g)| \leq a(f, f)^{1/2} \cdot a(g, g)^{1/2} = \|f\| \cdot \|g\|, \qquad (4.73)$$

and we obtain a bound on the left side via the energy norm of f and g.

Theoretically (4.73) "stands on shaky ground", because the energy norm of the influence function is usually infinitely large, but if we simply ignore this—success proves us right—it is the fastest way to decrease the error of the solution at one particular point and with respect to one particular functional.

The idea of goal-oriented adaptive refinement is so convincing, it makes you wonder why normal adaptive refinement, where the neighborhood of the source point is not refined, actually succeeds. The riddle is solved when we look at the two possible representations of the error, see Fig. 4.12,

$$\begin{aligned} u(\boldsymbol{x}) - u_h(\boldsymbol{x}) &= \int_\Omega G(\boldsymbol{y}, \boldsymbol{x})(p(\boldsymbol{y}) - p_h(\boldsymbol{y}))\, d\Omega_{\boldsymbol{y}} \\ &= \int_\Omega (G(\boldsymbol{y}, \boldsymbol{x}) - G_h(\boldsymbol{y}, \boldsymbol{x}))p(\boldsymbol{y})\, d\Omega_{\boldsymbol{y}}\,. \end{aligned} \qquad (4.74)$$

The refinement of the mesh near the load reduces the error $p - p_h$ and this error is weighted with the exact kernel which, since it is exact, is independent of the size h of the elements. The second equation explains why this effect is independent of the non-refinement at the source point, because if the solution lies in $\mathcal{V}_h$ or nearly so, if $p - p_h$ is small, the error in the Green's function is orthogonal to this load and then it does not matter if the Green's function is only a rough approximation.

For not to put the reader on the wrong track: The results of (4.74) on the two meshes in Fig. 4.12 are not the same, only on the same mesh do the two formulas in (4.74) agree.

Given these remarks one could assume that a smooth plot is a guarantee that the FE-solution is "exact", but there is no such guarantee, since **far-field errors**, keyword ***pollution***, can lead to a *drift* of the FE-solution, which goes unnoticed, if one only pays attention to the jumps [4].

4.11 Pollution

Pollution is the phenomenon that the solution in a part A of a structure is negatively affected by errors in a remote part B. We speak of *global errors*. In 2-D and 3-D the influence functions struggle with these effects, see Sect. 6.9, since the errors at the edge radiate into the domain.

The wall plate in Fig. 4.13 illustrates this phenomenon quite well. The influence function for the shear force N_{yx} in section $A - A$ is a sliding motion of the upper part by one meter to the right.

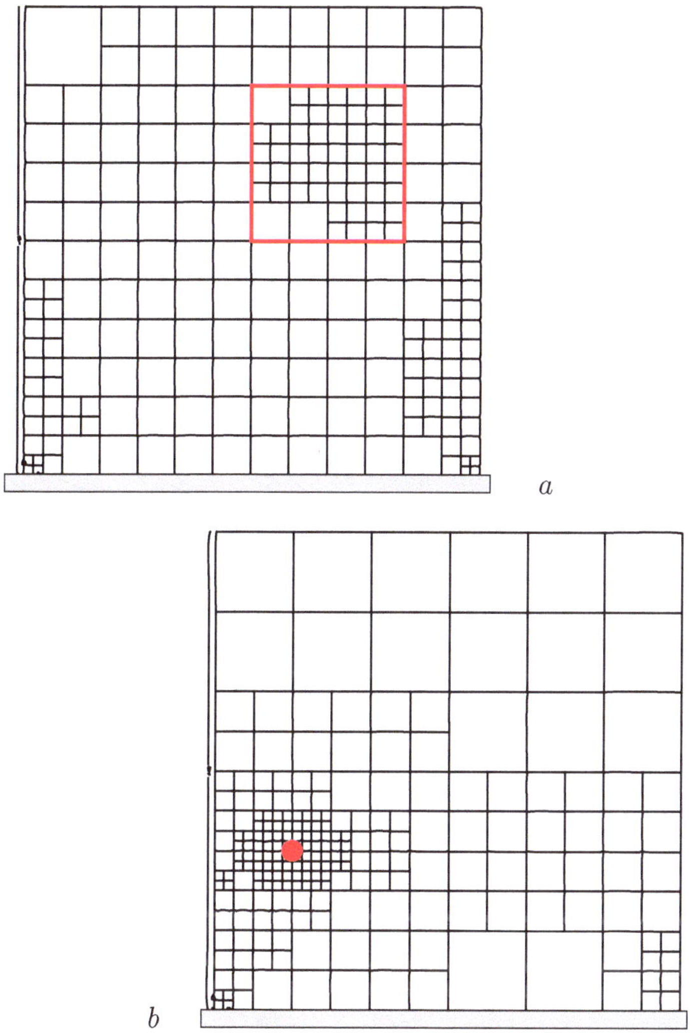

Fig. 4.12 Two possible strategies: **a** either the zone with the load, or **b** the neighborhood of the source point is refined

An FE-mesh, however, cannot perform this motion, since the elements would have to be sliced through and so an FE-program, which uses bilinear elements, produces the shape in Fig. 4.13b, where the upper edge of the plate moves 2.3 m to the right! So, a load of 1 kN at the upper edge produces a 2.3-fold shear force in section $A - A$. The difficulty which the FE-program has in producing the sliding motion in section $A - A$, leads to a large error at the upper edge of the plate.

An adaptive refinement of the mesh significantly reduces the error, see Fig. 4.13d.

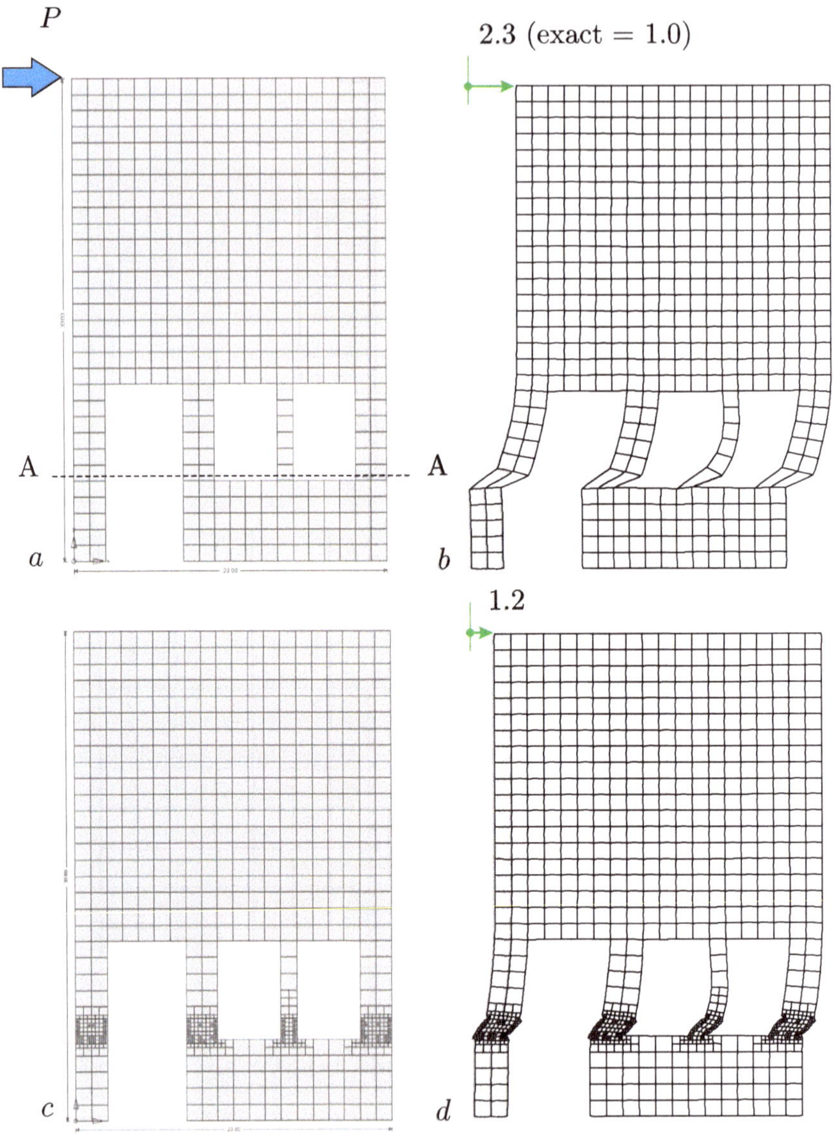

Fig. 4.13 Influence function for the shear force N_{yx} in section $A - A$, **a** plate and FE-mesh, **b** FE-influence function G_h, **c** improved model, **d** improved FE-solution G_h [5]

Also, if the cut passes exactly through the *center* of the element, the upper edge really moves 1 m to the right and the shear force N_{yx} will be exact.

In the case of the so-called *Wilson element* the horizontal cut can pass through any point on the element axis and the result is correct [4]. Only the sliding motion within the element itself, the abrupt jump in the horizontal displacement, cannot be modeled, since this would require a non-conforming element.

4.11.1 Causes

Possible causes of Pollution are [3]:

1. Discontinuities in the load, like jumps in the traffic load (a mild problem)
2. Large and small elements side by side; however, graded meshes can dampen *pollution*
3. Non-smooth influence functions, see Fig. 4.14
4. Singularities on the edge, see Sect. 6.9.

If the coefficients in the differential equation jump, because e.g. the thickness of the plate jumps, this affects also the influence functions, the **regularity** ("the smooth-

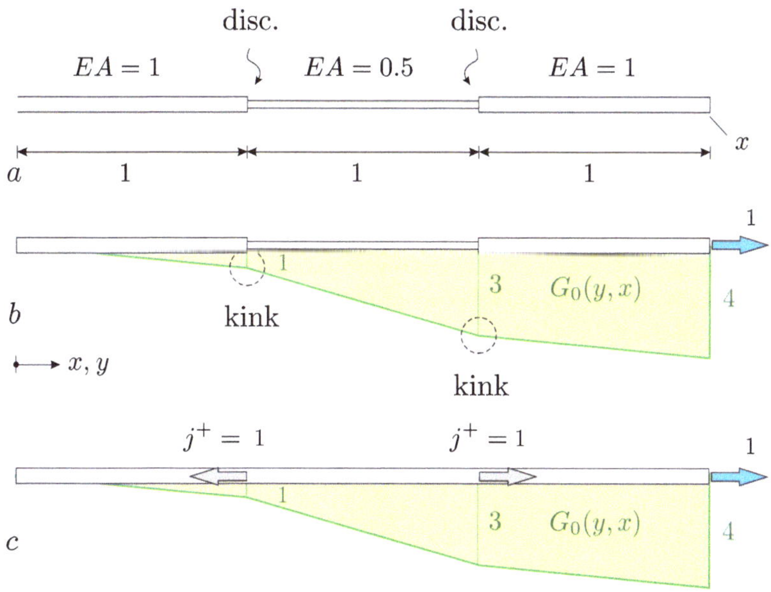

Fig. 4.14 Stretched bar, **a** system, **b** influence function for the displacement at the bar end, **c** the same shape can be generated in a uniform bar, $EA = 1$, with the help of auxiliary forces j^+, see Sect. 5.11

ness") of the influence functions decreases, kinks appear as in Fig. 4.14b and these tend to produce *pollution*.

All FE-solutions—except 1-D elements (EA and EI constant)—exhibit a global error, struggle with *pollution*.

The error manifests itself as ***drift*** at the nodes, since the exact solution and the FE-solution do not agree at the nodes [4]. Surprisingly, this is a "quality feature", since the interpolant is an inferior solution—the errors in the stresses are on average greater than those of the FE-solution.

Only in frames are the interpolant u_I and the FE-solution u_h (EA and EI const.) the same and so the FE-program must only add the local solutions

$$u = u_I + \text{local solutions}\,, \tag{4.75}$$

to produce the exact solution.

Interpolant: The function $u_I(x)$ in $\mathcal{V}_h$ which interpolates the exact solution u at the nodes.

The error of the FE-solution $\|u - u_h\|$ (measured in the energy norm) is less than the error $\|u - u_I\|$ of the interpolant. Meshes on which functions can be interpolated well are also good FE-meshes, since the FE-solution is still a little bit better than the interpolant.

4.11.2 Details

The subject *pollution and finite elements* goes back to *Babuška* [3]. What triggered his comments was, if we see it correctly, the remark by *Zienkiewicz et Zhu* [6], that it is only necessary to shrink the residual $p - p_h$ to get close to the exact solution. But in the case of non-smooth solutions there is also—this is the criticism—the ***drift*** of an FE-solution, which cannot be discovered by solely looking at the residual $p - p_h$ [7], p. 259.

Babuška defined as the ***global error*** or *pollution*, the distance between the interpolant u_I and the FE-solution

$$u_g = u_I - u_h \tag{4.76}$$

and as the ***local error*** the error of the interpolant on the elements

$$u_l = u - u_I\,. \tag{4.77}$$

In 1-D problems these are the small "arcs" which span from node to node (at the nodes $u_l = 0$ is zero).

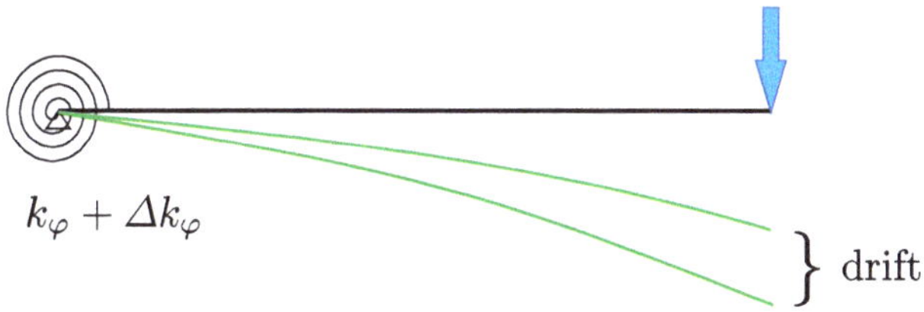

Fig. 4.15 Drift produced by an error in the stiffness of the spring

The sum is the error of the FE-solution [3], p. 186,

$$e = u_g + u_l = u_I - u_h + u - u_I = u - u_h\,. \tag{4.78}$$

The global error manifests itself as *drift* at the nodes and is therefore often called the ***far-field error***, since it is caused by disturbances $p - p_h$ in remote structural parts or by singularities on the edge, see Sect. 6.9. The latter is "overlooked" by an error estimator, which only looks at the residual $p - p_h$ in the domain.

The simplest illustration of this phenomenon is a cantilever beam attached to a torsional spring (k_φ), see Fig. 4.15. An error in the stiffness $k_\varphi + \Delta k_\varphi$ can be compared to a singularity on the edge and the further one moves out on the beam, the larger the *drift* becomes in the FE-solution—even if $p_h = p$.

4.12 Super Convergence

This subject is practically the counterpart to *pollution*. At superconvergent points an FE-solution is more accurate than the order of the shape functions would suggest. Superconvergent points are often the ***Gauss points*** of numerical quadrature. This can even be explained.

Let us study a beam, since in a beam the influence functions G_0^h of the nodal displacements agree with the exact influence functions G_0. In addition, the following applies: (i) the shape functions φ_i are homogeneous solutions, (ii) the influence functions are homogeneous solutions at all points—except the source point, (iii) homogeneous solutions are defined by their nodal values. So, the error in an FE-influence function G_0^h is zero outside of the element with the source point x, and a 1-D FE-solution[4] is exact at all points x lying on elements with zero load, see Fig. 4.16.

What happens, if x lies on an element, which carries a load p? In each element the exact deflection w can be split into a homogeneous solution w_0 and a particular solution w_p corresponding to fixed ends

---

[4]Before we add the local solutions.

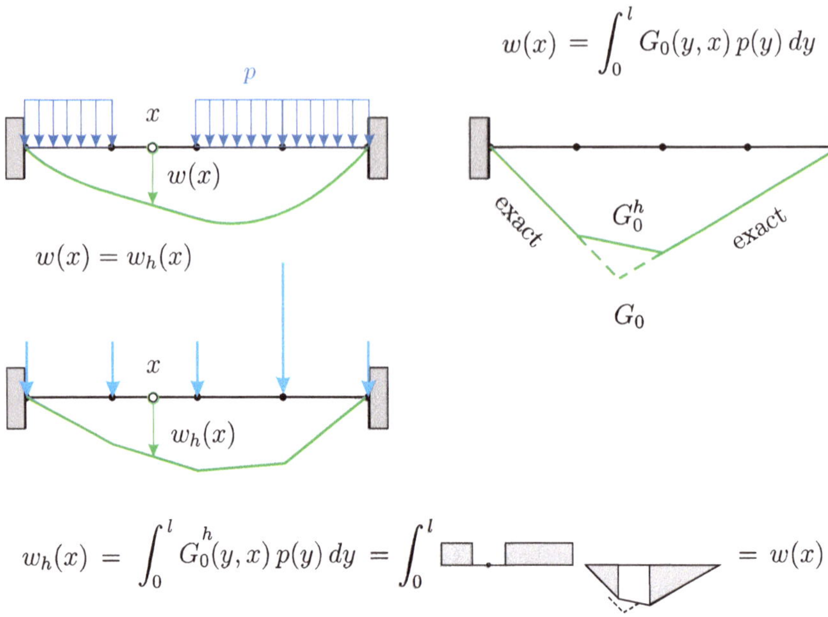

Fig. 4.16 In 1-D problems the error is zero in any element which carries no load

$$w(x) = w_0(x) + w_p(x) \qquad EI\, w_0^{IV}(x) = 0 \qquad EI\, w_p^{IV} = p\,, \tag{4.79}$$

and the homogeneous solution is just the FE-solution, $w_h = w_0$, since the load is only applied at the nodes. The error $e(x) = w(x) - w_h(x) = w_p(x)$ of the FE-solution is therefore identical with the particular solution and so the error in the bending moment is identical with the bending moment M_p of the particular solution, (= the moment if the element were clamped on both sides, l_e = element length)

$$M_p(x) = -EI\, w_p''(x) = \frac{p\, l_e^2}{2}\, (\frac{1}{6} - \frac{x}{l_e} + \frac{x^2}{l_e^2})\,. \tag{4.80}$$

Here is the surprise! The error is zero, where $M_p(x) = 0$, and these two points $x_1 = 0.21132$ and $x_2 = 0.78868$, the length of the element is $l_e = 1$, are exactly the Gauss points. How does this happen?

(i) The integral of M_p is zero, since the ends are fixed, $w_p'(0) = w_p'(l_e) = 0$, see Fig. 4.17,

$$\int_0^{l_e} M_p(x)\, dx = \int_0^{l_e} -EI\, w_p''(x)\, dx = -EI\, [w_p'(l_e) - w_p'(0)] = 0\,. \tag{4.81}$$

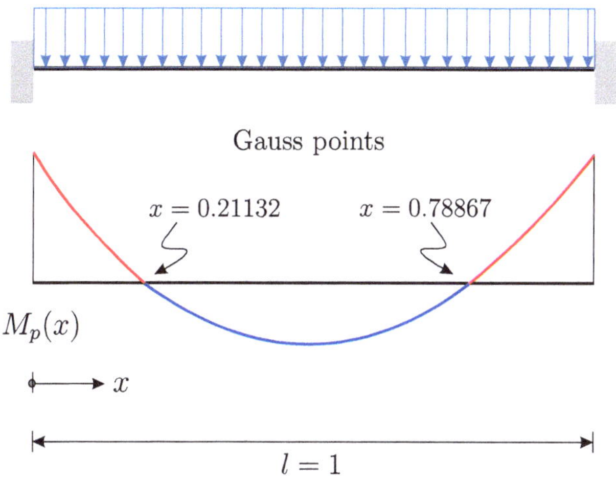

Fig. 4.17 The zeros of $M_p(x)$ are exactly the Gauss points

(ii) $M_p(x)$ is a second-degree polynomial, and therefore the function must vanish at the $n = 2$ Gauss points of a $2n - 1 = 3$ formula

$$\int_0^{l_e} M_p(x)\,dx = w_1\, M_p(x_1) + w_2\, M_p(x_2) = 2\, w_1\, M_p(x_1) = 0\,, \tag{4.82}$$

where the w_i are the weights of the quadrature.

Given other loads it is not guaranteed that M_p is zero at the Gauss points. Then one must search for the superconvergent points (= zeros of M_p) in the plots of the bending moments, although some rules still apply. For example, if the load on the interval $[-1, 1]$ is triangular, $p(x) = (1 + x)/2$, then because of

$$\begin{aligned} \int_{-1}^{+1} M_p(x)\,dx &= \int_{-1}^{+1} [-\frac{1}{12} - \frac{x}{20} + \frac{x^2}{4} + \frac{x^3}{12}]\,dx \\ &= 1.0 \times M_p(-0.5775) + 1.0 \times M_p(0.5775) = 0 \end{aligned} \tag{4.83}$$

M_p must have opposite values at the two Gauss points.

In the same sense the shear force $V(x) = -EI\,w'''(x)$ is zero at the midpoint of an element if the load p is constant—for the same reason as before: The integral of $V(x) = M'(x)$ is zero, since the two fixed end moments are the same, $M(l_e) - M(0) = 0$, and because $V(x)$ is a linear function, which can be integrated exactly with one Gauss point.

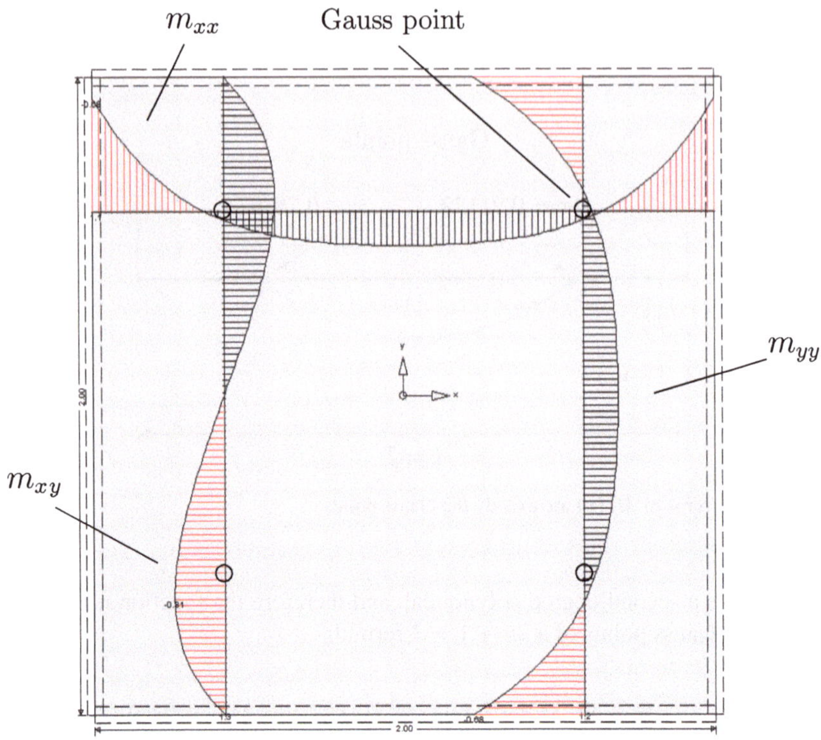

Fig. 4.18 Clamped plate and the position of the four Gauss points and plot of the bending moment m_{xx}, m_{xy}, and m_{yy}

Next, let us look at a rectangular, clamped plate element. Since the normal derivative $w_n = \nabla w \bullet \boldsymbol{n} = 0$ and the tangential derivative $w_t = \nabla w \bullet \boldsymbol{t} = 0$ on the edge are zero, the gradient $\nabla w = \{w,_x\,, w,_y\}^T$ is also zero and therefore the integral of m_{xx} must be zero

$$\int_{\Omega_e} m_{xx}\, d\Omega = \int_{\Omega_e} (w,_{xx} + \nu\, w,_{yy})\, d\Omega = \int_{\Gamma_e} (w,_x\, n_x + \nu\, w,_y\, n_y)\, ds = 0\,, \tag{4.84}$$

and the integrals of m_{yy} and m_{xy} as well.

In a beam which carries a uniform load p, the moment $M(x)$ is a second-degree polynomial. If the same were true in a plate, the moments m_{xx} and m_{yy} should be zero at the four Gauss points. But this seems to be only approximately true, see Fig. 4.18.

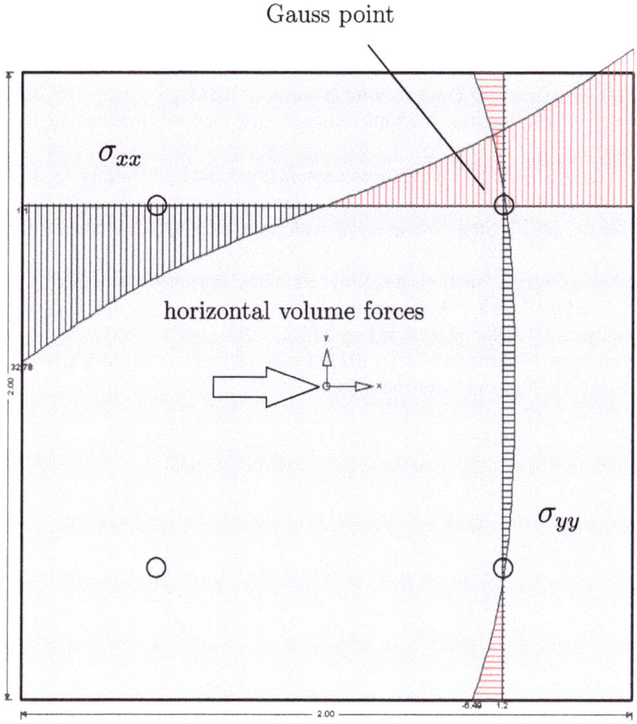

Fig. 4.19 Fixed plate element with horizontal forces $\boldsymbol{p} = \{1, 0\}^T$; σ_{xx} and σ_{xy} are approximately linear and σ_{yy} is quadratic

For the same reason as above in a rectangular plate element fixed on all sides, the integral of $\sigma_{xx} = \varepsilon_{xx} + \nu\,\varepsilon_{yy} = u_{x,x} + \nu\, u_{y,y}$

$$\int_{\Omega_e} \sigma_{xx}\, d\Omega = E \int_{\Omega_e} (u_{x,x} + \nu\, u_{y,y})\, d\Omega = E \int_{\Gamma_e} (u_x\, n_x + \nu\, u_y\, n_y)\, ds = 0\,, \tag{4.85}$$

and of σ_{yy} and σ_{xy} must be zero.

Horizontal surface loads $\boldsymbol{p} = \{1, 0\}^T$ will produce approximately linear stresses σ_{xx} and quadratic stresses σ_{yy}, see Fig. 4.19. Linear stresses σ_{xx} would be zero at the center of the element and quadratic stresses σ_{yy} would be zero at all four Gauss points.

References

1. Höllig K (2003) Finite element methods with B-splines. SIAM
2. Cottrell JA, Hughes TJR, Bazilevs Y (2009) Isogeometric analysis: toward integration of CAD and FEA. Wiley
3. Babuška I, Strouboulis T (2001) The finite element method and its reliability. Oxford University Press
4. Hartmann F, Katz C (2010) Structural analysis with finite elements, 2nd edn. Springer, Heidelberg
5. Hartmann F (2013) Green's functions and finite elements. Springer, Heidelberg
6. Zienkiewicz OC, Zhu JZ (1987) A simple error estimator and adaptive procedure for practical engineering analysis. Int J Num Methods Eng 24:337–357
7. Babuška I, Whiteman JR, Strouboulis T (2011) Finite elements. Oxford University Press, An Introduction to the Method and Error Estimation

Chapter 5
Stiffness Changes and Reanalysis

The subject of this chapter are changes in the stiffness of single structural members and we study how such shifts propagate through a structure, how they affect the stress distribution in a structure.

In terms of linear algebra such modifications correspond to an update of the stiffness matrix, $\boldsymbol{K} \rightarrow \boldsymbol{K} + \boldsymbol{\Delta K}$, and the solution $\boldsymbol{u} \rightarrow \boldsymbol{u}_c = \boldsymbol{u} + \boldsymbol{\Delta u}$

$$(\boldsymbol{K} + \boldsymbol{\Delta K})\,(\boldsymbol{u} + \boldsymbol{\Delta u}) = \boldsymbol{f} \qquad \text{or} \qquad \boldsymbol{K}_c\,\boldsymbol{u}_c = \boldsymbol{f}\,. \tag{5.1}$$

The approach in this chapter is that we compute the response $\boldsymbol{u}_c$ of the modified structure via the response $\boldsymbol{u}$ of the original, unmodified structure $\boldsymbol{K}\,\boldsymbol{u} = \boldsymbol{f}$, a technique which is called *reanalysis*.

The essential insight is: The new displacement vector $\boldsymbol{u}_c$ is the response of the original system when a vector $\boldsymbol{f}^+ = -\boldsymbol{\Delta K}\,\boldsymbol{u}_c$ is added to the right side

$$\boldsymbol{K}\boldsymbol{u}_c = \boldsymbol{f} + \boldsymbol{f}^+\,, \tag{5.2}$$

and $\boldsymbol{f}^+$ is orthogonal to all rigid-body motions $\boldsymbol{u}_0 = \boldsymbol{a} + \boldsymbol{x} \times \boldsymbol{b}$, which is the reason why the effects of local stiffness changes most often subside rapidly.

A stiffness change triggers a compensating displacement $\boldsymbol{\Delta u} = \boldsymbol{u}_c - \boldsymbol{u}$, and $\boldsymbol{\Delta u} = \boldsymbol{K}^{-1}\boldsymbol{f}^+$ is the reaction of the structure to additional self-equilibrated nodal forces $\boldsymbol{f}^+ = -\boldsymbol{\Delta K}\boldsymbol{u}_c$, that is $\boldsymbol{f}^+ \bullet (\boldsymbol{a} + \boldsymbol{b} \times \boldsymbol{x}) = 0$

The problem with this approach is of course that the new vector $\boldsymbol{u}_c$, on which $\boldsymbol{f}^+ = \boldsymbol{\Delta K}\,\boldsymbol{u}_c$ depends, is unknown, though eventually the vector $\boldsymbol{\Delta K}\boldsymbol{u} \simeq \boldsymbol{f}^+$ can serve as substitute. We will also present two techniques to calculate $\boldsymbol{u}_c$ directly, see Sect. 5.22, but our main concern is not to beat the computer, but we want insight into the effects of stiffness changes.

F. Hartmann and P. Jahn, *Statics and Influence Functions*, Springer Series in Solid and Structural Mechanics 13, https://doi.org/10.1007/978-3-030-55889-5_5

The algebra behind reanalysis is easily explained: To solve the equation

$$-(H + \Delta H)u_c'' = p \tag{5.3}$$

we place the term $-\Delta H\, u_c''$ on the right side,

$$-H\, u_c'' = p + \Delta H\, u_c'' \tag{5.4}$$

and so the equivalent nodal forces are

$$\int_0^l \varphi_i\, p\, dx + \int_0^l \varphi_i\, \Delta H\, u_c''\, dx = f_i + f_i^+ . \tag{5.5}$$

Back to $\boldsymbol{K}\,\boldsymbol{u}_c = \boldsymbol{f} + \boldsymbol{f}^+$. The new solution $\boldsymbol{u}_c$ is, like the old solution $\boldsymbol{u} = \sum_i f_i\, \boldsymbol{g}_i$, an expansion in terms of the influence functions of the nodal displacements (= the columns $\boldsymbol{g}_i$ of the **old** inverse $\boldsymbol{K}^{-1}$)

$$\boldsymbol{u}_c = \sum_i f_i\, \boldsymbol{g}_i + f_a^+\, \boldsymbol{g}_a + f_b^+\, \boldsymbol{g}_b + \cdots = \boldsymbol{u} + f_a^+\, \boldsymbol{g}_a + f_b^+\, \boldsymbol{g}_b + \cdots , \tag{5.6}$$

only that some columns appear twice, and then carry **additional weights** $f_a^+, f_b^+, \ldots$ corresponding to the modified values k_{ab}. If for example four entries $k_{3,3}$, $k_{3,7}$, $k_{7,3}$, $k_{7,7}$ change, two (initially unknown) weights f_3^+ and f_7^+ mark the difference between the new and the old displacement vector

$$\boldsymbol{u}_c - \boldsymbol{u} = f_3^+\, \boldsymbol{g}_3 + f_7^+\, \boldsymbol{g}_7 . \tag{5.7}$$

Nothing new is added, the old is only supplemented—with old.

This is the same approach as in the **force method**, since we do not change the stiffness matrix, but we change the right side, $\boldsymbol{f}$ becomes $\boldsymbol{f} + \boldsymbol{f}^+$.

The force method chooses a statically determinate structure as its *primary structure*, and in the sequence all calculations are done on this system. It corresponds to the undisturbed system $\boldsymbol{K}$ since the redundants X_i play the same role as the f_i^+. While the f_i^+ couple the added element to the structure, the X_i make that the gaps at the joints close. Both, the X_i and f_i^+, are *additional loads* which appear on the right side, while the proper analysis is done with the primary structure (matrix $\boldsymbol{K}$).

This approach has a further advantage: We do not need two sets of influence functions, one for the primary structure (matrix $\boldsymbol{K}$) and a separate set for the statically indeterminate structure (matrix $\boldsymbol{K}_c$), since the changes in any functional

$$J(\boldsymbol{e}) = J(\boldsymbol{u}_c) - J(\boldsymbol{u}) = \boldsymbol{g}^T \boldsymbol{f}^+ , \tag{5.8}$$

can be calculated with the influence functions $\boldsymbol{g}$ of the original system $\boldsymbol{K}$.

5.1 Parameter Identification

If you pull on a spring with a force f and measure the extension u, you can determine the stiffness $k = f/u$ of the spring. Parameter identification means fitting the elements k_{ij} of a stiffness matrix to measured data. Mathematically this counts as a (difficult) **inverse problem**. Since the columns $\boldsymbol{g}_i$ of $\boldsymbol{K}^{-1}$ are the influence functions of the nodal displacements, everything depends on the influence functions. How do the system responses change with corrections $k_{ij} \to k_{ij} + \Delta k_{ij}$? How to modify the k_{ij} to reproduce the measured data? How sensitive is the structure to such changes?

These are precisely the questions that also play an important role in reanalysis, because in Sect. 5.14 we will see that the derivative of the displacement vector with respect to an element k_{ij}

$$\frac{\partial \boldsymbol{u}}{\partial k_{ij}} = -\boldsymbol{K}^{-1} u_j \, \boldsymbol{e}_i = -u_j \, \boldsymbol{g}_i \,, \tag{5.9}$$

depends on the influence function $\boldsymbol{g}_i$ and the nodal displacement u_j. In this sense the title of the chapter could also have been *Reanalysis and Parameter Identification*. We could extend this list also further by adding *Structural Health Monitoring* which makes ample use of parameter identification techniques, but then mainly in the dynamic range [1].

5.2 Introductory Remarks

We start with a spring, $k\,u = f$, see Fig. 5.1. The reaction of the spring to a force $f = 1$ (the Dirac delta) is $G = 1/k$, and a shift in the stiffness, $k + \Delta\,k$, changes the response to $G_c = 1/(k + \Delta\,k)$, and so the reaction to any force f before, u, and after, u_c, is

$$u = \frac{1}{k} f \qquad u_c = \frac{1}{k + \Delta\,k} f \,. \tag{5.10}$$

A **Taylor series** of the updated influence function

$$\frac{1}{k + \Delta\,k} = \frac{1}{k} - \frac{1}{k^2} \Delta\,k + \cdots \tag{5.11}$$

illustrates, how the spring reacts to the shift in the stiffness

$$u_c \simeq \left[\frac{1}{k} - \frac{1}{k}\frac{\Delta\,k}{k}\right] f = u - \frac{1}{k} \underbrace{\Delta\,k \cdot u}_{force} \,. \tag{5.12}$$

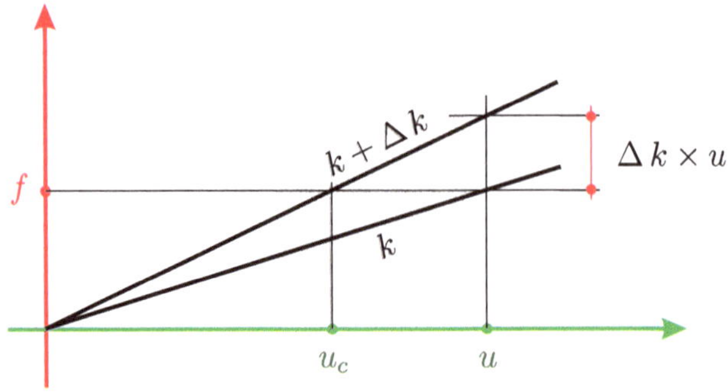

Fig. 5.1 A change in the stiffness means a change in the slope of $k\,u = f$

The increase, $k \to k + \Delta k$, lets the original u shoot beyond the target, the force in the spring, $(k + \Delta k)\,u = f + \Delta k \cdot u$, becomes too large, and to eliminate this too much an opposite displacement $\Delta u \simeq -\Delta k \cdot u/k$ must correct this.

Since the Taylor series of a stiffness matrix reads [2],

$$(\boldsymbol{K} + \boldsymbol{\Delta K})^{-1} = \boldsymbol{K}^{-1} - \boldsymbol{K}^{-1}\boldsymbol{\Delta K}\,\boldsymbol{K}^{-1} + \cdots \tag{5.13}$$

the analogy between (5.12) and

$$\boldsymbol{u}_c = (\boldsymbol{K} + \boldsymbol{\Delta K})^{-1}\boldsymbol{f} \simeq \boldsymbol{u} - \boldsymbol{K}^{-1}\boldsymbol{\Delta K}\,\boldsymbol{u} \tag{5.14}$$

is evident. If we multiply this equation from the left with $\boldsymbol{K}$

$$\boldsymbol{K}\boldsymbol{u}_c = \boldsymbol{K}\boldsymbol{u} - \boldsymbol{\Delta K}\boldsymbol{u}\,, \tag{5.15}$$

it has (almost) the form in which we treat it in this chapter. Almost, since we do not use a Taylor series, but the exact formula, i.e. we replace $-\boldsymbol{\Delta K}\boldsymbol{u}$ with the exact $-\boldsymbol{\Delta K}\boldsymbol{u}_c =: \boldsymbol{f}^+$

$$\boldsymbol{K}\boldsymbol{u}_c = \boldsymbol{K}\boldsymbol{u} - \boldsymbol{\Delta K}\boldsymbol{u}_c = \boldsymbol{f} + \boldsymbol{f}^+\,. \tag{5.16}$$

The approximation (5.14) means $\boldsymbol{u}_c - \boldsymbol{u} \simeq -\boldsymbol{K}^{-1}\boldsymbol{\Delta K}\boldsymbol{u}$. We can even understand this equation, if we apply it to a beam, $EI \to EI + \Delta EI$, and write it in integral form

$$w_c(x) - w(x) = -\int_0^l \underbrace{G_0(y,x)}_{\boldsymbol{K}^{-1}}\ \underbrace{\Delta EI\,\frac{d^4}{dy^4}}_{\boldsymbol{\Delta K}} \underbrace{\int_0^l G_0(y,z)\,p(z)\,dz}_{\boldsymbol{u}}\,dy \tag{5.17}$$

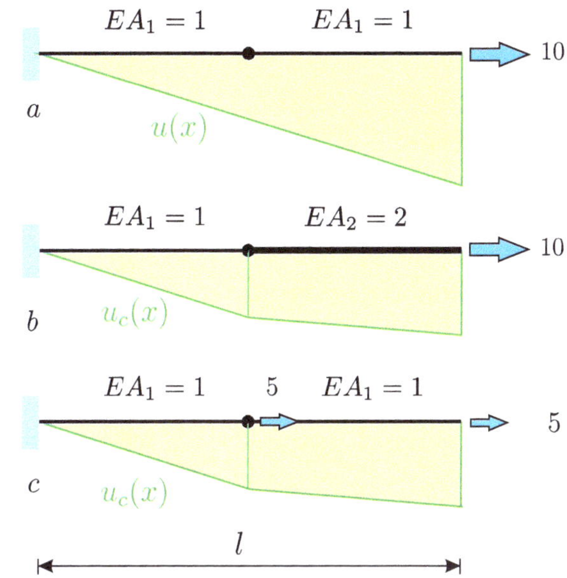

Fig. 5.2 In the second element the stiffness *EA* is doubled

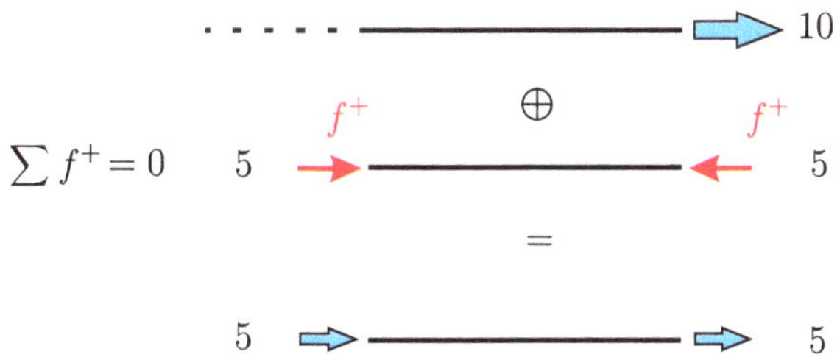

or

$$w_c(x) - w(x) = -\int_0^l G_0(y, x)\, \frac{\Delta EI}{EI} p(y)\, dy\,. \tag{5.18}$$

So, a decrease, $\Delta EI < 0$, is equivalent to an increase in the load p.[1]

Example 5.1 An elementary example may illustrate the technique. The bar in Fig. 5.2 consists of two elements with the same stiffness $EA = 1$ kN, and a force $f_2 = 10$ kN pulls at its right end

[1] It is $EI\, d^4/dy^4 G_0 = \delta_0$ and if you replace EI by ΔEI, this is $\Delta EI/EI\, \delta_0$ and so the integral gives $(\Delta EI/EI\, \Delta_0, p) = \Delta EI/EI \cdot p(y)$. Furthermore, $\boldsymbol{\Delta K u} = \delta W_i = \delta W_e$ and δW_e contains the operator $\Delta EI\, d^4/dy^4$.

$$\begin{bmatrix} 2 & -1 \\ -1 & 1 \end{bmatrix} \begin{bmatrix} u_1 \\ u_2 \end{bmatrix} = \begin{bmatrix} 0 \\ 10 \end{bmatrix}, \tag{5.19}$$

and so, $u_1 = 10\,\text{m}$, $u_2 = 20$ m.

On doubling the stiffness of the second element, $EA \to 2EA$,

$$\begin{bmatrix} 3 & -2 \\ -2 & 2 \end{bmatrix} \begin{bmatrix} u_1^c \\ u_2^c \end{bmatrix} = \begin{bmatrix} 0 \\ 10 \end{bmatrix}, \tag{5.20}$$

the nodal displacements become $u_1^c = 10\,\text{m}$, $u_2^c = 15$ m.

The question is then which vector $\boldsymbol{f}^+$ we must add to the right side $\boldsymbol{f}$ of the original system (5.19) to produce the same effect, the same displacements

$$\begin{bmatrix} 2 & -1 \\ -1 & 1 \end{bmatrix} \begin{bmatrix} u_1^c \\ u_2^c \end{bmatrix} = \begin{bmatrix} 0 \\ 10 \end{bmatrix} + X \begin{bmatrix} 1 \\ -1 \end{bmatrix}. \tag{5.21}$$

We assume the vector $\boldsymbol{f}^+$ to be an equilibrium vector (the sum of its components is zero), and the result justifies this assumption since with $X = 5$ we obtain a solution, a vector

$$\boldsymbol{f} + \boldsymbol{f}^+ = \begin{bmatrix} 0 \\ 10 \end{bmatrix} + \begin{bmatrix} 5 \\ -5 \end{bmatrix} = \begin{bmatrix} 5 \\ 5 \end{bmatrix}, \tag{5.22}$$

to which the original system reacts in the same way as the modified system, the nodal displacements $u_i = u_i^c$ are the same

$$\begin{bmatrix} 2 & -1 \\ -1 & 1 \end{bmatrix} \begin{bmatrix} 10 \\ 15 \end{bmatrix} = \begin{bmatrix} 5 \\ 5 \end{bmatrix}. \tag{5.23}$$

5.3 Adding or Subtracting Stiffness

A change in the stiffness of an element can be interpreted as placing an **additional element** in front of the original element, so that the two elements together have the target stiffness, see Fig. 5.3.

Since the additional element must be synchronized with the motions of the original structure **coupling forces** need to hold the nodes together, and these forces are just the $\boldsymbol{f}^+$. This explains why the forces $\boldsymbol{f}^+$ are equilibrium forces, because if they were not, the added element would fly off.

If the stiffness of the element increases, the coupling forces $\boldsymbol{f}^+$ normally have the tendency to hinder the deformation of the element, they stiffen, so to speak, the element.

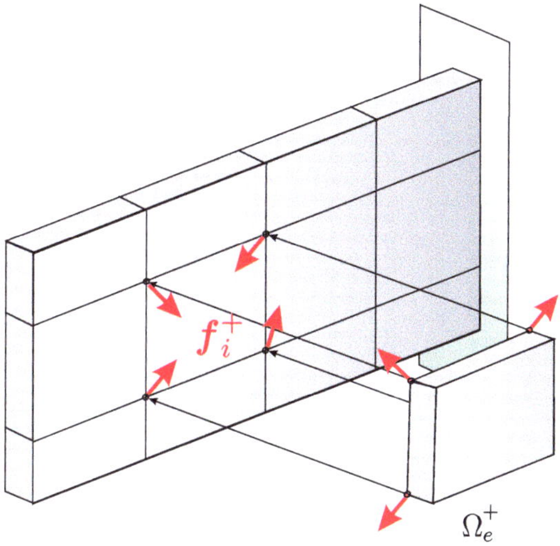

Fig. 5.3 A stiffness change $\boldsymbol{K} + \boldsymbol{\Delta K}$ is equivalent to attaching an element Ω_e^+ having the stiffness matrix $\boldsymbol{\Delta K}$ to the structure [3]

Conversely, if the stiffness in the element decreases, the coupling forces $\boldsymbol{f}^+$ will add to the deformations of the element, they act like additional weights at the nodes of the element.

5.4 Dipoles and Monopoles

Two opposite forces $f_i^+ = \pm 1/h$, a distance h apart, become a dipole when h tends to zero.

If, however, the two forces remain finite even in the limit $h = 0$, we call this a ***pseudodipole***. The proton (+) and the electron (−) in a hydrogen atom form such a pseudodipole, since the distance between the two opposite elementary charges is so small that their effects on a point charge outside the atom practically cancel out.

The same can be said about the forces f_i^+, since to each force f_i^+ corresponds an opposite force f_j^+, so that the two forces f^+, seen from a distance, resemble a pseudodipole, see Fig. 5.4.

The effect of the forces f_i^+ on any point x of the structure depends on how large the **runtime difference** from the point x to the force $+f_i^+$ and the counterforce $-f_i^+$ is. If two forces $\pm f_i^+$ are only a small distance apart, because the element Ω_e is small, their effects cancel each other out, since the influence function hardly changes on the element, $G' \simeq 0$.

Imagine a bar, which is modeled with seven linear elements, and in the second element the stiffness changes, $EA_c = EA + \Delta EA$, while we are about to calculate the influence function for the normal force in the last element by spreading the element at its midpoint by one unit.

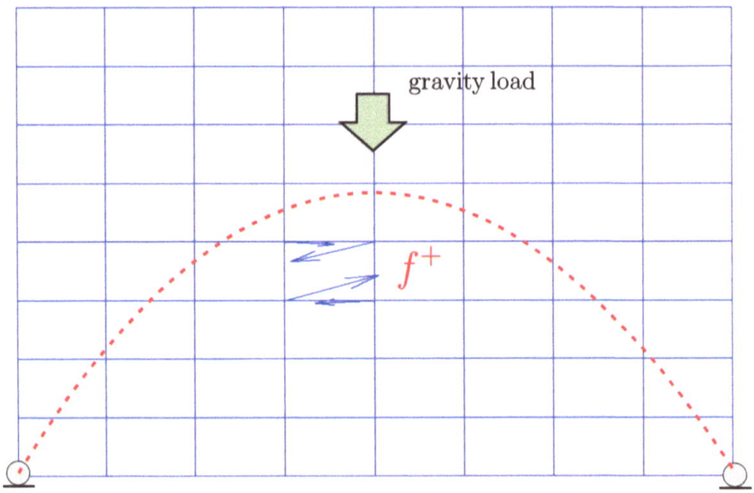

Fig. 5.4 A stiffness change in an element activates coupling forces f_i^+. These forces follow the directions of the principal stresses (- - -) and they are equilibrium forces, comparable to pseudodipoles

The effect of the spread propagates to the left, to element # 2, where two forces $\pm f^+$ (at the two nodes of the element) simulate the effect of the stiffness change, $EA_c = EA + \Delta EA$. One force f_i^+ pulls to the left and an opposite force f_{i+1}^+ pulls to the right. The influence function for the normal force

$$G(y, x) = \sum_j g_j(x)\, \varphi_j(y) \tag{5.24}$$

will have the value g_i at the left node and g_{i+1} at the right node of element # 2, and so the effect of the stiffness change on the normal force $N = J(u)$, *new–old*, in the last element

$$N_c - N = J(u_c) - J(u) = f_i^+ g_i - f_{i+1}^+ g_{i+1} = f_i^+ \cdot (g_i - g_{i+1}) \simeq f_i^+ \cdot G' \cdot l_e\,, \tag{5.25}$$

will only be noticeable if the influence function has a distinct slope, $G' \gg 0$, in the element #2. We may express this observation by saying:

The force pair $\pm f_i^+$ "differentiates" the influence function.

Remark 5.1 This picture of two forces f_i^+ and f_{i+1}^+ exactly opposite each other and of the same size, is only true in the 1-D model of a bar. In a beam element the f_i^+ would be two forces (which still add to zero) and two moments and all four balance in the sense that they perform zero work on acting through any rigid body motion $w_0 = a + b \cdot x$.

5.5 Displacements and Forces

Let us study this more systematically in easy steps. We begin with a displacement, $J(u) = u(x)$. The displacement before

$$J(u) = \boldsymbol{g}^T \boldsymbol{f} = \boldsymbol{u}^T \boldsymbol{j} \tag{5.26}$$

and after the shift in the stiffness is

$$J(u_c) = \boldsymbol{g}_c^T \boldsymbol{f} = \boldsymbol{u}_c^T \boldsymbol{j}_c\,, \tag{5.27}$$

where $\boldsymbol{g}_c$ is the influence function belonging to the modified system. Note, the two vectors $\boldsymbol{j}$ and $\boldsymbol{j}_c$,

$$\boldsymbol{j} = \{\varphi_1(x), \varphi_2(x), \ldots, \varphi_n(x)\}^T = \boldsymbol{j}_c \tag{5.28}$$

are the same, since the stiffness *EA* does not enter the definition of a **displacement-functional**, say $u(x)$, and so

$$\begin{aligned} J(u_c) - J(u) &= \boldsymbol{j}^T(\boldsymbol{u}_c - \boldsymbol{u}) = \boldsymbol{j}^T \boldsymbol{K}^{-1}(\boldsymbol{f} + \boldsymbol{f}^+) - \boldsymbol{j}^T \boldsymbol{K}^{-1} \boldsymbol{f} \\ &= \boldsymbol{j}^T \boldsymbol{K}^{-1} \boldsymbol{f}^+ = \boldsymbol{g}^T \boldsymbol{f}^+\,. \end{aligned} \tag{5.29}$$

This may be different with forces like $J(u) = N(x) = EA\, u'(x)$. If the source point x lies on the affected element, $EA \rightarrow EA_c$, the vectors $\boldsymbol{j}$

$$\boldsymbol{j} = \{EA\,\varphi_1'(x), EA\,\varphi_2'(x), \ldots, EA\,\varphi_n'(x)\}^T \tag{5.30}$$

and

$$\boldsymbol{j}_c = \{EA_c\,\varphi_1'(x), EA_c\,\varphi_2'(x), \ldots, EA_c\,\varphi_n'(x)\}^T \tag{5.31}$$

are not the same,[2] and so

$$\begin{aligned} J(u_c) - J(u) &= \boldsymbol{j}_c^T \boldsymbol{K}^{-1}(\boldsymbol{f} + \boldsymbol{f}^+) - \boldsymbol{j}^T \boldsymbol{K}^{-1} \boldsymbol{f} \\ &= (\boldsymbol{j}_c - \boldsymbol{j})^T \boldsymbol{u} + \boldsymbol{j}_c^T \boldsymbol{u}^+\,. \end{aligned} \tag{5.32}$$

[2] Only the functions $\varphi_i(x)$ whose support contains the source point x make non-zero contributions to the vectors $\boldsymbol{j}$ and $\boldsymbol{j}_c$, so that these vectors are relatively empty.

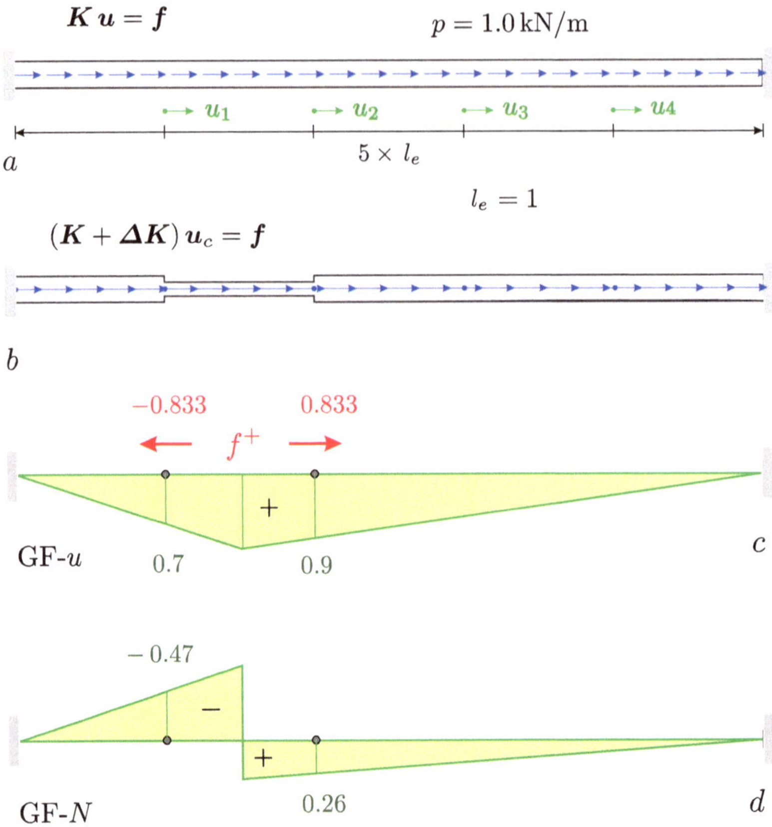

Fig. 5.5 Stiffness change in a stretched bar, the f_i^+ are antisymmetric, the GF-u (Green's function) is symmetric and the GF-N is antisymmetric

The first term on the right handles the step from EA to EA_c, and the second term is the influence of the additional displacement $\boldsymbol{u}^+ = \boldsymbol{K}^{-1}\boldsymbol{f}^+$ on the normal force.

The constriction of the second element (originally the bar had a uniform shape) in Fig. 5.5b means a drop from $EA = 1$ to $EA_c = 0.5$

$$\boldsymbol{K} = \begin{bmatrix} 2 & -1 & 0 & 0 \\ -1 & 2 & -1 & 0 \\ 0 & -1 & 2 & -1 \\ 0 & 0 & -1 & 2 \end{bmatrix} \qquad \boldsymbol{\Delta K} = \begin{bmatrix} -0.5 & 0.5 & 0 & 0 \\ 0.5 & -0.5 & 0 & 0 \\ 0 & 0 & 0 & 0 \\ 0 & 0 & 0 & 0 \end{bmatrix} \tag{5.33}$$

and the vectors are

$$\boldsymbol{u} = \begin{bmatrix} 2.0 \\ 3.0 \\ 3.0 \\ 2.0 \end{bmatrix} \quad \boldsymbol{u}_c = \begin{bmatrix} 1.833 \\ 3.500 \\ 3.333 \\ 2.167 \end{bmatrix} \quad \boldsymbol{f}^+ = \begin{bmatrix} -0.833 \\ 0.833 \\ 0.0 \\ 0.0 \end{bmatrix} \quad \boldsymbol{j} = \begin{bmatrix} -1.0 \\ 1.0 \\ 0.0 \\ 0.0 \end{bmatrix}. \tag{5.34}$$

Note, $\boldsymbol{j}_c = 0.5 \cdot \boldsymbol{j}$, since $EA_c = 0.5\, EA$, and thus the change in $N(x)$ at the midpoint of the second element is

$$N_c - N = J(\boldsymbol{u}_c) - J(\boldsymbol{u}) = -0.5 \cdot \boldsymbol{j}^T \boldsymbol{u} + 0.5 \cdot \boldsymbol{j}^T \boldsymbol{K}^{-1} \boldsymbol{f}^+ = -0.1665\,. \tag{5.35}$$

5.6 Symmetry and Antisymmetry

Influence functions for displacements are generated by **monopoles** and for normal forces (stresses) by **dipoles**. Since the f_i^+ are antisymmetric, stiffness changes should have a greater impact on the stresses (*antisymmetric* × *antisymmetric*) than on displacements (*antisymmetric* × *symmetric*). A quick check

$$\frac{N_c - N}{N} = \frac{0.8335 - 1.0}{1.0} = -0.1665 \qquad \frac{u_c - u}{u} = \frac{2.667 - 2.5}{2.5} = 0.0666 \tag{5.36}$$

confirms this tendency.

In statically determinate systems, however, the opposite is true: The displacements change, but the stresses do not. The reason is that the f_i^+ are equilibrium forces, which do zero work on acting through the piecewise linear influence functions of the section forces N, M, V, while the influence functions for displacements are "real" curves to which the nodal fores f_i^+ are not orthogonal.

Stiffness changes affect the displacements, but not the section forces in a statically determinate structure.

5.7 Orthogonality

A change $J(e) = J(u_c) - J(u)$ in a functional can be expressed as

$$J(e) = \boldsymbol{g}^T \boldsymbol{f}^+ = \sum_{i=1}^{N} \boldsymbol{g}_i^T \boldsymbol{f}_i^+\,, \tag{5.37}$$

and this can be split further into N scalar products of the nodal displacements $\boldsymbol{g}_i$ and the equivalent nodal forces $\boldsymbol{f}_i^+$ at the N nodes, see (3.230), and therefore the interplay

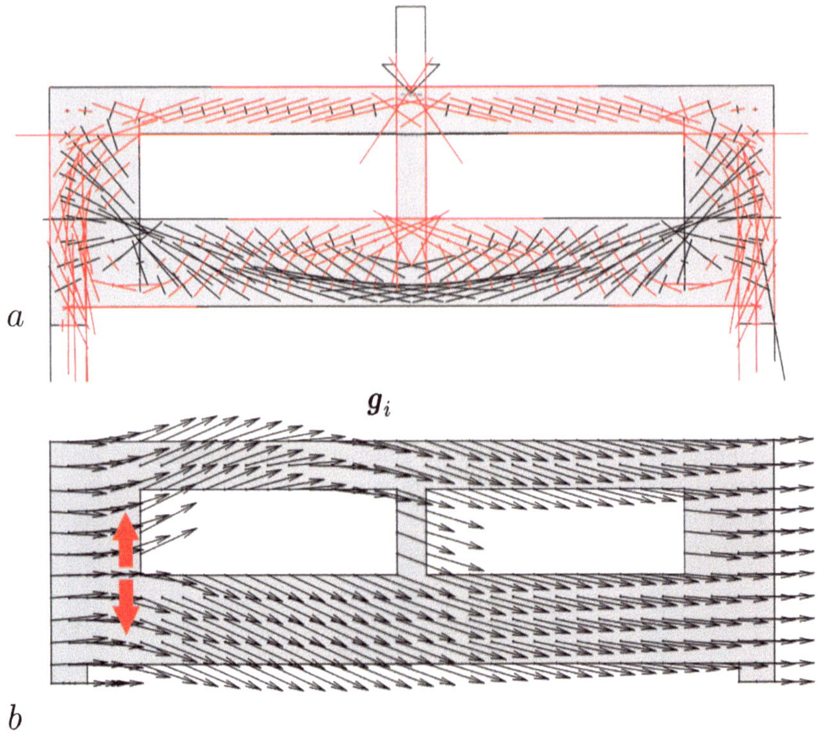

Fig. 5.6 Plate, **a** principal stresses under a vertical point load, **b** influence function for σ_{yy} near the corner. If the vector $\boldsymbol{g}_i$ points in the direction of the principal stresses, stiffness changes—which generate forces $\boldsymbol{f}_i^+$ in the direction of the principal stresses—have the greatest impact on the stress σ_{yy} at the corner

of two plots determines the size of $J(e)$: (i) the sensitivity plot of the functional and (ii) the plot of the principal stresses.

$$J(e) = \boldsymbol{g}^T \boldsymbol{f}^+ = \text{sensitivity plot} \times \text{principal stresses}.$$

The sensitivity plot determines the direction and the size of the $\boldsymbol{g}_i$, and from numerical tests we know that the principal stresses determine the **direction** of the forces $\boldsymbol{f}_i^+$, see Fig. 5.6, while the **size** of the $\boldsymbol{f}_i^+$ depends on the size of the element and on the size of the stiffness change in the element.

It is the same situation as in Sect. 3.32, only the nodal forces are now the $\boldsymbol{f}_i^+$.

5.8 The Effects Fade Away

The further we move away from the source point, the more "linear" an influence function becomes, see Fig. 5.7, since the vector $\boldsymbol{g}$ resembles more and more a vector $\boldsymbol{g} = \boldsymbol{a} + \boldsymbol{x} \times \boldsymbol{b}$, and this means, because the vector $\boldsymbol{f}^+$ is orthogonal to such vectors, that the influence of the forces $\boldsymbol{f}^+$ tends to zero with increasing distance from the source point at which we observe the effects which the $\boldsymbol{f}^+$ produce,

$$J(e) = J(u_c) - J(u) = \boldsymbol{g}^T \boldsymbol{f}^+ \simeq (\boldsymbol{a} + \boldsymbol{x} \times \boldsymbol{b})^T \boldsymbol{f}^+ = 0\,. \tag{5.38}$$

This is the reason, why it is possible to obtain accurate results with averaged material parameters.

If $J(u)$ is a force and the source point lies on the affected element, then, as shown above, the formula must be supplemented with a **correction term** due to the difference $EA_c - EA$

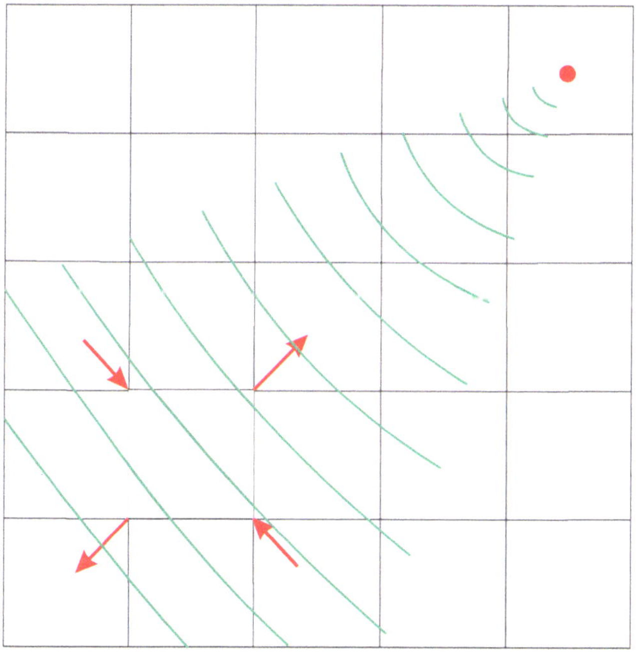

Fig. 5.7 A nearly uniform field hits on forces f^+

$$J(e) = J(u_c) - J(u) = \underbrace{(\boldsymbol{j}_c - \boldsymbol{j})^T \boldsymbol{u}}_{correction} + \boldsymbol{j}_c^T \boldsymbol{K}^{-1} \boldsymbol{f}^+ . \tag{5.39}$$

The first correction is purely local, happens only at the source point, while the second term captures the influence of the forces f_i^+, but these effects subside quickly with increasing distance.

5.9 The Relevance of These Results

The relevance of these results is better understood, when we look at how we average out material parameters to homogenize mechanical properties.

Concrete consists of a wide variety of gravel and cement stone. Each aggregate grain has a different modulus of elasticity, and therefore we should model each aggregate grain with its own finite element. But instead we calculate with an average modulus of elasticity and get quite satisfactory results.

The reason is that the nodal forces f_i^+ on the shell of the aggregate grain, with which we correct the deviations of the modulus of elasticity from the mean value, are equilibrium forces, and so the effects of these forces cancel in the distance.

Imagine a plate Ω consisting of a series of various elements, $\Omega = \Omega_1 \cup \Omega_2 \cup \ldots \Omega_n$, which all have their own modulus E_i, which deviates from the mean value E by an amount $\Delta E_i = E - E_i$, see Fig. 5.8. The exact displacement vector $\boldsymbol{u}_c$ would be the solution of the system

$$\boldsymbol{K}_c \boldsymbol{u}_c = \boldsymbol{f} , \tag{5.40}$$

where the matrix $\boldsymbol{K}_c$ is assembled from the different element matrices $\boldsymbol{K}_e(E_i)$. If, instead, we calculate with a simplified matrix $\boldsymbol{K}$ based on a uniform modulus E

$$\boldsymbol{K}\boldsymbol{u} = \boldsymbol{f} , \tag{5.41}$$

the error in a displacement

$$J(u_c) - J(u) = \boldsymbol{g}^T(\boldsymbol{f} + \boldsymbol{f}^+) - \boldsymbol{g}^T\boldsymbol{f} = \boldsymbol{g}^T\boldsymbol{f}^+ \tag{5.42}$$

will be relatively small, since the forces f_i^+, which originate from the error terms $\Delta E_i = E_i - E$

$$\boldsymbol{K}\boldsymbol{u}_c = \boldsymbol{f} + \boldsymbol{f}^+ , \tag{5.43}$$

on the one hand (1) form equilibrium groups and (2) positive deviations $\Delta E_i > 0$ and negative deviations $\Delta E_i < 0$ will approximately balance. These two effects together ensure that the far field errors will be small, see Fig. 5.8. You do not have to model

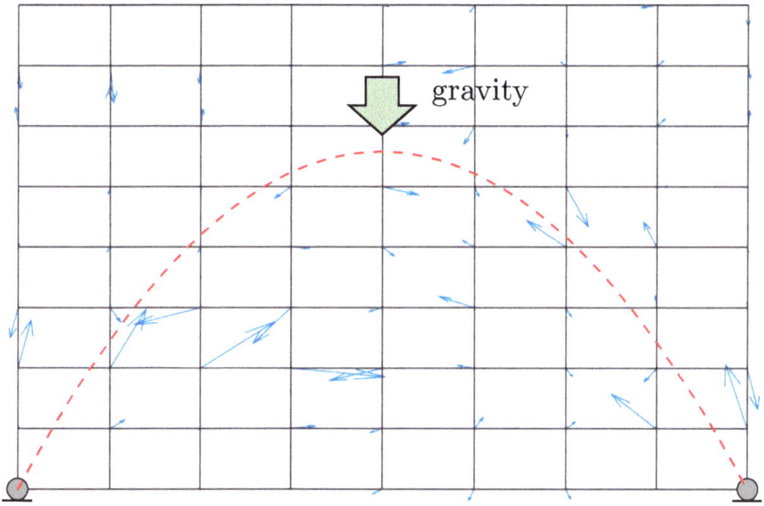

Fig. 5.8 Gravity load and forces f_i^+. First Young's modulus was the same in each element, $E = 1$, and next it was changed at random, $0.5 < E_i < 1.5$, and the forces f_i^+ were calculated. The response of the system $\boldsymbol{K}$ to the forces f_i plus f_i^+, $\boldsymbol{K}\boldsymbol{u}_c = \boldsymbol{f} + \boldsymbol{f}^+$, is the same vector $\boldsymbol{u}_c$ as in the refined model $\boldsymbol{K}_c\boldsymbol{u}_c = \boldsymbol{f}$ with the updated matrix $\boldsymbol{K}_c$, incorporating the different values E_i

each aggregate grain with its own element, mathematics makes sure that the errors cancel each other out.

In Fig. 5.8 the variation in the stiffness E_i was randomly distributed while in Fig. 5.9a, b the stiffness was reduced in two zones by 90%. Remarkable in Fig. 5.9 is the concentration of the forces f_i^+ along the edge of the two zones.

Since the forces f_i^+ are equilibrium forces, we tend to imagine the forces f_i^+ as coming in pairs $\pm f_i^+$, in particular in frames. If, as in Fig. 5.10, a column between two floors fails, the effects of the two opposing nodal forces f_i^+ will cancel each other out in the distance.

But if a ground column fails, the lower force is anchored to the ground and so only the upper force f_i^+ can make its influence felt, and because the antagonist is missing its effect will spread further than that of a pair $\pm f_i^+$.

The failure of a ground column is more critical than the failure of a floor column.

We think we must distinguish between ***internal*** and ***external*** stiffness changes (closer to the edge, to the supports). The **balancing motions** due to changes inside a structure are local affairs, subside rapidly, but stiffness changes closer to the edge—especially the supports—radiate more widely. The closer a compensating motion comes to a **rigid body motion** ("like a jerk in the whole structure"), the more dangerous it is.

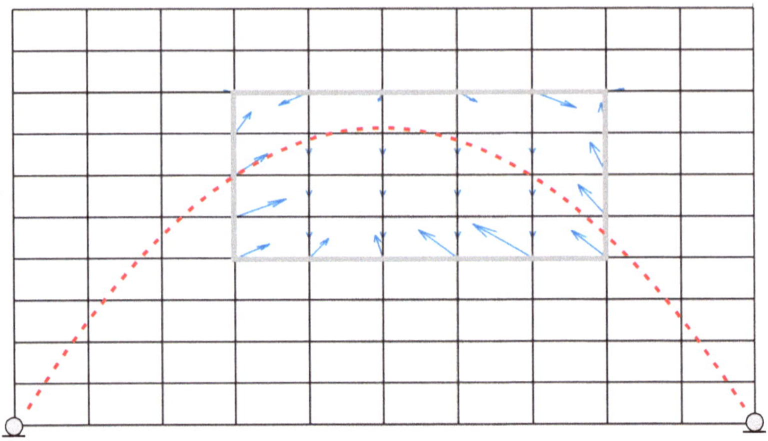

a

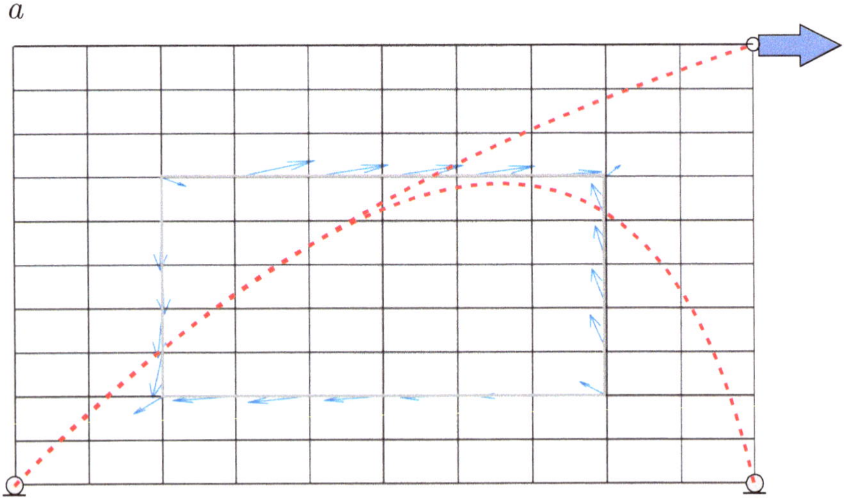

b

Fig. 5.9 Plate and forces f_i^+, **a** gravity load and **b** a corner force. In the two patches Young's modulus E was reduced by 90%. The analogy is obvious: A thin metal sheet welded to a steel tank may experience the same set of forces, only that the forces are distributed continuously along the perimeter and they will have the opposite direction, see Sect. 5.28

5.10 Frames

In Fig. 5.11 we see the bending moments in a frame before and after an edge column failed. The transition from Fig. 5.11a–c corresponds to the action of the additional force f_i^+ in Fig. 5.11b, (we only show the normal force)

$$(\boldsymbol{K} + \boldsymbol{\Delta K})\,\boldsymbol{u}_c = \boldsymbol{f} \qquad \boldsymbol{K}\,\boldsymbol{u}_c = \boldsymbol{f} + \boldsymbol{f}^+ . \tag{5.44}$$

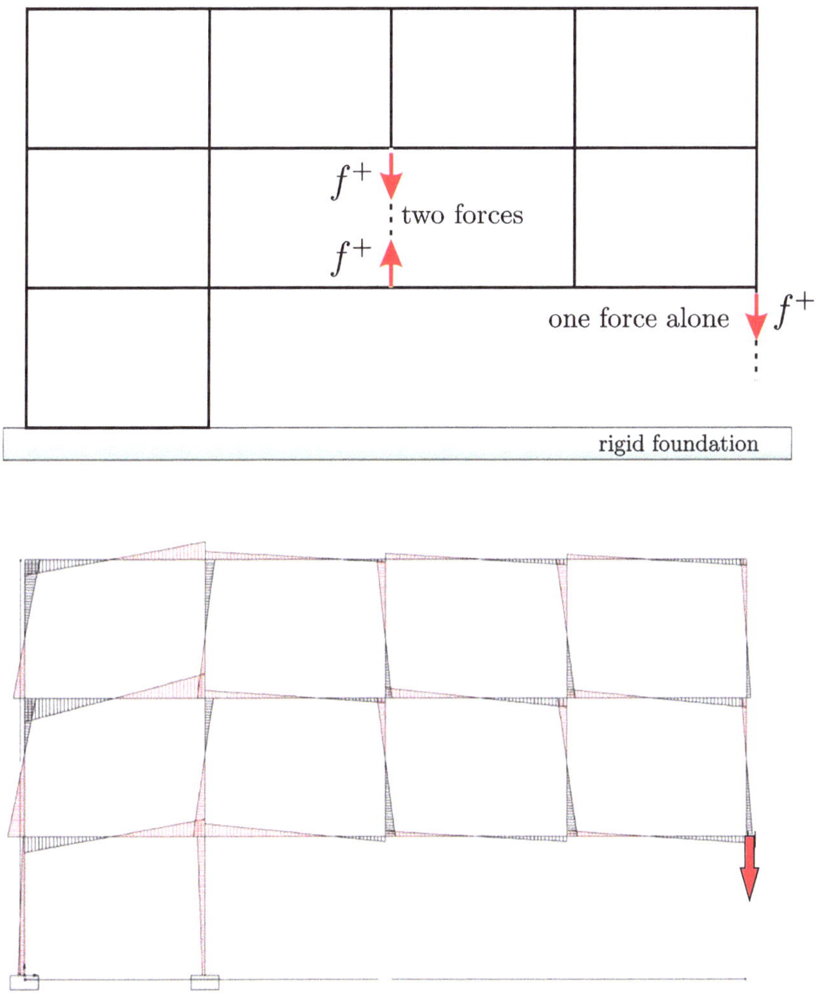

Fig. 5.10 Loss of a column in between two upper floors and in between the second floor and the foundation plate

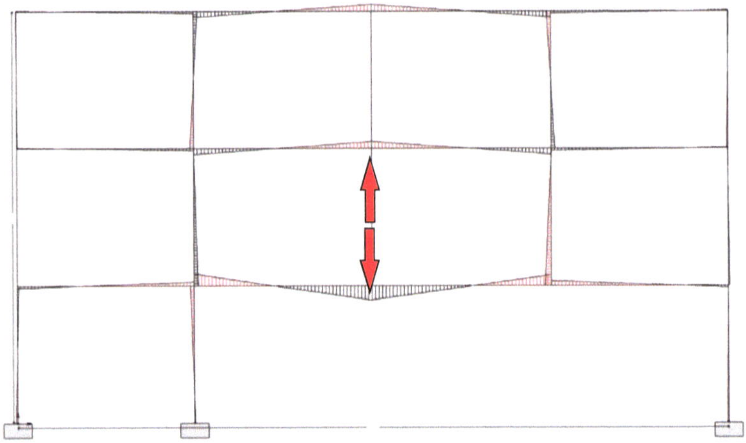

Fig. 5.10 (continued)

The vertical displacements of the edge node in Fig. 5.11b after the collapse of the column was

$$u_c = 21.3\,\text{mm}\,, \tag{5.45}$$

which resulted in forces

$$\pm f^+ = \frac{EA}{h}\,21.3 \cdot 10^{-3}\,\text{m} = 7.6 \cdot 10^3\,\text{kN} \tag{5.46}$$

corresponding to $\boldsymbol{\Delta K}\,\boldsymbol{u}_c = \boldsymbol{f}^+$.

In Fig. 5.12 two pin-ended diagonals are added to a Vierendeel frame. The new vector $\boldsymbol{u}_c$ is the solution to $\boldsymbol{K}\boldsymbol{u}_c = \boldsymbol{f} + \boldsymbol{f}^+$. The influence of the first pair $\pm\boldsymbol{f}^+$ on $M(x)$ will be noticeable since the relative displacements of the end points is relatively large, $\Delta \gg 1$, (large runtime difference), while the influence of the second pair will be small since the nodal displacements hardly differ, $\Delta \ll 1$, (small runtime difference).

5.11 Forces j^+

The formula $J(e) = \boldsymbol{g}^T\boldsymbol{f}^+$ makes us guess, whether there is not also a formula

$$J(e) = \boldsymbol{u}^T\boldsymbol{j}^+\,, \tag{5.47}$$

see Fig. 5.13.

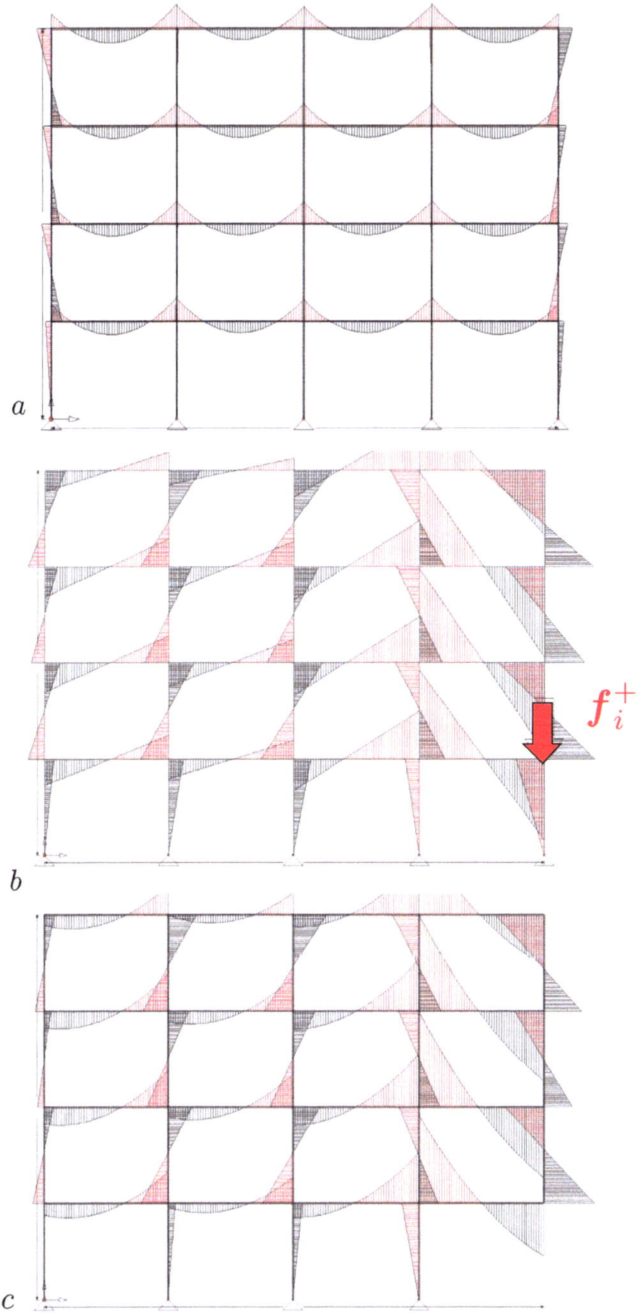

Fig. 5.11 Failure of an edge column, **a** moment distribution under gravity load, **b** force f^+ and moments generated by this force, **c** a + b = resulting moment distribution

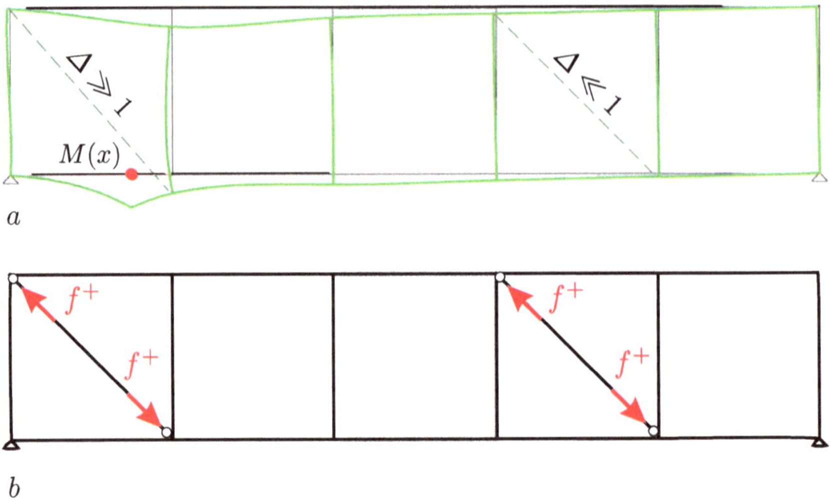

Fig. 5.12 Vierendeel frame, **a** influence function for a moment $M(x)$, **b** adding pin-ended diagonals to the frame

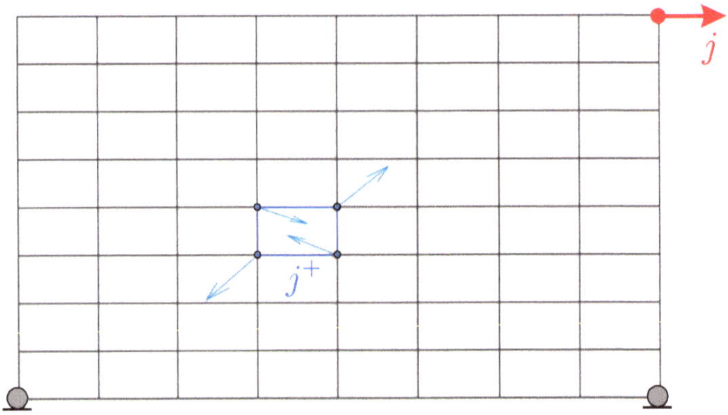

Fig. 5.13 Calculation of the influence function for a nodal displacement with a Dirac delta, i.e. a nodal force j. A stiffness change in an element activates additional nodal forces $\boldsymbol{j}^+$, so that $\boldsymbol{K}\boldsymbol{g}_c = \boldsymbol{j} + \boldsymbol{j}^+$ produces the modified vector $\boldsymbol{g}_c$, the modified influence function, see also Fig. 4.14

And there is indeed, because the displacement $u(x)$ at a point x can be calculated in two ways

$$u(x) = \int_0^l G(y, x)\, p(y)\, dy = \int_0^l u(y)\, \delta(y - x)\, dy\,. \tag{5.48}$$

In the linear algebra of finite elements, the second equation reads

$$J(u) = u(x) = \boldsymbol{u}^T \boldsymbol{j}(x)\,, \tag{5.49}$$

where the vector $\boldsymbol{j}(x)$ are the equivalent nodal forces, $j_i = \varphi_i(x)$, corresponding to a Dirac delta at the point x. The equation $(\boldsymbol{K} + \boldsymbol{\Delta K})\boldsymbol{g}_c = \boldsymbol{j}$ implies

$$\boldsymbol{K}\boldsymbol{g}_c = \boldsymbol{j} - \boldsymbol{\Delta K}\boldsymbol{g}_c = \boldsymbol{j} + \boldsymbol{j}^+\,, \tag{5.50}$$

and multiplying it from the left with $\boldsymbol{u}$ (it is $\boldsymbol{u}^T\boldsymbol{K} = \boldsymbol{f}^T$)

$$\underbrace{\boldsymbol{u}^T\boldsymbol{K}\boldsymbol{g}_c}_{J(\boldsymbol{u}_c)} = \underbrace{\boldsymbol{u}^T\boldsymbol{j}}_{J(\boldsymbol{u})} - \boldsymbol{u}^T\boldsymbol{\Delta K}\boldsymbol{g}_c = \boldsymbol{u}^T\boldsymbol{j} + \boldsymbol{u}^T\boldsymbol{j}^+\,, \tag{5.51}$$

results in

$$J(e) = u_c(x) - u(x) = \boldsymbol{u}^T\boldsymbol{j}^+\,, \tag{5.52}$$

where

$$\boldsymbol{j}^+ = -\boldsymbol{\Delta K}\boldsymbol{g}_c\,. \tag{5.53}$$

5.12 Replacement as Alternative

Instead of working with forces f_i^+, there are other ways to make ends meet. Imagine a situation where a pin-ended column in between two floors, see Fig. 5.14, must be replaced. Provided the new column has the same stiffness EA,

- in the building yard it must have the same length as the earlier unstressed column, and
- the new column must be prestressed prior to the installation, so that it has the same length as the column which it replaces.

After the installation, the restraints can be removed, and nothing has changed.

If instead the new column has a stiffness $EA_c > EA$, excess forces ($= f_i^+$) will press against the upper and lower floor as soon as the restraints are removed, see Fig. 5.14. Excess forces because to compress a column, when $EA_c > EA$, larger forces are necessary than in the case $EA_c = EA$. When the restraints are removed, these excess forces press against the floor plates.

If the new column is less stiff, $EA_c < EA$, the situation is the same, we only must turn the sign of the f_i^+ around.

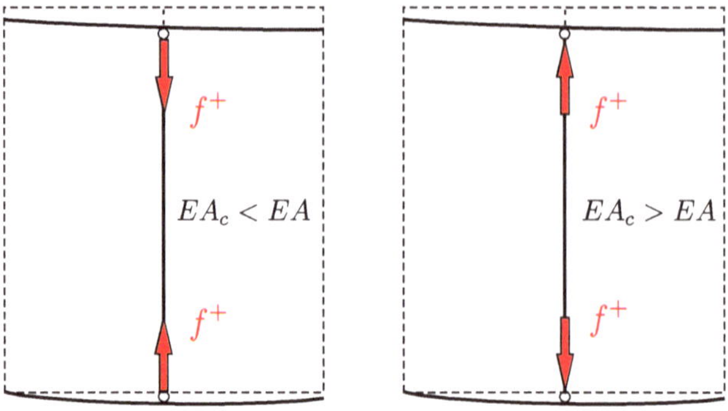

Fig. 5.14 Increase and decrease in the longitudinal stiffness *EA* of an interfloor column

5.13 The Derivative of the Inverse K^{-1}

If the stiffness matrix $\boldsymbol{K}$ is a function of a parameter s, the derivative of its inverse $\boldsymbol{K}^{-1}$ with respect to s is [4], p. 163,

$$\frac{d}{ds}\boldsymbol{K}^{-1} = -\boldsymbol{K}^{-1}\frac{d\boldsymbol{K}}{ds}\boldsymbol{K}^{-1}\,. \tag{5.54}$$

We test this formula with a bar as in Fig. 3.46 which consists of four linear elements. The stiffness in the first three elements is $k = EA/l_e = 1$ and $k = (1 + s)$ in the fourth element, and so the stiffness matrix is

$$\boldsymbol{K}(s) = \begin{bmatrix} 2 & -1 & 0 \\ -1 & 2 & -1 \\ 0 & -1 & 2+s \end{bmatrix}. \tag{5.55}$$

The two-terms Taylor expansion of the formula $\boldsymbol{u}(s) = \boldsymbol{K}^{-1}(s)\boldsymbol{f}$ reads

$$\boldsymbol{u}(s) \simeq [\boldsymbol{K}^{-1}(0) + \frac{d}{ds}\boldsymbol{K}^{-1}(0)\,ds]\boldsymbol{f} = \boldsymbol{u}(0) + \boldsymbol{du}(0)\,ds\,. \tag{5.56}$$

At $s = 0$ the inverse is

$$\boldsymbol{K}^{-1}(0) = \begin{bmatrix} 0.75 & 0.5 & 0.25 \\ 0.5 & 1.0 & 0.5 \\ 0.25 & 0.5 & 0.75 \end{bmatrix}, \tag{5.57}$$

and the derivative $d\boldsymbol{K}/ds$ contains only one non-zero entry, $k_{3,3} = 1$, and so—given a load $\boldsymbol{f} = \{1, 1, 1\}^T$—the increment $\boldsymbol{du}(0)$ is

$$\boldsymbol{du}(0) = -\boldsymbol{K}^{-1}(0)\,\frac{d}{ds}\,\boldsymbol{K}(0)\,\boldsymbol{K}^{-1}(0)\boldsymbol{f} = \begin{bmatrix} -0.3750 \\ -0.7500 \\ -1.1250 \end{bmatrix}. \tag{5.58}$$

If the stiffness k in the fourth element increases from $k = 1$ to double its value, $k = 1 + s = 2$, which means ds =1, the result is

$$\boldsymbol{u}(1.0) = \begin{bmatrix} 1.5 \\ 2.0 \\ 1.5 \end{bmatrix} + 1.0 \cdot \begin{bmatrix} -0.3750 \\ -0.7500 \\ -1.1250 \end{bmatrix} = \begin{bmatrix} 1.1250 \\ 1.2500 \\ 0.3750 \end{bmatrix}, \qquad \begin{bmatrix} 1.2857 \\ 1.5714 \\ 0.8571 \end{bmatrix} \text{(exact)} \tag{5.59}$$

The stiffness matrices of bar and beam elements carry the factors $k = EI/l^3$ and $k = EA/l$ respectively up front and therefore the derivative of the global stiffness matrix $\boldsymbol{K}$ with respect to the stiffness k of a single element is (in a large system) a vast array of empty space and embedded into this void is the element matrix without the k and the increment $d/dk\,\boldsymbol{K}\,\Delta k$ has—if we only concentrate on the non-zero part—the form

$$\frac{d}{dk}\boldsymbol{K}\,\Delta k = \Delta k \begin{bmatrix} 1 & -1 \\ -1 & 1 \end{bmatrix} \qquad \Delta k = \frac{\Delta EA}{l}. \tag{5.60}$$

This is the matrix $\boldsymbol{\Delta K}$ in $(\boldsymbol{K} + \boldsymbol{\Delta K})\boldsymbol{u}_c = \boldsymbol{f}$ after (5.60) is padded with zeros to match the size of the matrix $\boldsymbol{K}$.

5.14 The Derivatives $\partial u/\partial f_k$ and $\partial u/\partial k_{ij}$

The derivative of the solution $\boldsymbol{u} = \boldsymbol{K}^{-1}\boldsymbol{f}$ with respect to a nodal force f_k is easy

$$\frac{\partial}{\partial f_k}\,\boldsymbol{u} = \frac{\partial}{\partial f_k}\,(\boldsymbol{g}_1 f_1 + \boldsymbol{g}_2 f_2 + \cdots + \boldsymbol{g}_n f_n) = \boldsymbol{g}_k\,, \tag{5.61}$$

and to determine the derivative of $\boldsymbol{u}$ with respect to an entry k_{ij} we expand

$$\frac{\partial}{\partial k_{ij}}(\boldsymbol{K}\boldsymbol{u} = \boldsymbol{f}) \tag{5.62}$$

which results in

$$\frac{\partial}{\partial k_{ij}}\boldsymbol{K}\,\boldsymbol{u} + \boldsymbol{K}\,\frac{\partial}{\partial k_{ij}}\,\boldsymbol{u} = \boldsymbol{e}_i\,\boldsymbol{e}_j^T\,\boldsymbol{u} + \boldsymbol{K}\,\frac{\partial \boldsymbol{u}}{\partial k_{ij}} = \frac{\partial}{\partial k_{ij}}\boldsymbol{f} = 0\,, \tag{5.63}$$

and this means[3]

$$\boxed{\frac{\partial \boldsymbol{u}}{\partial k_{ij}} = -\boldsymbol{K}^{-1} u_j \, \boldsymbol{e}_i = -u_j \, \boldsymbol{g}_i} \tag{5.64}$$

So, given a change on the diagonal, $k_{ii} \to k_{ii} + \Delta k_{ii}$, the effect on the displacement vector is approximately

$$\boldsymbol{u}_c - \boldsymbol{u} \simeq -u_i \, \boldsymbol{g}_i \, \Delta k_{ii} \,. \tag{5.65}$$

This corresponds to Eq. (5.12) in the introduction, where we looked at the change in a single spring ($k\, u = f$) with $1/k \equiv g$

$$u_c - u \simeq -u \, \frac{1}{k} \, \Delta \, k. \tag{5.66}$$

The transition $\boldsymbol{u} \to \boldsymbol{u}_c$ is controlled by the columns $\boldsymbol{g}_i$ of the inverse

This is why influence functions play such an important role. In Fig. 5.15 a shift of the rotational stiffness $k_{ii} \to k_{ii} + \Delta k_{ii}$ of a node leads to a shift $\boldsymbol{u}_c - \boldsymbol{u} \simeq -u_i \, \boldsymbol{g}_i \, \Delta k_{ii}$ in the displacement vector and this shift is proportional to $\boldsymbol{g}_i$.

The same is true, of course, with regard to the internal forces, as the moments

$$M_c - M \simeq -u_i \, M(G_i) \, \Delta k_{ii} \,, \tag{5.67}$$

since the difference $M_c - M$ is a scaled-up, amplified version of the moment $M(G_i)$ of the influence function for the rotation u_i, see Fig. 5.15.

Normally, it will not be just one term on the diagonal, which changes, but a complete element matrix. In the case of the stepped bar in Fig. 5.16 which consists of three elements, with $EA = 3, 2, 1$, the stiffness in element two drops from $EA = 2$ to $EA = 1$

$$\boldsymbol{K}_2 = \begin{bmatrix} 2 & -2 \\ -2 & 2 \end{bmatrix} \quad \Rightarrow \quad \boldsymbol{K}_2^{new} = \begin{bmatrix} 1 & -1 \\ -1 & 1 \end{bmatrix} . \tag{5.68}$$

This means

$$\Delta k_{11} = \Delta k_{22} = -1 \qquad \Delta k_{12} = \Delta k_{21} = +1 \,, \tag{5.69}$$

and so the change in the displacement vector is approximately

[3] in the scalar case: $k\, u = f,\; du/dk = -1/k^2 \cdot f = -1/k \cdot u$.

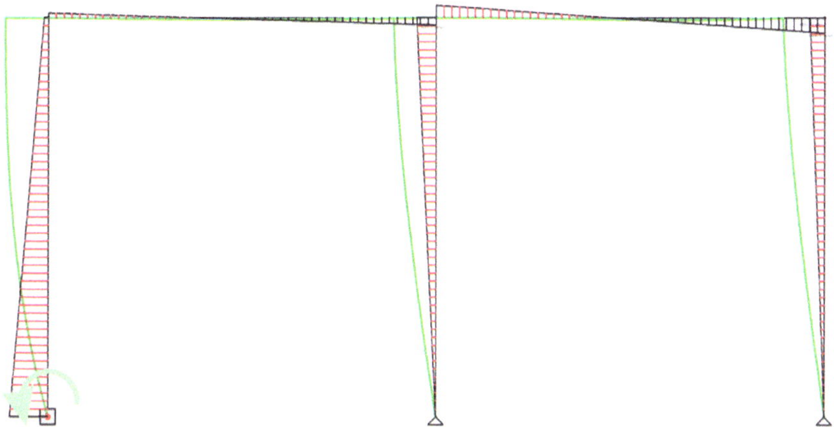

Fig. 5.15 Change in the rotational stiffness of a node. Influence function for the rotation u_i and the moments of the influence function

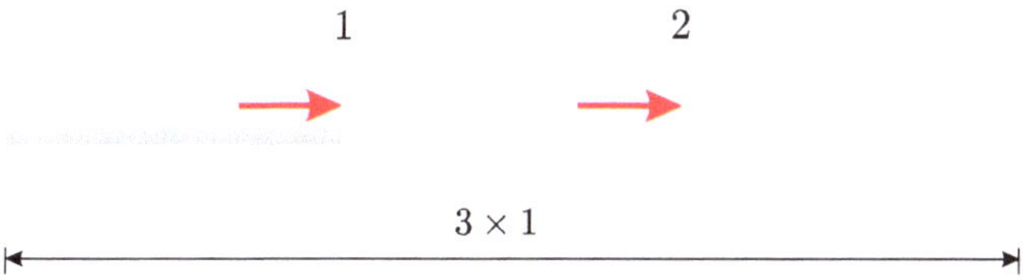

Fig. 5.16 Stepped bar

$$\boldsymbol{u}_c - \boldsymbol{u} \simeq \sum_{i,j=1}^{2} \frac{\partial \boldsymbol{u}}{\partial k_{ij}} \Delta k_{ij} = \sum_{i,j=1}^{2} -u_j \, \boldsymbol{g}_i \, \Delta k_{ij} \,. \tag{5.70}$$

The $\boldsymbol{g}_i$ are the columns of the inverse

$$\boldsymbol{K} = \begin{bmatrix} 5 & -2 \\ -2 & 3 \end{bmatrix} \quad \boldsymbol{K}^{-1} = \begin{bmatrix} 0.2727 & 0.1818 \\ 0.1818 & 0.4545 \end{bmatrix} \quad \boldsymbol{K}_c = \begin{bmatrix} 4 & -1 \\ -1 & 2 \end{bmatrix}, \tag{5.71}$$

and the u_j are the nodal displacements $\boldsymbol{u} = \{0.4545, 0.6364\}^T$ in the load case $\boldsymbol{f} = \{1, 1\}^T$, which gives

$$\begin{aligned} \boldsymbol{u}_c - \boldsymbol{u} &\simeq -\big[(u_1 \Delta k_{11} + u_2 \Delta k_{12}) \, \boldsymbol{g}_1 + (u_1 \Delta k_{21} + u_2 \Delta k_{22}) \, \boldsymbol{g}_2\big] \\ &= \begin{bmatrix} -0.0165 \\ 0.0496 \end{bmatrix} \qquad \begin{bmatrix} -0.0260 \\ 0.0779 \end{bmatrix} \text{(exact)} \,. \end{aligned} \tag{5.72}$$

We can also do a test: According to (3.166)

$$(\boldsymbol{I} + \boldsymbol{K}^{-1} \boldsymbol{\Delta K})\, \boldsymbol{u}_c = \boldsymbol{u}\,, \tag{5.73}$$

and substituting for $\boldsymbol{u}_c$ the approximation $\boldsymbol{u}_c \simeq \{0.4380, 0.6860\}^T$ the result is

$$(\boldsymbol{I} + \boldsymbol{K}^{-1} \boldsymbol{\Delta K})\, \boldsymbol{u}_c = \begin{bmatrix} 0.4606 \\ 0.6183 \end{bmatrix} \simeq \begin{bmatrix} 0.4545 \\ 0.6364 \end{bmatrix} = \boldsymbol{u}\,, \tag{5.74}$$

while the exact vector $\boldsymbol{u}_c = \{0.4286, 0.7143\}^T$ returns exactly $\boldsymbol{u}$.

To be complete we mention also the formula $\partial f_i / \partial u_j = k_{ij}$, which is a direct consequence of $\boldsymbol{K}\, \boldsymbol{u} = \boldsymbol{f}$ and so $\partial \boldsymbol{f} / \partial u_j = \boldsymbol{k}_j$ (column j of $\boldsymbol{K}$).

5.15 Integrating over the Defective Element

A change of one single entry k_{ij} in a stiffness matrix $\boldsymbol{K}$ will affect all entries of the inverse $\boldsymbol{K}^{-1}$. All influence functions, as for example the influence function for the transverse force $V(x)$ of the frame in Fig. 5.17, will take a new shape, $G_3(y, x) \to G_3^c(y, x)$, and so any such shift $k_{ij} \to k_{ij} + \Delta k_{ij}$ translates into a shift

$$V_c(x) - V(x) = \sum_e \int_0^{l_e} (G_3^c(y, x) - G_3(y, x))\, p(y)\, dy\,, \tag{5.75}$$

of the shear force at all points x. If the frame is ten stories high, we would have to integrate over many, many frame elements to update the shear force at one single point x.

But we can take a shortcut: **We need to integrate only over the defective element to determine** $J(e)$.

To explain this, we consider the two-span beam in Fig. 5.18. Given a uniform stiffness EI the **weak form** of the beam equation $EI\, w^{IV} = p$ is

$$\int_0^l EI\, w''\, \delta w''\, dx = \int_0^l p\, \delta w\, dx\,, \tag{5.76}$$

or

$$a(w, \delta w) = (p, \delta w) \tag{5.77}$$

for short. When the stiffness in the second span becomes $EI + \Delta EI$, the deflection of the beam changes, $w \to w_c$, and the weak form reads

$$\int_0^l EI\, w_c''\, \delta w''\, dx + \underbrace{\int_{l/2}^l \Delta EI\, w_c''\, \delta w''\, dx}_{d(w_c, \delta w)} = \int_0^l p\, \delta w\, dx\,, \tag{5.78}$$

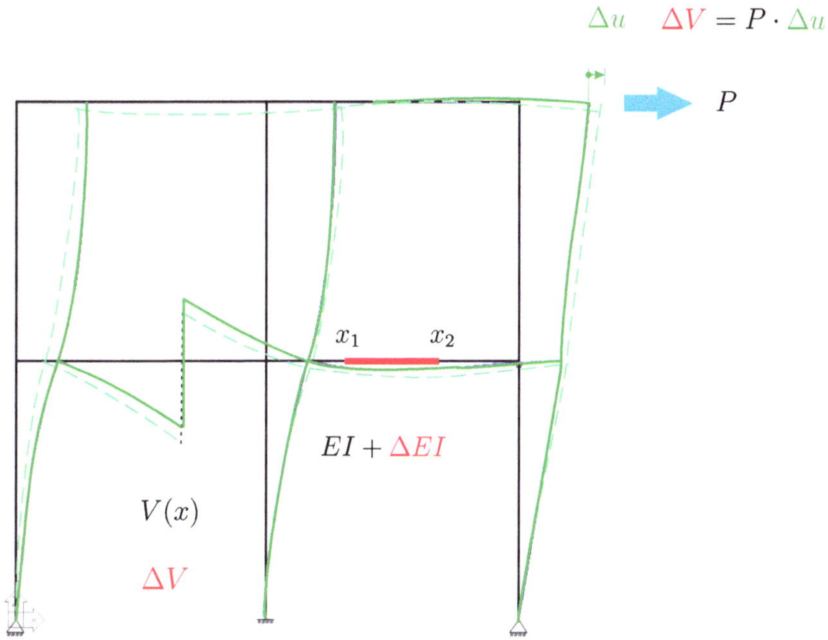

Fig. 5.17 A local stiffness change affects the influence function for the shear force $V(x)$ in all parts of the frame

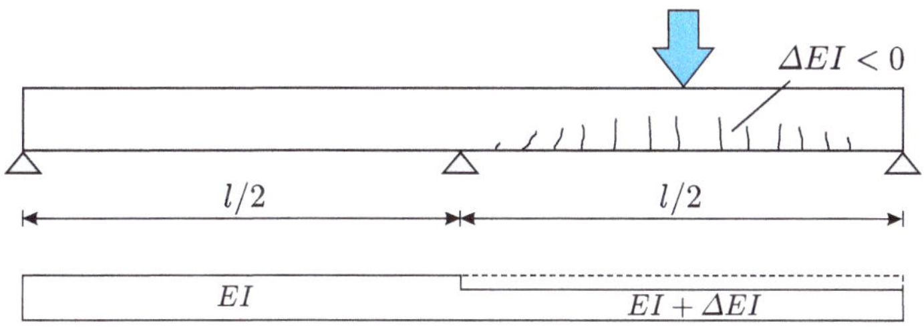

Fig. 5.18 Cracks in a beam

or

$$a(w_c, \delta w) + d(w_c, \delta w) = (p, \delta w)\,. \tag{5.79}$$

If we subtract the two equations this gives

$$a(w_c - w, \delta w) + d(w_c, \delta w) = 0\,, \tag{5.80}$$

or with $e = w_c - w$

$$a(e, \delta w) = -d(w_c, \delta w)\,. \tag{5.81}$$

If we choose for δw the influence function G of a functional $J(w)$, the result is the shift in the functional, $J(e) = a(e, G) = -d(w_c, G)$,

$$\boxed{J(e) = -d(w_c, G)} \tag{5.82}$$

Note,

$$d(w_c, \delta w) = \int_{l/2}^{l} \Delta EI\, w_c''\, \delta w''\, dx = \frac{\Delta EI}{EI} \int_{l/2}^{l} \frac{M_c\, M}{EI_c}\, dx\,. \tag{5.83}$$

In the notation of linear algebra, the formulation is even easier. With the transition $\boldsymbol{K} \rightarrow \boldsymbol{K} + \boldsymbol{\Delta K}$ the original weak formulation

$$\boldsymbol{\delta u}^T \boldsymbol{K u} = \boldsymbol{\delta u}^T \boldsymbol{f} \tag{5.84}$$

becomes

$$\boldsymbol{\delta u}^T (\boldsymbol{K} + \boldsymbol{\Delta K}) \boldsymbol{u}_c = \boldsymbol{\delta u}^T \boldsymbol{f}\,, \tag{5.85}$$

and so the vector $\boldsymbol{e} = \boldsymbol{u}_c - \boldsymbol{u}$ solves the equation

$$\boldsymbol{\delta u}^T \boldsymbol{K e} = -\boldsymbol{\delta u}^T \boldsymbol{\Delta K u}_c \qquad \text{for all } \boldsymbol{\delta u}\,. \tag{5.86}$$

Given a structure, which is in a state $\boldsymbol{u}$, we can take measurements by means of an influence function, the vector $\boldsymbol{g}$,

$$J(\boldsymbol{u}) = \boldsymbol{g}^T \boldsymbol{K u}\,, \tag{5.87}$$

and if we substitute in (5.86) for $\boldsymbol{\delta u}$ the vector $\boldsymbol{g}$, this gives

$$J(\boldsymbol{e}) = -\boldsymbol{g}^T \boldsymbol{\Delta K u}_c \qquad (= -d(w_c, G))\,, \tag{5.88}$$

which means, we only need to measure on the affected element(s), see Fig. 5.19; this equation corresponds to (5.82).

The expression (5.88) is the standard strain energy product on Ω_e

$$J(\boldsymbol{e}) = -d(w_c, G) = -\alpha \cdot a(G, w_c)_{\Omega_e} \tag{5.89}$$

between the (old) influence function G for $J(w)$ and the new solution w_c, where α is the ratio

$$\alpha = \frac{\Delta E}{E} = \frac{\Delta EI}{EI} = \frac{\Delta EA}{EA} = \frac{\Delta k}{k} = \frac{\Delta K}{K} \qquad \text{etc.} \tag{5.90}$$

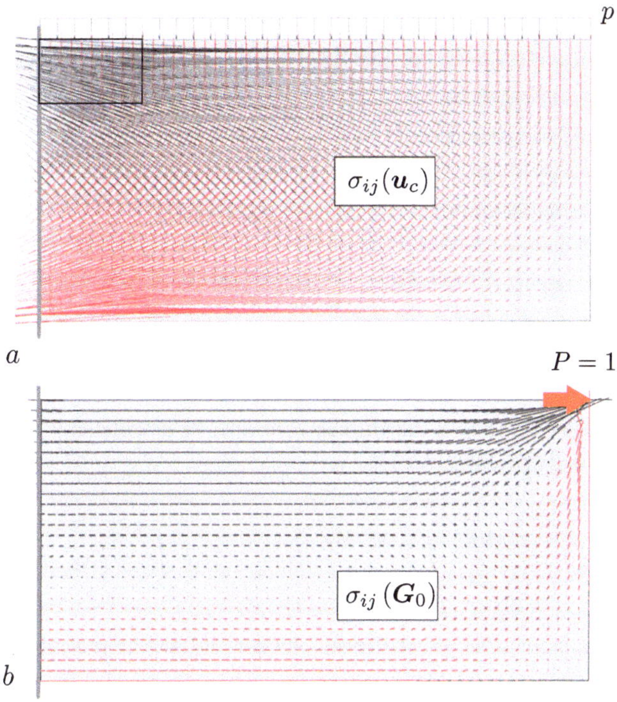

$$u_{cx}(\boldsymbol{x}) - u_x(\boldsymbol{x}) = -\frac{\Delta E}{E}\, a(\boldsymbol{G}_0, \boldsymbol{u})_{\Omega_e} = -\frac{\Delta E}{E} \int_{\Omega_e} \sigma_{ij}(\boldsymbol{u}_c)\, \varepsilon_{ij}(\boldsymbol{G}_0)\, d\Omega_{\boldsymbol{y}}$$

Fig. 5.19 Plate, **a** stress distribution in the plate after the uppermost element Ω_e cracked, **b** stresses resulting from the single force $P = 1$, that is the influence function for the displacement u_x

5.16 Mohr or the Weak Form of Influence Functions

Established that stiffness changes can be handled by a **local analysis**, we want to discuss the necessary steps in the following in more details.

Any displacement $u(x)$ can be calculated in two ways

$$u(x) = J(u) = \int_0^l G(y, x)\, p(y)\, dy = \int_0^l \frac{M(y)\, M_G(y, x)}{EI}\, dy = a(G, w) \quad (5.91)$$

either with a strong influence function or a weak influence function (Mohr). Here, and in the following M_G is the moment of the influence function.

Suppose $J(w)$ is the deflection of a beam at a point x

$$J(w) = \int_0^l \frac{M(y)\, M_G(y, x)}{EI}\, dy\,. \quad (5.92)$$

Then one element of the beam cracks, $EI \to EI + \Delta EI$, and so the moments $M \to M_c$ and $M_G \to M_G^c$ change, and we have to start all over again

$$J(w_c) = \int_0^l \frac{M_c(y)\, M_G^c(y, x)}{EI_c}\, dy\,. \tag{5.93}$$

The function EI_c in the denominator is the *step function*

$$EI_c = \begin{cases} EI + \Delta EI & \text{in interval } [x_a, x_b] \\ EI & \text{else} \end{cases}\,. \tag{5.94}$$

Now comes the interesting point.

As demonstrated above, (5.88), the shift in $J(w)$ can be calculated by integrating only over the defective element

$$\boxed{J(w_c) - J(w) = -\frac{\Delta EI}{EI} \int_{x_a}^{x_b} \frac{M_c\, M_G}{EI_c}\, dy = -\Delta EI \int_{x_a}^{x_b} w_c''\, G''\, dy} \tag{5.95}$$

M_c is the load moment *after* the stiffness change and M_G is the moment of the influence function *before* the change.

It also makes no difference whether we pair M_c with M_G or M with M_G^c

$$M \times M_G^c \equiv M_c \times M_G\,, \tag{5.96}$$

the result is the same

$$\boxed{J(w_c) - J(w) = -\frac{\Delta EI}{EI} \int_{x_a}^{x_b} \frac{M\, M_G^c}{EI_c}\, dy} \tag{5.97}$$

And the point x **can lie anywhere**. Imagine it is a point on the tenth floor, where the focus is on a value $J(w)$, a deflection $w(x)$ or a moment $M(x)$ or else, and we want to assess, which effects cracks in an element on the third floor will have on $J(w)$.

The change $J(e)$ at x can be calculated *solely* by integrating over the damaged element $[x_a, x_b]$ on the third floor. What matters is, how much of the action, the Dirac delta at x on the tenth floor, arrives on the third floor, how large the moment M_G of the influence function is in the cracked element $[x_a, x_b]$, and how large the load moment M_c is in the cracked element on the third floor.

The difficulty with (5.95) is the unknown load moment M_c, which forces us to substitute for M_c the original moment M

$$J(w_c) - J(w) \simeq -\frac{\Delta EI}{EI} \int_{x_a}^{x_b} \frac{M\, M_G}{EI_c}\, dy\,. \tag{5.98}$$

One additional remark: Equation(5.95), which looks like Mohr's integral, can also predict the effects of stiffness changes on forces, such as $V(x) \rightarrow V(x) + \Delta V$, while Mohr's original integral in the formulation as *principle of virtual forces*,

$$\mathcal{G}(G, w) = \delta W_e^* - \delta W_i^* = 0\,, \tag{5.99}$$

cannot. The riddle is resolved if we remember that (5.95) is a recast of (5.88).

5.17 Near and Far

The change $J(w_c) - J(w)$ will only be prominent, if the **strain energy product** $\delta W_i = a(G, w_c)_{\Omega_e}$ in the element "is large",[4] which means if M_c and M_G respectively are both large and are not orthogonal(!).

Since influence functions (usually) quickly subside, stiffness changes in distant regions will mostly have little impact on the internal forces in the "foreground".

And because δW_i is a **scalar product**, M and M_G can be orthogonal, as when M_c is antisymmetric, and M_G is symmetric. In such situations, regardless of the size of M and M_G, the change $J(w_c) - J(w)$ is zero.

Prior to any calculation the shapes of the influence functions allow therefore to estimate, which stiffness changes will have an effect.

In Fig. 5.20 a frame braces against a strong wind from the left. The strong wind may lead to cracks, $EI \rightarrow EI + \Delta EI$, in some frame elements, and we want to know how such cracks influence the fixed end moment at the foot of the leftmost column—that is at one single point (***focus on a point***).

First, we compute the influence function G for the moment and plot the bending moment distribution M_G of this influence function. Then we move from element to element and evaluate the expression

$$M_c - M = d(w_c, G) = -\frac{\Delta EI}{EI} \int_0^{l_e} \frac{M_c\, M_G}{EI_c}\, dy\,. \tag{5.100}$$

Each such integral is the influence, a crack ΔEI in an element $(0, l_e)$ has on the fixed end moment $M(x)$. The frame elements in which the moments M_G and $M_c \simeq M$ *both* are large (and not orthogonal) contribute the most. Not surprisingly are these the two columns at the ground floor.

A related question is the search for the maximum effect, which a stiffness change in one particular element has on the frame (***focus on an element***). This can be written as

[4] We omit the asterix on δW_i^*.

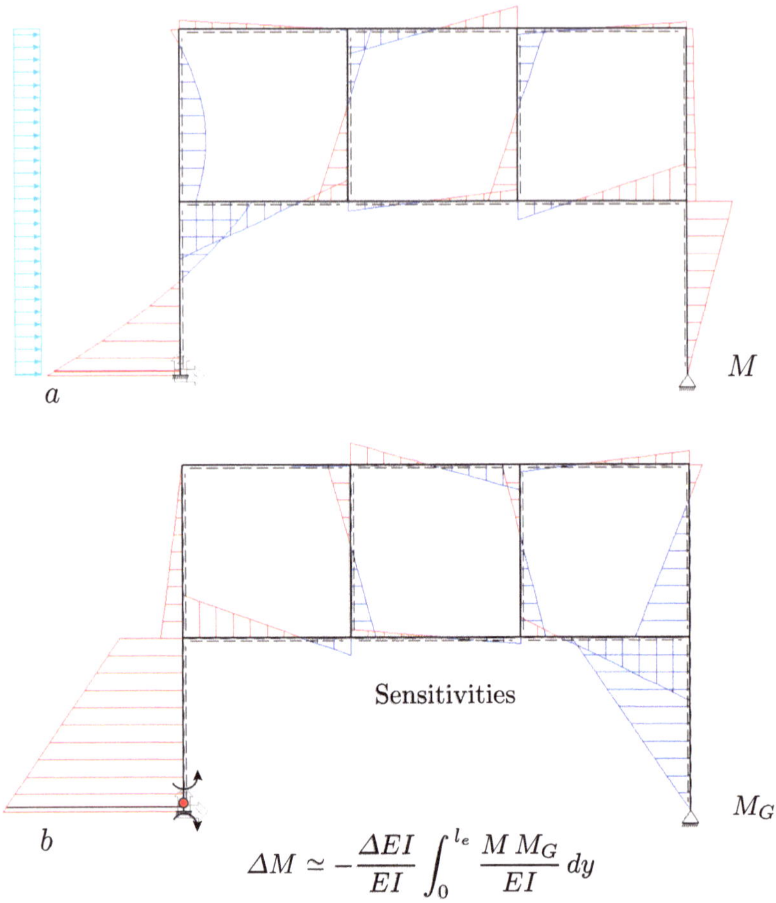

Fig. 5.20 Frame, **a** moment M from wind load, and **b** moment M_G produced by a unit rotation of the foot point, (= influence function for the moment at that point). Elementwise the integral of M and M_G is weighted with the ratio $\Delta EI/EI$ and this allows to assess approximately the effect of a local change ΔEI on the moment

$$max\; J(e) = \int_0^{l_e} \frac{M_c\, M_G}{EI_c}\, dy \qquad all\ relevant\ GF\,, \tag{5.101}$$

which means we search for the curve M_G which, when superposed with M_c in the cracked element $(0, l_e)$, returns the largest value $J(e)$, where M_G is the moment of the influence function, which belongs to the functional $J(w)$.

5.18 Supports

In case the stiffness of a support is modified, the change $J(e) = -d(G, w_c)$ in any functional **can be measured directly at the spring**. In particular for complex 3-D models with a large number of degrees of freedom, this is an advantage not to be missed.

Elastic Support

The weak form of the beam (equation) in Fig. 5.21 is

$$\int_0^l \frac{M\,\delta M}{EI}\,dx + \delta w'\,w'\,k = \int_0^l p\,\delta w\,dx\,. \tag{5.102}$$

Given a shift $k \to k + \Delta k$ in the spring stiffness it becomes

$$\underbrace{\int_0^l \frac{M_c\,\delta M}{EI}\,dx + \delta w'\,w_c'\,k}_{a(w_c,\delta w)} + \underbrace{\Delta k\,w_c'\,\delta w'}_{d(w_c,\delta w)} = \int_0^l p\,\delta w\,dx\,, \tag{5.103}$$

and so, the corresponding shift in any functional $J(w)$ is (M_G is the moment of the influence function in the spring before the shift)

$$J(e) = -d(w_c, w_G) = -\Delta k\,w_c'\,w_G' = -\Delta k\,\frac{M_c}{k_c}\,\frac{M_G}{k}\,. \tag{5.104}$$

The moment $M_c = k_c\,w_c'$ in the spring after the shift in the stiffness can either be determined directly or by iteration.

In the direct approach we apply (5.104) to the functional $J(w) = M$, the moment in the spring itself, since then M_c appears on both sides of the equation

$$J(e) = M_c - M = -\Delta k\,\frac{M_c}{k_c}\,\frac{M_G}{k}\,, \tag{5.105}$$

and so,

$$M_c = M\,\frac{k_c\,k}{k_c\,k + \Delta k\,M_G}\,. \tag{5.106}$$

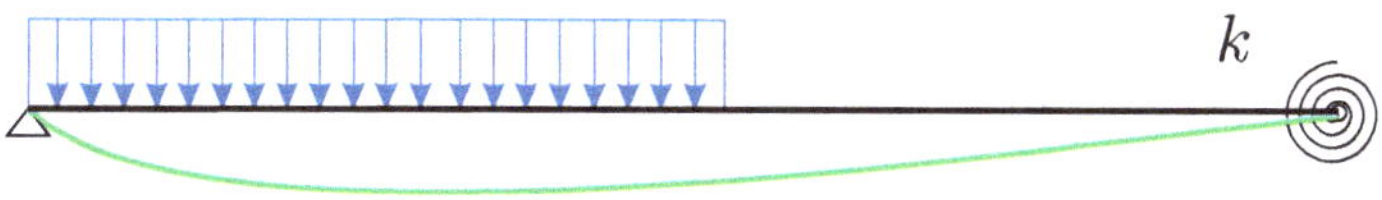

Fig. 5.21 Torsional spring

Or we iterate, we create a sequence $M_c^{(i)}$, which converges quickly to M_c

$$M_c^{(i+1)} = -\Delta k \, \frac{M_c^{(i)}}{k_c} \, \frac{M_G}{k} + M \qquad M_c^{(0)} = M \, . \tag{5.107}$$

Equation (5.104) allows then to calculate the shift in any functional

$$\boxed{J(e) = -\Delta k \frac{M \, M_G}{k_c \, k + \Delta k \, M_G}} \tag{5.108}$$

We only need the moment M_G of the influence function and the load moment M, both before the shift.

If m springs at m points x_i change, (5.104) becomes a linear system, which can be solved either directly or by iteration

$$M_c^{(i+1)}(x_j) = -\sum_{l=1}^{m} \Delta k(x_l) \, \frac{M_c^{(i)}(x_l)}{k_c(x_l)} \, \frac{M_G(x_l, x_j)}{k(x_l)} + M(x_j) \quad j \quad = 1, 2, \ldots m \, . \tag{5.109}$$

Here, $M_G(x_l, x_j)$ is the moment "ante" in the spring x_l of the influence function for the moment at the point x_j.

The two steel halls in Figs. 5.22 and 5.23 were analyzed with the help of these equations. In the first hall the focus was on the transition from rigid connections at the joints to elastic joints. In the second system, which we will discuss here, the analysis focused on how the transition from fixed joints to elastic joints at the foot of the columns would affect the horizontal displacement of the crane track at a specific point x_P. In an earlier analysis the columns were considered fully clamped.

First, a single horizontal force $P = 100$ kN was applied at x_p and the fixed end moments $M_G^P(x_i), i = 1, 2, \ldots 8$ at the eight supports were calculated. Then the eight end moments $M_c(x_i)$ due to the traffic load (*crane (25t lifting load)—vert. edge loads + slip forces in series 4*) were determined iteratively as in (5.109). Because of the original fixed clamping $k(x_i) = \infty$ with the modification as described in (5.127)

$$M_c^{(i+1)}(x_j) = -\sum_{l=1}^{m} \frac{1}{k_c(x_l)} \, M_c^{(i)}(x_l) \, M_G(x_l, x_j) + M(x_j) \quad j \quad = 1, 2, \ldots m \, . \tag{5.110}$$

With the moments $M_G^P(x_i)/100$ and the $M_c(x_i)$ the horizontal displacement $u_c(x_P)$ could then be calculated, see Fig. 5.24,

$$u_c(x_P) = -\sum_{l=1}^{8} \frac{1}{k_c(x_l)} M_c(x_l) \, M_G^P(x_l) \frac{1}{100} \, . \tag{5.111}$$

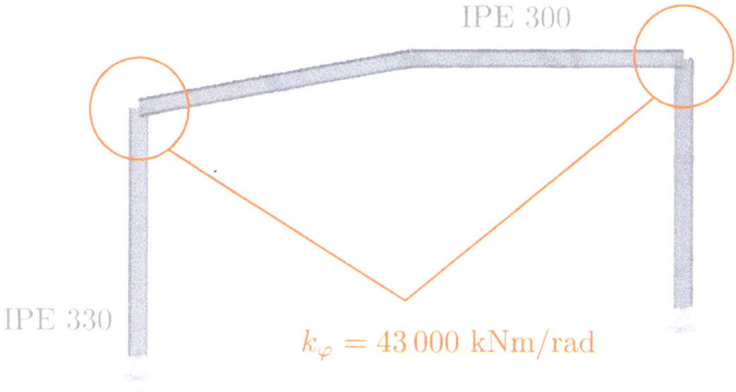

Fig. 5.22 3D-model of a steel frame consisting of 47 elements and with 480 degrees of freedom. Elastic joints reduced the bending moment from $M = -86.6$ kNm to $M_c = -75.0$ kNm [5]

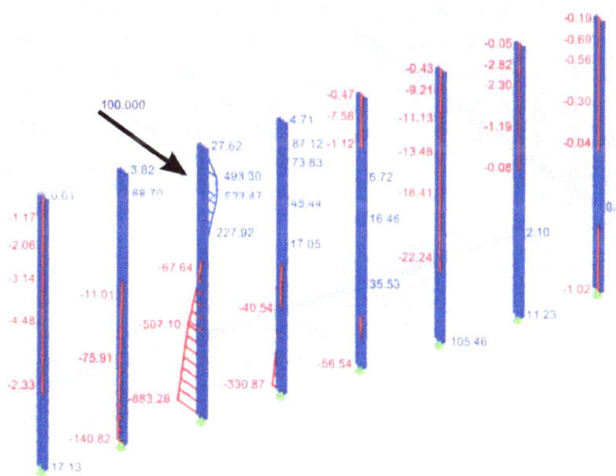

Fig. 5.23 Steel hall, 1 643 elements, 5 340 dof, 121 load cases. To be studied was the influence of the elastic support of the eight columns on the horizontal displacement of the crane runway. The technique introduced here reduces the problem to an 8×8 system which was solved by iteration [5]

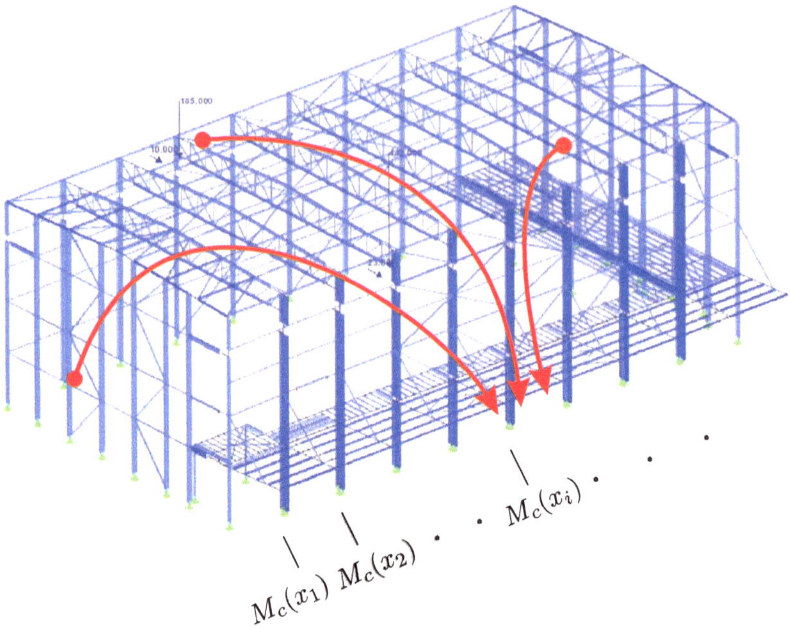

Fig. 5.24 Any influence function (G) generates moments (M_y^G) at the supports of the eight columns and the scalar product of this vector $\boldsymbol{M}_y^G$ with the vector $\boldsymbol{M}_c$ of the eight support moments $k_c^{-1}(x_i)M_c(x_i)$ provides the change $J(e) = \boldsymbol{M}_c^T \boldsymbol{M}_y^G$ of the relevant quantity in the transition from fixed supports to elastic supports

Direct Solution

The alternative to iteration is a direct solution. We illustrate this technique with a torsional spring whose stiffness $k \to k + \Delta k$ changes. If $J(w) = w'$ is the slope at the spring,

$$J(e) = w_c' - w' = -\Delta k\, w_c'\, w_G', \tag{5.112}$$

this can be solved for w_c'

$$w_c' = w' \frac{1}{1 + \Delta k\, w_G'}. \tag{5.113}$$

If the spring has, say, the degree of freedom u_7, w_G' would be the entry $g_{7,7}$ in the flexibility matrix $\boldsymbol{F} = \boldsymbol{K}^{-1}$.

Given w_c' it is possible to calculate the new moment $M_c = (k + \delta k)\, w_c'$ in the spring and the change $J(e)$ in any other functional, e.g. the displacement u_3 at a node. Since, as assumed, the spring has the degree of freedom u_7, w_G' is the entry $g_{3,7}$ in the flexibility matrix $\boldsymbol{F} = \boldsymbol{K}^{-1}$,

$$J(e) = u_{c3} - u_3 = -\Delta k\, w_c'\, g_{3,7}\,. \tag{5.114}$$

If two springs at two points x_1 and x_2 change their elastic properties, the weak form is

$$d(w_c, \delta w) = \Delta k_1\, w_c'(x_1)\, \delta w'(x_1) + \Delta k_2\, w_c'(x_2)\, \delta w'(x_2)\,, \tag{5.115}$$

and we must determine $w_c'(x_1)$ and $w_c'(x_2)$. These slopes are two functionals $J_1(w)$ and $J_2(w)$

$$J_i(w) = w'(x_i) = \int_0^l G_i(y, x_i)\, p(y)\, dy \qquad i = 1, 2\,, \tag{5.116}$$

to which belong influence functions $G_i(y, x_i)$. The shifts in the two functionals

$$\begin{aligned} J_1(e) &= w_c'(x_1) - w'(x_1) = -\Delta k_1\, w_c'(x_1)\, G_1'(x_1, x_1) - \Delta k_2\, w_c'(x_2)\, G_2'(x_1, x_2) \\ J_2(e) &= w_c'(x_2) - w'(x_2) = -\Delta k_1\, w_c'(x_1)\, G_1'(x_1, x_2) - \Delta k_2\, w_c'(x_2)\, G_2'(x_2, x_2)\,, \end{aligned} \tag{5.117}$$

lead to the system

$$\begin{bmatrix} 1 + a_{11} & a_{12} \\ a_{21} & 1 + a_{22} \end{bmatrix} \begin{bmatrix} w_c'(x_1) \\ w_c'(x_2) \end{bmatrix} = \begin{bmatrix} w'(x_1) \\ w'(x_2) \end{bmatrix} \tag{5.118}$$

where

$$a_{11} = \Delta k_1\, G_1'(x_1, x_1) \qquad a_{12} = \Delta k_2\, G_1'(x_1, x_2) \tag{5.119a}$$

$$a_{21} = \Delta k_1\, G_1'(x_1, x_2) \qquad a_{22} = \Delta k_2\, G_2'(x_2, x_2)\,. \tag{5.119b}$$

If the degrees of freedom of the two springs are u_7 and u_9, the terms $G_i'(x_j, x_i)$ would be the following entries in the inverse $\boldsymbol{F} = \boldsymbol{K}^{-1}$

$$G_1'(x_1, x_1) = g_{7,7} \qquad G_1'(x_1, x_2) = g_{7,9} \tag{5.120a}$$

$$G_2'(x_2, x_1) = g_{9,7} \qquad G_2'(x_2, x_2) = g_{9,9}\,, \tag{5.120b}$$

and extending this to any functional $J(w)$ means

$$J(e) = -\Delta k_1\, w_c'(x_1)\, G'(x_1, x) - \Delta k_2\, w_c'(x_2)\, G'(x_2, x)\,, \tag{5.121}$$

where $G(y, x)$ is the influence function for $J(w)$ and the $G'(x_i, x)$ are the slopes at the two springs.

Displacement Spring

To be complete, we also discuss a change $k \to k + \Delta k$ in a displacement spring, and we let $J(w) = w(x)$ be the deflection at a point x. According to Mohr's equation we have

$$w(x) = \int_0^l \frac{M\, M_G}{EI}\, dy + k \cdot w(l) \cdot G(x, l)\,, \tag{5.122}$$

where M_G is the moment of the influence function $G(x, y)$, see Fig. 5.25. To evaluate

$$J(e) = w_c(x) - w(x) = -\Delta k\, w_c(l)\, G(x, l) \tag{5.123}$$

we need $w_c(l)$ and so we apply the formula at the point $x = l$

$$w_c(l) - w(l) = -\Delta k\, w_c(l)\, G(l, l)\,, \tag{5.124}$$

where G is now the influence function for $w(l)$ and we solve for

$$w_c(l) = \frac{w(l)}{1 + \Delta k\, G(l, l)}\,, \tag{5.125}$$

and so (5.123) becomes

$$\boxed{J(e) = -\frac{\Delta k\, w(l)\, G(x, l)}{1 + \Delta k\, G(l, l)}} \tag{5.126}$$

This formula allows to track the change in any functional $J(w)$, if we know by how much the spring (stiffness k) is compressed (= $G(x, l)$) by the Dirac delta acting at the distant source point x.

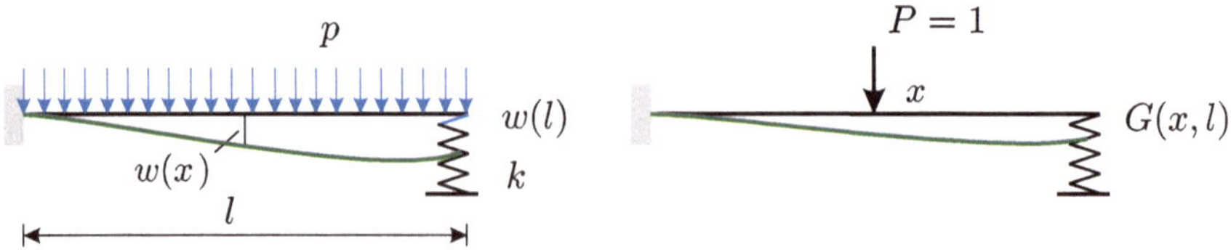

$$w(x) = \int_0^l \frac{M\,\bar{M}}{EI}\, dx + k\, w(l)\, G(x, l)$$

Fig. 5.25 Stiffness change in a spring

Rigid Supports

If the clamped end of a beam, $k_\varphi = \infty$, yields, if the support suddenly acts like a torsional spring with stiffness k^c_φ, Eq. (5.112) becomes

$$J(e) = -\Delta k_\varphi\, w'_c\, w'_G = -\frac{k^c_\varphi - k_\varphi}{k^c_\varphi\, k_\varphi} M_c\, M_G = (\frac{1}{k^c_\varphi} - \frac{1}{k_\varphi})\, M_c\, M_G = \frac{1}{k^c_\varphi}\, M_c\, M_G\,, \tag{5.127}$$

where M_G is the fixed end moment ($k_\varphi = \infty$) of the influence function and M_c is the load moment in the (then elastic) torsional spring.

Applied to $J(e) = M_c - M$ itself this results in

$$M_c - M = \frac{1}{k_c}\, M_c\, M_G\,, \tag{5.128}$$

where M_G is the moment, if the (released) fixed end is spread by $\tan\varphi = 1$. This can be solved for M_c and so any $J(e)$ can be determined with (5.127).

The same applies of course to displacement springs since

$$\boxed{J(e) = \frac{1}{k_c}\, R_c\, R_G} \tag{5.129}$$

where $R_c(k < \infty)$ and $R_G(k = \infty)$ are the corresponding support reactions.

We apply (5.129) to the beam in Fig. 5.26, to determine the change $J(e) = M - M_c$ in the fixed end moment, when the end support is removed, $k = \infty \to k = 0$. In this situation the force R_c is the inverted support reaction R, and the stiffness at the released end of the cantilever beam is

$$k_c = \frac{3EI}{l^3}\,. \tag{5.130}$$

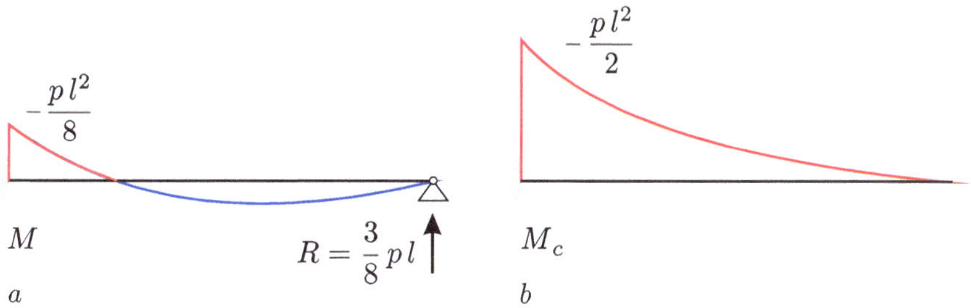

Fig. 5.26 Uniform load

The support reaction ($k = \infty$) of the influence function for the fixed end moment M is $R_G = 1/l$ and it is $R_c = 3/8 \cdot p\,l$, and so the change in the fixed end moment is

$$J(e) = M_c - M = \frac{l^3}{3\,EI} \cdot \frac{3}{8} p\,l \cdot \frac{1}{l} = p\,l^2 \,\frac{3}{8} \tag{5.131}$$

which agrees with the engineer's solution, $pl^2/2 - pl^2/8$.

5.19 Integral Bridges

A chance to apply these ideas provide integral bridges, where the foundation and the superstructure are monolithically connected in order to simplify the maintenance of the bridge.

The idea is new and when—in the course of the renewal of the autobahn A3—the *Fahrbachtalbrücke* near Aschaffenburg was to be designed as a **semi-integral bridge**, (no monolithic connection with the abutments), the authorities insisted, investigations were carried out on piles *in situ* in order to estimate the influence of variations $c \pm \Delta c$ in the soil stiffness c on the superstructure.

In a master thesis, the ideas developed here, were used to pursue this computationally [6]. The example is well suited, because there is a relatively large distance between the source points in the bridge and the piles, see Fig. 5.27, and the influence has to go the way practically twice, from the superstructure to the piles to generate the forces $\boldsymbol{f}^+$ and the forces $\boldsymbol{f}^+$ travel back the same way, to change the internal forces, $M(x) \rightarrow M_c(x)$, $V(x) \rightarrow V_c(x)$, etc.

In such structures influences play "ping-pong" after a stiffness change, the signals change between the super- and the substructure back and forth until the signals are balanced. The forces $\boldsymbol{f} + \boldsymbol{f}^+$ produce the new equilibrium position $\boldsymbol{u}_c$ and this must produce the matching forces $\boldsymbol{f}^+$.

Back to the bridge, where we focus on the moment $M(x)$ at a point x

$$M(x) = \int_0^l G_2(y, x)\, p(y)\, dy = \boldsymbol{g}^T \boldsymbol{f}\,. \tag{5.132}$$

The question of how much the moment is affected by variations in the modulus of subgrade reaction of the piles, $c \rightarrow c + \Delta c$, aims at the difference between the influence function G_2 (modulus c, matrix $\boldsymbol{K}$) and the influence function G_2^c, (modulus $c + \Delta c$, matrix $\boldsymbol{K}_c$)

$$M_c(x) = \int_0^l G_2^c(y, x)\, p(y)\, dy = \boldsymbol{g}_c^T \boldsymbol{f}\,. \tag{5.133}$$

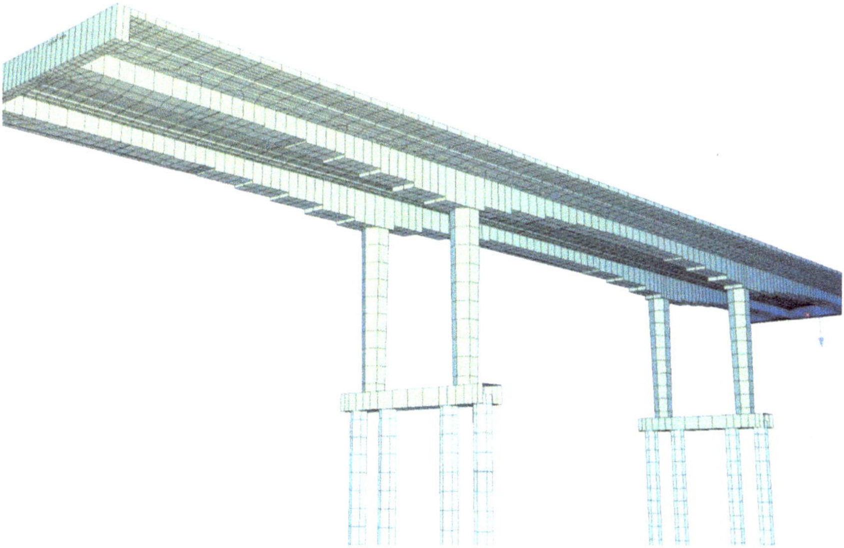

a

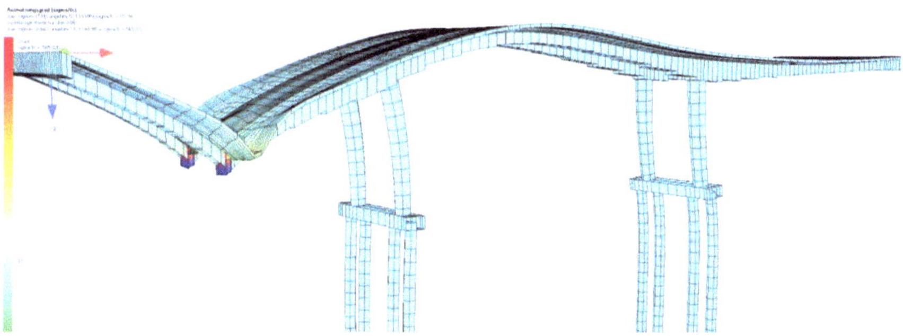

b

Fig. 5.27 Semi-integral bridge , **a** FE-model, **b** influence function for the moment $M\,(x)$

Because of

$$g_c^T f = g^T (f + f^+)\,, \tag{5.134}$$

we can trace this back to the effect of the vector $\boldsymbol{f}^+$ on the system $\boldsymbol{K}$, where the vectors $\boldsymbol{g}$ and $\boldsymbol{g}_c$ are the nodal values of the two influence functions and the vector $\boldsymbol{f}^+ = -\boldsymbol{\Delta K u}_c$ are the additional forces at the nodes of the piles due to the change of the soil stiffness, $c \to c + \Delta c$.

If we expand the deflection of a pile element

$$w_c(x) \simeq w_c(0) + w_c'(0) \cdot x + \frac{1}{2}\, w_c''(0) \cdot x^2\,, \tag{5.135}$$

only the quadratic terms will contribute to $\boldsymbol{f}^+ = -\boldsymbol{\Delta K u}_c$, since in each element the rows of $\boldsymbol{\Delta K}_e$ are (approximately) orthogonal to rigid body motions—the first two terms. However, we expect that the quadratic terms will be the smallest of the three terms, and so $\boldsymbol{f}^+$ will also be quite small. In addition, the piles are some distance away from the superstructure, so the influence of the pile forces $\boldsymbol{f}^+$ on the superstructure will be relatively small, regardless of their size. The numerical results confirmed this, the moments $M\,(x)$ and $M_c(x)$ hardly differ.

One can also fence it the other way around by directly studying the changes $G_2(y, x) \to G_2^c(y, x)$. The nodal values $\boldsymbol{g}$ of the influence function G_2 are the solution of the system $\boldsymbol{K g} = \boldsymbol{j}$ where $j_i = M\,(\varphi_i)(x)$, and the nodal values $\boldsymbol{g}_c$ of the influence function G_2^c are the solution of the system

$$\boldsymbol{K g}_c = \boldsymbol{j} + \boldsymbol{j}^+\,. \tag{5.136}$$

The displacement $g_c(x) = g_c(0) + g_c'(0)\,x + 0.5 * g_c''(0)^2/x \ldots$ generated by spreading the source point x should be subject to the same argument as above. The entries in the vector $\boldsymbol{j}^+ = -\boldsymbol{\Delta K g}_c$ should be small and because the nodes of the piles are relatively far away from the superstructure, the influence of the $\boldsymbol{j}^+$ will be small and therefore also the difference between $G_2(y, x)$ and $G_2^c(y, x)$.

Remark 5.2 Above, we simplified somewhat: Since the equation of a pile element is $EI\, w^{IV} + c\, w = p$, the stiffness matrix of a pile element is (in most FE-codes) a sum, $\boldsymbol{K} = \boldsymbol{K}_{Beam} + c \cdot \boldsymbol{M}$, of the normal beam matrix and the “mass matrix” $\boldsymbol{M}$, that is $m_{ij} = (\varphi_i, \varphi_j)$. Only the rows of $\boldsymbol{K}_{Beam}$ are orthogonal to the rigid body motions, while the mass matrix senses the rigid body motions in $\boldsymbol{u}_c$, but the overall effect seems to be rather mild.

5.20 Retrofitting

Influence functions are also well suited to find those elements, where retrofitting measures, $\boldsymbol{K} \to \boldsymbol{K} + \boldsymbol{\Delta K}$, tend to have the greatest impact, since the response $J(e) = J(u_c) - J(u)$ to any such measure can be studied directly on the element,

$$J(e) = J(u_c) - J(u) = \boldsymbol{g}^T \boldsymbol{f}^+ = \boldsymbol{g}^T \boldsymbol{\Delta K} \boldsymbol{u}_c\,. \tag{5.137}$$

In a beam element this equation has the form

$$J(e) = J(u_c) - J(u) = -\frac{\Delta EI}{EI} \int_{x_a}^{x_b} \frac{M_G\, M_c}{EI_c}\, dx\,, \tag{5.138}$$

where ΔEI is the added stiffness, $EI_c = EI + \Delta EI$, M_G is the moment of the influence function, and M_c is the load moment *after* the stiffness change.

Suppose we want to reduce the deflection of the Vierendeel beam in Fig. 5.28. To this end we apply a single force $P = 1$, and we weight the moments M_G of the force with the moment $M \sim M_c$. Where both are large (and not orthogonal), there we best increase the stiffness, $EI \to EI + \Delta EI$, and these zones are—not surprisingly—just the joints.

5.21 Classical Formulation

In statics influence functions are based on Betti's theorem and textbooks go a long way to explain to students how the superposition of two states, w and G, renders the deflection $w(x)$ of a beam. Technically though it can be expressed much simpler: Betti is duality, is adjoint analysis

$$\mathscr{B}(\boldsymbol{u}, \hat{\boldsymbol{u}}) = \hat{\boldsymbol{u}}^T \boldsymbol{K}\, \boldsymbol{u} - \boldsymbol{u}^T \boldsymbol{K} \hat{\boldsymbol{u}} = 0\,. \tag{5.139}$$

In optimization influence functions are simply the solution of the adjoint problems and adjoint formulations are seen as a clever mathematical "trick" to make problems easier to solve.

This approach is called ***adjoint method of analysis*** and it leads to the approximation

$$J(e) = J(u_c) - J(u) \simeq -\boldsymbol{g}^T \boldsymbol{\Delta K} \boldsymbol{u} \tag{5.140}$$

while the exact expression is

$$J(e) = J(u_c) - J(u) = -\boldsymbol{g}^T \boldsymbol{\Delta K} \boldsymbol{u}_c = \boldsymbol{g}^T \boldsymbol{f}^+\,. \tag{5.141}$$

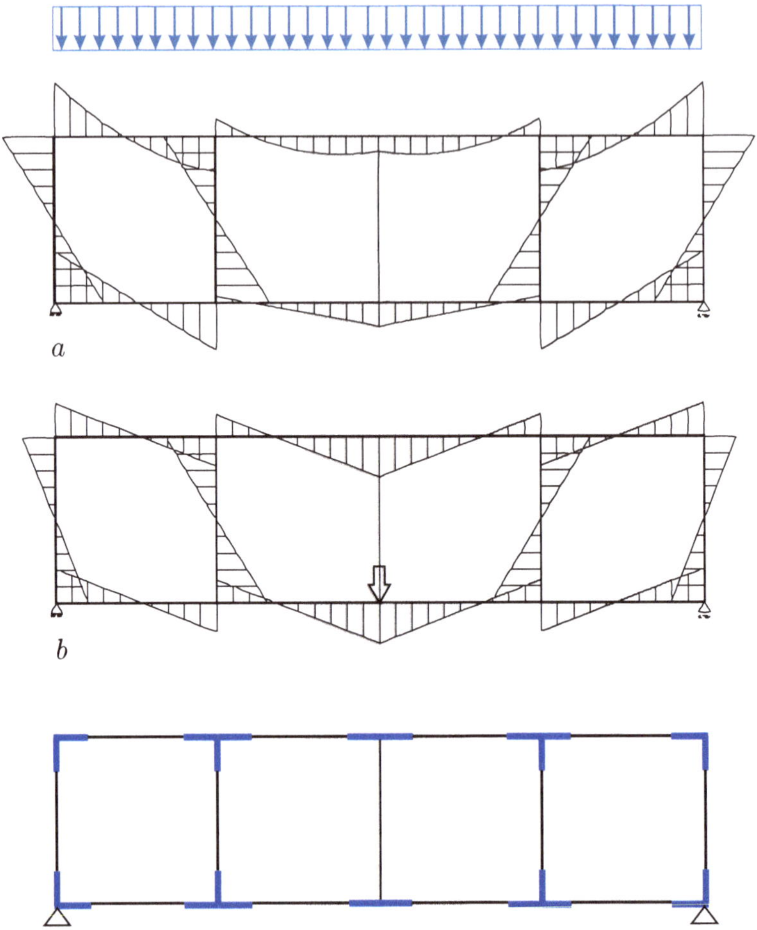

Fig. 5.28 Retroffitting a Vierendeel frame, **a** bending moments and load, **b** bending moment M_G from point load $P = 1$ (Dirac delta δ_0), **c** suggested modifications [7]

To keep the equation simple, we have presumed that the functional form of $J(u)$ does not depend on the stiffness EI, as is the case if $J(u) = u(x)$.

To derive the approximation (5.140), we assume that the stiffness matrix

$$\boldsymbol{K} = \boldsymbol{K}(p_1, p_2, \ldots, p_m) \tag{5.142}$$

and the vector $\boldsymbol{f}$ are a function of m model parameters p_i, and we calculate the derivative of the solution of $\boldsymbol{Ku} = \boldsymbol{f}$, with respect to the p_i.

Since the equation $\boldsymbol{K}\boldsymbol{u} - \boldsymbol{f} = \boldsymbol{0}$ is constant, its derivatives are zero

$$\frac{d}{dp_i}(\boldsymbol{K}\boldsymbol{u} - \boldsymbol{f}) = \frac{\partial \boldsymbol{K}}{\partial p_i}\boldsymbol{u} + \boldsymbol{K}\,\frac{\partial \boldsymbol{u}}{\partial p_i} - \frac{\partial \boldsymbol{f}}{\partial p_i} = 0\,, \tag{5.143}$$

which gives

$$\frac{\partial \boldsymbol{u}}{\partial p_i} = -\boldsymbol{K}^{-1}\,(\frac{\partial \boldsymbol{K}}{\partial p_i}\boldsymbol{u} - \frac{\partial \boldsymbol{f}}{\partial p_i}) = -\boldsymbol{K}^{-1}\boldsymbol{f}^p(i)\,. \tag{5.144}$$

The vector $\boldsymbol{f}^p(i)$ is called ***pseudoload vector***. There are m such vectors corresponding to the m parameters p_i. In the following we will simply write $\boldsymbol{f}^p$, since we know which (i) is meant.

Example 5.2 We study how the solution $\boldsymbol{u} = \{u_3, u_4\}^T$ of the cantilever beam in Fig. 5.29 depends on the length ℓ of the beam,

$$\boldsymbol{K} = EI\begin{bmatrix} 12/\ell^3 & 6/\ell^2 \\ 6/\ell^2 & 4/\ell \end{bmatrix} \qquad \boldsymbol{u} = \frac{p}{EI}\begin{bmatrix} \ell^4/8 \\ -\ell^3/6 \end{bmatrix} \qquad \boldsymbol{f} = p\begin{bmatrix} \ell/2 \\ \ell^2/12 \end{bmatrix}. \tag{5.145}$$

With

$$\boldsymbol{K}^{-1} = \frac{1}{EI}\begin{bmatrix} \ell^3/3 & -\ell^2/2 \\ -\ell^2/2 & \ell \end{bmatrix} \quad \frac{\partial \boldsymbol{K}}{\partial \ell} = EI\begin{bmatrix} -36/\ell^4 & -12/\ell^3 \\ -12/\ell^3 & -4/\ell^2 \end{bmatrix} \quad \frac{\partial \boldsymbol{f}}{\partial \ell} = p\begin{bmatrix} 1/2 \\ \ell/6 \end{bmatrix} \tag{5.146}$$

we obtain

$$\frac{\partial \boldsymbol{u}}{\partial \ell} = -\left[\boldsymbol{K}^{-1}\right]\left(\left[\frac{\partial \boldsymbol{K}}{\partial \ell}\right]\left[\boldsymbol{u}\right] - \left[\frac{\partial \boldsymbol{f}}{\partial \ell}\right]\right) = \frac{p}{2}\begin{bmatrix} \ell^3 \\ -\ell^2 \end{bmatrix}. \tag{5.147}$$

Since the gradient $\partial \boldsymbol{u}/\partial l$ is not constant, the expression

$$\boldsymbol{u}(\ell + \Delta\ell) - \boldsymbol{u}(\ell) \simeq \frac{\partial \boldsymbol{u}}{\partial \ell}\,\Delta\ell\,, \tag{5.148}$$

it is only a first-order approximation.

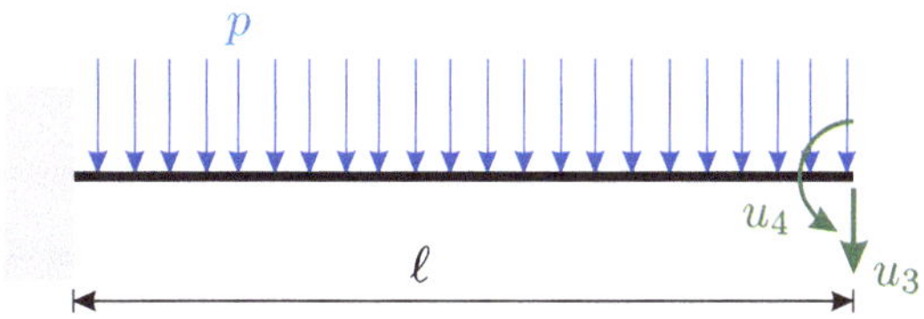

Fig. 5.29 Cantilever beam and uniform load

But the derivative of the solution $\boldsymbol{u} = \boldsymbol{K}^{-1}\boldsymbol{f}$ with respect to p is constant

$$\frac{\partial \boldsymbol{u}}{\partial p} = -\left[\boldsymbol{K}^{-1}\right]\left(\left[\ \boldsymbol{0}\ \right]\left[\boldsymbol{u}\right] - \left[\frac{\partial \boldsymbol{f}}{\partial p}\right]\right) = \frac{1}{EI}\begin{bmatrix} \ell^4/8 \\ -\ell^3/6 \end{bmatrix}, \tag{5.149}$$

and therefore

$$\boldsymbol{u}(p + \Delta p) - \boldsymbol{u}(p) = \frac{\partial \boldsymbol{u}}{\partial p}\,\Delta p = \frac{1}{EI}\begin{bmatrix} \ell^4/8 \\ -\ell^3/6 \end{bmatrix} \cdot \Delta p \tag{5.150}$$

is exact.

In contrast, the dependence of the solution $\boldsymbol{u}$ on EI is nonlinear, since when we solve for $\boldsymbol{u} = \boldsymbol{K}^{-1}\boldsymbol{f}$ we divide by the determinant (see Cramer's rule) and so the EI in the denominator becomes a $-EI^{-2}$ in the derivative.

It is

$$\frac{\partial \boldsymbol{K}}{\partial EI} = \begin{bmatrix} 12/\ell^3 & 6/\ell^2 \\ 6/\ell^2 & 4/\ell \end{bmatrix} \qquad \frac{\partial \boldsymbol{f}}{\partial EI} = \boldsymbol{0}\,, \tag{5.151}$$

and with $\boldsymbol{K}^{-1}\partial \boldsymbol{K}/\partial EI = EI^{-1} \cdot \boldsymbol{I}$ the gradient becomes

$$\frac{\partial \boldsymbol{u}}{\partial EI} = -\left[\boldsymbol{K}^{-1}\right]\left(\left[\frac{\partial \boldsymbol{K}}{\partial EI}\right]\left[\boldsymbol{u}\right] - \left[\frac{\partial \boldsymbol{f}}{\partial EI}\right]\right) = -\frac{1}{EI} \cdot \boldsymbol{u} = -\frac{p}{EI^2}\begin{bmatrix} \ell^4/8 \\ -\ell^3/6 \end{bmatrix}, \tag{5.152}$$

and so

$$\boldsymbol{u}(EI + \Delta EI) - \boldsymbol{u}(EI) \simeq \frac{\partial \boldsymbol{u}}{\partial EI} \cdot \Delta EI = -\frac{p}{EI^2}\begin{bmatrix} \ell^4/8 \\ -\ell^3/6 \end{bmatrix} \cdot \Delta EI\,, \tag{5.153}$$

is just the first term in the Taylor series.

Functionals

Next, we will look at functionals and study the **sensitivity** of a functional $J(\boldsymbol{u}) = \boldsymbol{j}^T\boldsymbol{u}$ with respect to the parameters p_i

$$\frac{dJ}{dp_i} = \frac{d}{dp_i}(\boldsymbol{j}^T\boldsymbol{u}) = \frac{d}{dp_i}\boldsymbol{j}^T\boldsymbol{u} + \boldsymbol{j}^T\,\frac{\partial \boldsymbol{u}}{\partial p_i}\,. \tag{5.154}$$

Here, we allow also the vector $\boldsymbol{j}$, the Dirac vector of the functional J, to depend on p_i, a situation which we neglected in the beginning.

With (5.144) follows

$$\frac{dJ}{dp_i} = \frac{d}{dp_i}\boldsymbol{j}^T\boldsymbol{u} - \boxed{\boldsymbol{j}^T\boldsymbol{K}^{-1}\boldsymbol{f}^p} = \frac{d}{dp_i}\boldsymbol{j}^T\boldsymbol{u} - \boldsymbol{\lambda}^T\boldsymbol{f}^p\,, \tag{5.155}$$

where we have introduced the ***adjoint variable***, the vector $\boldsymbol{\lambda} = \boldsymbol{K}^{-1}\boldsymbol{j}$. This is seen in the literature as a "trick", which allows to move the inverse $\boldsymbol{K}^{-1}$ away from $\boldsymbol{f}^p$ via an adjoint formulation.[5]

Adjoint simply means ordering the terms differently, because a scalar product of two vectors $\boldsymbol{a}$ and $\boldsymbol{M}\,\boldsymbol{b}$ can be read in two ways if the matrix $\boldsymbol{M} = \boldsymbol{M}^T$ is symmetric

$$\boldsymbol{a}^T\boldsymbol{M}\boldsymbol{b} = (\boldsymbol{M}\boldsymbol{a})^T\boldsymbol{b}\,, \tag{5.156}$$

or here

$$\boldsymbol{j}^T(\boldsymbol{K}^{-1}\boldsymbol{f}^p) = (\boldsymbol{K}^{-1}\boldsymbol{j})^T\boldsymbol{f}^p = \boldsymbol{\lambda}^T\boldsymbol{f}^p\,. \tag{5.157}$$

Since $\boldsymbol{K}^{-1}$ is an inverse we can write $\boldsymbol{\lambda}^T\boldsymbol{f}^p$, if $\boldsymbol{\lambda}$ is the solution to $\boldsymbol{K}\boldsymbol{\lambda} = \boldsymbol{j}$, and so the vector $\boldsymbol{\lambda}$ is exactly the vector $\boldsymbol{g}$ in $\boldsymbol{K}\boldsymbol{g} = \boldsymbol{j}$.

With the sensitivities dJ/dp_i we can then estimate the increment of a functional

$$J(p_i + \Delta p_i) - J(p_i) \simeq \Delta J_i = \frac{dJ}{dp_i}\,\Delta p_i\,, \tag{5.158}$$

a step which is central to optimization.

Comparison

To compare this result with the $\boldsymbol{f}^+$ approach, we switch to finite differences and write

$$\frac{d}{d\,p_i}\boldsymbol{j}^T\boldsymbol{u} \simeq \frac{1}{\Delta p_i}(\boldsymbol{j}(p_i + \Delta p_i) - \boldsymbol{j}(p_i))^T\boldsymbol{u} = \frac{1}{\Delta p_i}(\boldsymbol{j}_c - \boldsymbol{j})^T\,\boldsymbol{u} \tag{5.159}$$

and—assuming the load does not depend on p_i, $\partial f/\partial p_i = \boldsymbol{0}$,

$$\boldsymbol{f}^p = \frac{\partial \boldsymbol{K}}{\partial p_i}\boldsymbol{u} \simeq \frac{1}{\Delta p_i}(\boldsymbol{K}(p_i + \Delta p_i) - \boldsymbol{K}(p_i))\,\boldsymbol{u} = \frac{1}{\Delta p_i}\,\boldsymbol{\Delta K u}\,, \tag{5.160}$$

equation (5.158), actually (5.155), becomes

$$J(p_i + \Delta p_i) - J(p_i) \simeq (\boldsymbol{j}_c - \boldsymbol{j})^T\boldsymbol{u} - \boldsymbol{\lambda}^T\,\boldsymbol{\Delta K u}\,. \tag{5.161}$$

The exact formula instead is,[6] see (5.32) in Sect. 5.5,

$$J(u_c) - J(u) = (\boldsymbol{j}_c - \boldsymbol{j})^T\boldsymbol{u} + \boldsymbol{j}_c^T\boldsymbol{K}^{-1}\boldsymbol{f}^+ = (\boldsymbol{j}_c - \boldsymbol{j})^T\boldsymbol{u} - \bar{\boldsymbol{g}}_c^T\,\boldsymbol{\Delta K}\boldsymbol{u}_c\,. \tag{5.162}$$

---

[5] In this approach, the vector $\boldsymbol{g}$ (or $\boldsymbol{\lambda}$) appears in a purely algebraic way without mentioning the concept of Green's functions or Dirac deltas. The final result is of course exactly the same.

[6] We write $\bar{\boldsymbol{g}}_c = \boldsymbol{K}^{-1}\boldsymbol{j}_c$ to distinguish it from $\boldsymbol{g}_c = \boldsymbol{K}_c^{-1}\boldsymbol{j}_c$.

Technically, the only unknown in this equation is the vector $\boldsymbol{u}_c$ and if we let $\boldsymbol{u}_c \sim \boldsymbol{u}$ and $\bar{\boldsymbol{g}}_c = \boldsymbol{K}^{-1}\boldsymbol{j}_c \simeq \boldsymbol{K}^{-1}\boldsymbol{j} = \boldsymbol{g} = \boldsymbol{\lambda}$ the two formulas are the same, and so—on a numerical level—the two techniques are identical.

Application to Stiffness Matrices

In stiffness matrices the stiffnesses k (we switch from $p_i \rightarrow k_i$) are pre-factors

$$\boldsymbol{K} = k \cdot \boldsymbol{K}' \qquad k = \frac{EA}{l} \qquad k = \frac{EI}{l^3} \qquad (\boldsymbol{K}' = \frac{1}{k}\boldsymbol{K})\,. \tag{5.163}$$

Assuming that the right side $\boldsymbol{f}$ does not depend on k and k does not appear in the vector $\boldsymbol{j}$ either (for example, if $J(\boldsymbol{u}) = \boldsymbol{j}^T\boldsymbol{u} = \boldsymbol{e}_i^T\boldsymbol{u} = u_i$), then

$$\boldsymbol{f}^p = (\frac{\partial \boldsymbol{K}}{\partial k}\boldsymbol{u} - \frac{\partial \boldsymbol{f}}{\partial k}) = \boldsymbol{K}'\boldsymbol{u} \tag{5.164}$$

and since

$$\frac{dJ}{dk} = \frac{d}{dk}\boldsymbol{j}^T\boldsymbol{u} - \boldsymbol{j}^T\boldsymbol{K}^{-1}\boldsymbol{f}^p = -\boldsymbol{\lambda}^T\boldsymbol{f}^p = -\boldsymbol{\lambda}^T\boldsymbol{K}'\boldsymbol{u}\,, \tag{5.165}$$

we obtain

$$J(\boldsymbol{e}) = J(k + \Delta k) - J(k) \simeq \frac{dJ}{dk}\,\Delta k = -\boldsymbol{\lambda}^T\boldsymbol{K}'\boldsymbol{u}\,\Delta k = -\boldsymbol{\lambda}^T\,\boldsymbol{\Delta K}\boldsymbol{u}\,. \tag{5.166}$$

If we write $\boldsymbol{g}$ instead of $\boldsymbol{\lambda}$, and read $-\boldsymbol{\Delta K}\boldsymbol{u}$ as an approximation to $\boldsymbol{f}^+ = -\boldsymbol{\Delta K}\boldsymbol{u}_c$, this is exactly the expression $J(\boldsymbol{e}) = \boldsymbol{g}^T\boldsymbol{f}^+$.

Remark 5.3 The vector $\boldsymbol{f}^+$ is the vector which must be applied additionally to the original system, to produce the new response $\boldsymbol{u}_c$

$$\boldsymbol{u}_c = \boldsymbol{u} + \boldsymbol{K}\boldsymbol{f}^+\,, \tag{5.167}$$

and so, it can be considered a pseudoload vector. In the literature, however, it is the "gradient" $\boldsymbol{f}^p$ of $\boldsymbol{f}^+$ which is called the **pseudoload vector**

$$\boldsymbol{f}^+ \simeq -\boldsymbol{f}^p \cdot \Delta k\,. \tag{5.168}$$

The minus is due to the change of sign $-\boldsymbol{f}^p\,\Delta k = -\boldsymbol{\Delta K}\boldsymbol{u} \simeq \boldsymbol{f}^+$.

Remark 5.4 How we link the terms is basically the essential "trick" in adjoint formulations, and it is also behind the "two-way formula"

$$\boxed{J(\boldsymbol{u}) = \boldsymbol{g}^T\boldsymbol{f} = \boldsymbol{j}^T\boldsymbol{K}^{-1}\boldsymbol{f} = \boldsymbol{j}^T\boldsymbol{u}} \tag{5.169}$$

and its continuation

$$\boxed{J(\boldsymbol{e}) = \boldsymbol{g}^T\boldsymbol{f}^+ = \boldsymbol{j}^T\boldsymbol{K}^{-1}\boldsymbol{f}^+ = \boldsymbol{j}^T\boldsymbol{u}^+} \tag{5.170}$$

Example 5.3 The fixed end moment of the cantilever beam in Fig. 5.29 stays the same, if $EI \to EI + \Delta EI$ changes, and so the derivative of the functional $J(w) = M(0)$ must be zero, see (5.155),

$$\frac{dJ}{dEI} = \frac{d}{dEI}\boldsymbol{j}^T\boldsymbol{u} - \boldsymbol{j}^T\boldsymbol{K}^{-1}\boldsymbol{f}^p = \frac{d}{dEI}\boldsymbol{j}^T\boldsymbol{u} - \boldsymbol{g}^T\frac{\partial \boldsymbol{K}}{\partial EI}\boldsymbol{u} = 0\,. \tag{5.171}$$

This is a situation, where also the Dirac vector $\boldsymbol{j}$ depends on the parameter p_i, here EI, because the entries in the Dirac vector

$$\boldsymbol{j} = \{M(\varphi_3)(0), M(\varphi_4)(0)\}^T = -EI \cdot \{0.375, 0.5\}^T \tag{5.172}$$

are the moments of the two shape functions φ_3 and φ_4 at the source point $x = 0$, and since

$$\frac{d}{dEI}\boldsymbol{j}^T\boldsymbol{u} = \frac{1}{EI}\boldsymbol{j}^T\boldsymbol{K}^{-1}\boldsymbol{f} = \frac{1}{EI}\cdot\boldsymbol{g}^T\boldsymbol{f}\,, \tag{5.173}$$

and $\partial\boldsymbol{K}/\partial EI = 1/EI \cdot \boldsymbol{K}$ the derivative is indeed zero

$$\frac{dJ}{dEI} = \frac{1}{EI}\cdot\boldsymbol{g}^T\boldsymbol{f} - \frac{1}{EI}\cdot\boldsymbol{g}^T\boldsymbol{f} = 0\,. \tag{5.174}$$

5.22 Calculation of u_c

The problem with the formula $J(\boldsymbol{e}) = -\boldsymbol{g}^T\boldsymbol{\Delta K}\boldsymbol{u}_c$ is that we must substitute for the unknown vector $\boldsymbol{u}_c$ the previous vector $\boldsymbol{u}$, if we want to work with it. But there are two techniques, ***iteration*** and ***direct solution***, which allow to determine the vector $\boldsymbol{u}_c$ directly.

5.22.1 *Iteration*

We multiply the equation

$$(\boldsymbol{K} + \boldsymbol{\Delta K})\boldsymbol{u}_c = \boldsymbol{f} \tag{5.175}$$

from the left with $\boldsymbol{K}^{-1}$

$$(\boldsymbol{I} + \boldsymbol{K}^{-1}\boldsymbol{\Delta K})\boldsymbol{u}_c = \boldsymbol{u} \tag{5.176}$$

and write it in the form

$$\boldsymbol{u}_c = -\boldsymbol{K}^{-1}\boldsymbol{\Delta K}\boldsymbol{u}_c + \boldsymbol{u}\,, \tag{5.177}$$

and start the following *fixed-point iteration* for $\boldsymbol{u}_c$

$$\boldsymbol{u}_c^{i+1} = -\boldsymbol{K}^{-1}\boldsymbol{\Delta K}\boldsymbol{u}_c^i + \boldsymbol{u} \qquad i = 1, 2, \ldots \tag{5.178}$$

with $\boldsymbol{u}^0 = \boldsymbol{u}$ as start vector.

The convergence can be accelerated by mixing the vector $\boldsymbol{u}_c^{i+1}$, the result of (5.177), with the previous vector $\boldsymbol{u}_c^i$ to produce an improved vector

$$\boldsymbol{u}_c^{i+1} = \boldsymbol{u}_c^i \cdot c_{max} \cdot q + \boldsymbol{u}_c^{i+1} \cdot q = \frac{1}{1 + c_{max}} \cdot (\boldsymbol{u}_c^i \cdot c_{max} + \boldsymbol{u}_c^{i+1})\,, \tag{5.179}$$

where c_{max} is the largest scale factor c_e among the modifications $\boldsymbol{\Delta K}_e = c_e\boldsymbol{K}_e$ and

$$q = \frac{1}{1 + c_{max}}\,. \tag{5.180}$$

Theoretically, the iteration breaks down if the structure is kinematic, if the stiffness matrix $\boldsymbol{K}_c = \boldsymbol{K} + \boldsymbol{\Delta K}$ is singular. Then there exists a vector $\boldsymbol{u}_0 \neq \boldsymbol{0}$ such that

$$(\boldsymbol{K} + \boldsymbol{\Delta K})\boldsymbol{u}_0 = \boldsymbol{0}\,, \tag{5.181}$$

and if we multiply this equation from the left with the inverse $\boldsymbol{K}^{-1}$ we see,

$$\boldsymbol{K}^{-1}(\boldsymbol{K} + \boldsymbol{\Delta K})\boldsymbol{u}_0 = (\boldsymbol{I} - \boldsymbol{K}^{-1}\boldsymbol{K})\boldsymbol{u}_0 = \boldsymbol{0}\,, \tag{5.182}$$

that the vector $\boldsymbol{u}_0$ is an eigenvector of $\boldsymbol{K}^{-1}\boldsymbol{\Delta K}$ with the eigenvalue 1 and the error $\boldsymbol{e}_{i+1} = \boldsymbol{u}_{i+1} - \boldsymbol{u}_i$ therefore does not shrink

$$\boldsymbol{e}_{i+1} = -\boldsymbol{K}^{-1}\boldsymbol{\Delta K}\boldsymbol{e}_i \qquad i = 1, 2, \ldots\,, \tag{5.183}$$

since a fixed-point iteration only converges if the eigenvalues of the iteration matrix are less than 1.

Note,

$$\boldsymbol{e}_{i+1} = \boldsymbol{u}^{i+1} - \boldsymbol{u}^i = -\boldsymbol{K}^{-1}\boldsymbol{\Delta K}\boldsymbol{u}_i + \boldsymbol{u} + \boldsymbol{K}^{-1}\boldsymbol{\Delta K}\boldsymbol{u}_{i-1} - \boldsymbol{u} \tag{5.184}$$

$$= -\boldsymbol{K}^{-1}\boldsymbol{\Delta K}(\boldsymbol{u}_i - \boldsymbol{u}_{i-1}) = -\boldsymbol{K}^{-1}\boldsymbol{\Delta K}\boldsymbol{e}_i\,. \tag{5.185}$$

But with not too large perturbations the iteration converges, even if the removal of an element makes the structure unstable, since the program does not attempt to calculate the non-existing inverse of the singular stiffness matrix $\boldsymbol{K} + \boldsymbol{\Delta K}$.

5.22.2 Direct Solution

Let us repeat the equation to be solved,

$$(\boldsymbol{I} + \boldsymbol{K}^{-1}\boldsymbol{\Delta K})\boldsymbol{u}_c = \boldsymbol{u}\,. \tag{5.186}$$

The product $\boldsymbol{K}^{-1}\boldsymbol{\Delta K}$ is a **sparse matrix**, since the element matrix $\boldsymbol{\Delta K}$, when we expand it to full size $n \times n$, contains many zeros. Let the matrix $\boldsymbol{\Delta K}$ contain only four non-zero entries

$$\boldsymbol{\Delta K} = \begin{bmatrix} & & \dots & & \\ .\,0 & 0 & 0\,0 & 0\,0 & . \\ .\,0 & k_{33}^{\Delta} & 0\,0 & k_{35}^{\Delta}\,0 & . \\ .\,0 & 0 & 0\,0 & 0\,0 & . \\ .\,0 & k_{53}^{\Delta} & 0\,0 & k_{55}^{\Delta}\,0 & . \\ .\,0 & 0 & 0\,0 & 0\,0 & . \\ & & \dots & & \end{bmatrix}, \tag{5.187}$$

two in row 3 and two in row 5 (as would be the case if we change a spring element between u_3 and u_5).

The product of two $n \times n$ matrices $\boldsymbol{A}$ and $\boldsymbol{B}$ can be written as the sum of n matrices,

$$\boldsymbol{AB} = \boldsymbol{g}_1\boldsymbol{r}_1 + \boldsymbol{g}_2\boldsymbol{r}_2 + \dots + \boldsymbol{g}_n\boldsymbol{r}_n\,, \tag{5.188}$$

column $\boldsymbol{g}_1$ of $\boldsymbol{A}$ times row $\boldsymbol{r}_1$ of $\boldsymbol{B}$ plus column $\boldsymbol{g}_2$ of $\boldsymbol{A}$ times row $\boldsymbol{r}_2$ of $\boldsymbol{B}$ etc. Each product $\boldsymbol{g}_i\boldsymbol{r}_i$, *column* $\times$ *row*, is an $(n \times n)$ matrix.

Because $\boldsymbol{\Delta K}$ contains only two non-zero rows it is

$$\boldsymbol{K}^{-1}\boldsymbol{\Delta K} = \boldsymbol{g}_3\boldsymbol{r}_3 + \boldsymbol{g}_5\boldsymbol{r}_5 \qquad \text{(sum of two } n \times n \text{ matrices)} \tag{5.189}$$

which, if we write this in column form, results in

$$\boldsymbol{K}^{-1}\boldsymbol{\Delta K} = [\boldsymbol{0} \quad \boldsymbol{0} \quad \boldsymbol{c}_3 \quad \boldsymbol{0} \quad \boldsymbol{c}_5 \quad \boldsymbol{0} \cdots \boldsymbol{0}] \qquad \text{column vectors}\,, \tag{5.190}$$

where the two columns

$$\boldsymbol{c}_3 = k_{33}^{\Delta} \cdot \boldsymbol{g}_3 + k_{53}^{\Delta} \cdot \boldsymbol{g}_5 \qquad \boldsymbol{c}_5 = k_{35}^{\Delta} \cdot \boldsymbol{g}_3 + k_{55}^{\Delta} \cdot \boldsymbol{g}_5\,, \tag{5.191}$$

are linear combinations of the two columns $\boldsymbol{g}_3$ and $\boldsymbol{g}_5$ of $\boldsymbol{K}^{-1}$.

Doing it this way system (5.186) becomes

$$\boldsymbol{u}_c + \boldsymbol{c}_3\, u_{c3} + \boldsymbol{c}_5\, u_{c5} = \boldsymbol{u}\,. \tag{5.192}$$

We choose rows 3 and 5 of this system to determine first the two values u_{c3} and u_{c5}

$$\begin{bmatrix} 1 + c_{33} & c_{35} \\ c_{53} & 1 + c_{55} \end{bmatrix} \begin{bmatrix} u_{c3} \\ u_{c5} \end{bmatrix} = \begin{bmatrix} u_3 \\ u_5 \end{bmatrix}, \tag{5.193}$$

and then the complete solution

$$\begin{aligned} \boldsymbol{u}_c &= -\boldsymbol{K}^{-1}\boldsymbol{\Delta K}\boldsymbol{u}_c + \boldsymbol{u} = -u_{c3}\boldsymbol{c}_3 - u_{c5}\boldsymbol{c}_5 + \boldsymbol{u} \\ &= -(u_{c3}\, k_{33}^{\Delta} + u_{c5}\, k_{35}^{\Delta})\boldsymbol{g}_3 - (u_{c3}\, k_{53}^{\Delta} + u_{c5}\, k_{55}^{\Delta})\boldsymbol{g}_5 + \boldsymbol{u} \\ &= \alpha_3 \cdot \boldsymbol{g}_3 + \alpha_5 \cdot \boldsymbol{g}_5 + \boldsymbol{u}\,. \end{aligned} \tag{5.194}$$

The new FE-solution $\boldsymbol{u}_c$ is an expansion in terms of the old influence functions of the nodal displacements,

$$\boldsymbol{u}_c = \alpha_3 \cdot \boldsymbol{g}_3 + \alpha_5 \cdot \boldsymbol{g}_5 + \boldsymbol{u} = \alpha_3 \cdot \boldsymbol{g}_3 + \alpha_5 \cdot \boldsymbol{g}_5 + \textstyle\sum_i f_i\, \boldsymbol{g}_i \tag{5.195}$$

only that the influence functions of u_3 and u_5 appear twice in the sense of a correction.

Note, this is exactly the formula

$$\boldsymbol{u}_c = \boldsymbol{K}^{-1}(\boldsymbol{f}^+ + \boldsymbol{f}) = f_3^+ \cdot \boldsymbol{g}_3 + f_5^+ \cdot \boldsymbol{g}_5 + \sum_i f_i\, \boldsymbol{g}_i\,, \tag{5.196}$$

since $\alpha_3 = f_3^+$ and $\alpha_5 = f_5^+$.

To generalize this approach, we write the two matrices as a sequence of column vectors $\boldsymbol{g}_i$ and row vectors $\boldsymbol{r}_i$, respectively

$$\boldsymbol{K}^{-1} = [\,\boldsymbol{g}_1, \boldsymbol{g}_2, \ldots, \boldsymbol{g}_n\,] \qquad \boldsymbol{\Delta K} = [\,\boldsymbol{r}_1, \boldsymbol{r}_2, \ldots, \boldsymbol{r}_n\,]^T\,, \tag{5.197}$$

and so the system $\boldsymbol{K}^{-1}\boldsymbol{\Delta K} = [\,\boldsymbol{c}_1, \boldsymbol{c}_2, \ldots, \boldsymbol{c}_n\,]$ consists of n columns

$$\boldsymbol{c}_i = \sum_{j=1}^{n} \boldsymbol{g}_j\, r_{ji} \qquad (\text{vector } \boldsymbol{g}_j \times \text{entry } i \text{ of row } \boldsymbol{r}_j)\,, \tag{5.198}$$

and the system (5.186) reads, row by row,

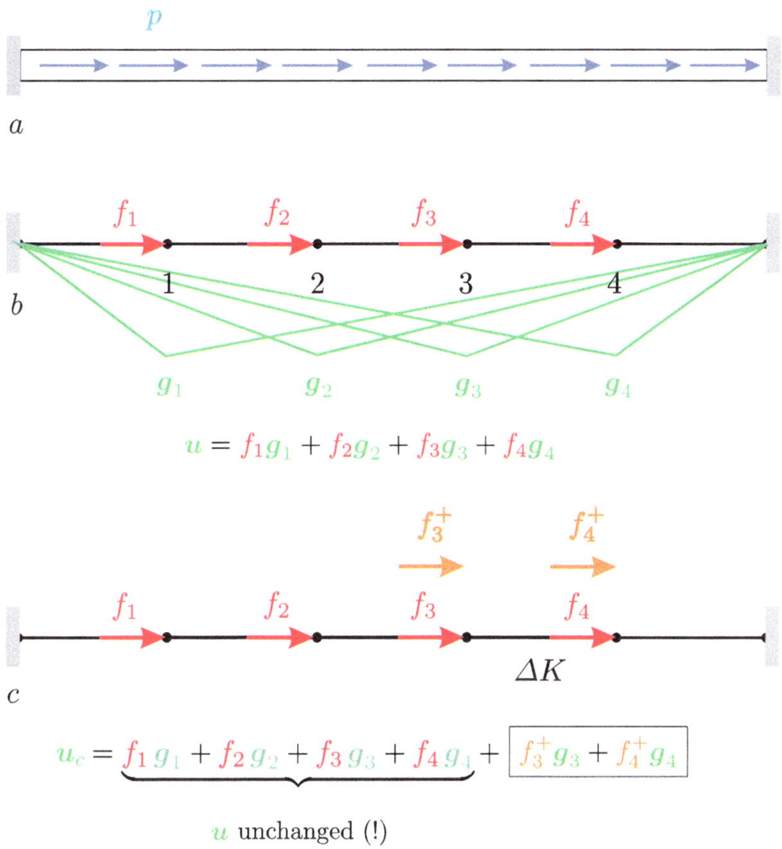

Fig. 5.30 Stretched bar and nodal forces, the vector $\boldsymbol{u}$ is the weighted sum of the columns $\boldsymbol{g}_i$ of the inverse. Stiffness changes ΔK generate additional forces f_i^+

$$u_{ci} + \sum_{j=1}^{n} c_{ij}\, u_{cj} = u_i \qquad i = 1, 2, \ldots, n\,. \tag{5.199}$$

If we delete all blank rows and columns, which contain only zeros, the result is a system of size $z \times z$, where z is the number of changes $k_{ii} \to k_{ii} + \Delta k_{ii}$ on the main diagonal; in the example above $z = 2$.

Let $\boldsymbol{u}_c^z$ be the solution. If we expand the solution $\boldsymbol{u}_c^z$ to a "full" vector, $\boldsymbol{u}_c^z \to \boldsymbol{u}_c^Z$, with n entries by padding $\boldsymbol{u}_c^z$ with zeros, then

$$\boldsymbol{u}_c = -\boldsymbol{K}^{-1} \boldsymbol{\Delta K} \boldsymbol{u}_c^Z + \boldsymbol{u} \tag{5.200}$$

is the new displacement vector $\boldsymbol{u}_c$.

Example 5.4 At the beginning the five elements of the bar in Fig. 5.30, had the same stiffness, $EA/l_e = 1$, but then in the fourth element the stiffness was doubled, $EA/l_e + \Delta EA/l_e = 1 + 1$. The original stiffness matrix $\boldsymbol{K}$ has 2 on the main diagonal and -1 on the secondary diagonals. Given the two matrices

$$\boldsymbol{K}^{-1} = \frac{1}{5}\begin{bmatrix} 4 & 3 & 2 & 1 \\ 3 & 6 & 4 & 2 \\ 2 & 4 & 6 & 3 \\ 1 & 2 & 3 & 4 \end{bmatrix} \qquad \boldsymbol{\Delta K} = \begin{bmatrix} 0 & 0 & 0 & 0 \\ 0 & 0 & 0 & 0 \\ 0 & 0 & 1 & -1 \\ 0 & 0 & -1 & 1 \end{bmatrix} \tag{5.201}$$

the system to be solved, $(\boldsymbol{I} + \boldsymbol{K}^{-1}\boldsymbol{\Delta K})\boldsymbol{u}_c = \boldsymbol{u}$, becomes

$$\begin{bmatrix} 1 & 0 & 0.2 & -0.2 \\ 0 & 1 & 0.4 & -0.4 \\ 0 & 0 & 1.6 & -0.6 \\ 0 & 0 & -0.2 & 1.2 \end{bmatrix} \begin{bmatrix} u_{c1} \\ u_{c2} \\ u_{c3} \\ u_{c4} \end{bmatrix} = \begin{bmatrix} u_1 \\ u_2 \\ u_3 \\ u_4 \end{bmatrix} = \begin{bmatrix} 2.0 \\ 3.0 \\ 3.0 \\ 2.0 \end{bmatrix} . \tag{5.202}$$

The solution of the (asymmetric) subsystem

$$\begin{bmatrix} 1.6 & -0.6 \\ -0.2 & 1.2 \end{bmatrix} \begin{bmatrix} u_{c3} \\ u_{c4} \end{bmatrix} = \begin{bmatrix} 3.0 \\ 2.0 \end{bmatrix} \tag{5.203}$$

is sufficient to determine the remaining terms

$$\begin{bmatrix} u_{c1} \\ u_{c2} \end{bmatrix} = -\begin{bmatrix} 0.2 & -0.2 \\ 0.4 & -0.4 \end{bmatrix} \begin{bmatrix} u_{c3} \\ u_{c4} \end{bmatrix} + \begin{bmatrix} 2.0 \\ 3.0 \end{bmatrix} , \tag{5.204}$$

and therewith the complete solution

$$\boldsymbol{u}_c = \{1.8889, 2.7778, 2.6667, 2.1111\}^T . \tag{5.205}$$

Given $\boldsymbol{u}_c$ we could, out of interest, calculate the f_i^+

$$-\boldsymbol{\Delta K u}_c = \boldsymbol{f}^+ = \{0, 0, -0.5556, 0.5556\}^T \tag{5.206}$$

and write the solution $\boldsymbol{u}_c$ in terms of the columns $\boldsymbol{g}_i$ of $\boldsymbol{K}^{-1}$

$$\boldsymbol{u}_c = \sum_{i=1}^{4} g_i f_i + g_3 f_3^+ + g_4 f_4^+ = \boldsymbol{u} - g_3 \cdot 0.5556 + g_4 \cdot 0.5556 . \tag{5.207}$$

5.22.3 Support Stiffness

A change in the stiffness of a support constitutes a one-term change on the diagonal, $k_{ii} \rightarrow k_{ii} + \Delta k_{ii}$, of the global stiffness matrix, and in such a case $\boldsymbol{\Delta K} = \boldsymbol{e}_i \, \boldsymbol{e}_i^T \, \Delta k_{ii}$ (= column × row) and

$$(\boldsymbol{I} + \boldsymbol{K}^{-1}(\boldsymbol{e}_i \, \boldsymbol{e}_i^T \, \Delta k_{ii})) \, \boldsymbol{u}_c = \boldsymbol{u} \tag{5.208}$$

becomes

$$(\boldsymbol{I} + \boldsymbol{g}_i \, \boldsymbol{e}_i^T \, \Delta k_{ii}) \, \boldsymbol{u}_c = \boldsymbol{u} \, . \tag{5.209}$$

The i-th equation

$$(1 + g_{ii} \, \Delta k_{ii}) u_{ci} \; = u_i \tag{5.210}$$

allows to determine u_{ci} and this one term is enough to calculate the complete solution $\boldsymbol{u}_c$

$$\boldsymbol{u}_c = \boldsymbol{u} - \boldsymbol{g}_i \, \boldsymbol{e}_i^T \, \boldsymbol{u}_c \, \Delta k_{ii} = \boldsymbol{u} - \boldsymbol{g}_i \, u_{ci} \, \Delta k_{ii} \, . \tag{5.211}$$

If in the two-span beam in Fig. 5.31 the spring support, $k = 10\,000$ kN/m fails, which means $\Delta k = -k$, we argue as follows: The compression of the spring under load was $u_2 = 5.5 \cdot 10^{-3}$ m, and a unit force $P = 1$ compresses the spring by $g_{22} = 88 \cdot 10^{-6}$ m, and so (5.210) becomes

$$(1 - 88 \cdot 10^{-6} \cdot 10^4) \, u_{c2} = 5.5 \cdot 10^{-3} \, , \tag{5.212}$$

which has the solution $u_{c2} = 45 \cdot 10^{-3}$ m.

5.23 Sherman-Morrison-Woodbury

If a matrix $\boldsymbol{K}_c$ has the shape $\boldsymbol{K}_c = \boldsymbol{K} - \boldsymbol{U} \, \boldsymbol{V}^T$, its inverse is

$$\boldsymbol{K}_c^{-1} = \boldsymbol{K}^{-1} + \boldsymbol{K}^{-1} \boldsymbol{U} \, (\boldsymbol{I} - \boldsymbol{V}^T \boldsymbol{K}^{-1} \boldsymbol{U})^{-1} \boldsymbol{V}^T \, \boldsymbol{K}^{-1} \, , \tag{5.213}$$

which is the *Sherman-Morrison-Woodbury formula* [4, 8]. For us, $-\boldsymbol{U} \, \boldsymbol{V}^T = \boldsymbol{\Delta K}$, and we can always write the split as in

$$\boldsymbol{\Delta K} = \frac{\Delta EA}{l} \begin{bmatrix} 1 & -1 \\ -1 & 1 \end{bmatrix} = \frac{\Delta EA}{l} \begin{bmatrix} 1 & 0 \\ 0 & 1 \end{bmatrix} \begin{bmatrix} 1 & -1 \\ -1 & 1 \end{bmatrix} = -\boldsymbol{U} \, \boldsymbol{V}^T \, . \tag{5.214}$$

The matrix $-\boldsymbol{U} = \Delta k \, \boldsymbol{I}$ is the unit matrix with a factor Δk, which here is $\Delta EA/l$.

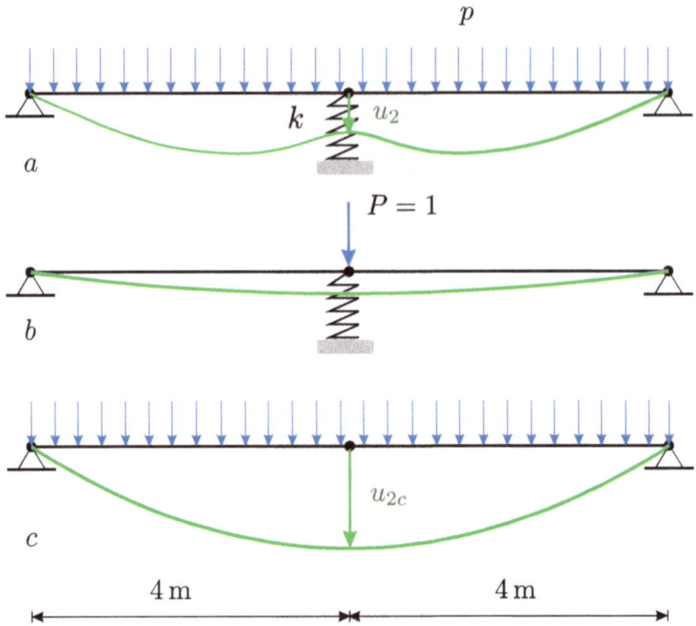

Fig. 5.31 Loss of a spring support, $k = 10\,000$ kN/m, in a two-span beam, $p = 10$ kN/m, $EI = 2.848 \cdot 10^4$ kNm^2

It remains to calculate the inverse $(\boldsymbol{I} - \boldsymbol{V}^T\boldsymbol{K}^{-1}\boldsymbol{U})^{-1}$. Suppose $\boldsymbol{K}$ is of size $N \times N$ and $\boldsymbol{V}^T\boldsymbol{K}^{-1}\boldsymbol{U}$ is of size $n \times n$ with $n \ll N$. Since the inverse of a unit matrix with a small sub-matrix on its diagonal is equal to the unit matrix plus the inverse of the sub-matrix, it suffices to calculate the inverse of the $n \times n$ matrix $\boldsymbol{I} - \boldsymbol{V}^T\boldsymbol{K}^{-1}\boldsymbol{U}$.

Example 5.5 To see this technique applied, we consider the bar in Fig. 5.30, where in the fourth element the stiffness $EA/l = 1$ doubles in value, $(\Delta k = 1)$, and so

$$\boldsymbol{K} = \begin{bmatrix} 2 & -1 & 0 & 0 \\ -1 & 2 & -1 & 0 \\ 0 & -1 & 2 & -1 \\ 0 & 0 & -1 & 2 \end{bmatrix}, \quad \boldsymbol{K}_c = \begin{bmatrix} 2 & -1 & 0 & 0 \\ -1 & 2 & -1 & 0 \\ 0 & -1 & 3 & -2 \\ 0 & 0 & -2 & 3 \end{bmatrix}. \tag{5.215}$$

Substituting the matrices $(\Delta k = 1)$

$$\boldsymbol{K}^{-1} = \frac{1}{5}\begin{bmatrix} 4 & 3 & 2 & 1 \\ 3 & 6 & 4 & 2 \\ 2 & 4 & 6 & 3 \\ 1 & 2 & 3 & 4 \end{bmatrix}, \quad -\boldsymbol{U} = \Delta k \cdot \boldsymbol{I}, \quad \boldsymbol{V} = \begin{bmatrix} 0 & 0 & 0 & 0 \\ 0 & 0 & 0 & 0 \\ 0 & 0 & 1 & -1 \\ 0 & 0 & -1 & 1 \end{bmatrix} \tag{5.216}$$

into (5.213) we obtain exactly the inverse of the matrix $\boldsymbol{K}_c$

$$\boldsymbol{K}_c^{-1} = \begin{bmatrix} 0.78 & 0.56 & 0.33 & 0.22 \\ 0.56 & 1.11 & 0.67 & 0.44 \\ 0.33 & 0.67 & 1.0 & 0.67 \\ 0.22 & 0.44 & 0.67 & 0.78 \end{bmatrix} = \boldsymbol{K}^{-1} + \boldsymbol{K}^{-1}\boldsymbol{U}(\boldsymbol{I} - \boldsymbol{V}^T\boldsymbol{K}^{-1}\boldsymbol{U})^{-1}\boldsymbol{V}^T\boldsymbol{K}^{-1}\,. \tag{5.217}$$

A comparison with the original $\boldsymbol{K}^{-1}$ confirms: Stiffness changes affect ***all*** influence functions, because no column of $\boldsymbol{K}^{-1}$ made it to $\boldsymbol{K}_c^{-1}$.

5.24 One-Click Reanalysis

In the program BE-FRAMES, see Chap. 10, both techniques to calculate $\boldsymbol{u}_c$, iteration and direct solution, are implemented as *One-Click Reanalysis*. The student can make changes to a frame by clicking on single elements and she or he immediately sees the effects, as in Fig. 5.32.

A click on an element means a change in the element matrix of the type $\boldsymbol{\Delta K}_e = c \cdot \boldsymbol{K}_e$ where c is a preset scaling factor; $c = 0$ would correspond to a complete loss of the element.

A series of clicks on, say, the elements 5, 7 and 9, simply means that $\boldsymbol{\Delta K}$ is a collection of scaled element matrices $\boldsymbol{\Delta K}_e$

$$\boldsymbol{u}_c = -\boldsymbol{K}^{-1}\,(\boldsymbol{\Delta K}_5 + \boldsymbol{\Delta K}_7 + \boldsymbol{\Delta K}_9)\boldsymbol{u}_c + \boldsymbol{u} \qquad \text{(direct sol.)} \tag{5.218}$$

5.24.1 When the Load "Is Hit"

If the student removes an element Ω_e, which carries a load, the new vector $\boldsymbol{u}_c$ is the solution of the system

$$(\boldsymbol{K} + \boldsymbol{\Delta K})\boldsymbol{u}_c = \boldsymbol{f} - \boldsymbol{f}_e\,, \tag{5.219}$$

where the entries in the vector $\boldsymbol{f}_e$ are the previous equivalent nodal forces of the dropped element Ω_e. In this case the program must also change the original right side, $\boldsymbol{f} \rightarrow \boldsymbol{f} - \boldsymbol{f}_e$.

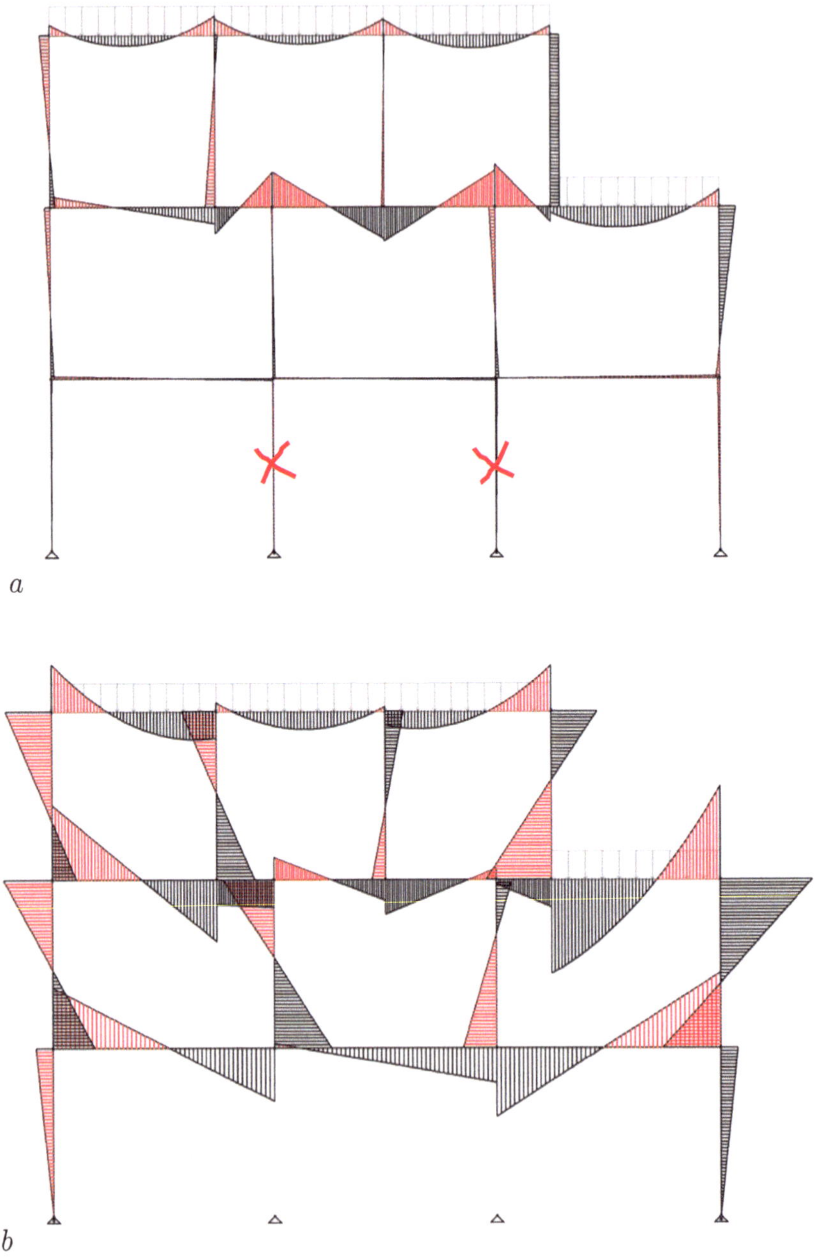

Fig. 5.32 One-click reanalysis (without modification of the stiffness matrix), **a** the crossed-out elements, **b** moment distribution without the two columns

5.24.2 Singular Stiffness Matrices

Since it is implemented in the program BE-FRAMES, we mention that the Gauss algorithm can also detect possible rigid body motions in frames. The Gauss algorithm transforms the stiffness matrix into an upper triangular matrix. If the algorithm is applied to the upper triangular matrix as well, the result is the unit matrix—for regular matrices. However, if the matrix is singular, the last columns are the eigenvectors to the eigenvalue(s) $\lambda = 0$, which can be displayed on the screen to notify the user of missing restraints, see Fig. 5.33.

The matrix created by this "twofold" Gauss is called the ***row reduced echelon form*** of a matrix and under this name the algorithm can be found in the literature.

5.25 Subsequent Installation of Joints

The student can also add joints to a frame and study how this affects the results. Mathematically a joint is a zero in $M(x) = 0$. If the student clicks at a point x_0 a Dirac delta δ_2 is applied at this point, the influence function for $M(x_0)$ is calculated and the moment $M_2(x_0)$ of the influence function itself at this point. Then the Dirac delta is scaled by a factor a to achieve $a \cdot M_2(x_0) + M(x_0) = 0$ and the moment distribution in the frame + hinge becomes $M_c(x) = M(x) + a \cdot M_2(x)$, see Fig. 5.34.

When multiple joints are added, a linear system controls the mutual influence of the different Dirac deltas.

5.26 Buckling Loads

The buckling loads of a frame are the eigenvalues of the second-order stiffness matrix $\boldsymbol{K}$. These too will be affected by modifications in the stiffness, $EI \to EI + \Delta EI$.

Here we only study the simplest case and assume the system to depend on a parameter t and therewith also the eigenvalues $\lambda(t)$ and the eigenvectors $\boldsymbol{u}(t)$

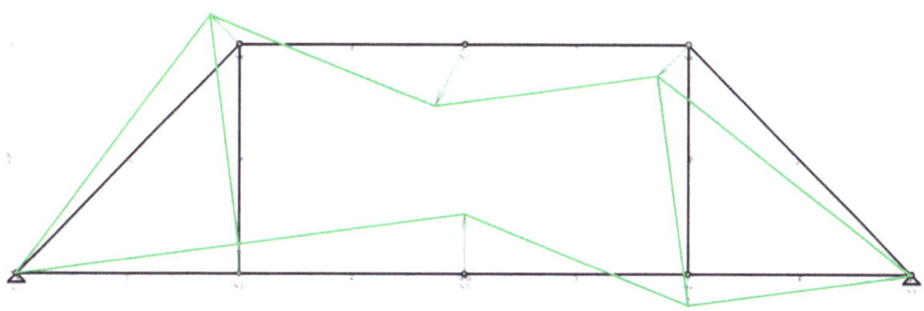

Fig. 5.33 Possible infinitesimal motions in a frame can be discovered with the "twofold" Gauss

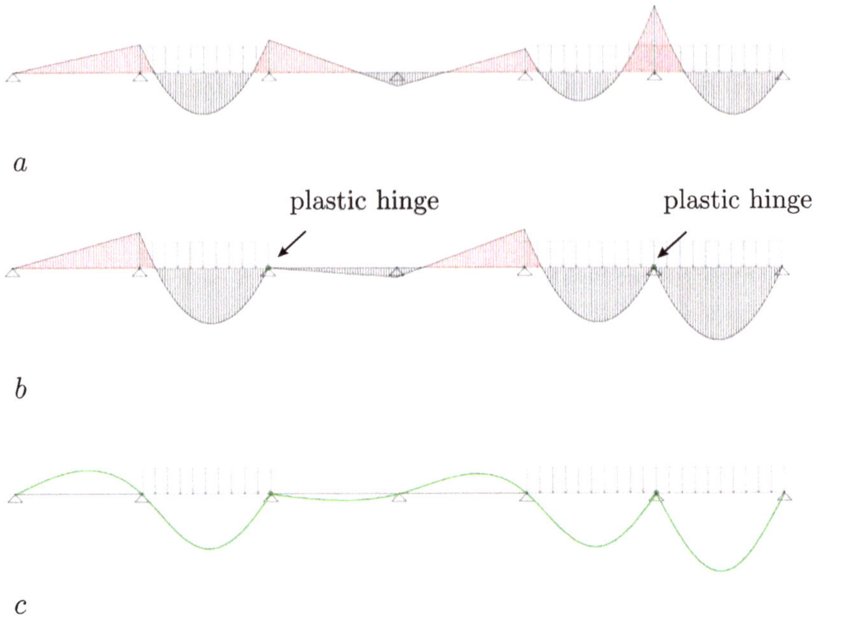

Fig. 5.34 Adding plastic hinges to a continuous beam, **a** original moment distribution, **b** after the hinges have been added, **c** deflection

$$\boldsymbol{K}(\lambda)\,\boldsymbol{u}(t) = \lambda(t)\,\boldsymbol{u}(t)\,. \tag{5.220}$$

The transposed matrix $\boldsymbol{K}^T$ has the same eigenvalues, but not necessarily the same eigenvectors $\boldsymbol{v}(t)$

$$\boldsymbol{K}^T(t)\boldsymbol{v}(t) = \lambda(t)\,\boldsymbol{v}(t)\,. \tag{5.221}$$

One can always normalize the eigenvectors of $\boldsymbol{K}$ and $\boldsymbol{K}^T$ belonging to $\lambda(t)$, so that $\boldsymbol{u}(t)^T\,\boldsymbol{v}(t) = 1$, and under this condition holds [4],

$$\frac{d\,\lambda(t)}{dt} = \boldsymbol{v}^T(t)\,\frac{d\,\boldsymbol{K}(t)}{dt}\,\boldsymbol{u}(t)\,. \tag{5.222}$$

For more details about the sensitivities of eigenvalues and eigenvectors see [9].

5.27 Dynamic Problems

Also harmonic oscillations, $\boldsymbol{K}\boldsymbol{u} - \omega^2 \boldsymbol{M}\boldsymbol{u} = \boldsymbol{0}$, lead to an eigenvalue problem $\boldsymbol{M}^{-1}\boldsymbol{K}\boldsymbol{u} = \lambda\,\boldsymbol{u}$. The oscillations of a chain of n masses are an expansion in terms of the eigenvectors of the matrix $\boldsymbol{M}^{-1}\boldsymbol{K}$

$$\boldsymbol{u}(t) = \sum_k \boldsymbol{u}_k\, e^{i\lambda_k t}\,, \tag{5.223}$$

where the λ_k are the associated eigenvalues.

In a static problem a shift $\boldsymbol{K} \to \boldsymbol{K} + \boldsymbol{\Delta K}$ produces additional forces f_i^+, but the effects will normally decrease rapidly—a local stiffness change often remains a local event—while in a dynamic problem **the amplitudes and the frequencies** change and this is quite a different situation: The whole frame oscillates to a different tune.

5.27.1 *Antisymmetry in the Compensating Motions*

Because of $u = 1/k \cdot f$, changes $k \pm \Delta k$ produce uneven effects, see Fig. 5.35. The gain in displacement Δu corresponding to a decrease $-\Delta k$ is larger than the reduction Δu corresponding to an increase $+\Delta k$. In other words: If you reduce the stiffness k of a spring by Δk, and then add Δk back to it, you are not where you started, but the mass f hangs deeper.

The reason is, you are on the curve $u = 1/k \cdot f$ when the stiffness decreases, $-\Delta k$ and on the curve $u = 1/(k - \Delta k) \cdot f$ when the stiffness increases, $+\Delta k$.

If you let one liter of air (= 20 %) escape out of a tire (5 l), and then pump it back in again, the car is at its old level. But based on the reduced volume (4 l) 1 liter of air

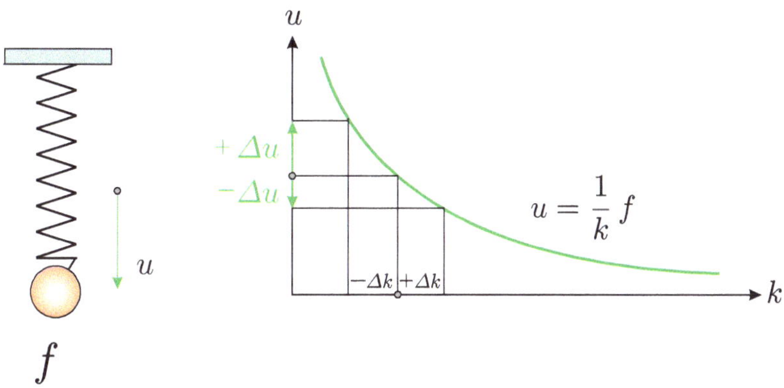

Fig. 5.35 The equation $u = 1/k \cdot f$ implies that a change $\pm\Delta k$ has different effects

is 25 %. Strictly speaking, you should have added only 0.8 l, which would not have been enough to reach the old level.

5.28 The Continuum

In Sect. 5.9 we studied a plate in which Young's modulus E was reduced in a rectangular region by 90 %, see Fig. 5.9, and we discovered that nodal forces f_i^+ were distributed along the edge of the inclusion, the less stiff zone. This suggested a comparison with a sheet metal which is welded to a steel tank to reinforce the tank. It is not just a picture. The nodal forces f_i^+ are only the discrete analog of line forces $\bar{t}$, which occur *numerically* (not physically(!)) where the stiffness of a patch changes. To explain this, we need to take a brief detour through potential theory.

Since the following pages are very technical, we start with a short summary.

What is the issue? The problem is that we have two domains with different parameters, and we are looking for an influence function for the solution $u(\boldsymbol{x})$, which is valid in both domains. Sure, one can always formulate two influence functions separately for Ω_a and Ω_b, but this is not the goal. It should be one.

This can be done in two ways. Either we add a layer $\bar{t}$ of **fictitious line forces** to the interface Γ_i between the two domains and this approach would result in a representation such as

$$(c_a(\boldsymbol{x}) + c_b(\boldsymbol{x}))\, u(\boldsymbol{x}) = \cdots + \int_{\Gamma_i} g\,\bar{t}\, ds_y \qquad \bar{t} = \frac{H_b - H_a}{H_a\, H_b}\, t \tag{5.224}$$

or we introduce a **new kernel** $\Delta H\ \partial g/\partial n$

$$(c_a(\boldsymbol{x}) H_a + c_b(\boldsymbol{x}) H_b)\, u(\boldsymbol{x}) = \cdots + \int_{\Gamma_i} \Delta H \frac{\partial g}{\partial n}\, u\, ds_y\,, \tag{5.225}$$

which is adjoint to the solution u on the interface. Since the characteristic functions $c_a(\boldsymbol{x})$ and $c_b(\boldsymbol{x})$ are zero in the respective other domain, this is like "one" influence function for $u(\boldsymbol{x})$. The left side of the first equation is $u(\boldsymbol{x})$ and the left side of the second equation is $H_a\, u(\boldsymbol{x})$ or $H_b\, u(\boldsymbol{x})$ respectively, depending on where $\boldsymbol{x}$ lies.

5.28.1 Potential Theory

Potential theory deals with the properties of fields, which are generated by point sources

$$g(\boldsymbol{y}, \boldsymbol{x}) = -\frac{1}{2\pi} \ln r \qquad 2-D \qquad g(\boldsymbol{y}, \boldsymbol{x}) = \frac{1}{4\pi} \frac{1}{r} \qquad 3-D\,. \tag{5.226}$$

These functions are at all points $\boldsymbol{y} \neq \boldsymbol{x}$ solutions of the ***Laplace equation***

$$\Delta g = \frac{\partial^2 g}{\partial y_1^2} + \frac{\partial^2 g}{\partial y_2^2} = 0 \qquad 2-D\,. \tag{5.227}$$

With these ***fundamental solutions*** we can create integral representations of arbitrary C^2 functions $u(\boldsymbol{x})$

$$c(\boldsymbol{x})\,u(\boldsymbol{x}) = \int_\Gamma [g\,\frac{\partial u}{\partial n} - \frac{\partial g}{\partial n}\,u]\,ds_y + \int_\Omega g\,(-\Delta u)\;d\Omega_y\,, \tag{5.228}$$

based on u and its normal derivative $\partial u/\partial n = \nabla u \bullet \boldsymbol{n}$ on the boundary Γ and the sum of the second derivatives $\Delta u = u,_{11} + u,_{22}$ in the domain Ω.

The function $c(\boldsymbol{x})$ is the so-called *characteristic function* of the domain Ω

$$c(\boldsymbol{x}) = \begin{cases} 1 & \boldsymbol{x} \in \Omega \\ \Delta\varphi/2\,\pi & \boldsymbol{x} \in \Gamma \\ 0 & \boldsymbol{x} \in \Omega^c \end{cases} \tag{5.229}$$

where $\Delta\varphi$ ($= \pi$ at a smooth point) is the angle of the boundary point $\boldsymbol{x}$ and Ω^c is the complement, everything which lies beyond the edge Γ.

Now, imagine a membrane, prestressed by a force H_a, and carrying a load p,

$$-H_a \Delta u = p \qquad u = 0 \;\text{ on}\, \Gamma\,. \tag{5.230}$$

Technically, the expression $H_a \partial u/\partial n$ on the edge is the support reaction, and the fundamental solution has the form

$$g_a(\boldsymbol{y}, \boldsymbol{x}) = \frac{1}{H_a}\,g \qquad g = -\frac{1}{2\,\pi}\,\ln\,r\,. \tag{5.231}$$

In a part Ω_b inside the membrane, however, the prestress has a different value H_b (how this can be realized is of no interest here). Technically the solution u satisfies different equations in Ω_a and Ω_b

$$-H_a\,\Delta u = p \quad \text{in}\, \Omega_a \qquad -H_b\,\Delta u = p \quad \text{in}\, \Omega_b\,. \tag{5.232}$$

The fundamental solution in the domain Ω_b is of course $1/H_b \cdot g$.

If we multiply (5.228) first with H_a and then with $1/H_a$ it is the integral representation with respect to the operator $-H_a \Delta u$ by means of g_a

$$c(\boldsymbol{x})\,u(\boldsymbol{x}) = \int_\Gamma [g_a\,H_a\,\frac{\partial u}{\partial n} - H_a\,\frac{\partial g_a}{\partial n}\,u]\,ds_y + \int_\Omega g_a\,(-H_a \Delta u)\;d\Omega_y\,. \tag{5.233}$$

Basically, nothing has changed. The expression is identical to (5.228).

If we formulate this equation also on Ω_b ($-H_b \Delta u = p$) and add the two representations, we obtain

$$
\begin{aligned}
(c_a(\boldsymbol{x}) + c_b(\boldsymbol{x}))\, u(\boldsymbol{x}) = u(\boldsymbol{x}) &= \int_\Gamma [g\, \frac{\partial u}{\partial n} - \frac{\partial g}{\partial n}\, u]\, ds_y \\
&+ \int_\Omega g\, (-\Delta u)\, d\Omega_y + \int_{\Gamma_i} g\, \bar{t}\, ds_y \qquad \bar{t} = \frac{H_b - H_a}{H_a\, H_b}\, t\,,
\end{aligned}
\tag{5.234}
$$

where Γ_i is the ***interface*** between the two domains, which is also the boundary of the inclusion Ω_b. The interface integral results as follows: On the interface

$$
t_a = H_a\, \partial u / \partial n = -H_b\, \partial u / \partial n = -t_b\,,
\tag{5.235}
$$

and thus the contributions from both domains add up to

$$
\int_{\Gamma_i} (g_a\, t_a - g_b\, t_a)\, ds_y = \int_{\Gamma_i} g\, t_a\, (\frac{H_b - H_a}{H_a\, H_b})\, ds_y\,.
\tag{5.236}
$$

So, the t in (5.234) is the t_a. The other two interface integrals

$$
\int_{\Gamma_i(\Omega_a)} \frac{\partial g}{\partial n}\, u\, ds_y + \int_{\Gamma_i(\Omega_b)} \frac{\partial g}{\partial n}\, u\, ds_y = 0
\tag{5.237}
$$

are zero, since u is constant, and the normal vector in $\partial g / \partial n = \nabla g \boldsymbol{\cdot} \boldsymbol{n}$ points to the respective opposite domain, and so the $u\, \partial g / \partial n$ cancel.

The expression (5.234) is the integral representation of a membrane with a uniform prestress $H = 1$ carrying a distributed load $-\Delta u$ and line forces $\bar{t}$ along the interface.

What can be done with this equation? Suppose we choose some line forces $\bar{t}$ and domain loads $p = -\Delta u$, then (5.234) formulates, if we let $\boldsymbol{x}$ circle the edge, the coupling condition between the boundary values u and $\partial u / \partial n$. Given two edge functions u and v ($= \partial u / \partial n$) which satisfy (5.234) $u(\boldsymbol{x})$ is the corresponding deflection.

Back to the differently prestressed membrane. For (5.234) to be the surface of the membrane, the function $\bar{t}$ must satisfy the equation

$$
\bar{t} = \frac{H_b - H_a}{Ha\, H_b}\, t_a\,,
\tag{5.238}
$$

where t_a is the normal derivative of the solution in Ω_a on the interface. The formulation of (5.234) is based on this property. To this corresponds that the vector $\boldsymbol{f}^+ = \boldsymbol{\Delta K}\, \boldsymbol{u}_c$ depends on the solution $\boldsymbol{u}_c$, which gives us trouble.

But the vector $\boldsymbol{f}^+$ also contains contributions from Ω_b. The entries in the original vector $\boldsymbol{f}$ are the integrals $f_i = (\varphi_i, -H_a \Delta u)$. If the prestress in Ω_b changes, the too much or the too little, the difference $p/H_a - p/H_b$, leads to additional nodal forces in Ω_b. These are part of the f_i^+. You see these "internal" f_i^+ in Fig. 5.9.

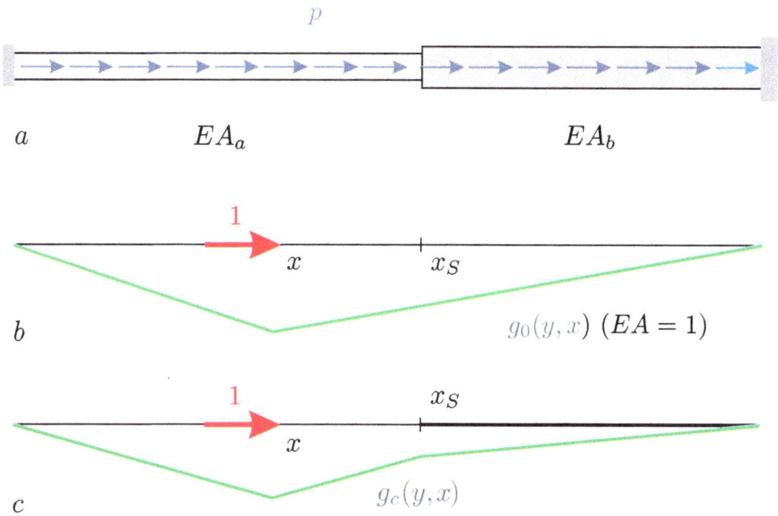

Fig. 5.36 Stepped bar, **a** system and loading, **b** influence function for a uniform bar, **c** exact influence function

Example 5.6 The bar in Fig. 5.36 a consists of two element with different values EA_a and EA_b. If $g(y, x)$ is a fundamental solution corresponding to a value $EA = 1$, then

$$g_a(y, x) = \frac{1}{EA_a} \cdot g(y, x) \qquad g_b(y, x) = \frac{1}{EA_b} \cdot g(y, x) \tag{5.239}$$

are fundamental solutions corresponding to EA_a and EA_b.

Next, we formulate *Betti's theorem*, first on $[0, x_S]$ and then on $[x_S, l]$

$$\mathscr{B}(g_a, u)_{[0,x_S]} = 0 \qquad \mathscr{B}(g_b, u)_{[x_S,l]} = 0\,. \tag{5.240}$$

This results in

$$c_a(x)u(x) = [g_a\, N - N(g_a)\, u]_0^{x_S} + \int_0^{x_S} g_a(y, x)\, p(y)\, dy \tag{5.241a}$$

$$c_b(x)u(x) = [g_b\, N - N(g_b)\, u]_{x_S}^{l} + \int_{x_S}^{l} g_b(y, x)\, p(y)\, dy\,. \tag{5.241b}$$

It is $N(g_a) = EA_a\, g_a' = g'$ and $N(g_b) = EA_b\, g_b' = g'$. The characteristic functions are[7]

[7] $(l - x_S)/l$ and x_S/l is the normal force N to the left and right of the point x_S on the homogeneous bar, if the force $P = 1$ is at the point x_S.

$$c_a(x) = \begin{cases} 0 & x \notin [0, x_s] \\ 1 & 0 \leq x < x_S \\ (l - x_S)/l & x = x_S \end{cases} \qquad c_b(x) = \begin{cases} x_S/l & x = x_S \\ 1 & x_S < x \leq l \\ 0 & x \notin [x_S, l] \end{cases} \tag{5.242}$$

Since u and g are zero at the ends of the bar, the sum of these two equations is

$$\begin{aligned} (c_a(x) + c_b(x))\, u(x) &= (g_a(x, x_S) - g_b(x, x_S))\, N(x_S) \\ &+ \int_0^{x_S} g_a(y, x)\, p(y)\, dy + \int_{x_S}^{l} g_b(y, x)\, p(y)\, dy \\ &= g(x, x_S) \cdot F + \int_0^{x_S} g_a(y, x)\, p(y)\, dy + \int_{x_S}^{l} g_b(y, x)\, p(y)\, dy \end{aligned} \tag{5.243}$$

where

$$F = \frac{EA_a - EA_b}{EA_a EA_b} N(x_S)\,. \tag{5.244}$$

Mathematically the bar carries additionally a single force F at the point x_S. Since the normal force is continuous at x_S, $EA_a\, \varepsilon_a = EA_b\, \varepsilon_b$, it holds

$$F = \frac{EA_a - EA_b}{EA_b} \varepsilon_a = \varepsilon_b - \varepsilon_a\,. \tag{5.245}$$

So, the "force" is the jump in the strain at the point x_S. A small transformation turns it into a real force

$$g(x, x_S) \cdot F = \frac{1}{EA_a} g(x, x_S)\, (\varepsilon_b - \varepsilon_a)\, EA_a\,. \tag{5.246}$$

That one can do without the force F shows Fig. 5.36c, where the stiffness change at the joint is built into the continuous Green's function, and thus the force F does not appear

$$u(x) = \int_0^l g_c(y, x)\, p(y)\, dy\,. \tag{5.247}$$

This would be the "one-term" influence function.

5.28.2 Kernels j^+

In Sect. 5.11 we guessed that there are not only forces f^+, but that there must also be forces j^+. Here we can clarify this.

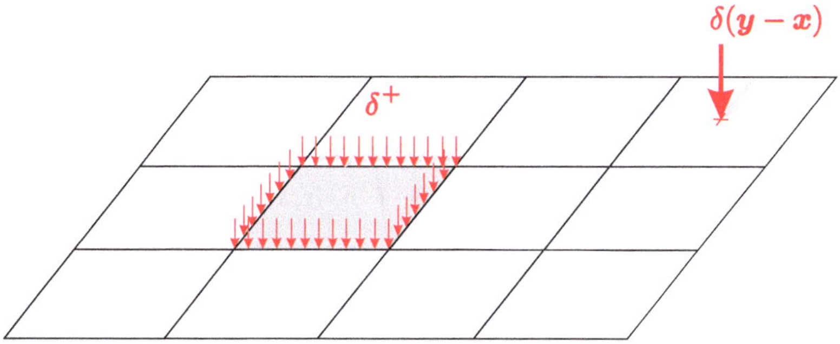

Fig. 5.37 To model a non-uniform domain requires additional Dirac deltas δ^+ at the interface of the inclusion

The membrane $\Omega = \Omega_a + \Omega_b$ consists, as above, of two differently prestressed zones Ω_a and Ω_b. When we formulate the representation (5.228) for both zones separately, multiply the respective representation with H_a or H_b and add the two equations, we arrive at

$$
\begin{aligned}
(c_a(\boldsymbol{x})H_a + c_b(\boldsymbol{x})H_b)\, u(\boldsymbol{x}) &= \int_{\Gamma_a} [g\, H_a \frac{\partial u}{\partial n} - H_a \frac{\partial g}{\partial n}\, u]\, ds_y \\
&+ \int_{\Gamma_b} [g\, H_b \frac{\partial u}{\partial n} - H_b \frac{\partial g}{\partial n}\, u]\, ds_y + \int_{\Omega} g\, p\, d\Omega_y \\
&+ \int_{\Gamma_i} (H_a - H_b) \frac{\partial g}{\partial n}\, u\, ds_y\,. \qquad (5.248)
\end{aligned}
$$

If, as in Fig. 5.37, the part Ω_b is embedded into Ω_a, we can treat Ω_b like a single finite element whose stiffness differs, $H \to H + \Delta H$, and the integral over Γ_b can be dropped

$$
\begin{aligned}
(c_a(\boldsymbol{x})H + c_b(\boldsymbol{x})(H + \Delta H))\, u(\boldsymbol{x}) &= \int_{\Gamma} [g\, H \frac{\partial u}{\partial n} - H \frac{\partial g}{\partial n}\, u]\, ds_y \\
&+ \int_{\Omega} g\, p\, d\Omega_y + \int_{\Gamma_i} \Delta H \frac{\partial g}{\partial n}\, u\, ds_y\,. \qquad (5.249)
\end{aligned}
$$

If the source point $\boldsymbol{x}$ lies e.g. in Ω_a, $c_b(\boldsymbol{x}) = 0$ and the result is

$$
H\, u(\boldsymbol{x}) = \int_{\Gamma} [g\, H \frac{\partial u}{\partial n} - H \frac{\partial g}{\partial n}\, u]\, ds_y + \int_{\Omega} g\, p\, d\Omega_y + \int_{\Gamma_i} \Delta H \frac{\partial g}{\partial n}\, u\, ds_y \qquad (5.250)
$$

which, after dividing by H, is the deflection $u(\boldsymbol{x})$.

Hence, a stiffness change leads to additional terms δ^+ (additional "Dirac deltas") on the edge Γ_i of the element

$$\delta(y - x) \quad \rightarrow \quad \delta(y - x) + \delta^+(y, x) = \delta(y - x) + \frac{\Delta H}{H}\frac{\partial g}{\partial n}\,, \tag{5.251}$$

against which the solution u has to work

$$J(u_c) = \int_\Omega \delta(y - x)\, u(y)\, d\Omega_y + \int_{\Gamma_i} \delta^+(y, x) u(y)\, ds_y\,, \tag{5.252}$$

to produce the value $u(x)$. It is classical potential theory.

5.28.3 The Two Approaches

And this brings us to the point where we need to make it explicit that there are *two* approaches. One leads to the additional forces $\bar{t}$ along the interface Γ_i and the other to the additional kernels δ^+ along Γ_i. If *Betti's theorem* is first formulated separately on Ω_a and Ω_b with the kernels g_a and g_b resp. and the equations are added, this produces an ***additional layer*** $\bar{t}$

$$(c_a(x) + c_b(x))\, u(x) = \cdots \int_{\Gamma_i} g\,\bar{t}\, ds_y \qquad \bar{t} = \frac{H_b - H_a}{H_a\, H_b}\, t\,. \tag{5.253}$$

But if you multiply the two integral representations for $u(x)$, before you add, with the coefficients H_a and H_b respectively, an additional *kernel* appears on the interface Γ_i

$$(c_a(x)H_a + c_b(x)H_b)\, u(x) = \cdots + \int_{\Gamma_i} \Delta H \frac{\partial g}{\partial n}\, u\, ds_y\,, \tag{5.254}$$

which we have called δ^+.

Physically the forces $\bar{t}$ have no business, since the internal forces balance at the interface Γ_i. But via these auxiliary forces $\bar{t}$ (or f^+) potential theory manages to calculate the exact value $u(x)$. They are, if you will, a mathematical artifact. Just as 1 + 1 = 2, so too the sum of the influence functions of two adjoining regions produces the layer $\bar{t}$

$$\cdots \int_{\Gamma_i} g_a \frac{\partial u}{\partial n}\, ds_y + \int_{\Gamma_i} g_b \frac{\partial u}{\partial n}\, ds_y = \cdots + \int_{\Gamma_i} g\,\bar{t}\, ds_y \tag{5.255}$$

or the kernel δ^+ respectively

$$\cdots + \int_{\Gamma_i} H_a \frac{\partial g}{\partial n}\, u\, ds_y + \int_{\Gamma_i} H_b \frac{\partial g}{\partial n}\, u\, ds_y = \cdots + \int_{\Gamma_i} \Delta H \frac{\partial g}{\partial n}\, u\, ds_y\,. \tag{5.256}$$

Remark 5.5 The *boundary elements* are based on potential theory, and in the program BE-SLABS, see Chap. 10, this technique is used to analyze plates with piecewise different stiffness. The influence functions are supplemented with two integrals, (the plate equation is of fourth order), which weight the deflection w and the slope $\partial w/\partial n$ (perpendicular to the edge of the plates) with two such kernels j^+ which are conjugated to w and $\partial w/\partial n$, i.e. which result from the jumps of the Kirchhoff kernel (shear force) and the moment kernel [10], p. 289,

$$\cdots + \int_{\Gamma_i} [M_\nu(g_0^a(\mathbf{y},\mathbf{x}))(K_a - K_b)\frac{\partial w}{\partial \nu}(\mathbf{y}) - V_\nu(g_0^a(\mathbf{y},\mathbf{x}))(K_a - K_b)w(\mathbf{y})]\, ds_{\mathbf{y}} + \cdots \tag{5.257}$$

References

1. Chen H-P, Ni, Y-Q (2018) Structural health monitoring of large civil engineering structures. Wiley Blackwell
2. Deif A (1986) Sensitivity analysis in linear systems. Springer, Heidelberg
3. Hartmann F (2013) Green's functions and finite elements. Springer, Heidelberg
4. Strang G (2019) Linear algebra and learning from data. Wellesley-Cambridge Press
5. Carl O, Hartman F (March 2017) Zhang C (2017) Schnelle Berechnung von Änderungen und Varianten bei komplexen Tragsystemen (3D-Modellen) - Neue Ansätze in der Baustatik unter Verwendung von Einflussfunktionen. Stahlbau 217–224
6. Sopoth M, Sopoth G (2008) Sensitivitätsanalyse an einem Brückenbauwerk in semi-integraler Bauweise. Master thesis University of Kassel
7. Hartmann F, Katz C (2010) Structural analysis with finite elements, 2nd edn. Springer, Heidelberg
8. Hager WW (1989) Updating the inverse of a matrix. SIAM Rev 31, 2:221–239
9. Fox RL, Kapoor MP (1968) Rates of changes of eigenvalues and eigenvectors. AIAA J 6(12):2426–2429
10. Hartmann F (1989) Introduction to boundary elements. Springer, Heidelberg

Chapter 6
Singularities

This chapter deals with the question when and where stresses peak, and why these singularities reduce the quality of the influence functions and therewith the solution itself, not only in the vicinity of the singular points, but at all points.

6.1 Singular Stresses

Stresses are proportional to the strains, $\sigma = E \cdot \varepsilon$, which means proportional to the first derivatives of the displacements, see Fig. 6.1a, and so we will always see singular stresses, if a displacement changes "on the spot", if it shoots straight up as in Fig. 6.1a.

The ***brachystochrone***, ($\beta\rho\alpha\chi\upsilon\varsigma$ = short), best illustrates the situation. This is the curve which starting from a given point A, and descending solely by the force of gravity, will reach another given point B in a shorter time than it could by any other path. This curve of quickest descent is a *cycloid*, see Fig. 6.1b. The trick is to start with an infinite slope for to gain as much speed as possible at the beginning.

This also happens in structures, where the stressed material tries to escape on the fastest route out from the dangerous zone, see Fig. 6.2, since the vertical displacement u_y starts with an infinite slope at the crack tip and this rapid start produces infinite stresses σ_{yy}.

The best illustration of this phenomenon can be found in traffic accident research where it is said: *"If the braking distance is zero, the force is infinite"*. What for a speeding car is the acceleration $a = dv/dt$ are in structures the strains $\varepsilon = du/dx$ and the curvatures $\kappa = d^2w/dx^2$.

In a crack the strain is infinite because in the uncracked concrete the two faces of the crack were the distance $dx = 0$ apart, and so even a very tiny crack opening du results in infinite stresses, $du/dx = du/0 = \infty$.

F. Hartmann and P. Jahn, *Statics and Influence Functions*,
Springer Series in Solid and Structural Mechanics 13,
https://doi.org/10.1007/978-3-030-55889-5_6

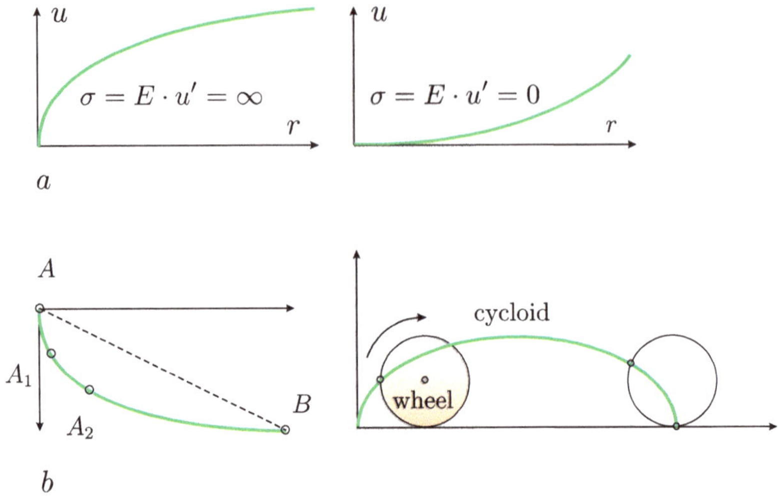

Fig. 6.1 Stress σ, **a** depending on how the displacement approaches zero, the stresses are either infinite or finite, **b** the fastest track from A to B in the gravitational field of a planet is not the shortest connection (− − −), but a cycloid. Because the initial acceleration at points A_1 or A_2 is less than at A (smaller slope), the travel time is the same as that from A

6.2 Singular Support Reactions

Two supports placed side by side as in Fig. 6.3 should lessen the strain on the beam, but the opposite is true since a point load in the first span produces a zig-zag shaped moment distribution, and the smaller the distance between the two inner supports is, the greater the shear force will be, since the shear force is the derivative of the zig-zag moment

$$V = \frac{dM}{dx}\,. \tag{6.1}$$

The influence function for the shear force confirms this tendency. The dislocation ± 0.5, which triggers the influence function, remains the same, but the flanks of the influence function become steeper and steeper when the distance shrinks, $h \to 0$, and so the influence function bulges out in the neighboring spans.

A similar effect was observed in the plate in Fig. 6.4, where a slight offset of two supporting walls produced a twisting moment, which required large support reactions to stabilize the plate. Such situations are not so rare.

Fig. 6.2 The stresses σ_{yy} at the crack tip are infinite, since u_y starts with an infinite slope at the crack tip ($\nu = 0$)

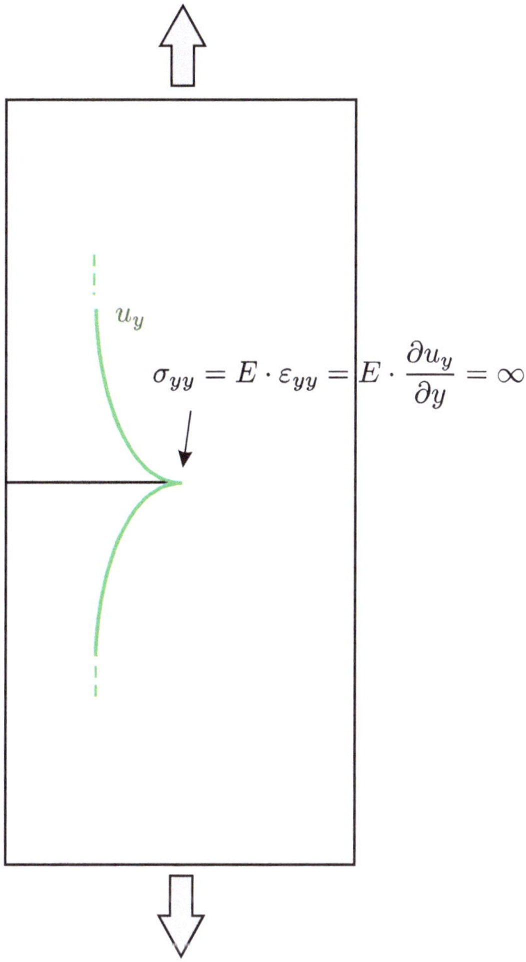

6.3 Single Forces

If we place a single force P on a plate, see Fig. 6.5, and we draw circles with radius r around the source point, the horizontal stresses integrated over the circle must be equal to the point load, and this also in the limit $r \to 0$.

What we integrate, though, are not the horizontal stresses but the horizontal tractions. If $\boldsymbol{S}$ is the stress tensor in the plate

$$\boldsymbol{S} = \begin{bmatrix} \sigma_{xx} & \sigma_{xy} \\ \sigma_{yx} & \sigma_{yy} \end{bmatrix}, \tag{6.2}$$

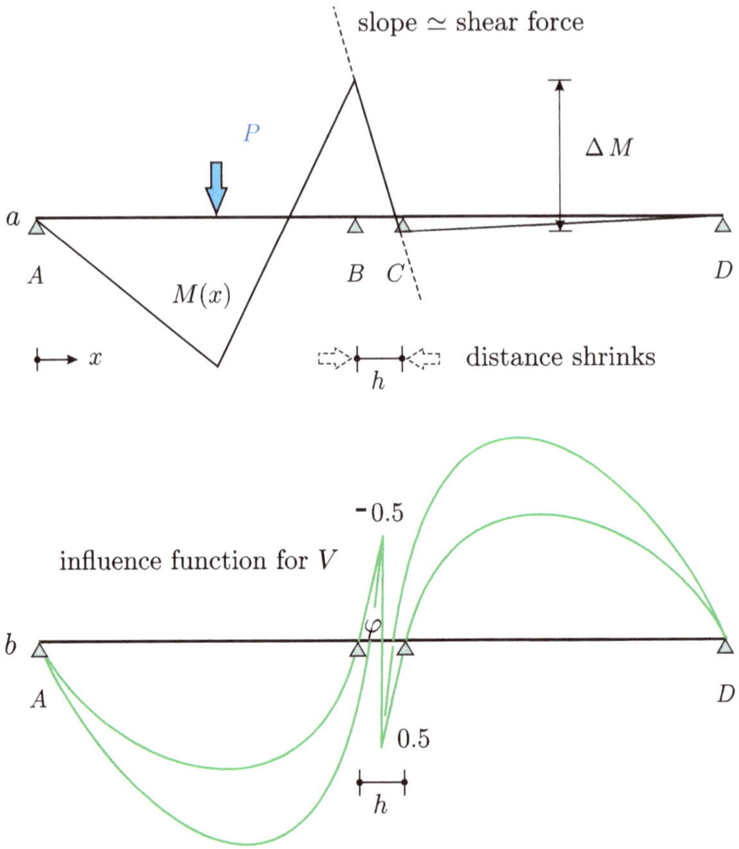

Fig. 6.3 Continuous beam, **a** the shorter the length of the span in the middle is, the steeper the bending moment and therewith its slope $V = M'$, **b** influence function for the shear force at the midpoint

and we are on the circle, the **tractions** $\boldsymbol{t}$ in the direction of the normal vector $\boldsymbol{n} = \{n_x, n_y\}^T$ are

$$\boldsymbol{t} = \boldsymbol{S}\,\boldsymbol{n} = \begin{bmatrix} \sigma_{xx} & \sigma_{xy} \\ \sigma_{yx} & \sigma_{yy} \end{bmatrix} \begin{bmatrix} \cos\varphi \\ \sin\varphi \end{bmatrix} \qquad \text{on the circle}\,, \tag{6.3}$$

and the equilibrium conditions require that the integral of the traction vector over the circle Γ equals the point load inside the circle[1]

$$\int_\Gamma \boldsymbol{t}\, ds + P \cdot \boldsymbol{e}_1 = \int_\Gamma \begin{bmatrix} t_x \\ t_y \end{bmatrix} ds + \begin{bmatrix} 1 \\ 0 \end{bmatrix} = \begin{bmatrix} 0 \\ 0 \end{bmatrix}. \tag{6.4}$$

[1]For more details see page 446.

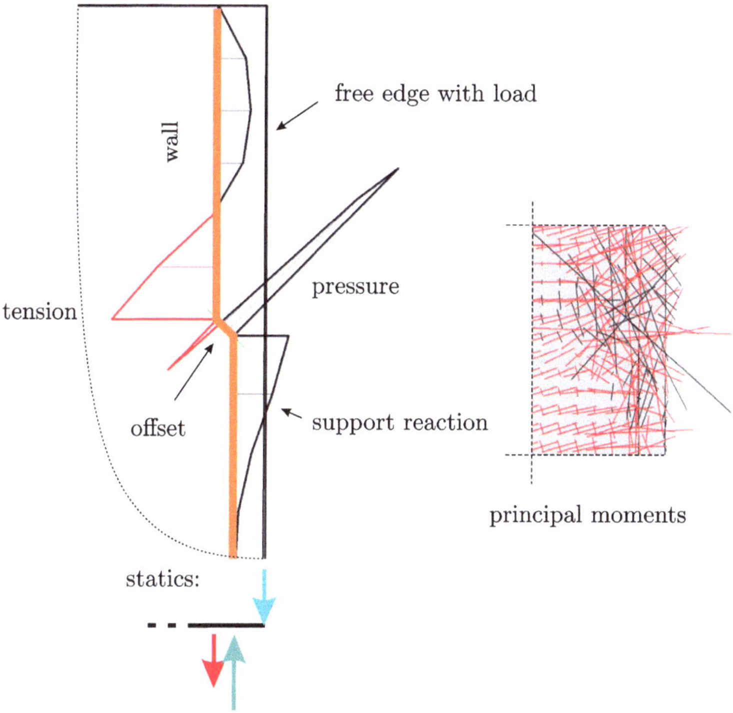

Fig. 6.4 Plate (detail)—a small offset in the supporting walls and the hefty consequences

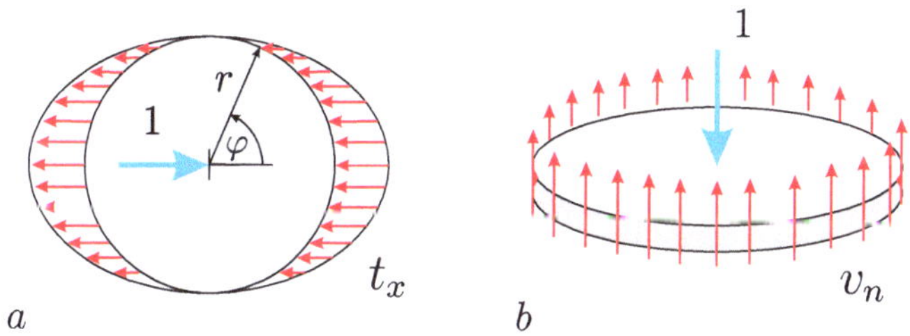

Fig. 6.5 Single force acting on a plate and on a slab. The stresses in the plate, as well as the shear force (v_n), behave as $1/r$, but since the plate deflection w is the threefold integral of v_n, the deflection $w = r^2 \ln r$ is finite at $r = 0$

If the radius r shrinks, the circumference of the circle, $|\Gamma| = 2\pi\, r$, shrinks as well, and to counteract this tendency the horizontal traction t_x must behave like $-1/(2\pi\, r)$

$$\lim_{r \to 0} \int_\Gamma t_x\, ds = \int_0^{2\pi} t_x\, r\, d\varphi = -\int_0^{2\pi} \frac{1}{2\pi\, r}\, r\, d\varphi = -1\,, \tag{6.5}$$

and become infinitely large in the limit, $r \to 0$.

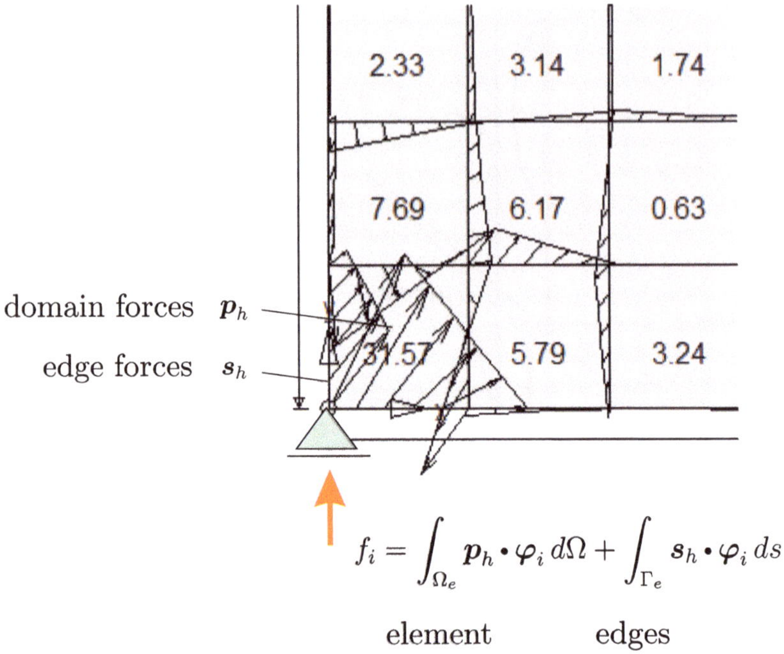

Fig. 6.6 Restraining forces = surface forces + line forces near the fixed node. The numbers are the integrals of the surface forces $\boldsymbol{p}_h$, see (6.7)

By how much does the source point get displaced? To find that out we integrate the strains. If we set Poisson's ratio $\nu = 0$, the strain $\varepsilon_{xx} = 1/E \cdot \sigma_{xx}$ depends only on the horizontal stress, and since

$$\sigma_{xx} = -\frac{1}{2\,\pi\, r} = E \cdot \varepsilon_{xx} = E \cdot \frac{\partial u}{\partial x} \simeq -\frac{1}{r}\,, \tag{6.6}$$

we conclude that the horizontal displacement u behaves like $-\ln r$, which is the integral of $-1/r$, and so the displacement will be infinitely large, since $-\ln 0 = \infty$.

If this sounds convincing, we must also accept that it is not possible to place a plate on point supports, since the plate would simply slip by.

The reason that in FE-analysis it is possible to keep a node fixed is of course that there is no 1:1 correspondence between the FE-solution and the physical model. At a fixed node of a mesh the displacement is indeed zero, $u_i = 0$, but it is not a single point force, which achieves this, but rather the combined effort of domain and line forces near the support, see Fig. 6.6. The numbers in the plot are the integrated domain forces of the FE-solution in the element (element resultants)

$$\int_{\Omega_e} (p_{hx}^2 + p_{hy}^2)\, d\Omega\,. \tag{6.7}$$

The nodal force f_i in the printout is the equivalent nodal support reaction, it signals how much work is done by these restraining forces on acting through the unit nodal displacement of the support.

6.4 Decay of Stresses

For a similar reason the stresses and strains in a structure must exhibit the opposite tendency and tend to zero with increasing distance from the load. Though, in this case it is not the equilibrium condition, but the **principle of conservation of energy**, which dictates this behavior.

When we draw a circle of radius R around the center of the load, the internal energy inside the circle (we leave the factor $1/2$ of eigenwork off)

$$\begin{aligned} W_i &= \int_{\Omega} \sigma_{ij}\, \varepsilon_{ij}\, d\Omega = \int_0^R \int_0^{2\pi} \sigma_{ij}\, \varepsilon_{ij}\, r\, dr\, d\varphi \\ &= \int_0^R \int_0^{2\pi} (\sigma_{11}\, \varepsilon_{11} + 2\, \sigma_{12}\, \varepsilon_{12} + \sigma_{22}\, \varepsilon_{22})\, r\, dr\, d\varphi \end{aligned} \tag{6.8}$$

must be equal to the external work $W_e(P)$ of the load—which in simplified terms is the displacement caused by the load $\times$ the load itself plus the work done by the section forces on the circumference Γ of the circle, which we call $W_e(R)$. From a certain radius R onwards, the circle is large enough to encompass all of the load and from this moment on the external work W_e will only change because $W_e(R)$ changes. The energy balance dictates

$$W_e = W_e(P) + W_e(R) = a(\boldsymbol{u}, \boldsymbol{u})_{\Omega} = W_i\,, \tag{6.9}$$

and so the expression

$$a(\boldsymbol{u}, \boldsymbol{u})_{\Omega} - W_e(R) = W_e(P) \tag{6.10}$$

must be constant, which means, the strain energy density $\sigma_{ij}\, \varepsilon_{ij}$ and the section forces must tend to zero with increasing radius R to counteract the increase in the size πR^2 of Ω and the circumference $2\pi R$ of Γ.

When we apply this logic to the continuous beam in Fig. 6.7, we realize that the internal energy W_i in each tail end $(x', \ell]$ must be equal to the external work (we integrate from the point x' up to the end of the beam)

$$W_i = \int_{x'}^{\ell} \frac{M^2}{EI}\, dx = P \cdot w(\ell) + W_e(x') = W_e\,, \tag{6.11}$$

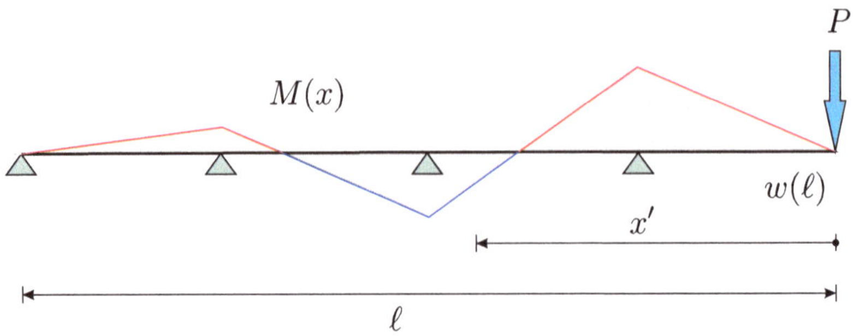

Fig. 6.7 The bending moment must ebb away

and this requires $W_i - W_e(x')$ to have the same value at each point x', and so $M(x)$ and $W_e(x')$ must shrink if x' moves to the left.

The requirement that energy is conserved is the reason why influence functions usually decay rapidly.

The exception are influence functions for forces in statically determinate structures, since such structures become kinematic when a hinge is added, and therefore even very wide swings of an influence function do not violate the law of conservation of energy, see Fig. 2.46.

Remark 6.1 The greater the distance R of an observer from the sun, the weaker the light appears, since the radiated energy E is distributed over an ever-larger sphere S

$$E = \int_S q\, dS = q \cdot 4\pi\, R^2\,, \tag{6.12}$$

and the energy density $q = E/S$ per m^2 drops like $1/R^2$.

This argument is also implicitly adopted by an engineer, who ignores loads which are added in remote regions of a plate, since he knows that the additional strain energy will decay rapidly—like the light of the sun.

However, this rule cannot be adhered to blindly. The shape of the structure and the location of the supports play a major role, as in the case of the "cantilever like" frame in Fig. 6.8. The larger the lever arm of the point load, the larger the fixed end moments.

In other words, if the additional load goes a long way, if its eigenwork is large, one must be careful, while in all other cases one can assume that the effects "trickle away".

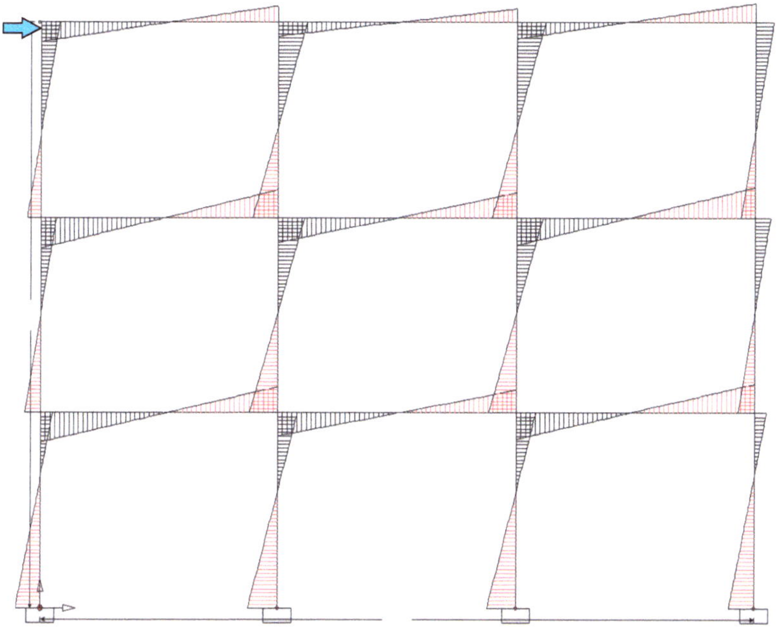

Fig. 6.8 Single force and multi-story frame

If you direct a very strong laser beam from the Earth to the Moon, and you apply a tiny rotation to the beam, the light spots on the moon travel with a speed, which is larger than the speed of light (theoretically, because the photons only travel with speed c).

So, rigid body rotations require attention. If these are possible, we must expect everything...

6.5 Infinite Stresses

Singular points are typically found at corner points as in Fig. 6.9. If we are sure, influence functions can predict also these stresses—perhaps not directly at the corner point itself, but close by—we face a problem: How can the dislocation of a single point (= influence function for the stress σ_{yy} at the crack tip) cause infinitely large displacements at the upper and lower edge of the plate, see Fig. 6.10? No other solution is possible if influence functions are also applicable at or near such singular points

$$\sigma_{yy}(\boldsymbol{x}) = \int_{\Gamma} G(\boldsymbol{y}, \boldsymbol{x}) \bullet \boldsymbol{p}(\boldsymbol{y}) ds_{y} = \infty\,. \tag{6.13}$$

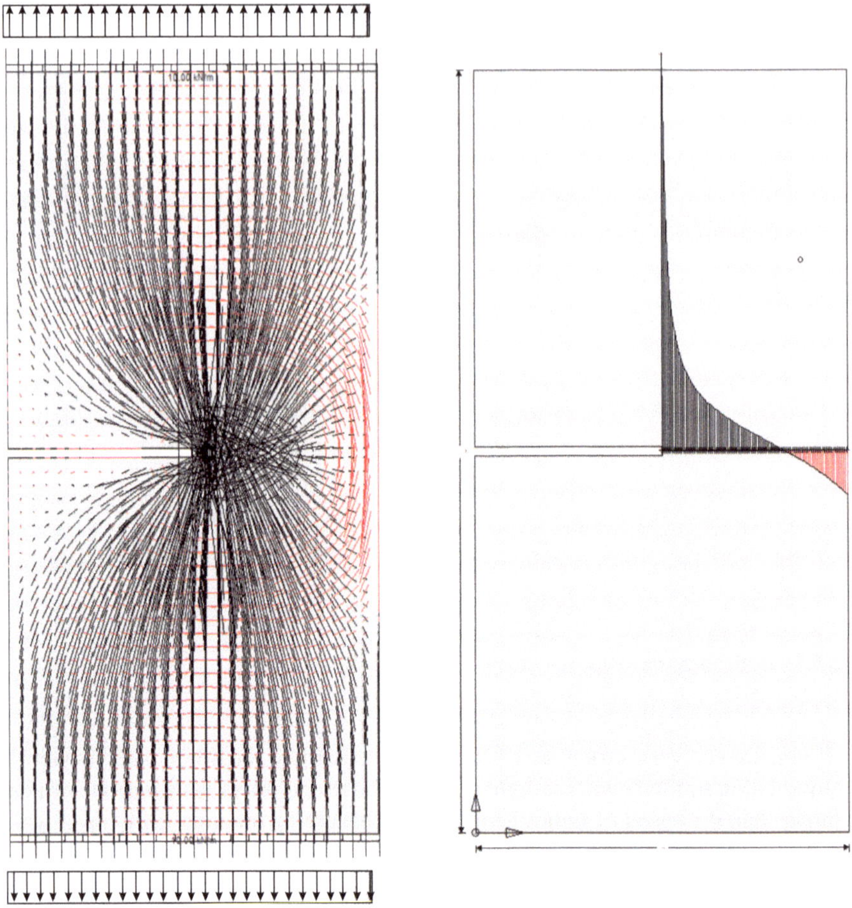

Fig. 6.9 Principal stresses and stresses σ_{yy} in a plate with a slit

In principle we have met the phenomenon before, in the case of the influence function for the shear force in Fig. 6.3. Imagine we analyze the cracked plate, see Fig. 6.11, with a mesh of triangular elements, Poisson's ratio ν is zero (for simplicity), and we compute the influence function for the sum of the two stresses $\sigma_{yy} + \sigma_{yy}$ in the two elements in the immediate vicinity of the crack tip. By taking the sum we preserve the symmetry of the problem.

To generate the influence function we need to apply the stresses σ_{yy} of the shape functions as nodal forces, see Fig. 6.11a (the mutual opposite nodal forces at the right end of the two elements cancel). At the rearward nodes on the symmetry axis, the vertical displacements are zero, and so if the two forces drive the two faces of the crack apart, they practically rotate the two sides of the crack about these rearward nodes, and if h tends to zero, the sides will rotate by 90°, which means the vertical displacements will become infinitely large.

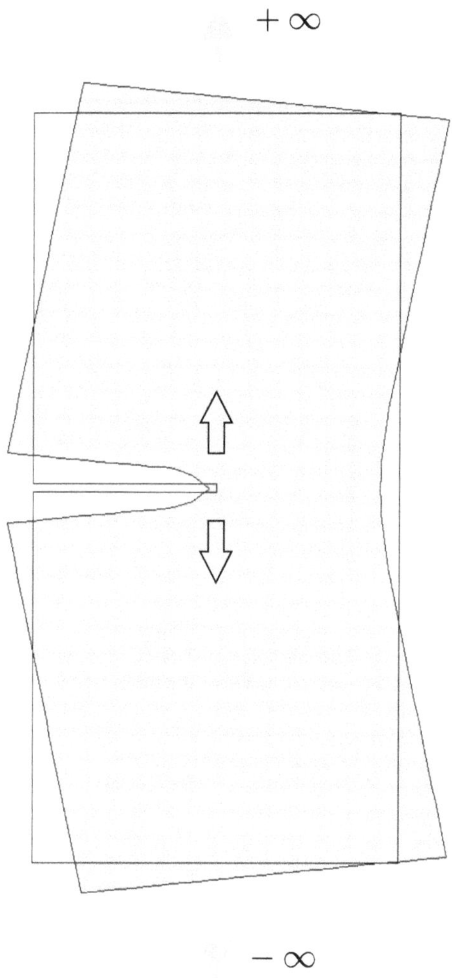

Fig. 6.10 A dislocation at the crack tip must result in infinite displacements

But why are the results different, when we repeat this maneuver at the smooth edge of a plate? If we now compute the influence function for $\sigma_{yy} + \sigma_{yy}$, no part of the plate protrudes beyond the edge, see Fig. 6.11c, and so nothing can rotate, and the displacements at distant points remain bounded.

Also the singular stresses of the L-shaped plate in Fig. 6.12 are due to the fact that the two nodal forces, which generate the influence function for the sum of the two stresses $2 \cdot \sigma_{\xi\xi}$ rotate in the limit, $h \to 0$, the faces by 90°.

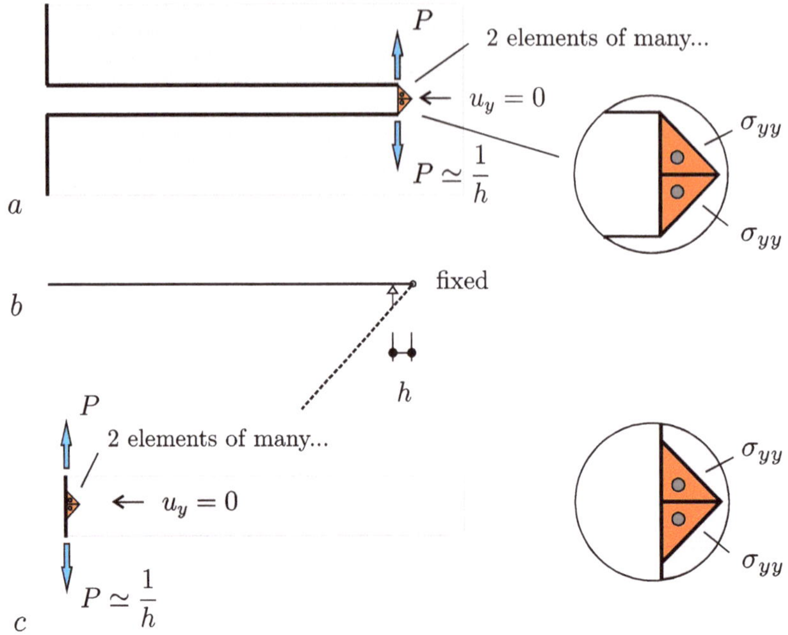

Fig. 6.11 Plate and triangular elements, **a** generating the influence function for $2 \cdot \sigma_{yy}$ at the crack tip, **b** kinematics, **c** action of the forces at the outer edge

6.6 Symmetry of Adjoint Effects

Some theory can help to understand these effects. Let us return to the cracked plate in Fig. 6.10. Two influence functions are at work here: The influence function for the stress σ_{yy} at the crack tip and the influence function for the sum of the vertical displacements at the upper and lower edge. To simplify the situation, we compute the vertical displacements at one point on the upper and one point on the lower edge simultaneously. (Instead of measuring at two points we could also take the integral of the displacements). This arrangement preserves the symmetry of the problem.

So, we operate with two functionals

$$J_1(\boldsymbol{u}) = \sigma_{yy}(\boldsymbol{u}) \qquad J_2(\boldsymbol{u}) = u_y(upper\ edge) + u_y(lower\ edge)\,. \tag{6.14}$$

To the first functional belongs the influence function $\boldsymbol{G}_1$ and to the second functional the influence function $\boldsymbol{G}_2$ and the two functionals must agree “crosswise”

$$J_1(\boldsymbol{G}_2) = J_2(\boldsymbol{G}_1)\,. \tag{6.15}$$

The influence function $\boldsymbol{G}_2$ is generated by two single forces $P = \pm 1$, which pull at the upper and lower edge, and produce infinitely large stresses σ_{yy} at the crack tip

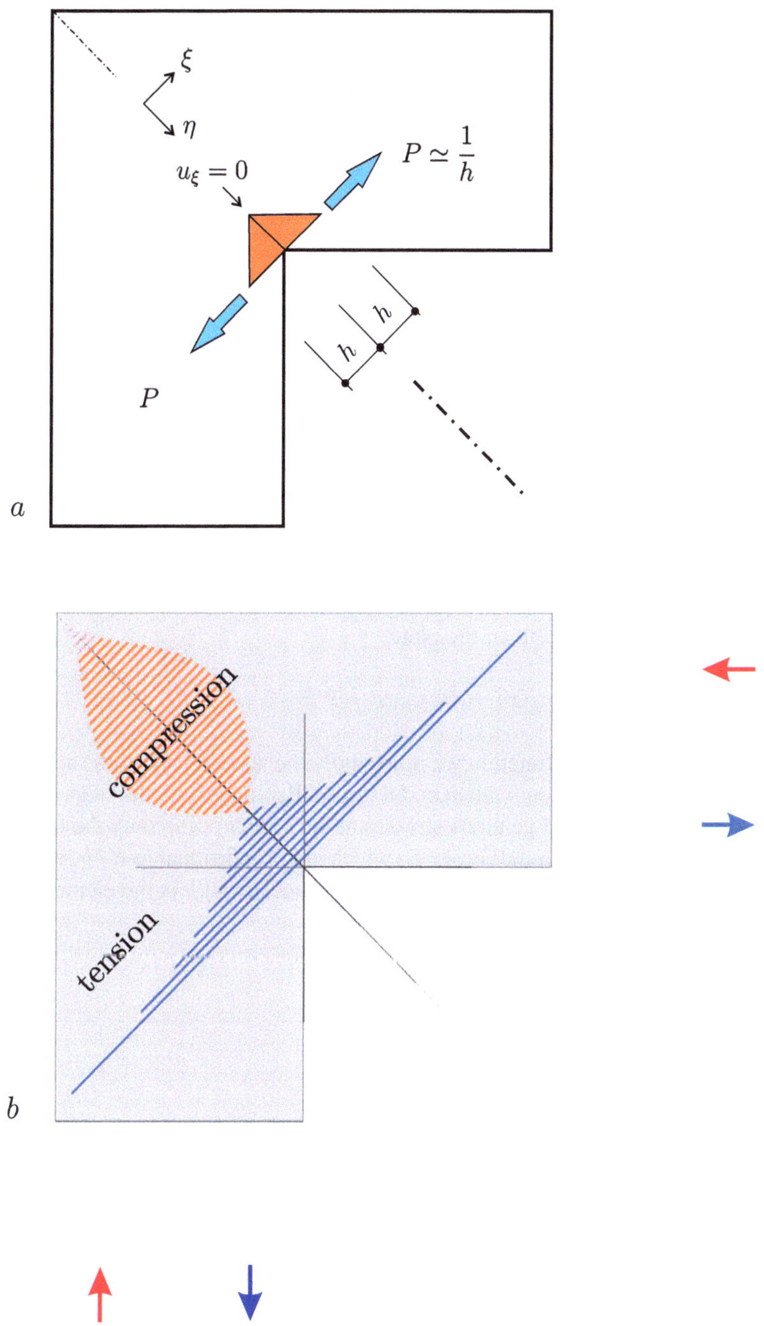

Fig. 6.12 At the reentrant corner the stresses become infinite

$$J_1(\boldsymbol{G}_2) = \infty \qquad J_1 \text{ measures} \sigma_{yy} \text{ of} \boldsymbol{G}_2 \,. \tag{6.16}$$

Vice versa, does the dislocation at the crack tip lead to infinitely large displacements at the upper and lower edge of the plate

$$J_2(\boldsymbol{G}_1) = \infty \qquad J_2 \text{ measures} u_y \text{ (at the edges) of} \boldsymbol{G}_1 \,, \tag{6.17}$$

and according to the symmetry built into Betti, see (3.170), these two values must be the same. If one is infinite, the other must be infinite as well.

6.7 Cantilever Wall Plate

But even in standard, no-fancy situations such as the self-weight of the plate in Fig. 6.13 we observe infinitely large stresses, here in the extreme fibers. We may assume, this would also happen if we replace the gravity load by a single force P, which acts at an arbitrary interior point $\boldsymbol{y}_P$.

If this is correct, the influence function for the stress σ_{xx} at the upper edge must be infinite at almost all points of the plate

$$\sigma_{xx}(\boldsymbol{x}) = G(\boldsymbol{y}_P, \boldsymbol{x}) \cdot P = \infty \cdot P \,. \tag{6.18}$$

To generate the influence function we must apply the stresses σ_{xx} of the shape functions at the corner point as equivalent nodal forces. But since only the shape functions of the corner element itself generate stresses at the corner point, only the four nodes near the corner will carry nodal forces f_i, which are proportional to E/h, where E is the modulus of elasticity, $E = 2.1 \cdot 10^5$ N/mm^2 (steel), and h is the element length.

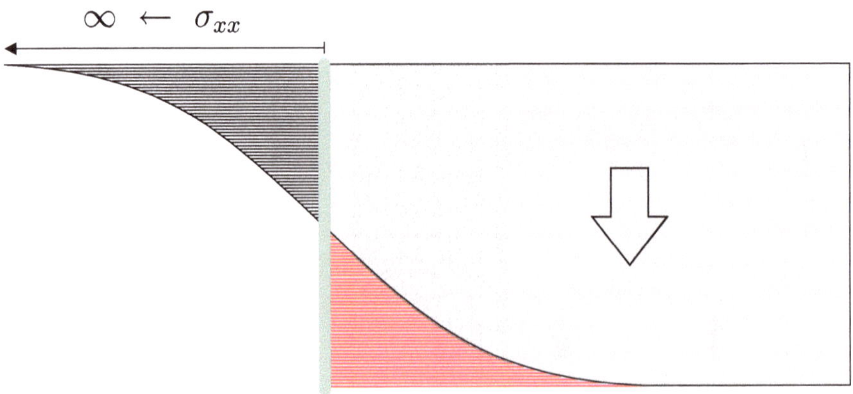

Fig. 6.13 Gravity load produces infinitely large stresses in the extreme fibres

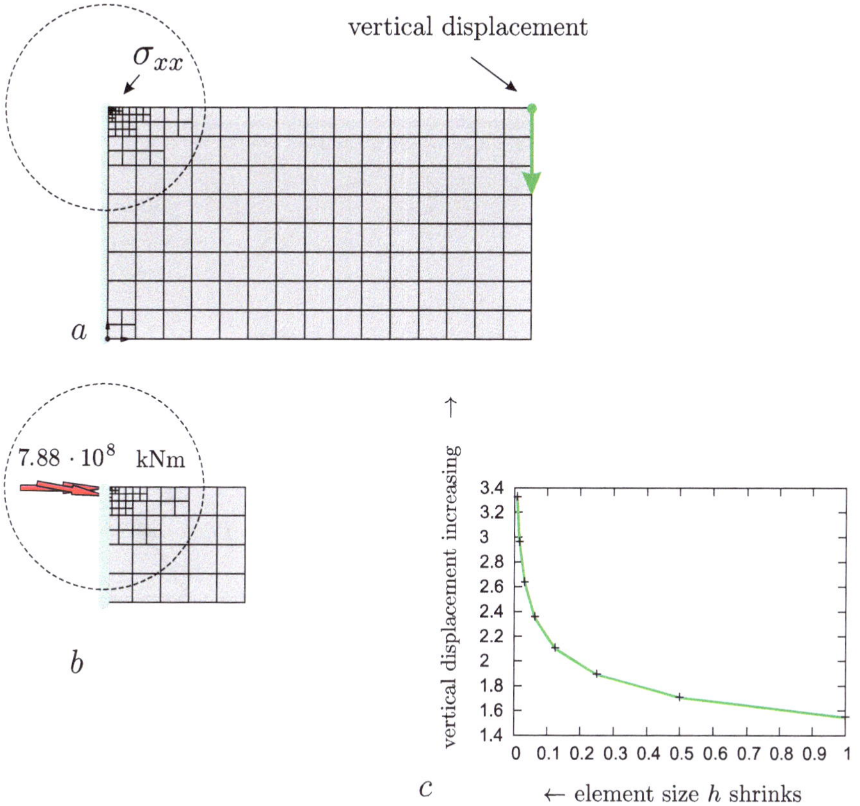

Fig. 6.14 Calculating the influence function for the stress σ_{xx} at the upper left corner, **a** mesh, **b** (two of the four) equivalent nodal forces, **c** vertical displacement of the upper right node

Numerical tests confirmed, indeed, an exponential increase in the vertical displacement of the opposite free corner, when the element size h approached zero, $h \to 0$, see Fig. 6.14c.

Imagine, we replace the main stresses by pairs of orthogonal arrows ("stream lines"), see Figs. 6.15a, b. In each vertical cross section, the vector sum of these arrows must balance the load vector. Now, the more these streamlines approach the left edge, the more they flatten, since the edge is fixed in vertical direction and so the streamlines must stretch ever more and more, for their shrinking vertical components to balance the applied load. This is the reason behind the infinite stresses.

It is the same situation as with a street lantern, which hangs between two walls. Before you manage to pull the rope taut, the rope rips apart.

Better to round off the corners of the plate, since then the streamlines can rotate and thus they have it easier to balance the vertical load, see Figs. 6.16 and 6.17. No need to develop infinite stresses.

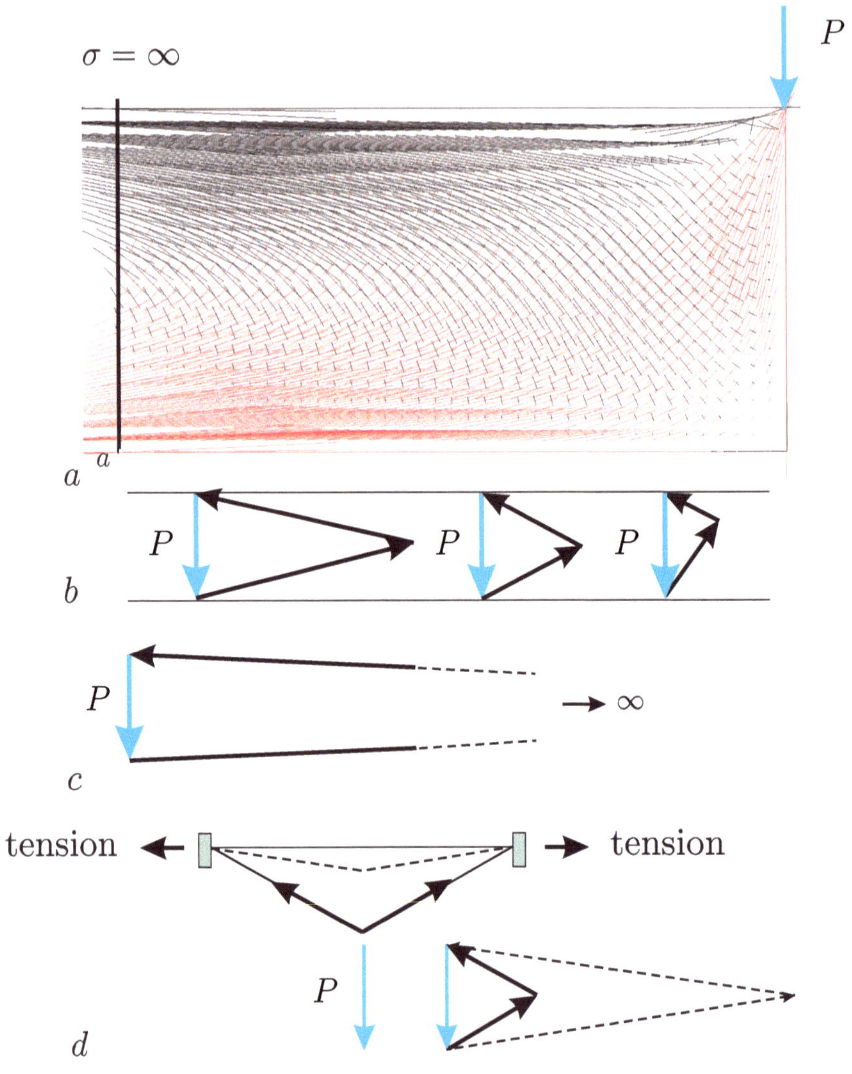

Fig. 6.15 Cantilever plate, **a** principal stresses ("stream lines"), **b** force polygon in different vertical sections, **c** close to the left edge the polygon becomes nearly infinitely flat and infinitely long, **d** street lantern—the same principle

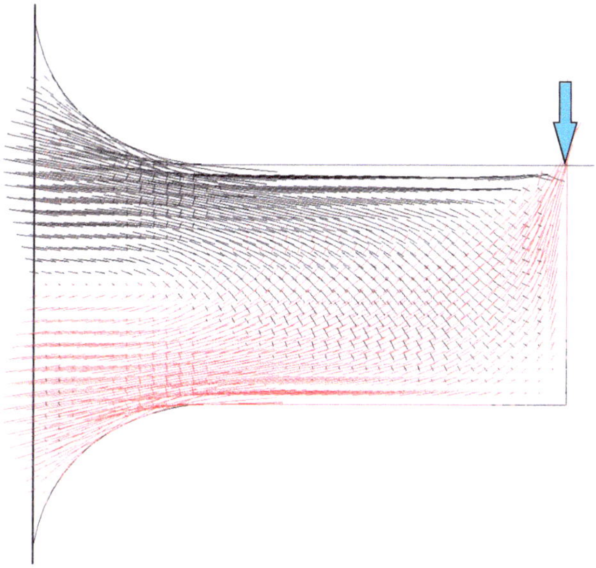

Fig. 6.16 By rounding out the corners the "stream lines" (= principal stresses) can rotate and so they have it easier to balance the vertical load

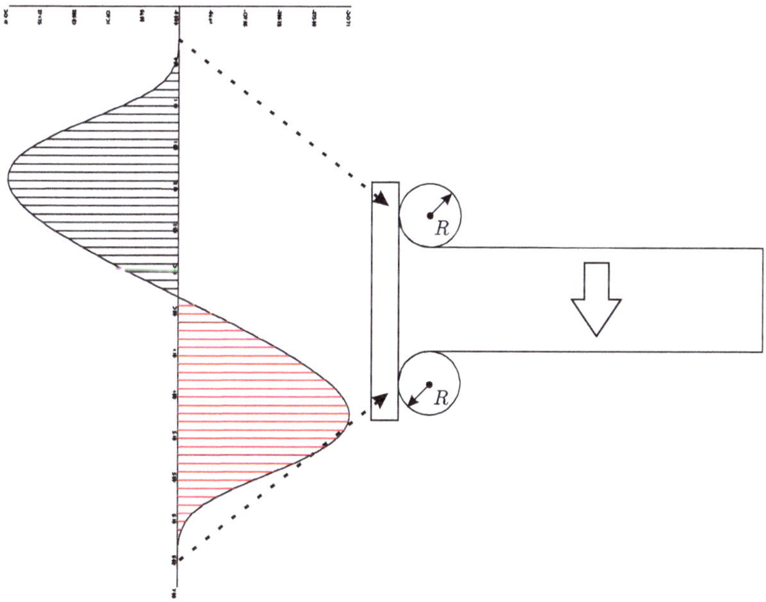

Fig. 6.17 Stress distribution (σ_{xx}) at the fixed edge if corners are rounded out

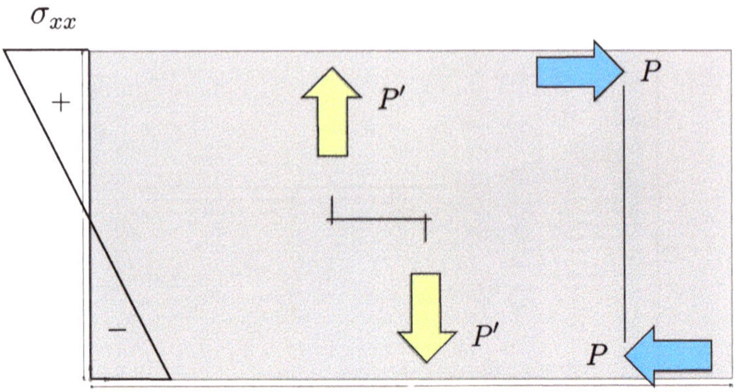

Fig. 6.18 The stresses σ_{xx} remain bounded in these load cases

Remark 6.2 Numerical tests prove that force couples do not produce singular stresses, see Fig. 6.18.

6.8 Standard Situations

But a cantilever plate is not so special. Even at such seemingly harmless points as the reentrant corners of openings in a wall plate, see Fig. 6.19, singularities develop. Normally the meshes are too coarse for the singularities to shine through, but when one really goes all the way in, as in Fig. 6.20, one notices the infinite stresses. In highly stressed mechanical parts as turbine blades such stress singularities can be relevant for the design (stress intensity factors).

6.9 Singularities in Influence Functions

Also influence functions must cope with the singularities on the boundary, because these singularities pollute the FE-solution via an unavoidable "boundary element" mechanism: **Also FE-solutions are potentials**, since they are the superposition of the load with influence functions, see Sect. 9.18.

To understand this, we consider a membrane, which covers an L-shaped opening, see Fig. 6.21. When the wind p presses down on the membrane, the slope of the membrane at the re-entrant corner will be very steep, we may even assume infinite in value, $w,_x = \infty$ and $w,_y = \infty$, and this means, the shear forces v_x and v_y

$$v_x = H\, w,_x \qquad v_y = H\, w,_y \qquad H = \text{prestress}\,, \tag{6.19}$$

will be infinite at the re-entrant corner.

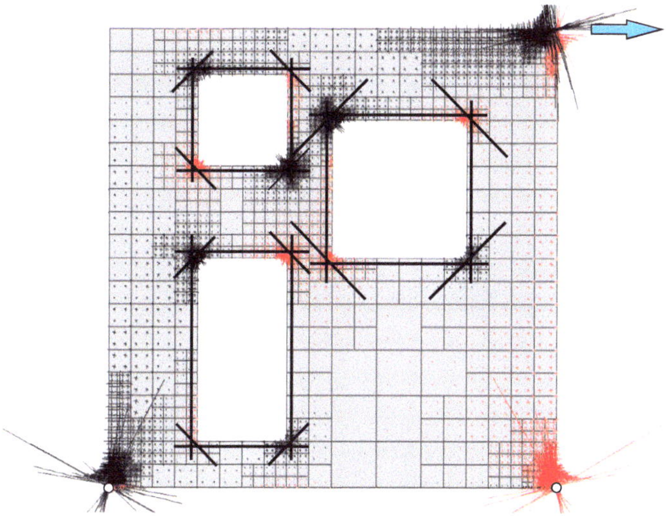

Fig. 6.19 Wall plate with adaptively refined mesh, and constructive reinforcement at the corners

But even in this dramatic situation the influence function for the FE-solution

$$w_h(\boldsymbol{x}) = \int_{\Omega} G_h(\boldsymbol{y}, \boldsymbol{x})\, p(\boldsymbol{y})\, d\Omega_{\boldsymbol{y}}\,, \tag{6.20}$$

looks the same and nothing seems to indicate that the kernel $G_h(\boldsymbol{y}, \boldsymbol{x})$ suffers from a singularity at the re-entrant corner. Where do the negative effects of the singularity hide?

We only want to sketch the answer. Readers who are familiar with potential theory or the boundary element method know that the solution of the equation $-\Delta w = p$ can be written as a boundary integral plus a domain integral, see Sect. 9.17,

$$w(\boldsymbol{x}) = \int_{\Gamma} (g(\boldsymbol{y}, \boldsymbol{x})\, \frac{\partial w}{\partial n}(\boldsymbol{y}) - \frac{\partial g(\boldsymbol{y}, \boldsymbol{x})}{\partial n}\, w(\boldsymbol{y}))\, ds_{\boldsymbol{y}} + \int_{\Omega} g(\boldsymbol{y}, \boldsymbol{x})\, p(\boldsymbol{y})\, d\Omega_{\boldsymbol{y}}\,. \tag{6.21}$$

That is, we can calculate the shape of the surface w via the auxiliary function $g(\boldsymbol{y}, \boldsymbol{x})$ from its boundary values w, and $\partial w/\partial n$ on the edge Γ, and from the way the load p is distributed in the domain Ω.

The function

$$g(\boldsymbol{y}, \boldsymbol{x}) = -\frac{1}{2\pi} \ln |\boldsymbol{y} - \boldsymbol{x}|\,, \tag{6.22}$$

is called a ***fundamental solution***, since it satisfies the equation $-\Delta g(\boldsymbol{y}, \boldsymbol{x}) = \delta(\boldsymbol{y} - \boldsymbol{x})$.

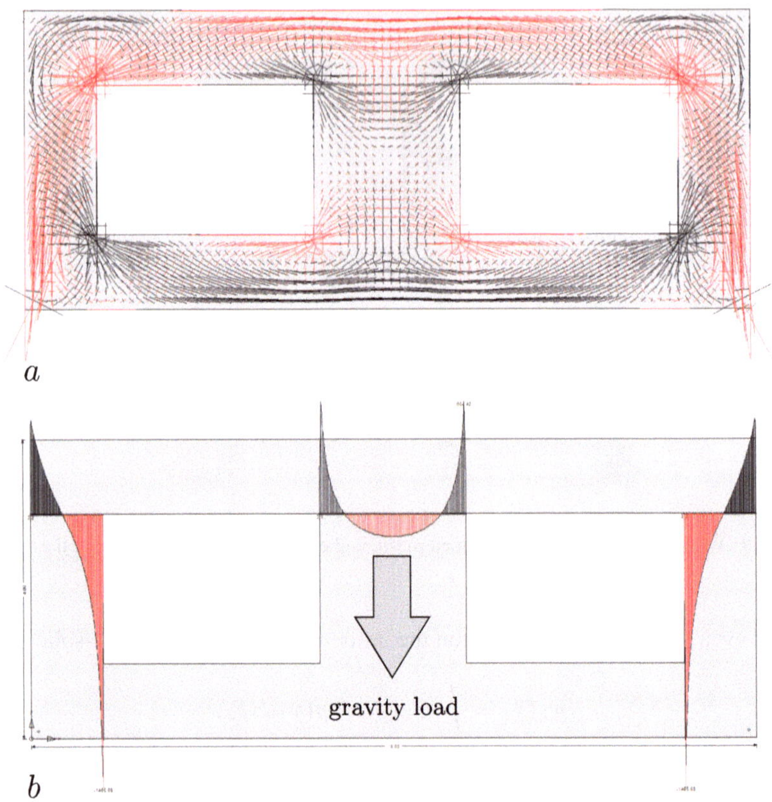

Fig. 6.20 Gravity load in a wall plate on two point supports, **a** principal stresses, **b** stresses σ_{yy} in horizontal sections (BE-solution)

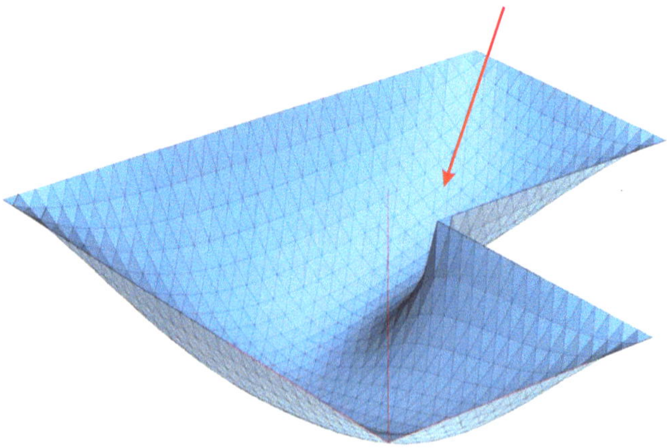

Fig. 6.21 L-shaped membrane under uniform pressure

This integral representation can also be applied to the influence function $G(\boldsymbol{y}, \boldsymbol{x})$. In this situation, the load to which $G(\boldsymbol{y}, \boldsymbol{x})$ corresponds, is a Dirac delta $\delta(\boldsymbol{y} - \boldsymbol{x})$, and since $G(\boldsymbol{y}, \boldsymbol{x})$ is zero on the edge Γ, the equation reduces to

$$\begin{aligned} G(\boldsymbol{y}, \boldsymbol{x}) &= \int_\Gamma [g(\boldsymbol{\xi}, \boldsymbol{y}) \frac{\partial G_h}{\partial n}(\boldsymbol{\xi}, \boldsymbol{x}) - \frac{\partial g(\boldsymbol{\xi}, \boldsymbol{y})}{\partial n} G(\boldsymbol{\xi}, \boldsymbol{x})]\, ds_{\xi} \\ &\quad + \underbrace{\int_\Omega g(\boldsymbol{\xi}, \boldsymbol{y})\, \delta(\boldsymbol{\xi} - \boldsymbol{x})\, d\Omega_{\xi}}_{= g(\boldsymbol{y}, \boldsymbol{x})} \\ &= \int_\Gamma g(\boldsymbol{\xi}, \boldsymbol{y}) \frac{\partial G}{\partial n}(\boldsymbol{\xi}, \boldsymbol{x})\, ds_{\xi} + g(\boldsymbol{y}, \boldsymbol{x})\,. \end{aligned} \tag{6.23}$$

Next, we repeat this with the FE-approximation $G_h(\boldsymbol{y}, \boldsymbol{x})$, which is the solution to the boundary value problem

$$-\Delta G_h(\boldsymbol{y}, \boldsymbol{x}) = \delta_h(\boldsymbol{y}, \boldsymbol{x}) \qquad G_h = 0 \qquad \text{at the edge } \Gamma\,, \tag{6.24}$$

where $\delta_h(\boldsymbol{y}, \boldsymbol{x})$ is a patchwork of distributed forces (and line forces), which try to simulate the action of a true point load, of a Dirac delta $\delta(\boldsymbol{y} - \boldsymbol{x})$. Therefore $G_h(\boldsymbol{y}, \boldsymbol{x})$ has the integral representation

$$G_h(\boldsymbol{y}, \boldsymbol{x}) = \int_\Gamma g(\boldsymbol{\xi}, \boldsymbol{y}) \underset{\uparrow}{\frac{\partial G_h}{\partial n}}(\boldsymbol{\xi}, \boldsymbol{x})\, ds_{\xi} + \int_\Omega g(\boldsymbol{\xi}, \boldsymbol{y})\, \delta_h(\boldsymbol{\xi}, \boldsymbol{x})\, d\Omega_{\xi}\,, \tag{6.25}$$

and here we see the critical point. At the re-entrant corner **the slope** $\partial G_h/\partial n$ of the membrane will probably be infinite (we are talking now about the load case $\delta_h(\boldsymbol{y}, \boldsymbol{x})$) and such singular shapes cannot be modeled accurately with finite elements, which means the normal derivative $\partial G_h/\partial n$ of the FE-solution will be inaccurate at the corner, and because this inaccuracy affects, see (6.25), also $G_h(\boldsymbol{y}, \boldsymbol{x})$, the influence function $G_h(\boldsymbol{y}, \boldsymbol{x})$ will be less accurate—not only at the corner—but in the whole domain, at any point.

Singularities on the edge propagate through this mechanism into the interior and reduce the quality of the influence function and therefore of the FE-solution w_h itself.

$$\frac{\partial G}{\partial n} \quad \rightarrow \quad \frac{\partial G_h}{\partial n} \quad \rightarrow \quad G_h \quad \rightarrow \quad w_h = \int_\Omega G_h\, p\, d\Omega_y$$

Nothing can break this chain.

This is like a canvas throwing wrinkles when the suspension on the edge is faulty.

To be precise we would have to add to the domain integral in (6.25) also the contributions of the line loads l_h (= jumps in the slope of $G_h(\boldsymbol{y}, \boldsymbol{x})$)

$$\int_\Omega g(\boldsymbol{\xi}, \boldsymbol{y})\, \delta_h(\boldsymbol{\xi}, \boldsymbol{x})\, d\Omega_\xi + \sum_i \int_{\Gamma_i} g(\boldsymbol{\xi}, \boldsymbol{y})\, l_h(\boldsymbol{\xi}, \boldsymbol{x})\, ds_\xi\,, \tag{6.26}$$

but we read the domain integral in (6.25) as a shorthand for both contributions. Here the focus is on the slope on the boundary and its effect. ***The slope is the critical part***.

Wall Plates

What in a membrane is the slope, are the tractions $\boldsymbol{t} = \boldsymbol{S}\,\boldsymbol{n}$ on the edge of a plate, and the equation analogous to (6.21) is the integral representation (*in tensor notation, summation over repeated indices is implied*)

$$u_i(\boldsymbol{x}) = \int_\Gamma (G_{ij}^F(\boldsymbol{y}, \boldsymbol{x})\, \underset{\uparrow}{t_j(\boldsymbol{y})} - T_{ij}^F(\boldsymbol{y}, \boldsymbol{x})\, u_j(\boldsymbol{y}))\, ds_y + \int_\Omega G_{ij}^F(\boldsymbol{y}, \boldsymbol{x})\, p_j(\boldsymbol{y})\, d\Omega_y$$

of a displacement field $\boldsymbol{u}(\boldsymbol{x}) = \{u_1, u_2\}^T$.

The functions $G_{ij}^F(\boldsymbol{y}, \boldsymbol{x})$ are the displacements ($j = 1, 2$) (x- and y-direction) of the elastic plane of infinite extent, if a single force $P = 1$ pushes the node $\boldsymbol{x}$ into the directions $i = 1, 2$ and the functions $T_{ij}^F(\boldsymbol{y}, \boldsymbol{x})$ are the tractions ($j = 1, 2$) (x- and y-direction) of these fields on the curve Γ.

The 2×2 matrix $\boldsymbol{G}^F$ is called the ***Somigliana matrix***, and it plays the same role as the function $g(\boldsymbol{y}, \boldsymbol{x})$ in (3.41), since it is the fundamental solution of the governing 2×2 system of equations (7.27).

When we apply this integral representation to Green's function this leads to

$$\begin{aligned} G_{ij}(\boldsymbol{y}, \boldsymbol{x}) &= \int_\Gamma (G_{jk}^F(\boldsymbol{\xi}, \boldsymbol{y})\, T_{ik}(\boldsymbol{\xi}, \boldsymbol{x}) - T_{jk}^F(\boldsymbol{\xi}, \boldsymbol{y})\, G_{ik}(\boldsymbol{\xi}, \boldsymbol{x}))\, ds_\xi \\ &+ \int_\Omega G_{jk}^F(\boldsymbol{\xi}, \boldsymbol{y})\, \delta_{ik}(\boldsymbol{\xi} - \boldsymbol{x})\, d\Omega_\xi\,, \end{aligned} \tag{6.27}$$

and in the case of the FE-solution to

$$\begin{aligned} G_{ij}^h(\boldsymbol{y}, \boldsymbol{x}) &= \int_\Gamma (G_{jk}^F(\boldsymbol{\xi}, \boldsymbol{y})\, \underset{\uparrow}{T_{ik}^h(\boldsymbol{\xi}, \boldsymbol{x})} - T_{jk}^F(\boldsymbol{\xi}, \boldsymbol{y})\, G_{ik}^h(\boldsymbol{\xi}, \boldsymbol{x}))\, ds_\xi \\ &+ \int_\Omega G_{jk}^F(\boldsymbol{\xi}, \boldsymbol{y})\, \delta_{ik}^h(\boldsymbol{\xi}, \boldsymbol{x})\, d\Omega_\xi\,. \end{aligned} \tag{6.28}$$

This suffices to recognize that the sensitive part in the approximation of the FE-influence function $G^h_{ij}(\boldsymbol{y}, \boldsymbol{x})$ are **the tractions** T^h_{ik} due to the approximate point load, the Dirac delta δ^h_{ik}. At singular points on the boundary the accuracy of these tractions will be questionable.

Kirchhoff Plates

What in a membrane is the slope are the moments and the Kirchhoff shear (support reactions) along the edge of the plate. Their singularities have a decisive influence on the accuracy of the FE-influence functions, see Fig. 6.22. In very simplified terms does the influence function of a plate look like,

$$w(\boldsymbol{x}) = \int_\Gamma (g\, w''' + g' w'' + g'' w' + g''' w)\, ds_y + \int_\Omega g\, p\, d\Omega_y\,, \tag{6.29}$$

where

$$g(\boldsymbol{y}, \boldsymbol{x}) = \frac{1}{2\pi K} r^2 \ln r \qquad K = \frac{E\, h^3}{12\,(1-\nu^2)} \tag{6.30}$$

is the fundamental solution, and K is the plate stiffness.

In this notation w' is the slope on the boundary, w'' is the moment and w''' is the Kirchhoff shear. The exact influence function G has the boundary values $G = 0$ and $G'' = 0$ (hinged edge), and so the formula reduces to

$$G(\boldsymbol{y}, \boldsymbol{x}) = \int_\Gamma (g\, G''' + g''\, G')\, ds_y + g(\boldsymbol{y}, \boldsymbol{x})\,. \tag{6.31}$$

The FE-influence function satisfies the zero-moment condition $G'' = 0$ on the hinged edge only approximately, and so we are not allowed to drop its contribution

$$G_h(\boldsymbol{y}, \boldsymbol{x}) = \int_\Gamma (g\, G_h''' + g' G_h'' + g'' G_h')\, ds_y + \int_\Omega g\, \delta_h\, d\Omega_y\,. \tag{6.32}$$

Consequently, the quality of the FE-influence function depends on the slope G_h' and the moment

$$G_h'' \equiv m_{xx}\, n_x^2 + 2\, m_{xy}\, n_x\, n_y + m_{yy}\, n_y^2 \qquad \text{(2nd deriv.)} \tag{6.33}$$

and the shear force

$$G_h''' \equiv m_{ij,j}\; n_i + \frac{d}{ds} m_{nt} \qquad \text{(3rd deriv.)}\,. \tag{6.34}$$

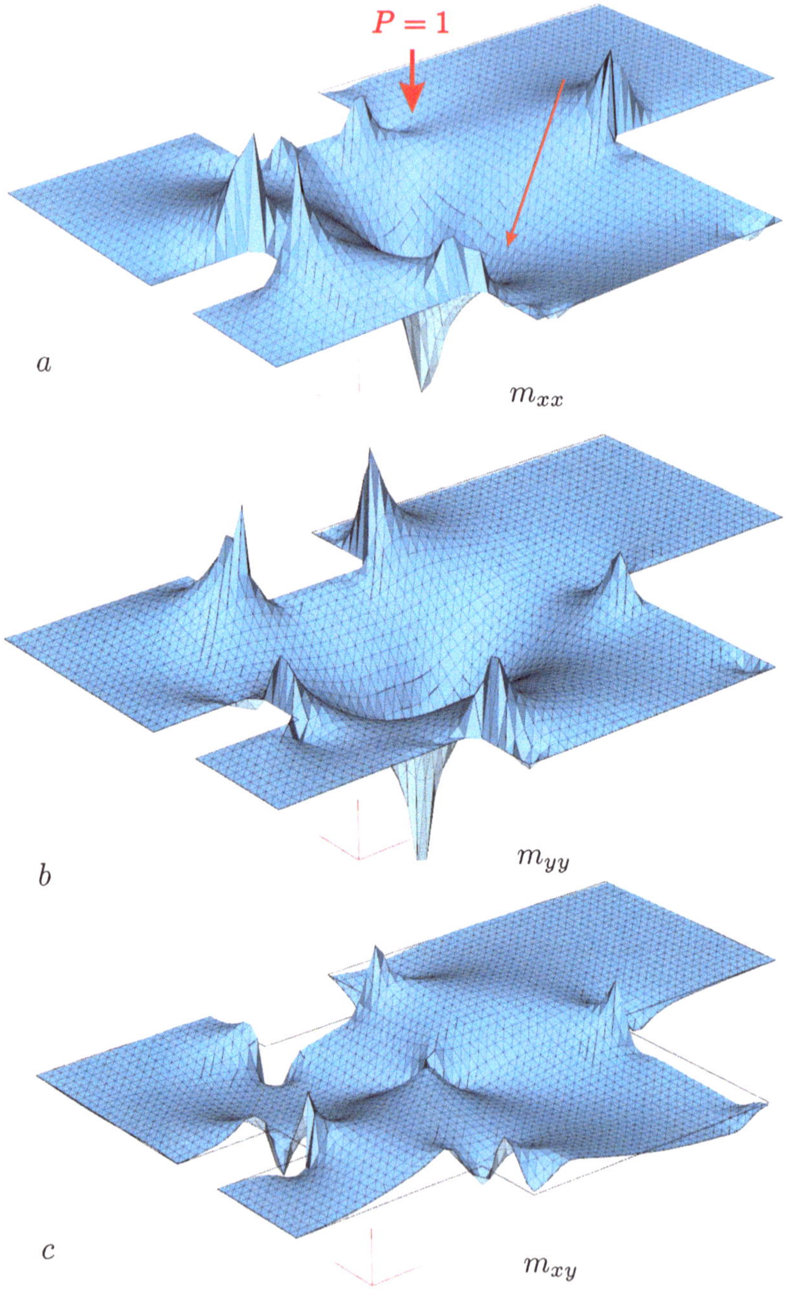

Fig. 6.22 Hinged plate and the bending moments produced by the Dirac delta $P = 1$, which generates the influence function $G_0(\boldsymbol{y}, \boldsymbol{x})$ for the deflection $w(\boldsymbol{x})$. At reentrant corners the moments become singular and this has a negative effect on the quality of the FE-influence function

If these functions are critical at a corner point, it will obviously diminish the quality of the influence function.

All this applies to the influence functions of the internal forces as well

$$m_{xx}(\boldsymbol{x}) = \int_\Gamma (G_2\, w''' + \cdots \qquad v_x(\boldsymbol{x}) = \int_\Gamma (G_3\, w''' + \cdots \tag{6.35}$$

and these react even more sensitively to singularities since they calculate second- or third-order derivatives. This is better seen, if we write (6.29) for all four quantities w, w', w'', w''' and keep only the **characteristic singularities** of the kernels

$$\begin{bmatrix} w \\ w' \\ w'' \\ w''' \end{bmatrix} = \int_\Gamma \begin{bmatrix} \boxed{r^{-1}} & \ln r & r \ln r & r^2 \ln r \\ r^{-2} & \boxed{r^{-1}} & \ln r & r \ln r \\ r^{-3} & r^{-2} & \boxed{r^{-1}} & \ln r \\ r^{-4} & r^{-3} & r^{-2} & \boxed{r^{-1}} \end{bmatrix} \begin{bmatrix} w \\ w' \\ w'' \\ w''' \end{bmatrix} ds_y + \int_\Omega \begin{bmatrix} G_0 \\ G_1 \\ G_2 \\ G_3 \end{bmatrix} p\, d\Omega_y\,. \tag{6.36}$$

Column #1 is the Kirchhoff shear of the G_i, column #2 are the moments, next come the normal derivatives and finally, in column #4, the G_i itself.

So much for the theory. In practice, however, the effects of singularities are unlikely to be so dramatic, since engineering accuracy is not that demanding, and the experienced engineer has a well-developed sense of what is credible and what is not.

Finite elements in structural analysis are both, modeling and "slide rule", and the engineer is therefore very flexible—not to say: indulgent—in the interpretation of FE-results.

In the case of the wall plate in Fig. 6.20 the tensile stress at the bottom increased from $\sigma_{xx} = 168$ kN/m^2 to $\sigma_{xx} = 220$ kN/m^2 after an adaptive refinement, see Fig. 6.23, which is an unusually large increase. Not in all cases will the difference be so large; unfortunately, no fixed rules can be established.

We think stress peaks are "harmless", if the plastic zone, which may form, does not cause massive compensating motions. In the case of the wall plate in Fig. 6.23 it is probably different, since it is only held fixed by two "tiny" point supports.

Perhaps a word about the separate roles of finite elements in mathematics and in engineering is in order here. For a mathematician finite elements are functions, shape functions, while for an engineer finite elements are structural elements, and the engineer therefore is not only interested in the approximation error, but also in the **modeling error**.

Both errors are interrelated. In contrast to the construction of a house, finite elements need to improve the foundations, when the roof truss is already in place, and the model still is open to adjustments. The analysis of the modeling error is as impor-

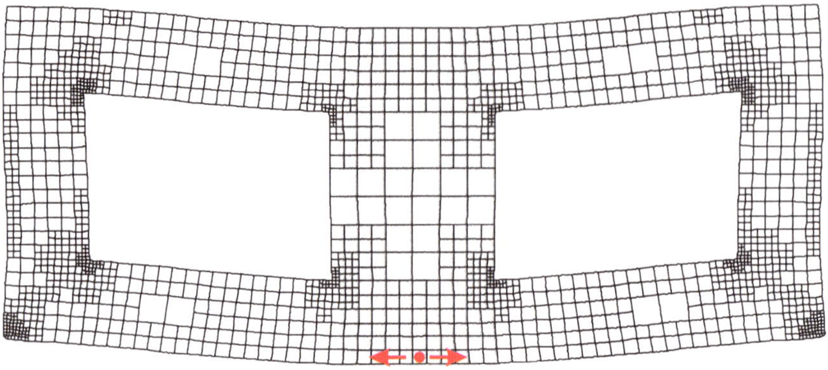

Fig. 6.23 Wall plate on point supports and gravity load. Adaptive refinement let the stress at the bottom increase from $\sigma_{xx} = 168$ kN/m^2 to $\sigma_{xx} = 220$ kN/m^2

tant as the analysis of the approximation error and here, above all, the expertise of the engineer is required.

For (too) long a time the focus was primarily on the numerical errors, on the development of sophisticated asymptotic error estimators, expressions of the type $O(h^n)$, but now the focus is also on the modeling error. Influence functions are well suited for this task, see the previous chapter, since they directly represent the sensitivities of a structure.

The keywords at this point are ***verification and validation***. Was the equation solved correctly—*verification*—and is the model able to provide the desired answer—*validation*? Answering this question is the engineer's job.

Chapter 7
Mixed Formulations

Variational formulations with different fields are called mixed formulations. They are based on a split of the displacement operator into a system of three equations according to the triad of ***strains, constitutive equations*** and ***equilibrium***.

7.1 Bernoulli Beam

To the Bernoulli beam corresponds the system

$$\begin{aligned} &\textbf{curvature} && \kappa - w'' = \kappa_0 && (7.1a)\\ &\textbf{constitutive equ.} && EI\,\kappa + M = M_0 && (7.1b)\\ &\textbf{equilibrium} && -M'' = m' + p\,, && (7.1c) \end{aligned}$$

where the terms on the right side are an *initial curvature* κ_0, e.g. from temperature $\kappa_0 = \alpha_T \Delta T / h$, an *initial moment* M_0, a distributed moment m [Nm/m] and a distributed load p [N/m].

We can read the system as the application of an **operator** $\boldsymbol{A}(\boldsymbol{\Sigma})$ to the triplet $\boldsymbol{\Sigma} = \{w, \kappa, M\}$. Given a **test function** $\boldsymbol{\delta\Sigma} = \{\delta w, \delta\kappa, \delta M\}$ (a virtual displacement) the virtual external work is the expression

$$< \boldsymbol{A}(\boldsymbol{\Sigma}), \boldsymbol{\delta\Sigma} >:= \int_0^l [(\kappa - w'')\,\delta M + (EI\,\kappa + M)\,\delta\kappa - M''\,\delta w]\,dx\,, \tag{7.2}$$

and integration by parts leads to Green's first identity

$$\mathscr{G}(\boldsymbol{\Sigma}, \boldsymbol{\delta\Sigma}) = < \boldsymbol{A}(\boldsymbol{\Sigma}), \boldsymbol{\delta\Sigma} > + [w'\,\delta M + M'\,\delta w]_0^l + a(\boldsymbol{\Sigma}, \boldsymbol{\delta\Sigma}) = 0\,, \tag{7.3}$$

F. Hartmann and P. Jahn, *Statics and Influence Functions*,
Springer Series in Solid and Structural Mechanics 13,
https://doi.org/10.1007/978-3-030-55889-5_7

where

$$a(\boldsymbol{\Sigma}, \boldsymbol{\delta\Sigma}) = \int_0^l [EI\, \kappa\, \delta\kappa + \kappa\, \delta M + M\, \delta\kappa + w'\, \delta M' + M'\, \delta w']\, dx \tag{7.4}$$

is the symmetric strain energy product, and one step more takes us to Green's second identity, to ***Betti's theorem***,

$$\begin{aligned}\mathscr{B}(\boldsymbol{\Sigma}, \boldsymbol{\delta\Sigma}) &= <\boldsymbol{A}(\boldsymbol{\Sigma}), \boldsymbol{\delta\Sigma}> + [w'\, \delta M + M'\, \delta w]_0^l \\ &\quad - [\delta w'\, M + \delta M'\, w]_0^l - <\boldsymbol{A}(\boldsymbol{\delta\Sigma}), \boldsymbol{\Sigma}> = 0\,.\end{aligned} \tag{7.5}$$

7.2 Timoshenko Beam

The displacements of a Timoshenko beam are the deflection w, and the rotation φ, see Fig. 7.1, which form a coupled system

$$-EI\, \theta'' + GA_s\, (w' + \theta) = 0 \tag{7.6a}$$

$$-GA_s\, (w'' + \theta') = p\,, \tag{7.6b}$$

that can be read as $-\boldsymbol{Lu} = \{0, p\}^T$. To it belongs the identity

$$\mathscr{G}(\boldsymbol{u}, \boldsymbol{\delta u}) = \int_0^l -\boldsymbol{Lu} \cdot \boldsymbol{\delta u}\, dx + [V\, \delta w + M\, \delta\theta]_0^l - a(\boldsymbol{u}, \boldsymbol{\delta u}) = 0\,, \tag{7.7}$$

where

$$\begin{aligned}a(\boldsymbol{u}, \boldsymbol{\delta u}) &= \int_0^l [V\, \delta\gamma + M\, \delta\kappa]\, dx \\ &= \int_0^l [GA_s(w' + \theta)\, (\delta w' + \delta\theta) + EI\, \theta'\, \delta\theta'\,]\, dx\,,\end{aligned} \tag{7.8}$$

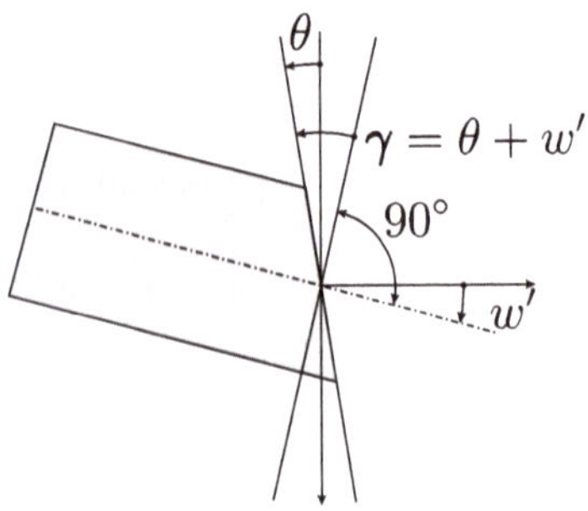

Fig. 7.1 Timoshenko beam

is the strain energy product. This is the **displacement formulation** of a Timoshenko beam.

The original mixed system behind (7.6) is

$$\begin{aligned} &\textbf{strains} && \theta' - \kappa = 0 && w' + \theta - \gamma = 0 && (7.9a)\\ &\textbf{material law} && EI\,\kappa - M = 0 && GA_s\gamma - V = 0 && (7.9b)\\ &\textbf{equilibrium} && -M' + V = 0 && -V' = p\,. && (7.9c) \end{aligned}$$

It can be read as the application of an operator $\boldsymbol{A}(\boldsymbol{\Sigma})$ to the triplet $\boldsymbol{\Sigma} = \{w, \theta; \kappa, \gamma; M, V\}$ and the **virtual external work** is the integral

$$\begin{aligned} <\boldsymbol{A}(\boldsymbol{\Sigma}), \boldsymbol{\delta\Sigma}>:= \int_0^l [(\theta' - \kappa)\,\delta M + (w' + \theta - \gamma)\,\delta V + (EI\,\kappa - M)\,\delta\kappa \\ + (GA_s\gamma - V)\,\delta\gamma + (-M' + V)\,\delta\theta - V'\,\delta w]\,dx\,. \qquad (7.10) \end{aligned}$$

Integration by parts of the last two terms, the equilibrium condition, leads to Green's first identity

$$\mathscr{G}(\boldsymbol{\Sigma}, \boldsymbol{\delta\Sigma}) = <\boldsymbol{A}(\boldsymbol{\Sigma}), \boldsymbol{\delta\Sigma}> + [-M\,\delta\theta - V\,\delta w]_0^l - a(\boldsymbol{\Sigma}, \boldsymbol{\delta\Sigma}) = 0 \qquad (7.11)$$

with the symmetric strain energy product

$$\begin{aligned} a(\boldsymbol{\Sigma}, \boldsymbol{\delta\Sigma}) = \int_0^l [(\theta' - \kappa)\,\delta M + (\delta\theta' - \delta\kappa) M + (w' + \theta - \gamma)\,\delta V \\ + (\delta w' + \delta\theta - \delta\gamma) V + EI\,\kappa\,\delta\kappa + GA_s\gamma\,\delta\gamma]\,dx\,. \qquad (7.12) \end{aligned}$$

7.3 Poisson Equation

The Poisson equation

$$-\Delta u(\boldsymbol{x}) = p(\boldsymbol{x}) \qquad (7.13)$$

can be split into a first-order system

$$\begin{aligned} \nabla u - \boldsymbol{\sigma} &= \boldsymbol{0}_{(2)} && (7.14a)\\ -\mathrm{div}\,\boldsymbol{\sigma} &= p_{(1)} && (7.14b) \end{aligned}$$

for two functions u and $\boldsymbol{\sigma}$ or $\boldsymbol{v} = \{u, \boldsymbol{\sigma}\}^T$, to which belongs the identity

$$\mathcal{G}(\boldsymbol{v}, \boldsymbol{\delta v}) = \int_{\Omega} [(\nabla u - \boldsymbol{\sigma}) \bullet \boldsymbol{\delta\sigma} - \text{div}\boldsymbol{\sigma}\, \delta u]\, d\Omega + \int_{\Gamma} \boldsymbol{\sigma} \bullet \boldsymbol{n}\, \delta u\, ds$$
$$- \underbrace{\int_{\Omega} (\nabla u \bullet \boldsymbol{\delta\sigma} + \nabla \delta u \bullet \boldsymbol{\sigma} - \boldsymbol{\sigma} \bullet \boldsymbol{\delta\sigma})\, d\Omega}_{a(\boldsymbol{v},\boldsymbol{\delta v})} = 0\,. \tag{7.15}$$

7.4 The Plate Equation

The constitutive equations are

$$\boldsymbol{E}(\boldsymbol{u}) - \boldsymbol{E} = \boldsymbol{E}_0 \tag{7.16a}$$
$$\boldsymbol{C}[\boldsymbol{E}] - \boldsymbol{S} = \boldsymbol{0} \tag{7.16b}$$
$$-\text{div}\boldsymbol{S} = \boldsymbol{p}\,, \tag{7.16c}$$

where $\boldsymbol{E} = [\varepsilon_{ij}]$ and $\boldsymbol{S} = [\sigma_{ij}]$ are the strain and stress tensor, respectively, $\boldsymbol{E}_0$ are initial strains (e.g. from temperature) and $\boldsymbol{E}()$ is the operator

$$\boldsymbol{E}(\boldsymbol{u}) = \frac{1}{2}(\nabla \boldsymbol{u} + \nabla \boldsymbol{u}^T) = \frac{1}{2}\begin{bmatrix} 2\,u_{1,1} & u_{1,2} + u_{2,1} \\ u_{2,1} + u_{1,2} & 2\,u_{2,2} \end{bmatrix}. \tag{7.17}$$

The divergence of the stress tensor $\boldsymbol{S}$, see Fig. 7.2, is the vector

$$\text{div}\, \boldsymbol{S} = \begin{bmatrix} \sigma_{11,1} + \sigma_{12,2} \\ \sigma_{21,1} + \sigma_{22,2} \end{bmatrix}. \tag{7.18}$$

The operator

$$\boldsymbol{C}[\boldsymbol{E}] = 2\mu \cdot \boldsymbol{E} + \lambda\, (\text{tr}\boldsymbol{E}) \cdot \boldsymbol{I} \tag{7.19}$$

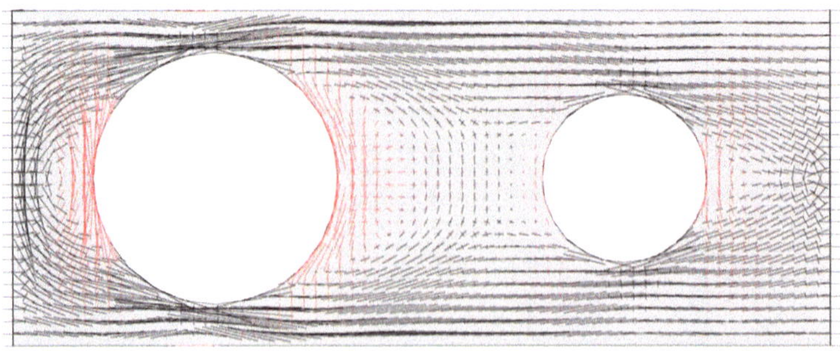

Fig. 7.2 Stressed plate

transforms the strain tensor into the stress tensor $\boldsymbol{S}$, where

$$\lambda = \frac{2\,\mu\,\nu}{1-2\,\nu} \qquad \text{tr}\boldsymbol{E} = \varepsilon_{11} + \varepsilon_{22} \qquad \text{(trace)} \tag{7.20}$$

and $\boldsymbol{I}$ is the unit tensor (unit matrix).

The left side of the system (7.16) is the application of an operator $\boldsymbol{A}(\boldsymbol{\Sigma})$ to the triplet $\boldsymbol{\Sigma} = \{\boldsymbol{u}, \boldsymbol{E}, \boldsymbol{S}\}$ and

$$\begin{aligned} < \boldsymbol{A}(\boldsymbol{\Sigma}), \boldsymbol{\delta\Sigma} >= &\int_\Omega (\boldsymbol{E}(\boldsymbol{u}) - \boldsymbol{E}) \boldsymbol{\cdot} \boldsymbol{\delta S}\, d\Omega + \int_\Omega (\boldsymbol{C}[\boldsymbol{E}] - \boldsymbol{S}) \boldsymbol{\cdot} \boldsymbol{\delta E}\, d\Omega \\ &+ \int_\Omega -\text{div}\boldsymbol{S} \boldsymbol{\cdot} \boldsymbol{\delta u}\, d\Omega \end{aligned} \tag{7.21}$$

is the associated virtual work with $\boldsymbol{\delta\Sigma} = \{\boldsymbol{\delta u}, \boldsymbol{\delta E}, \boldsymbol{\delta S}\}$ as "virtual displacementc" or "test field".

The dot $\boldsymbol{\cdot}$ denotes here also the scalar product of two matrices

$$\boldsymbol{S} \boldsymbol{\cdot} \boldsymbol{E} = \sigma_{11}\,\varepsilon_{11} + \sigma_{12}\,\varepsilon_{12} + \sigma_{21}\,\varepsilon_{21} + \sigma_{22}\,\varepsilon_{22}\,. \tag{7.22}$$

If $\boldsymbol{S} \in C^1(\Omega)$ is a symmetric matrix and $\boldsymbol{\delta u} \in C^1(\Omega)$ an arbitrary displacement field, then holds (integration by parts)

$$\int_\Omega -\text{div}\boldsymbol{S} \boldsymbol{\cdot} \boldsymbol{\delta u}\, d\Omega = -\int_\Gamma \boldsymbol{Sn} \boldsymbol{\cdot} \boldsymbol{\delta u}\, ds + \int_\Omega \boldsymbol{S} \boldsymbol{\cdot} \boldsymbol{E}(\boldsymbol{\delta u})\, d\Omega\,, \tag{7.23}$$

and so

$$\begin{aligned} < \boldsymbol{A}(\boldsymbol{\Sigma}), \boldsymbol{\delta\Sigma} > = &\int_\Omega (\boldsymbol{E}(\boldsymbol{u}) - \boldsymbol{E}) \boldsymbol{\cdot} \boldsymbol{\delta S}\, d\Omega + \int_\Omega (\boldsymbol{C}[\boldsymbol{E}] - \boldsymbol{S}) \boldsymbol{\cdot} \boldsymbol{\delta E}\, d\Omega \\ &+ \int_\Omega \boldsymbol{S} \boldsymbol{\cdot} \boldsymbol{E}(\boldsymbol{\delta u})\, d\Omega - \int_\Gamma \boldsymbol{Sn} \boldsymbol{\cdot} \boldsymbol{\delta u}\, ds\,. \end{aligned} \tag{7.24}$$

The three domain integrals in this equation form a symmetric, bilinear form since $\boldsymbol{C}[\boldsymbol{E}] \boldsymbol{\cdot} \boldsymbol{\delta E} = \boldsymbol{E} \boldsymbol{\cdot} \boldsymbol{C}[\boldsymbol{\delta E}]$

$$\begin{aligned} a(\boldsymbol{\Sigma}, \boldsymbol{\delta\Sigma}) := &\int_\Omega (\boldsymbol{E}(\boldsymbol{u}) - \boldsymbol{E}) \boldsymbol{\cdot} \boldsymbol{\delta S}\, d\Omega + \int_\Omega \boldsymbol{C}[\boldsymbol{E}] \boldsymbol{\cdot} \boldsymbol{\delta E}\, d\Omega \\ &+ \int_\Omega \boldsymbol{S} \boldsymbol{\cdot} (\boldsymbol{E}(\boldsymbol{\delta u}) - \boldsymbol{\delta E})\, d\Omega\,, \end{aligned} \tag{7.25}$$

which we call the strain energy product of $\boldsymbol{\Sigma}$ and $\boldsymbol{\delta\Sigma}$.

Green's first identity of the operator $\boldsymbol{A}(\boldsymbol{\Sigma})$ is

$$\mathcal{G}(\boldsymbol{\Sigma}, \delta\boldsymbol{\Sigma}) = < \boldsymbol{A}(\boldsymbol{\Sigma}), \delta\boldsymbol{\Sigma} > + \int_{\Gamma} \boldsymbol{Sn} \cdot \delta\boldsymbol{u}\, ds - a(\boldsymbol{\Sigma}, \delta\boldsymbol{\Sigma}) = 0\,, \tag{7.26}$$

and this expression is the basis of all the different variation principles in elasticity theory, as the ***Hu-Washizu principle*** or the ***Hellinger-Reissner principle***, [1].

The system $\boldsymbol{A}(\boldsymbol{\Sigma})$ can be transformed into a system for the displacement field $\boldsymbol{u}$ alone

$$-\boldsymbol{L}\boldsymbol{u} := -[\mu\, \Delta\boldsymbol{u} + \frac{\mu}{1-2\nu}\, \nabla\, \mathrm{div}\boldsymbol{u}] = \boldsymbol{p} - \mathrm{div}\, \boldsymbol{C}[\boldsymbol{E}_0] \tag{7.27}$$

to which belongs the identity

$$\mathcal{G}(\boldsymbol{u}, \delta\boldsymbol{u}) = \underbrace{\int_{\Omega} -\boldsymbol{L}\boldsymbol{u} \cdot \delta\boldsymbol{u}\, d\Omega + \int_{\Gamma} \boldsymbol{\tau}(\boldsymbol{u}) \cdot \delta\boldsymbol{u}\, ds}_{\delta W_e} \underbrace{- \int_{\Omega} \boldsymbol{E}(\boldsymbol{u}) \cdot \boldsymbol{C}[\boldsymbol{E}(\delta\boldsymbol{u})]\, d\Omega}_{\delta W_i} = 0\,, \tag{7.28}$$

where $\boldsymbol{\tau}(\boldsymbol{u})$ is the stress vector $\boldsymbol{Sn}$ of the field $\boldsymbol{u}$ on the edge Γ.

7.5 Kirchhoff Plate

The system consists of the three equations

$$\boldsymbol{K} - \boldsymbol{K}(w) = \boldsymbol{K}_0 \tag{7.29a}$$

$$\boldsymbol{C}[\boldsymbol{M}] + \boldsymbol{M} = \boldsymbol{0} \tag{7.29b}$$

$$-\mathrm{div}^2 \boldsymbol{M} = p\,, \tag{7.29c}$$

which is the application of an operator $\boldsymbol{A}(\boldsymbol{\Sigma})$ to the triplet $\boldsymbol{\Sigma} = \{w, \boldsymbol{K}, \boldsymbol{M}\}$. The tensor $\boldsymbol{K}_0$ stands for possible **initial curvatures** and it is

$$\mathrm{div}^2\, \boldsymbol{M} = \mathrm{div} \begin{bmatrix} m_{xx},_x + m_{xy},_y \\ m_{yx},_x + m_{yy},_y \end{bmatrix} = m_{xx},_{xx} + 2\, m_{xy},_{xy} + m_{yy},_{yy}\ . \tag{7.30}$$

The expression $\boldsymbol{K}(w)$ are of course the "curvatures"

$$\boldsymbol{K}(w) = \begin{bmatrix} w,_{xx} & w,_{xy} \\ w,_{yx} & w,_{yy} \end{bmatrix}\ . \tag{7.31}$$

Given a symmetric matrix $\boldsymbol{M} \in C^2(\Omega)$ and a function $\delta w \in C^2(\Omega)$ we have (via integration by parts)

$$\int_\Omega -\mathrm{div}^2 \boldsymbol{M}\, \delta w \, d\Omega = -\int_\Gamma (V_n\, \delta w - M_n \frac{\partial \delta w}{\partial n})\, ds - [[M_{nt}\, \delta w]] \\ - \int_\Omega \boldsymbol{M} \boldsymbol{\cdot} \boldsymbol{K}(\delta w)\, d\Omega \,, \tag{7.32}$$

where (in tensor notation)

$$V_n = \frac{d}{ds} M_{nt} + Q_n \quad M_{nt} = M_{ij}\, n_i\, t_j \quad M_n = M_{ij}\, n_i\, n_j \quad Q_n = M_{ij,i}\, n_j \,, \tag{7.33}$$

and $\boldsymbol{n} = \{n_1, n_2\}^T$ and $\boldsymbol{t} = \{t_1, t_2\}^T$ are the normal and tangent vectors on the edge (each is a unit vector).

The system (7.29) is the application of an operator $\boldsymbol{A}()$ to the triplet $\boldsymbol{\Sigma} = \{w, \boldsymbol{K}, \boldsymbol{M}\}$ and

$$< \boldsymbol{A}(\boldsymbol{\Sigma}), \boldsymbol{\delta\Sigma} >= \int_\Omega (\boldsymbol{K}(w) - \boldsymbol{K}) \boldsymbol{\cdot} \boldsymbol{\delta M}\, d\Omega + \int_\Omega (\boldsymbol{C}[\boldsymbol{K}] - \boldsymbol{M}) \boldsymbol{\cdot} \boldsymbol{\delta} K\, d\Omega \\ + \int_\Omega -\mathrm{div}^2 \boldsymbol{M}\, \delta w\, d\Omega \tag{7.34}$$

is the corresponding "pairing" with $\boldsymbol{\delta\Sigma} = \{\delta w, \boldsymbol{\delta K}, \boldsymbol{\delta M}\}$, a "virtual displacement". Integration by parts of the last integral leads to the identity

$$\mathscr{G}(\boldsymbol{\Sigma}, \boldsymbol{\delta}\Sigma) = < \boldsymbol{A}(\boldsymbol{\Sigma}), \boldsymbol{\delta\Sigma} > + \int_\Gamma (V_n\, \delta w - M_n \frac{\partial \delta w}{\partial n})\, ds + [[M_{nt}\, \delta w]] \\ - a(\boldsymbol{\Sigma}, \boldsymbol{\delta\Sigma}) = 0 \tag{7.35}$$

with the symmetric **strain energy product**

$$a(\boldsymbol{\Sigma}, \boldsymbol{\delta\Sigma}) = \int_\Omega (\boldsymbol{K}(w) - \boldsymbol{K}) \boldsymbol{\cdot} \boldsymbol{\delta M}\, d\Omega + \int_\Omega \boldsymbol{C}[\boldsymbol{K}] \boldsymbol{\cdot} \hat{\boldsymbol{K}}\, d\Omega \\ + \int_\Omega \boldsymbol{M} \boldsymbol{\cdot} (\boldsymbol{K}(\delta w) - \boldsymbol{\delta K})\, d\Omega \,. \tag{7.36}$$

The expression

$$[[M_{nt}\, \delta w]] = \sum_i F_i\, \delta w(\boldsymbol{x}_i) \tag{7.37}$$

is the virtual work of the corner forces F_i.

The system $\boldsymbol{A}(\boldsymbol{\Sigma})$ for the triplet $\boldsymbol{\Sigma} = \{w, \boldsymbol{K}, \boldsymbol{M}\}$ can be reduced to an equation for the deflection w alone

$$K\,\Delta\Delta w = p\,. \tag{7.38}$$

To the left side belongs the identity

$$\mathcal{G}(w, \delta w) = \underbrace{\int_\Omega K\,\Delta\Delta w\,\delta w\,d\Omega + \int_\Gamma (V_n\,\delta w - M_n\,\frac{\partial \delta w}{\partial n})\,ds + [[M_{nt}\,\delta w]]}_{\delta W_e}$$
$$- \underbrace{\int_\Omega \boldsymbol{K}(w) \cdot \boldsymbol{C}[\boldsymbol{K}(\delta w)]\,d\Omega}_{\delta W_i} = 0\,, \tag{7.39}$$

which formulates the ***principle of virtual displacements***, and

$$\mathcal{B}(w, \hat{w}) = \mathcal{G}(w, \hat{w}) - \mathcal{G}(\hat{w}, w) = 0 \tag{7.40}$$

or

$$\mathcal{B}(w, \hat{w}) = \int_\Omega K\,\Delta\Delta w\,\hat{w}\,d\Omega + \int_\Gamma (V_n\,\hat{w} - M_n\,\frac{\partial \hat{w}}{\partial n})\,ds + [[M_{nt}\,\hat{w}]]$$
$$- \int_\Gamma (\hat{V}_n\,w - \hat{M}_n\,\frac{\partial w}{\partial n})\,ds - [[\hat{M}_{nt}\,w]] - \int_\Omega K\,\Delta\Delta \hat{w}\,w\,d\Omega = 0 \tag{7.41}$$

is ***Betti's theorem***.

7.6 Reissner-Mindlin Plate

The displacements are the deflection and the rotations about the two axes

$$w(\boldsymbol{x}) \qquad \boldsymbol{\varphi} = \{\varphi_1, \varphi_2\}^T\,. \tag{7.42}$$

The curvatures and strains are defined as

$$\boldsymbol{K}(\boldsymbol{\varphi}) - \boldsymbol{K} = \boldsymbol{0}_{\,(2\times 2)} \tag{7.43a}$$
$$\boldsymbol{\gamma}(\boldsymbol{\varphi}, w) - \boldsymbol{\gamma} = \boldsymbol{0}_{\,(2)}\,. \tag{7.43b}$$

The constitutive equations are

$$C[\boldsymbol{K}] - \boldsymbol{M} = \boldsymbol{0}_{\,(2\times2)} \tag{7.44a}$$

$$a\boldsymbol{\gamma} - \boldsymbol{q} = \boldsymbol{0}_{\,(2)} \tag{7.44b}$$

and the equilibrium conditions read

$$-\text{div}\boldsymbol{M} + \boldsymbol{q} = b\,\nabla\, p_{\,(2)} \tag{7.45a}$$

$$-\text{div}\boldsymbol{q} = p_{\,(1)}\,. \tag{7.45b}$$

It is

$$\boldsymbol{K}(\boldsymbol{\varphi}) = \begin{bmatrix} \varphi_{1,1} & \frac{1}{2}(\varphi_{1,2} + \varphi_{2,1}) \\ \text{sym.} & \varphi_{2,2} \end{bmatrix}, \qquad \boldsymbol{\gamma}(\boldsymbol{\varphi}, w) = \begin{bmatrix} \varphi_1 + w_{,1} \\ \varphi_2 + w_{,2} \end{bmatrix}, \tag{7.46}$$

and

$$C[\boldsymbol{K}] = K\,(1-\nu)\boldsymbol{K} + \nu\, K\,(\text{tr}\boldsymbol{K})\boldsymbol{I}\,, \tag{7.47}$$

where

$$K = \frac{E\,h^3}{12\,(1-\nu^2)} \qquad a = K\,\frac{1-\nu}{2}\,\bar{\lambda}^2 \qquad b = \frac{\nu}{1-\nu}\,\frac{1}{\bar{\lambda}^2} \qquad \bar{\lambda}^2 = \frac{10}{h^2}\,. \tag{7.48}$$

The triplet $\boldsymbol{\Sigma}$ consists of the terms

$$\boldsymbol{\Sigma} = \{w, \boldsymbol{\varphi}_{(2)};\, \boldsymbol{K}_{\,(2\times2)}, \boldsymbol{\gamma}_{(2)};\, \boldsymbol{M}_{\,(2\times2)}, \boldsymbol{q}_{\,(2)}\} \tag{7.49}$$

and the virtual work is the expression

$$\begin{aligned} < \boldsymbol{A}(\boldsymbol{\Sigma}), \delta\boldsymbol{\Sigma} > = \int_\Omega & [(\boldsymbol{K}(\boldsymbol{\varphi}) - \boldsymbol{K}) \cdot \delta\boldsymbol{M} + (\boldsymbol{\gamma}(\boldsymbol{\varphi}, w) - \boldsymbol{\gamma}) \cdot \delta\boldsymbol{q} \\ & + (C[\boldsymbol{K}] - \boldsymbol{M}) \cdot \delta\boldsymbol{K} + (a\,\boldsymbol{\gamma} - \boldsymbol{q}) \cdot \delta\boldsymbol{\gamma} + (-\text{div}\,\boldsymbol{M} + \boldsymbol{q}) \cdot \delta\boldsymbol{\varphi} \\ & + (-\text{div}\,\boldsymbol{q})\,\delta w]\,d\Omega\,. \end{aligned} \tag{7.50}$$

Integration by parts, see (7.23),

$$\int_\Omega -\text{div}\,\boldsymbol{M} \cdot \delta\boldsymbol{\varphi}\,d\Omega = -\int_\Gamma \boldsymbol{M}\boldsymbol{n} \cdot \delta\boldsymbol{\varphi}\,ds + \int_\Omega \boldsymbol{M} \cdot \boldsymbol{K}(\delta\boldsymbol{\varphi})\,d\Omega \tag{7.51}$$

and

$$\int_\Omega -\text{div}\,\boldsymbol{q}\,\delta w\,d\Omega = -\int_\Gamma \boldsymbol{q} \cdot \boldsymbol{n}\,\delta w\,ds + \int_\Omega \boldsymbol{q} \cdot \nabla\delta w\,d\Omega \tag{7.52}$$

leads to **Green's first identity**

$$\mathscr{G}(\boldsymbol{\Sigma}, \boldsymbol{\delta\Sigma}) = <\boldsymbol{A}(\boldsymbol{\Sigma}), \boldsymbol{\delta\Sigma}> + \int_{\Gamma} (\boldsymbol{Mn} \cdot \boldsymbol{\delta\varphi} + \boldsymbol{q} \cdot \boldsymbol{n}\, \delta w)\, ds - a(\boldsymbol{\Sigma}, \boldsymbol{\delta\Sigma}) = 0 \tag{7.53}$$

with the symmetric strain energy product[1]

$$\begin{aligned} a(\boldsymbol{\Sigma}, \boldsymbol{\delta\Sigma}) = \boldsymbol{K}(\varphi) \cdot \boldsymbol{\delta M} + \boldsymbol{M} \cdot \boldsymbol{K}(\delta\varphi) + \boldsymbol{C}[\boldsymbol{K}] \cdot \boldsymbol{\delta K} - \boldsymbol{K} \cdot \boldsymbol{\delta M} - \boldsymbol{M} \cdot \boldsymbol{\delta K} \\ + \gamma(\varphi, w) \cdot \boldsymbol{\delta q} + \gamma(\delta\varphi, \delta w) \cdot \boldsymbol{q} - \gamma \cdot \boldsymbol{\delta q} - \boldsymbol{q} \cdot \boldsymbol{\delta\gamma}\,. \end{aligned} \tag{7.54}$$

Alternatively, these equations can be reformulated as a system of three differential equations for the three displacements w, φ_1, φ_2

$$-\text{div}\, \boldsymbol{C}[\boldsymbol{K}(\varphi)] + a\gamma(\varphi, w) = b\, \nabla\, p_{\,(2)} \tag{7.55a}$$

$$-\text{div}\,(a\gamma(\varphi, w)) = p_{\,(1)}\,. \tag{7.55b}$$

The associated identity is

$$\begin{aligned} \mathscr{G}(\varphi, w; \delta\varphi, \delta w) = & \int_{\Omega} [-\text{div}\, \boldsymbol{C}[\boldsymbol{K}(\varphi)] \cdot \boldsymbol{\delta\varphi} - a\, \text{div}\, \gamma(\varphi, w)\, \delta w]\, d\Omega \\ + & \int_{\Gamma} (\boldsymbol{C}[\boldsymbol{K}(\varphi)]\boldsymbol{n} \cdot \boldsymbol{\delta\varphi} + a\gamma(\varphi, w) \cdot \boldsymbol{n}\, \delta w)\, ds - a(\varphi, w; \delta\varphi, \delta w) = 0 \end{aligned} \tag{7.56}$$

where

$$a(\varphi, w; \boldsymbol{\delta\varphi}, \delta w) = \int_{\Omega} [\boldsymbol{C}[\boldsymbol{K}(\varphi)] \cdot \boldsymbol{K}(\boldsymbol{\delta\varphi}) + a\gamma(\varphi, w) \cdot \gamma(\boldsymbol{\delta\varphi}, \delta w)]\, d\Omega \tag{7.57}$$

is the symmetric strain energy product.

7.7 Influence Functions

To keep the formulations simple let us look at a bar $\boldsymbol{\Sigma} = \{u, \varepsilon, N\}$ with given initial strains ε^+ and an initial normal force N^+

$$\begin{aligned} u' - \varepsilon &= \varepsilon^+ \\ EA\,\varepsilon - N &= N^+ \\ -N' &= p\,. \end{aligned} \tag{7.58}$$

---

[1] it is $\boldsymbol{C}[\boldsymbol{K}] \cdot \boldsymbol{\delta K} = \boldsymbol{C}[\boldsymbol{\delta K}] \cdot \boldsymbol{K}$.

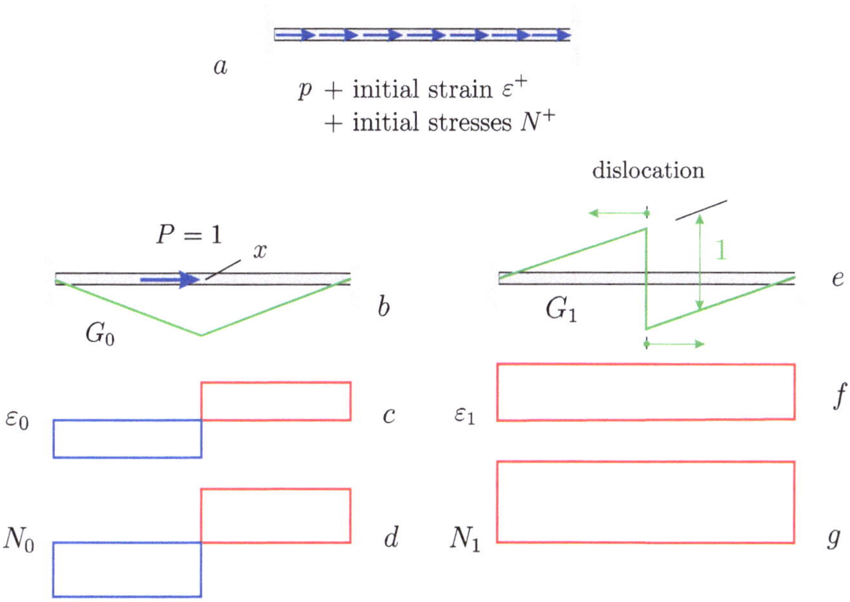

Fig. 7.3 Bar, **a** system and loading, **b** influence function G_0, **c** influence function G_1 and associated strains and normal forces

To this system $\boldsymbol{A}(\boldsymbol{\Sigma})$ belongs the identity

$$\mathscr{G}(\boldsymbol{\Sigma}, \boldsymbol{\delta\Sigma}) = \langle \boldsymbol{A}(\boldsymbol{\Sigma}), \boldsymbol{\delta\Sigma} \rangle + [N\, \delta u]_0^l - a(\boldsymbol{\Sigma}, \boldsymbol{\delta\Sigma}) = 0 \tag{7.59}$$

with

$$a(\boldsymbol{\Sigma}, \boldsymbol{\delta\Sigma}) = \int_0^l (u' - \varepsilon)\, \delta N\, dx + \int_0^l EA\, \varepsilon\, \delta\varepsilon\, dx + \int_0^l N\, (\delta u' - \delta\varepsilon)\, dx \tag{7.60}$$

which leads to Green's second identity

$$\mathscr{B}(\boldsymbol{\Sigma}, \hat{\boldsymbol{\Sigma}}) = \langle \boldsymbol{A}(\boldsymbol{\Sigma}), \hat{\boldsymbol{\Sigma}} \rangle + [N\, \hat{u}]_0^l - [u\, \hat{N}]_0^l - \langle \boldsymbol{\Sigma}, \boldsymbol{A}(\hat{\boldsymbol{\Sigma}}) \rangle = 0\,. \tag{7.61}$$

Next, let $\boldsymbol{\Sigma} = \{u, \varepsilon, N\}$ be the solution to $\boldsymbol{A}(\boldsymbol{\Sigma}) = [\varepsilon^+, N^+, p\,]^T$, see Fig. 7.3a, and let $\boldsymbol{\Sigma}_0 = \{G_0, \varepsilon_0, N_0\}$ be the solution to a force $P = 1$ at a point x, see Fig. 7.3b, and

$$\varepsilon_0 = \frac{d}{dy}\, G_0(y, x) \qquad N_0 = EA\, \varepsilon_0\,. \tag{7.62}$$

We formulate Green's second identity (7.61) on the split interval $I_\varepsilon := [0, x - \varepsilon] \cup [x + \varepsilon, l]$ and let $\varepsilon \to 0$. The right side of the influence function is zero on I_ε, $\boldsymbol{A}(\boldsymbol{\Sigma}_0) = [0, 0, 0]^T$, and so

$$\lim_{\varepsilon \to 0} \mathscr{B}(\boldsymbol{\Sigma}, \boldsymbol{\Sigma}_0)_{I_\varepsilon} = \langle \boldsymbol{A}(\boldsymbol{\Sigma}), \boldsymbol{\Sigma}_0 \rangle + [N\, G_0]_0^l - [u\, N_0]_0^l - u(x) = 0\,, \tag{7.63}$$

where the single term $u(x)$ is the limit of

$$\lim_{\varepsilon \to 0} \{N_0(x - \varepsilon)\, u(x - \varepsilon) - N_0(x + \varepsilon)\, u(x + \varepsilon)\} = 1 \cdot u(x)\,. \tag{7.64}$$

Considering the boundary conditions on u and $\boldsymbol{\Sigma}_0$, it follows

$$u(x) = \langle \boldsymbol{A}(\boldsymbol{\Sigma}), \boldsymbol{\Sigma}_0 \rangle = \int_0^l [N_0\, \varepsilon^+ + \varepsilon_0\, N^+ + G_0\, p\,]\, dx\,, \tag{7.65}$$

and in an analogous manner we obtain the influence function for the normal force

$$N(x) = \langle \boldsymbol{A}(\boldsymbol{\Sigma}), \boldsymbol{\Sigma}_1 \rangle = \int_0^l [N_1\, \varepsilon^+ + \varepsilon_1\, N^+ + G_1\, p\,]\, dx\,. \tag{7.66}$$

7.8 Betti Extended

All FE-methods for mixed formulations, be it the *Hu-Washizu principle* or the *Hellinger-Reissner Principle* or else are based on the *Galerkin orthogonality*

$$a(\boldsymbol{\Sigma} - \boldsymbol{\Sigma}_h, \boldsymbol{\Sigma}_i) = 0 \qquad \boldsymbol{\Sigma}_i \in V_h\,. \tag{7.67}$$

It is the same algebra as before and so also the formulation of **Betti extended** for mixed problems is a repetition. Let $\boldsymbol{\Sigma}_1$ and $\boldsymbol{\Sigma}_2$ be two equilibrium positions corresponding to two right sides $\boldsymbol{p}_1$ and $\boldsymbol{p}_2$, which can now also include initial strains and initial stresses, $\boldsymbol{p} = \{\varepsilon^+, N^+, p\}$,

$$\boldsymbol{A}(\boldsymbol{\Sigma}_1) = \boldsymbol{p}_1 \qquad \boldsymbol{A}(\boldsymbol{\Sigma}_2) = \boldsymbol{p}_2\,, \tag{7.68}$$

then in *Betti's theorem* $<\boldsymbol{\Sigma}_1, \boldsymbol{p}_2> \; = \; <\boldsymbol{\Sigma}_2, \boldsymbol{p}_1>$ the FE-solutions may replace the originals

$$<\boldsymbol{\Sigma}_{1h}, \boldsymbol{p}_2> \; = \; <\boldsymbol{\Sigma}_{2h}, \boldsymbol{p}_1> \qquad \text{Betti extended}\,. \tag{7.69}$$

If $\boldsymbol{\Sigma}_1 = \boldsymbol{G}_0$ is the influence function for $u(x)$, $\boldsymbol{A}(\boldsymbol{G}_0) = \boldsymbol{\delta}_0 = \{0, 0, \delta_0\}^T$, the switch $\boldsymbol{G}_0 \to \boldsymbol{G}_{0h}$ and $\boldsymbol{\Sigma} \to \boldsymbol{\Sigma}_h$

$$< \boldsymbol{G}_{0h}, \boldsymbol{p} > \, = \, < \boldsymbol{\Sigma}_h, \boldsymbol{\delta}_0 > \tag{7.70}$$

results in the influence function for the displacement of the FE-solution

$$u_h(x) = \int_0^l [N_0^h \, \varepsilon^+ + \varepsilon_0^h \, N^+ + G_0^h \, p \,] \, dx \, . \tag{7.71}$$

Reference

1. Hartmann F (1985) The mathematical foundation of structural mechanics. Springer, Heidelberg

Chapter 8
Nonlinear Problems

In this chapter we discuss nonlinear problems. The focus is on the questions, which parts of the algebra of finite elements can be directly transferred to the nonlinear theory and where the nonlinear theory differs from the linear theory and what are the characteristic features of nonlinear theories.

To be as illustrative as possible, we have included selected examples to detail the numerical analysis of nonlinear problems.

The transition from linear theory to nonlinear theory, unfortunately, means a loss of transparency in the formulations, since nonlinear finite element formulations consist to a large part in a seemingly endless application of the product and chain rules of calculus, coordinate transformations, first-order approximations, and integration by parts and all this spiced with linear algebra. (For help with the technical issues see [1]).

We have tried to make the walk through the subject as transparent as possible.

8.1 Introduction

The key point in any nonlinear formulation are the three steps $u \to \varepsilon \to \sigma \to p$. If these are understood and formulated correctly Green's first identity

$$\mathscr{G}(u, \delta u) = a_u(u, \delta u) - (p, \delta u) = 0\,, \tag{8.1}$$

formulates itself automatically and the "rest" is just algebra and a fast computer.

As in the linear theory we let the FE-solution $u_h = \sum_j u_j\, \varphi_j(x)$ and we find the nodal displacements u_i by solving the n equations

$$a_u(u_h, \varphi_i) - (p, \varphi_i) = 0 \qquad i = 1, 2, \ldots, n\,, \tag{8.2}$$

F. Hartmann and P. Jahn, *Statics and Influence Functions*, Springer Series in Solid and Structural Mechanics 13,
https://doi.org/10.1007/978-3-030-55889-5_8

$$\mathcal{G}(u, \delta u) = \int_0^l -\mathcal{N}' \,\delta u \, dx + [\mathcal{N}\,\delta u]_0^l - \underbrace{\int_0^l \varepsilon_u(u, \delta u)\,\sigma\, A \, dx}_{a_u(u,\delta u)} = 0 \tag{8.14}$$

with $\varepsilon_u(u, \delta u) := (1 + u')\,\delta u'$ as the Gateaux derivative

$$\frac{d}{d\eta}\varepsilon(u + \eta\,\delta u)|_{\eta=0} = \frac{d}{d\eta}\big(u' + \eta\,\delta u' + \frac{1}{2}\,(u' + \eta\,\delta u')^2\big)|_{\eta=0} \tag{8.15}$$

of $\varepsilon(u)$ in the direction of δu.

8.3.1 Newton's Method

Let $u(x)$ be the longitudinal displacement of a bar fixed on the left, $u(0) = 0$, and with a free end, $N(l) = 0$. Given the solution

$$u_h = \sum_j u_j\,\varphi_j(x)\,, \tag{8.16}$$

we determine the nodal displacements u_j by solving the n equations

$$a_u(u_h, \varphi_i) - \int_0^l p\,\varphi_i \, dx = k_i(\boldsymbol{u}) - f_i = 0 \qquad i = 1, 2, \ldots, n\,. \tag{8.17}$$

This is a set $\boldsymbol{k}(\boldsymbol{u}) = \boldsymbol{f}$ of n nonlinear equations, which a computer solves iteratively with ***Newton's method***

$$\boldsymbol{u}_{i+1} = \boldsymbol{u}_i - (\nabla \boldsymbol{k}(\boldsymbol{u}_i))^{-1}(\boldsymbol{k}(\boldsymbol{u}_i) - \boldsymbol{f}) \tag{8.18}$$

or

$$\boldsymbol{u}_{i+1} = \boldsymbol{u}_i - \boldsymbol{K}_T^{-1}(\boldsymbol{u}_i)\,(\boldsymbol{k}(\boldsymbol{u}_i) - \boldsymbol{f})\,, \tag{8.19}$$

where $\boldsymbol{K}_T(\boldsymbol{u}_i)$ is the ***tangential stiffness matrix***.

8.4 Geometrically Nonlinear Beam

The bending stiffness EI and longitudinal stiffness EA are constant, and the distributed loads are p_x and p_z. The displacements are the longitudinal displacement u and the deflection w, and this pair can also be written as a vector $\boldsymbol{v} = \{u, w\}^T$,

$$\varepsilon = u' + \frac{1}{2}\,(w')^2 \quad \kappa = w'' \tag{8.20a}$$

$$N = EA\,\varepsilon \quad M = -EI\,\kappa \tag{8.20b}$$

$$-N' = p_x \quad -M'' - (N\,w')' = p_z\,. \tag{8.20c}$$

The "displacement formulation" of this system is

$$-EA\,(u' + \frac{1}{2}\,(w')^2)' = p_x \tag{8.21a}$$

$$EI\,w^{IV} - (EA\,(u' + \frac{1}{2}\,(w')^2)\,w')' = p_z\,, \tag{8.21b}$$

or in a slightly more transparent formulation

$$-N' = p_x \tag{8.22a}$$

$$EI\,w^{IV} - (N\,w')' = p_z\,. \tag{8.22b}$$

Let

$$N = N(\boldsymbol{v}) = EA\,(u' + \frac{1}{2}\,(w')^2)\,, \qquad M = M(w) = -EI\,w''\,, \tag{8.23}$$

and let $\boldsymbol{Lv}$ the left side of (8.21), integration by parts of the work integral results in

$$\begin{aligned}\int_0^l \boldsymbol{Lv} \bullet \boldsymbol{\delta v}\,dx &= \int_0^l ((Eq_1)\cdot\delta u + (Eq_2)\cdot\delta w)\,dx \\ &= \int_0^l [(-N'\,\delta u - (M'' + (N\,w')')\,\delta w]\,dx \\ &= -[N\,\delta u + (M' + N\,w')\,\delta w - M\,\delta w']_0^l + a_{\boldsymbol{v}}(\boldsymbol{v}, \boldsymbol{\delta v})\,,\end{aligned} \tag{8.24}$$

where

$$\begin{aligned}a_{\boldsymbol{v}}(\boldsymbol{v}, \boldsymbol{\delta v}) &= \int_0^l (-M\,\delta w'' + N\,(\delta u' + w'\,\delta w'))\,dx \\ &= \int_0^l (\frac{M(w)\,M_w(w, \delta w)}{EI} + \frac{N(\boldsymbol{v})\,N_{\boldsymbol{v}}(\boldsymbol{v}, \boldsymbol{\delta v})}{EA})\,dx\end{aligned} \tag{8.25}$$

with the ***Gateaux derivatives*** of M and N

$$M_w(w, \delta w) = [\frac{d}{d\varepsilon} M(w + \varepsilon\,\delta w)]_{|\varepsilon=0} = \delta M \tag{8.26}$$

$$N_{\boldsymbol{v}}(\boldsymbol{v}, \boldsymbol{\delta v}) = [\frac{d}{d\varepsilon} N(\boldsymbol{v} + \varepsilon\,\boldsymbol{\delta v})]_{|\varepsilon=0} = EA(\delta u' + w'\delta w')\,. \tag{8.27}$$

So, Green's first identity is the expression

$$\mathcal{G}(\boldsymbol{v}, \boldsymbol{\delta v}) = \int_0^l \boldsymbol{L u} \cdot \boldsymbol{\delta u}\, dx + [N\,\delta u + (M' + N\,w')\,\delta w - M\,\delta w']_0^l \\ - a_v(\boldsymbol{v}, \boldsymbol{\delta v}) = 0\,. \tag{8.28}$$

If the normal force N is constant, we arrive at the well-known equation

$$EI\,w^{IV} - N\,w'' = p_z \tag{8.29}$$

of second-order theory.

8.4.1 Conservation of Energy

For a linear bar conservation of energy means

$$\frac{1}{2}\,\mathcal{G}(u, u) = \frac{1}{2}\int_0^l p\,u\,dx - \frac{1}{2}\int_0^l \frac{N^2}{EA}\,dx = W_e - W_i = 0\,. \tag{8.30}$$

In the case of the cantilever beam in Fig. 8.1, with

$$\varepsilon = l\sqrt{\frac{P_x}{EI}}\,, \qquad P_x = P_z = 1\,, \qquad l = EI = 1\,, \tag{8.31}$$

the displacements according to **second-order theory** are

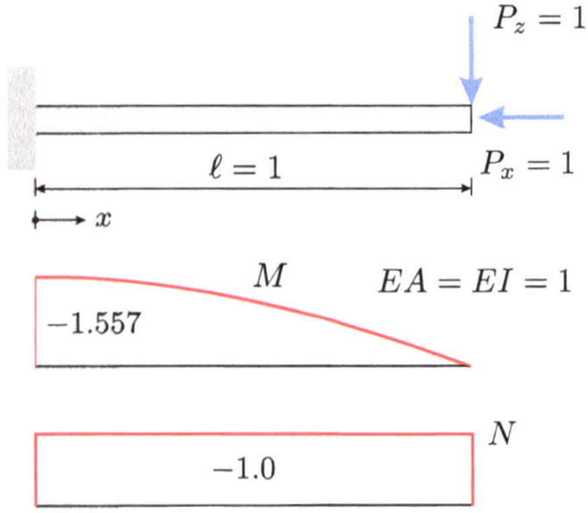

Fig. 8.1 Second-order theory

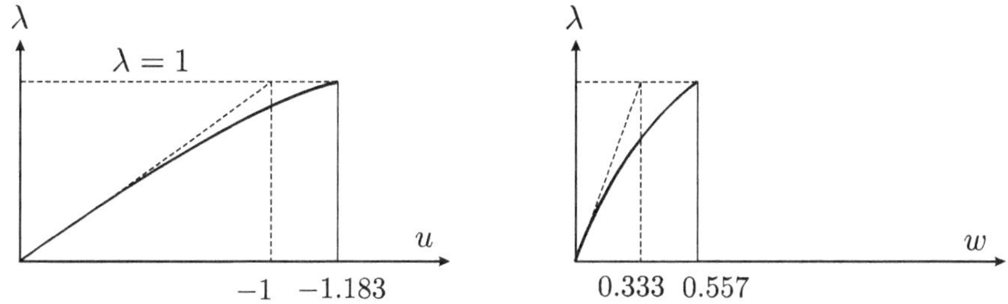

Fig. 8.2 Displacements at the end of the beam

$$u(x) = \frac{1}{\varepsilon(\sin^2(\varepsilon) - 1)} [-0.5\sin(\varepsilon(x-2)) + 3\sin(2\varepsilon) - \sin(2\varepsilon(x-1)) \\ + 4\sin(\varepsilon x) - 6\,\varepsilon x - 8\,\varepsilon^3 x + 4\,\varepsilon x \sin^2(\varepsilon) + 8\,\varepsilon^3 x \sin^2(\varepsilon)] \tag{8.32a}$$

$$w(x) = \frac{P_z}{\varepsilon^3} [\tan(\varepsilon) \cdot (1 - \cos(\varepsilon x)) + \sin(\varepsilon x) - \varepsilon x] \,. \tag{8.32b}$$

The statement $W_e = W_i$ is still true, but we cannot calculate the external work W_e as in the linear theory, see Fig. 8.2, by simply superposing the load with the displacements

$$W_e \neq \frac{1}{2}(P_x\, u(l) + P_z\, w(l)) = \frac{1}{2}(1.0 \cdot 1.183 + 1.0 \cdot 0.557) = 0.87\,, \tag{8.33}$$

but instead we must count the increments. Let $\boldsymbol{dv} = \{du, dw\}$ be the displacement increment with respect to a load increment, then we read

$$dW_i = a_v(\boldsymbol{v}, \boldsymbol{dv}) = \int_0^l (\frac{M(w)\, M_w(w, dw)}{EI} + \frac{N(\boldsymbol{v})\, N_v(\boldsymbol{v}, \boldsymbol{dv})}{EA})\, dx \tag{8.34}$$

as the increase in internal energy and Conservation of energy

$$\int dW_i = \int_0^l [\int_0^M \frac{\bar{M}\, d\bar{M}}{EI} + \int_0^N \frac{\bar{N}\, d\bar{N}}{EA}]\, dx \\ = \frac{1}{2} \int_0^l (\frac{N^2}{EA} + \frac{M^2}{EI})\, dx = 0.9670 \tag{8.35}$$

is therefore the internal energy at the end of the load path.

In order to calculate the external work, we assume the load to slowly increase,

$$P_x(\lambda) = \lambda\, P_x = \lambda \cdot 1 \qquad P_z(\lambda) = \lambda\, P_z = \lambda \cdot 1 \qquad 0 \leq \lambda \leq 1\,. \tag{8.36}$$

so that the external work is [2], p. 333,

$$\int dW_e = \int_0^u \lambda \cdot 1\, d\bar{u} + \int_0^w \lambda \cdot 1\, d\bar{w} = \int_0^1 (\bar{u}_{,\lambda} + \bar{w}_{,\lambda})\, d\lambda\,. \qquad (8.37)$$

This integral is best evaluated numerically, $\Delta = 0.05$,

$$\int dW_e = \int_0^1 [\frac{\bar{u}(\lambda + \Delta) - \bar{u}(\lambda)}{\Delta} + \frac{\bar{w}(\lambda + \Delta) - \bar{w}(\lambda)}{\Delta}] d\lambda \qquad (8.38)$$

and with *Simpson's rule* we obtain

$$\int dW_e = 0.9669\,, \qquad (8.39)$$

which matches the internal energy quite well.[1]

8.5 Geometrically Nonlinear Kirchhoff Plate

The formulation is the same as for the geometrically nonlinear beam, it only consists of a more extensive set of equations, and so we refer interested readers to pp. 325–328 in [2].

8.6 Nonlinear Elasticity Theory

In the triplet $\{\boldsymbol{u}, \boldsymbol{E}, \boldsymbol{S}\}$ the tensor $\boldsymbol{E}$ is the Green-Lagrange strain tensor and $\boldsymbol{S}$ is the second Piola-Kirchhoff stress tensor. We assume the material to be hyperelastic, i.e. there exists a strain energy function $\boldsymbol{W}$ such that $\boldsymbol{S} = \partial \boldsymbol{W}/\partial \boldsymbol{E}$.

In the presence of volume loads $\boldsymbol{p}$ the elastic state $\boldsymbol{\Sigma} = \{\boldsymbol{u}, \boldsymbol{E}, \boldsymbol{S}\}$ at each point $\boldsymbol{x}$ of the undeformed body satisfies the system[2]

$$\boldsymbol{E}(\boldsymbol{u}) - \boldsymbol{E} = \boldsymbol{0} \qquad \frac{1}{2}(u_{i,j} + u_{j,i} + u_{k,i}\, u_{k,j}) - \varepsilon_{ij} = 0 \qquad (8.40a)$$

$$\boldsymbol{W}'(\boldsymbol{E}) - \boldsymbol{S} = \boldsymbol{0} \qquad \frac{\partial W}{\partial \varepsilon_{ij}} - \sigma_{ij} = 0 \qquad (8.40b)$$

$$-\text{div}(\boldsymbol{S} + \nabla \boldsymbol{u} \boldsymbol{S}) = \boldsymbol{p} \qquad -(\sigma_{ij} + u_{i,k}\, \sigma_{kj})_{,j} = p_i \qquad (8.40c)$$

---

[1] Since $EA = 1$ the example is of course academic because the compression exceeds the length of the bar.

[2] In linear 1-D problems $\sigma = E\,\varepsilon$, $\boldsymbol{W}(\boldsymbol{E}) = 0.5\, E\, \varepsilon^2$.

with matching displacement boundary conditions $\boldsymbol{u} = \bar{\boldsymbol{u}}$ on the part Γ_D of the boundary and stress boundary conditions $\boldsymbol{t}(\boldsymbol{S}, \boldsymbol{u}) = \bar{\boldsymbol{t}}$ on the remaining part Γ_N of the boundary, where

$$\boldsymbol{t}(\boldsymbol{S}, \boldsymbol{u}) := (\boldsymbol{S} + \nabla \boldsymbol{u} \boldsymbol{S})\boldsymbol{n} \tag{8.41}$$

is the stress vector at a boundary point with an outward pointing unit normal vector $\boldsymbol{n}$.

Symmetric tensors $\boldsymbol{S}$ satisfy the identity (int. by parts)

$$\begin{aligned}\int_\Omega &-\mathrm{div}(\boldsymbol{S} + \nabla \boldsymbol{u} \boldsymbol{S}) \cdot \delta \boldsymbol{u}\, d\Omega \\ &= -\int_\Gamma \boldsymbol{t}(\boldsymbol{S}, \boldsymbol{u}) \cdot \delta \boldsymbol{u}\, ds + \int_\Omega \boldsymbol{E}_{\boldsymbol{u}}(\boldsymbol{u}, \delta \boldsymbol{u}) \cdot \boldsymbol{S}\, d\Omega\,,\end{aligned} \tag{8.42}$$

where

$$\boldsymbol{E}_{\boldsymbol{u}}(\boldsymbol{u}, \delta \boldsymbol{u}) := \frac{1}{2}\,(\nabla \delta \boldsymbol{u} + \nabla \delta \boldsymbol{u}^T + \nabla \boldsymbol{u}^T\, \nabla \delta \boldsymbol{u} + \nabla \delta \boldsymbol{u}^T\, \nabla \boldsymbol{u}) \tag{8.43}$$

is the Gateaux derivative of the tensor $\boldsymbol{E}(\boldsymbol{u})$ at $\boldsymbol{u}$,

$$\frac{d}{d\varepsilon}[\boldsymbol{E}(\boldsymbol{u} + \varepsilon \delta \boldsymbol{u})]_{|_{\varepsilon=0}} = \boldsymbol{E}_{\boldsymbol{u}}(\boldsymbol{u}, \delta \boldsymbol{u})\,. \tag{8.44}$$

We so arrive at Green's first identity of the operator $\boldsymbol{A}(\boldsymbol{\Sigma})$, the system (8.40),

$$\mathscr{G}(\boldsymbol{\Sigma}, \delta \boldsymbol{\Sigma}) = \underbrace{\langle \boldsymbol{A}(\boldsymbol{\Sigma}), \delta \boldsymbol{\Sigma} \rangle + \int_\Gamma \boldsymbol{t}(\boldsymbol{S}, \boldsymbol{u}) \cdot \delta \boldsymbol{u}\, ds}_{\delta W_e} - \underbrace{a_{\boldsymbol{\Sigma}}(\boldsymbol{\Sigma}, \delta \boldsymbol{\Sigma})}_{\delta W_i} = 0\,, \tag{8.45}$$

where

$$\begin{aligned}\langle A(\boldsymbol{\Sigma}), \delta \boldsymbol{\Sigma} \rangle := &\int_0^l (\boldsymbol{E}(\boldsymbol{u}) - \boldsymbol{E}) \cdot \delta \boldsymbol{S}\, d\Omega + \int_\Omega (\boldsymbol{W}'(\boldsymbol{E}) - \mathbf{S}) \cdot \delta \boldsymbol{E}\, d\Omega \\ &+ \int_\Omega -(\mathrm{div} \boldsymbol{S} + \nabla \boldsymbol{u} \boldsymbol{S}) \cdot \delta \boldsymbol{u}\, d\Omega\end{aligned} \tag{8.46}$$

and

$$\begin{aligned}a_{\boldsymbol{\Sigma}}(\boldsymbol{\Sigma}, \delta \boldsymbol{\Sigma}) = &\int_\Omega (\boldsymbol{E}(\boldsymbol{u}) - \boldsymbol{E}) \cdot \delta \boldsymbol{S}\, d\Omega \\ &+ \int_\Omega (\boldsymbol{W}'(\boldsymbol{E}) - \mathbf{S}) \cdot \delta \boldsymbol{E}\, d\Omega + \int_\Omega \boldsymbol{E}_{\boldsymbol{u}}(\boldsymbol{u}, \delta \boldsymbol{u}) \cdot \boldsymbol{S}\, d\Omega\,.\end{aligned} \tag{8.47}$$

In a pure displacement formulation, where everything is formulated in terms of $\boldsymbol{u}$, $\boldsymbol{\Sigma} = \{\boldsymbol{u}, \boldsymbol{E}(\boldsymbol{u}), \boldsymbol{W}'(\boldsymbol{E}(\boldsymbol{u}))\}$, and if $\boldsymbol{\delta u} = \boldsymbol{0}$ on Γ_D, the system (8.45) reduces to ($\bar{\boldsymbol{t}}$ are edge forces on Γ_N)

$$\mathscr{G}(\boldsymbol{u}, \boldsymbol{\delta u}) = \int_\Omega \boldsymbol{p} \cdot \boldsymbol{\delta u}\, d\Omega + \int_{\Gamma_N} \bar{\boldsymbol{t}} \cdot \boldsymbol{\delta u}\, ds - \int_\Omega \boldsymbol{E}_{\boldsymbol{u}}(\boldsymbol{u}, \boldsymbol{\delta u}) \cdot \boldsymbol{S}\, d\Omega = 0\,, \quad (8.48)$$

where $\boldsymbol{S} = \boldsymbol{W}'(\boldsymbol{E}(\boldsymbol{u}))$. In the following we write

$$a_{\boldsymbol{u}}(\boldsymbol{u}, \boldsymbol{\delta u}) := \int_\Omega \boldsymbol{E}_{\boldsymbol{u}}(\boldsymbol{u}, \boldsymbol{\delta u}) \cdot \boldsymbol{S}\, d\Omega\,. \quad (8.49)$$

The FE-solution $\boldsymbol{u}_h = \sum_j u_j \boldsymbol{\varphi}_j$ is then the solution of the system

$$\int_\Omega \boldsymbol{p} \cdot \boldsymbol{\varphi}_i\, d\Omega + \int_{\Gamma_N} \bar{\boldsymbol{t}} \cdot \boldsymbol{\varphi}_i\, ds - a_{\boldsymbol{u}}(\boldsymbol{u}_h, \boldsymbol{\varphi}_i) = 0 \qquad i = 1, 2, \ldots n\,, \quad (8.50)$$

or $\boldsymbol{f} - \boldsymbol{k}(\boldsymbol{u}) = \boldsymbol{0}$ for short, where $\boldsymbol{u}$ is the vector of the nodal displacements.[3]

8.6.1 Linearization

To solve $\boldsymbol{k}(\boldsymbol{u}) = \boldsymbol{f}$, we have to linearize the equation. Let us assume that the function $\boldsymbol{u} = \boldsymbol{u}_h + \boldsymbol{u}_\Delta$ can be split into an approximation $\boldsymbol{u}_h$ plus an increment $\boldsymbol{u}_\Delta$, and so the identity (8.48) reads

$$\mathscr{G}(\boldsymbol{u}, \boldsymbol{\delta u}) = \int_\Omega \boldsymbol{p} \cdot \boldsymbol{\delta u}\, d\Omega + \int_{\Gamma_N} \bar{\boldsymbol{t}} \cdot \boldsymbol{\delta u}\, ds - a_{\boldsymbol{u}}(\boldsymbol{u}_h + \boldsymbol{u}_\Delta, \boldsymbol{\delta u}) = 0\,. \quad (8.51)$$

The Gateaux derivative of the expression

$$a_{\boldsymbol{u}}(\boldsymbol{u}, \boldsymbol{\delta u}) := \int_\Omega \boldsymbol{E}_{\boldsymbol{u}}(\boldsymbol{u}, \boldsymbol{\delta u}) \cdot \boldsymbol{W}'(\boldsymbol{E}(\boldsymbol{u}))\, d\Omega \quad (8.52)$$

with respect to a displacement increment $\boldsymbol{u}_\Delta$ is

$$\begin{aligned} a_T(\boldsymbol{u}, \boldsymbol{u}_\Delta, \boldsymbol{\delta u}) &:= \left[\frac{d}{d\varepsilon}\, a_{\boldsymbol{u}}(\boldsymbol{u} + \varepsilon\, \boldsymbol{u}_\Delta, \boldsymbol{\delta u})\right]_{\varepsilon=0} \\ &= \int_\Omega [\nabla \boldsymbol{u}_\Delta \nabla \boldsymbol{\delta u} \cdot \boldsymbol{W}'(\boldsymbol{E}(\boldsymbol{u})) + \boldsymbol{E}_{\boldsymbol{u}}(\boldsymbol{u}, \boldsymbol{\delta u}) \cdot \boldsymbol{C}[\boldsymbol{E}_{\boldsymbol{u}}(\boldsymbol{u}, \boldsymbol{u}_\Delta)]]\, d\Omega\,, \end{aligned} \quad (8.53)$$

[3] We use the same letter for the vector $\boldsymbol{u}$ as for the exact solution $\boldsymbol{u}(\boldsymbol{x})$.

where

$$C = \frac{\partial^2 W}{\partial E\, \partial E} = \frac{\partial}{\partial E}\, S\,, \tag{8.54}$$

which here is the second derivative of W. This derivative is to be taken at the position $E(u)$. Note, $a_T(u, u_\Delta, \delta u)$ is linear in the second and third argument, u_Δ and δu. Therefore, the system of equations for u_Δ later will be linear, see (8.57), $K_T(u)\, u_\Delta = \ldots$

With

$$a_u(u + u_\Delta, \delta u) \simeq a_u(u, \delta u) + a_T(u, u_\Delta, \delta u)\,, \tag{8.55}$$

equation (8.50) becomes

$$\int_\Omega p \cdot \varphi_i \, d\Omega + \int_{\Gamma_N} \bar{t} \cdot \varphi_i \, ds - a_u(u, \varphi_i) - a_T(u, u_\Delta, \varphi_i) = 0 \quad i = 1, 2, \ldots n\,, \tag{8.56}$$

or

$$f - k(u) - K_T(u)\, u_\Delta = 0\,, \tag{8.57}$$

where u_Δ is the nodal displacement vector of the displacement field u_Δ and K_T is the tangential stiffness matrix

$$(K_T)_{ij} = a_T(u, \varphi_j, \varphi_i)\,. \tag{8.58}$$

The iteration rule then reads

$$u_{i+1} = u_i - K_T^{-1}(u_i)(k(u_i) - f)\,. \tag{8.59}$$

8.6.2 A Truss Element in 3-D

We consider a truss element which extends along the x axis, see Fig. 8.3. The motion of its axis is

$$\varphi(x, y, z) = (x + u(x))\, e_1 + (y + v(x))\, e_2 + (z + w(x))\, e_3\,, \tag{8.60}$$

where $u(x) = \{u(x), v(x), w(x)\}^T$ is the displacement vector. The **deformation gradient** is

Fig. 8.3 Truss element

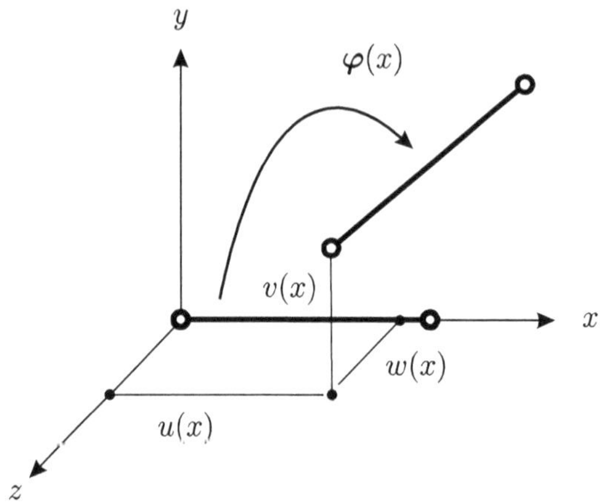

$$F = \nabla \varphi = \begin{bmatrix} 1 + u' & 0 & 0 \\ v' & 1 & 0 \\ w' & 0 & 1 \end{bmatrix} = I + \nabla u \,, \tag{8.61}$$

and the **Green-Lagrange strain tensor** is

$$E(u) = \frac{1}{2} \, (F^T \, F - I) = \begin{bmatrix} \varepsilon_x & \frac{1}{2} \, v' & \frac{1}{2} \, w' \\ \frac{1}{2} \, v' & 0 & 0 \\ \frac{1}{2} \, w' & 0 & 0 \end{bmatrix} , \tag{8.62}$$

where

$$\begin{aligned} \varepsilon(u) := \varepsilon_x(u) &= \frac{1}{2} \, ((1 + u')^2 + (v')^2 + (w')^2 - 1) \\ &= u' + \frac{1}{2} \, ((u')^2 + (v')^2 + (w')^2) \qquad (\,)' = \frac{d}{dx} \,. \end{aligned} \tag{8.63}$$

Green's first identity reads, since all terms are only functions of x,

$$\begin{aligned} \mathscr{G}(u, \delta u) = \int_0^l -\text{div}\,(S + \nabla\, u\, S) \cdot \delta u \, A \, dx + [(S + \nabla\, u\, S) \, e_1 \cdot \delta u \, A\,]_0^l \\ - \int_0^l E_u(u, \delta u) \cdot S \, A \, dx = 0 \,. \end{aligned} \tag{8.64}$$

We let $\sigma = \sigma_{xx} = E \, \varepsilon$ and all other $\sigma_{ij} = 0$, and so with $\mathscr{N} = A(\sigma + u' \, \sigma)$ the identity reads

$$\mathscr{G}(\boldsymbol{u}, \boldsymbol{\delta u}) = \int_0^l -\mathcal{N}'\,\hat{u}\,dx + [\mathcal{N}\,\hat{u}\,]_0^l - \underbrace{\int_0^l \varepsilon_{\boldsymbol{u}}(\boldsymbol{u}, \boldsymbol{\delta u})\,\sigma\,A\,dx}_{a_{\boldsymbol{u}}(\boldsymbol{u}, \boldsymbol{\delta u})} = 0\,, \tag{8.65}$$

where

$$\varepsilon_{\boldsymbol{u}}(\boldsymbol{u}, \boldsymbol{\delta u}) = (1 + u')\,\delta u' + v'\,\delta v' + w'\,\delta w'\,. \tag{8.66}$$

The **Gateaux derivative** of the strain energy product $a_{\boldsymbol{u}}(\boldsymbol{u}, \boldsymbol{u}, \boldsymbol{\delta u})$ is

$$\begin{aligned} a_T(\boldsymbol{u}, \boldsymbol{u}_\Delta, \boldsymbol{\delta u}) &:= \left[\frac{d}{d\varepsilon}\,a_{\boldsymbol{u}}(\boldsymbol{u} + \varepsilon \boldsymbol{u}_\Delta, \boldsymbol{\delta u})\right]_{\varepsilon=0} \\ &= \int_0^l [\varepsilon_{\boldsymbol{u}}(\boldsymbol{u}_{\boldsymbol{\Delta}}, \boldsymbol{\delta u})\,\sigma + \varepsilon_{\boldsymbol{u}}(\boldsymbol{u}, \boldsymbol{\delta u})\,\sigma_{\boldsymbol{u}}]\,A\,dx\,, \end{aligned} \tag{8.67}$$

where

$$\varepsilon_{\boldsymbol{u}}(\boldsymbol{u}_\Delta, \boldsymbol{\delta u}) = \left[\frac{d}{d\varepsilon}\,\varepsilon_{\boldsymbol{u}}(\boldsymbol{u} + \varepsilon\,\boldsymbol{u}_\Delta, \boldsymbol{\delta u})\right]_{\varepsilon=0} = u'_\Delta \delta u' + v'_\Delta\,\delta v' + w'_\Delta\,\delta w' \tag{8.68}$$

and

$$\sigma_{\boldsymbol{u}} = \left[\frac{d}{d\varepsilon}\,\sigma(\boldsymbol{u} + \varepsilon\,\boldsymbol{u}_\Delta)\right]_{\varepsilon=0} = E\,\varepsilon_{\boldsymbol{u}}(\boldsymbol{u}, \boldsymbol{u}_\Delta)\,. \tag{8.69}$$

The elements of the **tangential stiffness matrix** are therefore

$$(K_T(\boldsymbol{u}))_{ij} = a_T(\boldsymbol{u}, \boldsymbol{\varphi}_i, \boldsymbol{\varphi}_j)\,. \tag{8.70}$$

The vector-valued functions $\boldsymbol{\varphi}_i(\xi)$ are the nodal displacements (three at each node) where ξ is the coordinate on the master element.

With linear shape functions on an element Ω_e of length l_e we have

$$u_e(x) = \sum_{i=1}^{2} u_i^e\,\varphi_i^e(\xi)\,,\; v_e(x) = \sum_{i=1}^{2} v_i^e\,\varphi_i^e(\xi)\,,\;\; w_e(x) = \sum_{i=1}^{2} w_i^e\,\varphi_i^e(\xi)\,, \tag{8.71}$$

and the nodal displacement vector is $\boldsymbol{u}^e = [u_1, v_1, w_1, u_2, v_2, w_2]^T$ and the 6×6 tangential stiffness matrix becomes [3],

$$\boldsymbol{K}_T^e(\boldsymbol{u}) = \begin{bmatrix} (\boldsymbol{A}_1 + \boldsymbol{A}_2) & -(\boldsymbol{A}_1 + \boldsymbol{A}_2) \\ -(\boldsymbol{A}_1 + \boldsymbol{A}_2) & (\boldsymbol{A}_1 + \boldsymbol{A}_2) \end{bmatrix}, \tag{8.72}$$

with

$$A_1 = \frac{E\,A}{l_e}\begin{bmatrix} (1+u_e')^2 & (1+u_e')\,v_e' & (1+u_e')\,w_e' \\ (1+u_e')\,v_e' & (v_e')^2 & v_e'\,w_e' \\ (1+u_e')\,w_e' & v_e'\,w_e' & (w_e')^2 \end{bmatrix} \tag{8.73}$$

and

$$A_2 = \frac{\sigma\,A}{l_e}\begin{bmatrix} 1 & 0 & 0 \\ 0 & 1 & 0 \\ 0 & 0 & 1 \end{bmatrix}. \tag{8.74}$$

This leads to the system

$$\boldsymbol{K}_T(\boldsymbol{u})\,\boldsymbol{u}_\Delta = \boldsymbol{f} - \boldsymbol{k}(\boldsymbol{u})\,, \tag{8.75}$$

which is solved iteratively

$$\boldsymbol{u}_{i+1} = \boldsymbol{u}_i - \boldsymbol{K}_T^{-1}(\boldsymbol{u}_i)(\boldsymbol{k}(\boldsymbol{u}_i) - \boldsymbol{f})\,. \tag{8.76}$$

Regarding the extension to 3-D beams we refer to [4].

8.6.3 Planar Problem

Let $\boldsymbol{X}$ be the coordinates in the initial position and $\boldsymbol{x}$ in the current configuration, see Fig. 8.4,

$$\boldsymbol{x} = \boldsymbol{X} + \boldsymbol{u}\,, \quad \text{or} \quad x_i = X_i + u_i\,, \tag{8.77}$$

where $\boldsymbol{u}$ is the displacement vector.

Fig. 8.4 Original and displaced position

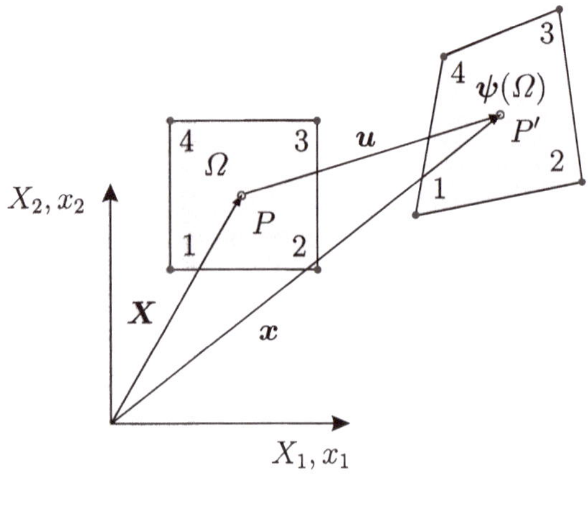

The deformation gradient is

$$F = \frac{\partial x}{\partial X} = I + \nabla u = \begin{bmatrix} 1 + u_{1,1} & u_{1,2} & 0 \\ u_{2,1} & 1 + u_{2,2} & 0 \\ 0 & 0 & 1 \end{bmatrix}, \tag{8.78}$$

and so the **Green-Lagrange strain tensor**

$$E = \frac{1}{2}(F^T F - I) = \begin{bmatrix} E_{11} & E_{12} \\ \text{sym.} & E_{22} \end{bmatrix} \tag{8.79}$$

has the components

$$\begin{aligned} E_{11} &= u_{1,1} + \frac{1}{2}\,(u_{1,1}^2 + u_{2,1}^2) \\ E_{22} &= u_{2,2} + \frac{1}{2}\,(u_{2,2}^2 + u_{1,2}^2) \\ E_{12} &= \frac{1}{2}\,(u_{1,2} + u_{2,1}) + \frac{1}{2}\,(u_{1,1}\,u_{1,2} + u_{2,2}\,u_{2,1})\,. \end{aligned} \tag{8.80}$$

The FE-displacement field can be written in two ways

$$u_h = \sum_{i=1}^{\text{DOFS}} u_i\,\varphi_i(x) \equiv \sum_i u_i \begin{bmatrix} \cdot \\ \cdot \end{bmatrix}, \qquad u_h = \sum_{i=1}^{\text{NODES}} u_i\,\psi_i(x) \equiv \sum_i \begin{bmatrix} \cdot \\ \cdot \end{bmatrix} \psi_i\,, \tag{8.81}$$

either as a sum over the degrees of freedom u_i or as the sum over the nodes. The $\varphi_i(x)$ are the displacement fields of the u_i, while the ψ_i in the second formula are the scalar-valued shape functions of the nodes ($\psi_i(x_j) = \delta_{ij}$) and the vectors $u_i = \{u_1^{(i)}, u_2^{(i)}\}^T$ are the displacements at the nodes x_i.

Here we use the second notation, so that for example the virtual external work of the volume forces p is[4]

$$\begin{aligned} \delta W_e &= \int_\Omega p^T \hat{u}\, d\Omega \\ &= \sum_{i=1}^n \int_\Omega p^T \hat{u}_i\,\psi_i\, d\Omega = \sum_{i=1}^n \hat{u}_i^T \int_\Omega p\,\psi_i\, d\Omega = \sum_{i=1}^n \hat{u}_i^T f_i\,. \end{aligned} \tag{8.82}$$

The gradient of the FE-displacement field u reads

$$\nabla u = \sum_{i=1}^n u_i \otimes \nabla\psi_i\,, \tag{8.83}$$

[4]In order not to overload the notation, we write a virtual displacement $\hat{u}$ and not δu and we write the FE-solution without index h.

and the deformation gradient is

$$\boldsymbol{F} = \boldsymbol{I} + \nabla \boldsymbol{u} = \boldsymbol{I} + \sum_{i=1}^{n} \boldsymbol{u}_i \otimes \nabla \psi_i \,. \tag{8.84}$$

The gradient of the virtual displacement field $\hat{\boldsymbol{u}} = \sum_i \hat{\boldsymbol{u}}_i \, \psi_i$ depends on (8.83), and the Gateaux derivative is

$$\begin{aligned} \boldsymbol{E}_{\boldsymbol{u}}(\boldsymbol{u}, \hat{\boldsymbol{u}}) :&= \frac{1}{2} (\nabla \hat{\boldsymbol{u}} + \nabla \hat{\boldsymbol{u}}^T + \nabla \boldsymbol{u}^T \, \nabla \hat{\boldsymbol{u}} + \nabla \hat{\boldsymbol{u}}^T \, \nabla \boldsymbol{u}) \\ &= \frac{1}{2} (\boldsymbol{F}^T \, \nabla \hat{\boldsymbol{u}} + \nabla \hat{\boldsymbol{u}}^T \, \boldsymbol{F}) \\ &= \frac{1}{2} \sum_{i=1}^{n} [\boldsymbol{F}^T (\hat{\boldsymbol{u}}_i \otimes \nabla \psi_i) + (\nabla \psi_i \otimes \hat{\boldsymbol{u}}_i) \boldsymbol{F}] \,. \end{aligned} \tag{8.85}$$

As in the linear theory, where $\boldsymbol{E} \bullet \boldsymbol{S} = \boldsymbol{\varepsilon} \bullet \boldsymbol{\sigma}$, the symmetry of $\boldsymbol{E}_{\boldsymbol{u}}(\boldsymbol{u}, \hat{\boldsymbol{u}})$ is reason to introduce the "Gateaux strain vector"

$$\varepsilon_{\boldsymbol{u}}(\boldsymbol{u}, \hat{\boldsymbol{u}}) := \begin{bmatrix} (\boldsymbol{E}_{\boldsymbol{u}}(\boldsymbol{u}, \hat{\boldsymbol{u}}))_{11} \\ (\boldsymbol{E}_{\boldsymbol{u}}(\boldsymbol{u}, \hat{\boldsymbol{u}}))_{22} \\ 2\,(\boldsymbol{E}_{\boldsymbol{u}}(\boldsymbol{u}, \hat{\boldsymbol{u}}))_{12} \end{bmatrix} = \sum_{i=1}^{n} \boldsymbol{B}_{L_i} \, \hat{\boldsymbol{u}}_i \,, \tag{8.86}$$

where

$$\boldsymbol{B}_{L_i} = \begin{bmatrix} F_{11} \, \psi_{i,1} & F_{21} \, \psi_{i,1} \\ F_{12} \, \psi_{i,2} & F_{22} \, \psi_{i,2} \\ F_{11} \, \psi_{i,2} + F_{12} \, \psi_{i,1} & F_{21} \, \psi_{i,2} + F_{22} \, \psi_{i,1} \end{bmatrix} . \tag{8.87}$$

Given a St. Venant type material

$$\boldsymbol{S} = \boldsymbol{C}[\boldsymbol{E}] = 2\mu \boldsymbol{E} + \lambda (\mathrm{tr}\, \boldsymbol{E}) \, \boldsymbol{I} \,, \tag{8.88}$$

the vector representation of the second Piola-Kirchhoff stress tensor is

$$\boldsymbol{\sigma} = \begin{bmatrix} S_{11} \\ S_{22} \\ S_{12} \end{bmatrix} = \begin{bmatrix} \lambda + 2\mu & \lambda & 0 \\ \lambda & \lambda + 2\mu & 0 \\ 0 & 0 & \mu \end{bmatrix} \begin{bmatrix} E_{11} \\ E_{22} \\ 2E_{12} \end{bmatrix} , \tag{8.89}$$

with

$$\mu = \frac{E}{2(1+\nu)} \,, \qquad \lambda = \frac{E\nu}{(1+\nu)(1-2\nu)} \,. \tag{8.90}$$

The weak form of the equilibrium is

$$\mathscr{G}(\boldsymbol{u}, \hat{\boldsymbol{u}}) = \sum_{i=1}^{n} \hat{\boldsymbol{u}}_i^T [\underbrace{\int_\Omega \boldsymbol{B}_{L_i}^T \cdot \boldsymbol{\sigma}\, d\Omega}_{\boldsymbol{k}_i} - \underbrace{\int_\Omega \boldsymbol{p}\, \psi_i\, d\Omega - \int_{\Gamma_N} \bar{\boldsymbol{t}}\, \psi_i\, ds}_{\boldsymbol{f}_i}] = 0 \quad (8.91)$$

or $\boldsymbol{k}(\boldsymbol{u}) - \boldsymbol{f} = \boldsymbol{0}$. As before, we linearize this equation by forming the Gateaux derivative in the direction of $\boldsymbol{u}_\Delta$

$$a_T(\boldsymbol{u}, \boldsymbol{u}_\Delta, \hat{\boldsymbol{u}}) = \int_\Omega [\nabla \boldsymbol{u}_\Delta\, \boldsymbol{S} \cdot \nabla \hat{\boldsymbol{u}} + \boldsymbol{E}_{\boldsymbol{u}}(\boldsymbol{u}, \hat{\boldsymbol{u}}) \cdot \boldsymbol{C}[\boldsymbol{E}_{\boldsymbol{u}}(\boldsymbol{u}_\Delta)]]\, d\Omega\,. \quad (8.92)$$

The discretization of the first term in (8.92) leads with

$$\nabla \boldsymbol{u}_{\Delta_h} = \sum_{j=1}^{n} \boldsymbol{u}_{\Delta j} \otimes \nabla \psi_j\,, \qquad \nabla \hat{\boldsymbol{u}} = \sum_{i=1}^{n} \hat{\boldsymbol{u}}_i \otimes \nabla \psi_i \quad (8.93)$$

to the so-called **stress-stiffness matrix**

$$\int_\Omega \nabla \boldsymbol{u}_\Delta\, \boldsymbol{S} \cdot \nabla \hat{\boldsymbol{u}}\, d\Omega = \sum_{i=1}^{n} \sum_{j=1}^{n} \hat{\boldsymbol{u}}_i^T \int_\Omega G_{ij}\, \boldsymbol{I}\, d\Omega\, \boldsymbol{u}_{\Delta j}\,, \quad (8.94)$$

where

$$G_{ij} := \nabla^T \psi_i\, \boldsymbol{S}\, \nabla \psi_j = \begin{bmatrix} \psi_{i,1} & \psi_{i,2} \end{bmatrix} \begin{bmatrix} S_{11} & S_{12} \\ S_{21} & S_{22} \end{bmatrix} \begin{bmatrix} \psi_{j,1} \\ \psi_{j,2} \end{bmatrix}. \quad (8.95)$$

With

$$\boldsymbol{E}_{\boldsymbol{u}}(\boldsymbol{u}, \boldsymbol{u}_\Delta) = \frac{1}{2} \sum_{j=1}^{n} [\boldsymbol{F}^T (\boldsymbol{u}_{\Delta j} \otimes \nabla \psi_j) + (\nabla \psi_j \otimes \boldsymbol{u}_{\Delta j}) \boldsymbol{F}] = \sum_{j=1}^{n} \boldsymbol{B}_{L_j} \boldsymbol{u}_{\Delta j} \quad (8.96)$$

the second term in (8.92) becomes

$$\int_\Omega \boldsymbol{E}_{\boldsymbol{u}}(\boldsymbol{u}, \hat{\boldsymbol{u}}) \cdot \boldsymbol{C}[\boldsymbol{E}_{\boldsymbol{u}}(\boldsymbol{u}, \boldsymbol{u}_\Delta)]\, d\Omega = \sum_{i=1}^{n} \sum_{j=1}^{n} \hat{\boldsymbol{u}}_i^T \int_\Omega \boldsymbol{B}_{L_i}^T \boldsymbol{C} \boldsymbol{B}_{L_j}\, d\Omega\, \boldsymbol{u}_{\Delta j}\,. \quad (8.97)$$

Adding all terms results in

$$a_T(\boldsymbol{u}, \boldsymbol{u}_\Delta, \hat{\boldsymbol{u}}) = \sum_{i=1}^{n} \sum_{j=1}^{n} \hat{\boldsymbol{u}}_i^T \boldsymbol{K}_{T_{ij}} \boldsymbol{u}_{\Delta j} \quad (8.98)$$

with

$$\boldsymbol{K}_{T_{ij}} := \int_{\Omega} \left[G_{ij}\, \boldsymbol{I} + \boldsymbol{B}_{L_i}^T \boldsymbol{C} \boldsymbol{B}_{L_j} \right] d\Omega\,. \tag{8.99}$$

In the case of a bilinear element, the (2×2) submatrices $\boldsymbol{K}_{T_{ij}}$ of the tangential stiffness matrix of a single element are arranged as follows

$$\boldsymbol{K}_T^e = \begin{bmatrix} \boldsymbol{K}_{T_{11}} & \boldsymbol{K}_{T_{12}} & \boldsymbol{K}_{T_{13}} & \boldsymbol{K}_{T_{14}} \\ \boldsymbol{K}_{T_{21}} & \boldsymbol{K}_{T_{22}} & \boldsymbol{K}_{T_{23}} & \boldsymbol{K}_{T_{24}} \\ \boldsymbol{K}_{T_{31}} & \boldsymbol{K}_{T_{32}} & \boldsymbol{K}_{T_{33}} & \boldsymbol{K}_{T_{34}} \\ \boldsymbol{K}_{T_{41}} & \boldsymbol{K}_{T_{42}} & \boldsymbol{K}_{T_{43}} & \boldsymbol{K}_{T_{44}} \end{bmatrix}. \tag{8.100}$$

After the complete system is assembled, the system is solved iteratively

$$\boldsymbol{u}_{i+1} = \boldsymbol{u}_i - \boldsymbol{K}_T^{-1}(\boldsymbol{u}_i)(\boldsymbol{k}(\boldsymbol{u}_i) - \boldsymbol{f})\,. \tag{8.101}$$

8.7 Nonlinear Functionals

Problems with constraints are the domain of **Lagrange multipliers**. As such a problem we can also read the evaluation of a functional $J(\boldsymbol{u}) = \boldsymbol{j}^T \boldsymbol{u}$ (we switch directly to finite elements), since the vector $\boldsymbol{u}$ must solve the system $\boldsymbol{K}\,\boldsymbol{u} = \boldsymbol{f}$. With a third vector $\boldsymbol{\lambda}$ we form the *Lagrange functional*

$$\mathcal{L}(\boldsymbol{u}, \boldsymbol{\lambda}) = J(\boldsymbol{u}) - \boldsymbol{\lambda}^T (\boldsymbol{K}\,\boldsymbol{u} - \boldsymbol{f}) \tag{8.102}$$

and the point $\{\boldsymbol{u}, \boldsymbol{\lambda}\}$ at which the functional $\mathcal{L}$ is stationary

$$d\mathcal{L} = \mathcal{L},_{u_i}\, du_i - \mathcal{L},_{\lambda_i}\, d\lambda_i = 0\,, \tag{8.103}$$

is found by solving the two equations

$$\boldsymbol{K}\,\boldsymbol{u} = \boldsymbol{f} \qquad \boldsymbol{K}\,\boldsymbol{\lambda} = \boldsymbol{j} \tag{8.104}$$

which results in

$$J(\boldsymbol{u}) = \boldsymbol{\lambda}^T \boldsymbol{f}\,, \tag{8.105}$$

and so the Lagrange multiplier is identical with the vector $\boldsymbol{g}$ of the influence function.

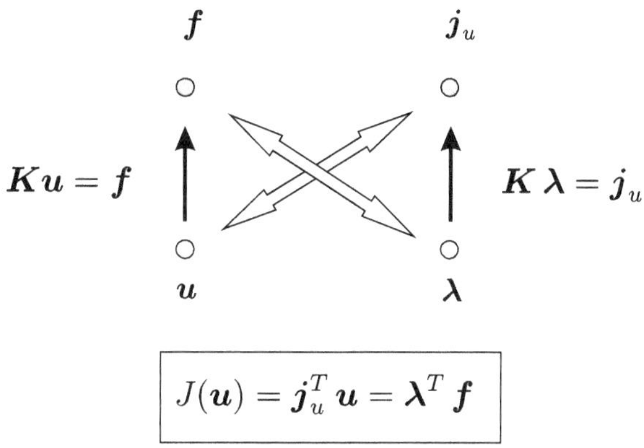

Fig. 8.5 Here, the Lagrange multiplier is the Gateaux derivative $\boldsymbol{j}_u$ of the non-linear functional $J(\boldsymbol{u})$

According to (8.105) the vector $\boldsymbol{\lambda}$ is the gradient of the functional $J(\boldsymbol{u})$ with respect to the right side $\boldsymbol{f}$ [5],

$$\frac{\partial J}{\partial f_i} = \lambda_i \,, \tag{8.106}$$

and its plot serves as a sensitivity plot of the functional, see section 3.32.

If the functional $J(u)$ is nonlinear, the equations read

$$\boldsymbol{K}\,\boldsymbol{u} = \boldsymbol{f} \qquad \boldsymbol{K}\,\boldsymbol{\lambda} = \boldsymbol{j}_u \,, \tag{8.107}$$

where

$$j_{ui} := \int_0^1 J'(\boldsymbol{u} - s\,\boldsymbol{u}; \varphi_i)\, ds \,, \tag{8.108}$$

and J' is the Gateaux derivative of $J(\boldsymbol{u})$ at the point $\boldsymbol{u}$, see Fig. 8.5. So, the vector $\boldsymbol{g} = \boldsymbol{\lambda}$ of the functional depends on $\boldsymbol{u}$, depends on where we are on the equilibrium path.

This way "influence functions" can also be formulated for nonlinear problems, more precisely for the tangential stiffness matrix at a particular equilibrium point $\boldsymbol{u}$, if one moves away from this point in "tangential direction". This is also how *goal oriented refinement* methods have been applied successfully to nonlinear problems; for details, see [6, 7].

References

1. Dhatt G, Touzot G, Lefrançois (2012) Finite element method. Wiley, ISTE
2. Hartmann F (1985) The mathematical foundation of structural mechanics. Springer, Heidelberg
3. Wriggers P (2010) Nonlinear finite element methods. Springer, Heidelberg
4. Gruttmann F, Sauer R, Wagner W (2000) Theory and numerics of three-dimensional beams with elastoplastic behaviour. Int J Num Methods Eng 48(12):1675–1702
5. Strang G (2019) Linear algebra and learning from data. Wellesley-Cambridge Press
6. Hartmann F (2013) Green's functions and finite elements. Springer, Heidelberg
7. Bangerth W, Rannacher R (2003) Adaptive finite element methods for differential equations. Birkhäuser Basel Boston Berlin

Chapter 9
Addenda

9.1 Basics

Before we concentrate on details, let us briefly describe the approach to the equations of statics in this book.

Euler Equation

The starting point are the ***Euler equations***, as the equation $-EAu'' = p$, which formulates the equilibrium of an infinitesimal element dx, see Fig. 9.1,

$$-N + N + dN + p\,dx = 0 \qquad (N = EAu')\,. \tag{9.1}$$

We superpose the left side of $-E\,A\,u'' = p$ with a test function δu

$$\int_0^l -E\,A\,u''\,\delta u\,dx\,, \tag{9.2}$$

apply integration by parts and arrive at Green's first identity

$$\mathscr{G}(u, \delta u) = \underbrace{\int_0^l -E\,A\,u''(x)\,\delta u(x)\,dx + [N\,\delta u]_0^l}_{\text{virt. external work}} - \underbrace{\int_0^l \frac{N\,\delta N}{EA}\,dx}_{\text{virt. internal energy}} = 0\,, \tag{9.3}$$

and therewith at the variational principles associated with $-EAu'' = p$.

F. Hartmann and P. Jahn, *Statics and Influence Functions*,
Springer Series in Solid and Structural Mechanics 13,
https://doi.org/10.1007/978-3-030-55889-5_9

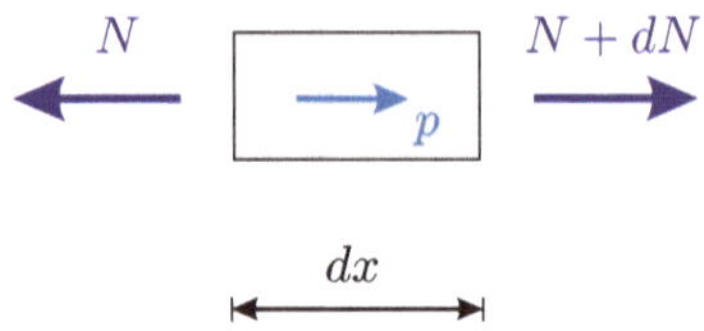

Fig. 9.1 Bar element

Principle of Virtual Displacements

The reverse formulation starts with the *principle of virtual displacements*, defines the virtual external and internal energy, δW_e and δW_i, and requires these two expressions to be the same for all δu

$$\delta W_e = \int_0^l p\,\delta u\,dx = \int_0^l \frac{N\,\delta N}{EA}\,dx = \delta W_i\,. \tag{9.4}$$

With Green's first identity the authors have a proven and tested template for how to define δW_e and δW_i, and so in the end everything fits nicely together. The actual challenge arises when several motions "overlap", such as the stretching, bending and twisting (3-D) of an open steel tube or the displacements of a shell element, where above all the kinematics must be mastered in order to arrive at a consistent formulation of $\delta W_e = \delta W_i$.

What favors the access $\delta W_e = \delta W_i$ in the eyes of the engineer is, that it is also the basis of the finite element method and so one can start directly the discretization without formulating a differential equation. But the differential equation is still "there", it still pulls the strings. ***Weak solutions do not make differential equations meaningless.***

The equation comes in through the back door, because if we integrate (9.4) by parts, $(\delta u(0) = \delta u(l) = 0)$, it is equivalent to

$$\int_0^l \underbrace{(-EA\,u'' - p)}_{Euler\ Equ.}\,\delta u\,dx = 0 \qquad \text{for all } \delta u\,, \tag{9.5}$$

which means:

The statement $\delta W_e = \delta W_i$ is *synonymous* with the fact that u is the solution of a differential equation, the Euler equation.

The Euler equations, together with the associated Green's identities, form the basis of statics and one can rightly claim:

An FE-program is Green's first identity in *bits* and *bytes*

9.2 Notation

The **gradient** of a scalar function u is a vector and the gradient of a vector-valued function $\boldsymbol{u} = \{u_1, u_2\}^T$ is a matrix,

$$\nabla u = \begin{bmatrix} u,_1 \\ u,_2 \end{bmatrix} \qquad \nabla \boldsymbol{u} = \begin{bmatrix} u_1,_1 & u_1,_2 \\ u_2,_1 & u_2,_2 \end{bmatrix} \qquad u_i,_j := \frac{\partial u_i}{\partial x_j}\,. \tag{9.6}$$

The opposite is the operator div, since the **divergence** of a matrix-valued function is a vector-valued function and the divergence of a vector-valued function $\boldsymbol{q} = \{q_1, q_2\}^T$ is a scalar function

$$\text{div}\, \boldsymbol{S} = \begin{bmatrix} \sigma_{11},_1 + \sigma_{12},_2 \\ \sigma_{21},_1 + \sigma_{22},_2 \end{bmatrix} \qquad \text{div}\boldsymbol{q} = q_1,_1 + q_2,_2\,. \tag{9.7}$$

The link between these two operators is the identity

$$\int_\Omega \text{div}\boldsymbol{S} \cdot \boldsymbol{\delta u}\, d\Omega = \int_\Gamma \boldsymbol{Sn} \cdot \boldsymbol{\delta u}\, ds - \int_\Omega \boldsymbol{S} \cdot \nabla \boldsymbol{\delta u}\, d\Omega\,, \tag{9.8}$$

and if $\boldsymbol{S}$ is a symmetric tensor, we have as well

$$\begin{aligned} \int_\Omega -\text{div}\boldsymbol{S} \cdot \boldsymbol{\delta u}\, d\Omega + \int_\Gamma \boldsymbol{Sn} \cdot \boldsymbol{\delta u}\, ds &= \int_\Omega \boldsymbol{S} \cdot \nabla \boldsymbol{\delta u}\, d\Omega \\ &= \int_\Omega \boldsymbol{S} \cdot \frac{1}{2}(\nabla \boldsymbol{\delta u} + \nabla \boldsymbol{\delta u}^T)\, d\Omega\,, \end{aligned} \tag{9.9}$$

which is the *principle of virtual displacements* for a plate, $\delta W_e = \delta W_i$.

Vector-valued functions $\boldsymbol{u} = \{u_x, u_y\}^T$ satisfy a similar equation

$$\int_\Omega \text{div}\boldsymbol{u}\, \delta u\, d\Omega = \int_\Gamma (\boldsymbol{u} \cdot \boldsymbol{n})\, \delta u\, ds - \int_\Omega \boldsymbol{u} \cdot \nabla \delta u\, d\Omega\,, \tag{9.10}$$

and in 1-D problems div = $()'$ and $\nabla = ()'$ are the same

$$\int_0^l u'\, \delta u\, dx = [u\, \delta u]_0^l - \int_0^l u\, \delta u'\, dx\,. \tag{9.11}$$

Vectors are column vectors and a dot denotes the scalar product between two vectors

$$\boldsymbol{f} \cdot \boldsymbol{u} = f_x\, u_x + f_y\, u_y\,. \tag{9.12}$$

Occasionally we also use the notation $\boldsymbol{f}^T \boldsymbol{u}$. The **dot** also denotes the **scalar product of two matrices**, like the strain and stress tensor

$$W_i = \frac{1}{2}\int_\Omega \boldsymbol{E} \cdot \boldsymbol{S}\, d\Omega$$
$$= \frac{1}{2}\int_\Omega \underbrace{[\varepsilon_{xx}\,\sigma_{xx} + \varepsilon_{xy}\,\sigma_{xy} + \varepsilon_{yx}\,\sigma_{yx} + \varepsilon_{yy}\,\sigma_{yy}]}_{scalarproduct}\, d\Omega\,. \tag{9.13}$$

In the literature you see also the notation

$$\boldsymbol{E} \cdot \boldsymbol{S} = \text{tr}\,(\boldsymbol{E} \otimes \boldsymbol{S}) = \boldsymbol{E} : \boldsymbol{S} \qquad (\text{tr} = \text{trace})\,, \tag{9.14}$$

where $\boldsymbol{E} \otimes \boldsymbol{S}$ is the ***direct product*** of the two tensors $\boldsymbol{E}$ and $\boldsymbol{S}$. The direct product of two vectors is a matrix

$$\boldsymbol{f} \otimes \boldsymbol{u} = \begin{bmatrix} f_x \\ f_y \end{bmatrix} \otimes \begin{bmatrix} u_x \\ u_y \end{bmatrix} = \begin{bmatrix} f_x \cdot u_x & f_x \cdot u_y \\ f_y \cdot u_x & f_y \cdot u_y \end{bmatrix} = \boldsymbol{f}\boldsymbol{u}^T = \boldsymbol{A} \tag{9.15}$$

with elements $a_{ij} = f_i \cdot u_j$; such matrices always have rank 1.

The multiplication table $\boldsymbol{T}_{(10\times 10)} = \boldsymbol{z} \otimes \boldsymbol{z}$ from 1 to 10 is the direct product of the vector $\boldsymbol{z} = \{1, 2, \ldots, 10\}^T$ with itself.

9.3 FE-Notation

A column vector $\boldsymbol{a}$ times a row vector $\boldsymbol{b}^T$ is a direct product, is a matrix

$$\boldsymbol{M} = \boldsymbol{a} \otimes \boldsymbol{b} = \boldsymbol{a}\boldsymbol{b}^T = \begin{bmatrix} a_1 \cdot b_1 & a_1 \cdot b_2 \\ a_2 \cdot b_1 & a_2 \cdot b_2 \end{bmatrix}. \tag{9.16}$$

In the FE-literature the stiffness matrix of a linear bar element with its two linear shape functions $\varphi_1(x)$ and $\varphi_2(x)$ is usually written in this way

$$\boldsymbol{K}^e_{(2\times 2)} = \int_0^{l_e} \begin{bmatrix} \varphi_1' \\ \varphi_2' \end{bmatrix} \underbrace{[\,EA\,]}_{\boldsymbol{E}} [\,\varphi_1'\ \varphi_2'\,]\, dx = \int_0^{l_e} \boldsymbol{B}^T_{(2\times 1)} \boldsymbol{E}_{(1\times 1)}\, \boldsymbol{B}_{(1\times 2)}\, dx \tag{9.17}$$

which means

$$k^e_{ij} = a(\varphi_i, \varphi_j) = \int_0^{l_e} \varphi_i'\, EA\, \varphi_j'\, dx\,. \tag{9.18}$$

In a beam element the entries of the matrix $\boldsymbol{B}_{(1\times 4)}$ are the second derivatives of the four shape functions

$$\boldsymbol{B}_{(1\times 4)} = \{-\frac{6}{l_e^2} + \frac{12\,x}{l_e^3}, \quad -\frac{4}{l_e} + \frac{6\,x}{l_e^2}, \quad \frac{6}{l_e}^2 - \frac{12\,x}{l_e^3}, \quad -\frac{2}{l_e} + \frac{6\,x}{l_e^2}\}, \tag{9.19}$$

and the stiffness matrix is

$$\boldsymbol{K}^e_{(4\times 4)} = \int_0^{l_e} \boldsymbol{B}^T_{(4\times 1)}\, EI_{(1\times 1)}\, \boldsymbol{B}_{(1\times 4)}\, dx\,. \tag{9.20}$$

Given a plate element with n nodal displacement fields, e.g.. $n = 3 \cdot 2 = 6$ for a *CST element* with its three nodes, the n columns of the matrix

$$\boldsymbol{B}_{(3\times n)} = \{\boldsymbol{\varepsilon}(\boldsymbol{\varphi}_1), \boldsymbol{\varepsilon}(\boldsymbol{\varphi}_2), \ldots, \boldsymbol{\varepsilon}(\boldsymbol{\varphi}_n)\} \tag{9.21}$$

are the strains of the n fields

$$\boldsymbol{\varepsilon}(\boldsymbol{\varphi}_i) = \{\varepsilon^{(i)}_{xx}, \varepsilon^{(i)}_{yy}, \varepsilon^{(i)}_{xy}\}^T \qquad \text{(column)} \tag{9.22}$$

and the product

$$\boldsymbol{E}_{(3\times 3)}\boldsymbol{B}_{(3\times n)} = \{\boldsymbol{\sigma}(\boldsymbol{\varphi}_1), \boldsymbol{\sigma}(\boldsymbol{\varphi}_2), \ldots, \boldsymbol{\sigma}(\boldsymbol{\varphi}_n)\} \qquad \boldsymbol{E} = \frac{E}{1-\nu^2}\begin{bmatrix} 1 & \nu & 0 \\ \nu & 1 & 0 \\ 0 & 0 & 2(1-\nu) \end{bmatrix} \tag{9.23}$$

are the stresses

$$\boldsymbol{\sigma}(\boldsymbol{\varphi}_i) = \{\sigma^{(i)}_{xx}, \sigma^{(i)}_{yy}, 2\sigma^{(i)}_{xy}\}^T \qquad \text{(column)}, \tag{9.24}$$

of the shape functions, and so

$$\begin{aligned} k^e_{ij} &= a(\boldsymbol{\varphi}_i, \boldsymbol{\varphi}_j) = \int_{\Omega_e} \boldsymbol{\sigma}^{(i)} \bullet \boldsymbol{\varepsilon}^{(j)}\, d\Omega \\ &= \int_{\Omega_e} (\sigma^{(i)}_{xx}\, \varepsilon^{(j)}_{xx} + \sigma^{(i)}_{yy}\, \varepsilon^{(j)}_{yy} + 2 \cdot \sigma^{(i)}_{xy}\, \varepsilon^{(j)}_{xy})\, d\Omega\,. \end{aligned} \tag{9.25}$$

9.4 Vectors and Functions

A function $u(x)$ is an "infinitely long" vector $\boldsymbol{u}$ and a linear operator L equals a matrix $\boldsymbol{L}$, and the inverse of a matrix

$$\boldsymbol{L}\,\boldsymbol{u} = \boldsymbol{f} \qquad \Rightarrow \qquad \boldsymbol{u} = \boldsymbol{L}^{-1}\boldsymbol{f} \tag{9.26}$$

equals the inverse L^{-1} of a linear operator

$$L\,u = f \quad \Rightarrow \quad u = \int_0^l L^{-1} f\, dx\,. \tag{9.27}$$

The scalar product of vectors, $\boldsymbol{u}^T \boldsymbol{\delta v} = (\boldsymbol{u}, \boldsymbol{\delta v})$, has its continuation in the L_2-scalar product

$$\int_0^l u(x)\,\delta v(x)\,dx = (u, \delta v) \tag{9.28}$$

and to the transposed matrix $\boldsymbol{L}^T$ corresponds the **adjoint operator** L^*

$$(L u, v) = (u, L^* v)\,. \tag{9.29}$$

In linear algebra you transpose and in calculus you integrate by parts.

Linear Functions

The derivative of $f(x) = a \cdot x$ is $f'(x) = a$. Each linear function (also each polynomial) is identical with its **Taylor expansion** about the point $x = 0$

$$f(x) = f(0) + f'(0) \cdot x = 0 + a \cdot x\,. \tag{9.30}$$

A linear function carries its derivative with it in its backpack, so to speak, it knows with which slope it started.

The gradients of the linear functional $J(\boldsymbol{u}) = \boldsymbol{g}^T \boldsymbol{f} = \boldsymbol{j}^T \boldsymbol{u}$ in the direction of $\boldsymbol{f}$ and $\boldsymbol{u}$, respectively, are the vectors $\boldsymbol{g}$ and $\boldsymbol{j}$, and the stiffness matrix $\boldsymbol{K}$ is the gradient of the linear vector-valued function $\boldsymbol{f} = \boldsymbol{K}\,\boldsymbol{u}$.

9.5 The Algebra of the Identities

The subject is best explained with matrices. Green's second identity is not limited to square matrices

$$\mathscr{B}(\boldsymbol{u}, \boldsymbol{v}) = \boldsymbol{u}_n^T \boldsymbol{A}_{n\times m}^T \boldsymbol{v}_m - \boldsymbol{v}_m^T \boldsymbol{A}_{m\times n} \boldsymbol{u}_n = 0\,. \tag{9.31}$$

Let $C(\boldsymbol{A}) \subset \mathbb{R}^m$ denote the column space, $R(\boldsymbol{A}) = C(\boldsymbol{A}^T) \subset \mathbb{R}^n$ the row space and $N(\boldsymbol{A}) \subset \mathbb{R}^n$ resp. $N(\boldsymbol{A}^T) \subset \mathbb{R}^m$ the null space of $\boldsymbol{A}$ resp. $\boldsymbol{A}^T$.

- column space of $\boldsymbol{A}$, the set of all $\boldsymbol{y}$ in $\mathbb{R}^m$ resulting from $\boldsymbol{y} = \boldsymbol{A}\boldsymbol{x}$,
- row space of $\boldsymbol{A}$, the set of all $\boldsymbol{x}$ in $\mathbb{R}^n$ resulting from $\boldsymbol{x} = \boldsymbol{A}^T \boldsymbol{y}$,

- null space of $\boldsymbol{A}$, the set of all $\boldsymbol{x}$ in $\mathbb{R}^n$ for which $\boldsymbol{A}\boldsymbol{x} = \boldsymbol{0}$,
- left null space of $\boldsymbol{A}$, the set of all $\boldsymbol{y}$ in $\mathbb{R}^m$ for which $\boldsymbol{A}^T \boldsymbol{y} = \boldsymbol{0}$

Subspace	Subspace of	Symbol	Dimension
Column space $C(\boldsymbol{A})$	$\mathbb{R}^m$	im($\boldsymbol{A}$)	r = rank
Nullspace (kernel) $N(\boldsymbol{A})$	$\mathbb{R}^n$	ker($\boldsymbol{A}$)	$n - r$
Row space $C(\boldsymbol{A}^T)$	$\mathbb{R}^n$	im($\boldsymbol{A}^T$)	r
Left nullspace (kernel) $N(\boldsymbol{A}^T)$	$\mathbb{R}^m$	ker($\boldsymbol{A}^T$)	$m - r$

The above identity implies: The vectors $\boldsymbol{v} \in N(\boldsymbol{A}^T)$ are orthogonal to the vectors in $C(\boldsymbol{A})$[1]

$$\mathscr{B}(\boldsymbol{u}, \boldsymbol{v}) = \boldsymbol{u}^T \boldsymbol{0} - \boldsymbol{v}^T \boldsymbol{A}\boldsymbol{u} = 0\,, \tag{9.32}$$

and the vectors $\boldsymbol{u} \in N(\boldsymbol{A})$ are orthogonal to the vectors in $C(\boldsymbol{A}^T)$

$$C(\boldsymbol{A}) \perp N(\boldsymbol{A}^T) \qquad\qquad C(\boldsymbol{A}^T) \perp N(\boldsymbol{A})\,. \tag{9.33}$$

Each vector $\boldsymbol{x} = \boldsymbol{x}_C + \boldsymbol{x}_{N'} \in \mathbb{R}^m$ is a part $\boldsymbol{x}_C$ from $C(\boldsymbol{A})$ and a part $\boldsymbol{x}_{N'}$ from $N(\boldsymbol{A}^T)$ and an analog splitting of the $\mathbb{R}^n$ is possible, $\boldsymbol{x} = \boldsymbol{x}_R + \boldsymbol{x}_N \in \mathbb{R}^n$, *The four fundamental subspaces* [1].

In statics, $\boldsymbol{K} = \boldsymbol{A}$ is the symmetric stiffness matrix and $C(\boldsymbol{K}) = R(\boldsymbol{K})$ and $N(\boldsymbol{K}) = N(\boldsymbol{K}^T)$ contain only the zero vector $\boldsymbol{0}$ if the structure is supported. If it is kinematic an equilibrium position $\boldsymbol{u}$ is only possible, if the forces $\boldsymbol{f}$ are orthogonal to the rigid body motions, $\boldsymbol{u}_0^T \boldsymbol{f} = 0$, if they are equilibrium forces, since this guarantees that $\boldsymbol{f}$ lies in $C(\boldsymbol{K})$, and a solution $\boldsymbol{K}\boldsymbol{u} = \boldsymbol{f}$ exists. The rule is, see (9.33),

$$\boldsymbol{f} \in C(\boldsymbol{K}) \qquad \Leftrightarrow \qquad \boldsymbol{f} \perp N(\boldsymbol{K}^T)\,. \tag{9.34}$$

Take the singular stiffness matrix of a bar element

$$\boldsymbol{K} = \frac{EA}{l}\begin{bmatrix} 1 & -1 \\ -1 & 1 \end{bmatrix}. \tag{9.35}$$

Each vector $\boldsymbol{u} = \boldsymbol{u}_R + \boldsymbol{u}_N \in \mathbb{R}^2$ can be split into a translation $\boldsymbol{u}_N = c\,\{1, 1\}^T$, $c =$ any number, and a stretch $\boldsymbol{u}_R \in R(\boldsymbol{K})$ of the element. Given any such $\boldsymbol{u}_R$ the forces $\boldsymbol{f} = \boldsymbol{K}\boldsymbol{u}_R$ balance, they are orthogonal to the one-vector $\boldsymbol{u}_0 = \{1, 1\}^T \in N(\boldsymbol{K}^T) = N(\boldsymbol{K})$.

The vector $\boldsymbol{f} = \{1, 0\}^T$ is not orthogonal to $N(\boldsymbol{K}^T)$, so it is not in $C(\boldsymbol{K})$ and therefore $\boldsymbol{K}\boldsymbol{u} = \boldsymbol{f}$ has no solution $\boldsymbol{u}$.

We mention these examples, since Green's identities are also such zero sums and many important results, just think of the Galerkin orthogonality, $a(u - u_h\,, \varphi_i) = 0$, are algebraic in nature.

---

[1] Each of the m columns $\boldsymbol{c}_i$ of $\boldsymbol{A}$ is orthogonal to $\boldsymbol{v}$, so $\boldsymbol{v}^T \boldsymbol{c}_i = 0$.

The equilibrium condition for a bar element

$$\mathcal{G}(u, 1) = \int_0^l -EA\, u''(x) \cdot 1\, dx + N(l) \cdot 1 - N(0) \cdot 1 = 0 \tag{9.36}$$

is practically a continuation of the statement $\boldsymbol{u}_0^T \boldsymbol{f} = 0$ of linear algebra.

The moments M_i of the redundants X_i in the force method are orthogonal to the final moment distribution, $(M_i, M) = 0$, and if a support of a beam settles, the deflection w is orthogonal to the virtual displacements, $a(w, \delta w) = 0$.

Or take the singular stiffness matrix of a bar consisting of two elements. The product $\boldsymbol{K}\boldsymbol{u}$ is the vector

$$\frac{EA}{l_e} \begin{bmatrix} 1 & -1 & 0 \\ -1 & 2 & -1 \\ 0 & -1 & 1 \end{bmatrix} \begin{bmatrix} u_1 \\ u_2 \\ u_3 \end{bmatrix} = \frac{EA}{l_e} \begin{bmatrix} u_1 - u_2 \\ -u_1 + 2 \cdot u_2 - u_3 \\ -u_2 + u_3 \end{bmatrix}, \tag{9.37}$$

and the row sum of this vector is zero

$$(u_1 - u_2) + (-u_1 + 2 \cdot u_2 - u_3) + (-u_2 + u_3) = 0 \tag{9.38}$$

whatever the u_i, and this means the singular system $\boldsymbol{K}\boldsymbol{u} = \boldsymbol{f}$ has only a solution if $f_1 + f_2 + f_3 = 0$, which is of course the equilibrium condition.

It is the requirement $\boldsymbol{f} \perp N(\boldsymbol{K}^T) = N(\boldsymbol{K})$ or $\boldsymbol{f}^T \boldsymbol{u}_0 = f_1 + f_2 + f_3 = 0$, since $\boldsymbol{u}_0 = \{1, 1, 1\}^T$ is the vector, which spans $\boldsymbol{N}(\boldsymbol{K}^T)$.

Each deflection curve w of a beam satisfies the equilibrium conditions

$$\mathcal{G}(w, a\,x + b) = 0\,. \tag{9.39}$$

Similarly, the product of the beam matrix $\boldsymbol{K}$ with any vector $\boldsymbol{w}$ results in nodal forces $\boldsymbol{f} = \boldsymbol{K}\boldsymbol{w}$, which are orthogonal to the zero eigenvectors $\boldsymbol{\delta w}_0$ of $\boldsymbol{K}$, translations and rotations, since $\boldsymbol{\delta w}_0^T \boldsymbol{f} = 0$.

To the split of the equation $-EA\,u'' = p$ into the three parts

$$u' = \varepsilon \qquad EA\,\varepsilon = N \qquad -N' = p\,. \tag{9.40}$$

corresponds the split $\boldsymbol{K} = \boldsymbol{A}^T \boldsymbol{C} \boldsymbol{A}$ of the stiffness matrix of a bar (and other elements as well). The matrix $\boldsymbol{A} = [\ldots 0 \;\; 1/l_i \;\; -1/l_i \;\; 0 \ldots]$ with $1/l_i$ on the diagonal and $-1/l_i$ on the right secondary diagonal and its transpose $\boldsymbol{A}^T = -\boldsymbol{A}$ differentiate, like d/dx and $-d/dx$ and the diagonal matrix $\boldsymbol{C}$ lists on its diagonal the stiffness EA_i of the individual elements i with length l_i. This scheme, see Fig. 9.2, is universal [1].

A truss is *statically determinate*, if $\boldsymbol{A}$ is square, since one can solve $\boldsymbol{A}^T \boldsymbol{N} = \boldsymbol{f}$ for $\boldsymbol{N}$ and then determine $\boldsymbol{u}$ via $\boldsymbol{C}\boldsymbol{A}\boldsymbol{u} = \boldsymbol{N}$.

If the truss is *statically indeterminate*, there are more bars than nodal displacements, $\boldsymbol{A}$ is a rectangular matrix, and hand methods like the *Ritter method* are not

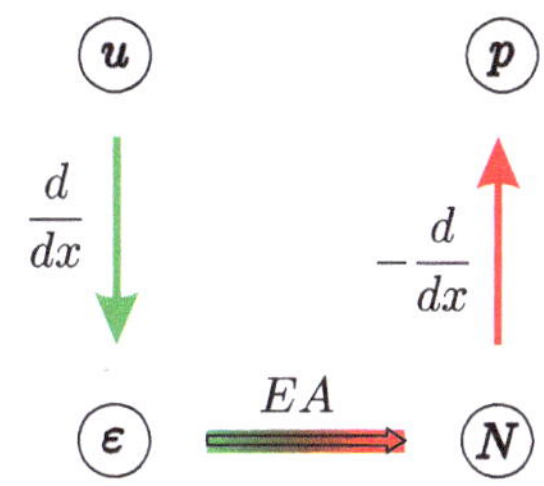

Fig. 9.2 The three steps in the case of a bar, $-d/dx\,(EA\,d/dx)\,u = p$

applicable, and so one needs a computer, to first solve $\boldsymbol{K}\boldsymbol{u} = \boldsymbol{f}$ for $\boldsymbol{u}$ and to then calculate the normal forces N_i in the single members via the strains in the members.

9.6 The Algebra of Finite Elements

The FE-solution $u_h = \sum_i u_i \varphi_i(x)$ is a variational solution, which satisfies the system

$$a(u_h, \varphi_i) - (p, \varphi_i) = 0 \qquad i = 1, 2, \ldots, n\,, \tag{9.41}$$

but many more equations and inequalities can be formulated with FE-solutions of linear problems. We present an excerpt.

The **solution space** $\mathcal{V}$ contains all functions u, which satisfy the geometric boundary conditions (Dirichlet) and $\mathcal{V}_h \subset \mathcal{V}$ is the **trial space**. In the following u is the exact solution, u_h is the FE-solution, $e = u - u_h$ is the error and v_h is a test function in $\mathcal{V}_h$ and p is the load.

- It is

$$a(e, v) = (p, v) - a(u_h, v) = (p, v) - (p_h, v) \qquad v \in \mathcal{V}\,. \tag{9.42}$$

- In particular

$$(p, u) - (p, u_h) = (p, e) = a(e, u) = (p, u) - (p_h, u) \tag{9.43}$$

  and therefore

$$(p, u_h) = (p_h, u) \qquad symmetry\,. \tag{9.44}$$

- Galerkin orthogonality

$$a(e, v_h) = 0 \quad v_h \in \mathcal{V}_h\,. \tag{9.45}$$

- The error on the right side, $p - p_h$, is orthogonal to the test functions

$$a(e, v_h) = (p, v_h) - (p_h, v_h) = 0 \quad v_h \in \mathcal{V}_h. \tag{9.46}$$

- The FE-load p_h is orthogonal to e

$$a(u_h, e) = (p_h, e) = 0\,. \tag{9.47}$$

- The shape forces p_i are orthogonal to e

$$a(\varphi_i, e) = (p_i, e) = 0. \tag{9.48}$$

- The FE-solution minimizes the error in the strain energy

$$a(e, e) \leq a(u - v_h, u - v_h) \quad v_h \in \mathcal{V}_h\,, \tag{9.49}$$

since for every $v_h \in \mathcal{V}_h$ holds

$$a(e + v_h, e + v_h) = a(e, e) + 2\underbrace{a(e, v_h)}_{=0} + \underbrace{a(v_h, v_h)}_{>0}\,. \tag{9.50}$$

The energy $a(e, e)$ necessary to correct the FE-solution, $u_h + e = u$, is the smallest possible compared to all other functions v_h in $\mathcal{V}_h$.
- The strain energy of the FE-solution is less than the exact strain energy (here without the factor 1/2)

$$a(u_h, u_h) \leq a(u, u) \qquad \text{in a LC } p\,, \tag{9.51}$$

because

$$\begin{aligned} 0 < a(u, u) &= a(u_h + e, u_h + e) \\ &= a(u_h, u_h) + 2\, \underbrace{a(e, u_h)}_{=0} + \underbrace{a(e, e)}_{>0}\,. \end{aligned} \tag{9.52}$$

- It is

$$\Pi(u) \leq \Pi(u_h)\,, \tag{9.53}$$

since

$$\begin{aligned} \Pi(u_h) = \Pi(u - e) &= \frac{1}{2}\, a(u, u) - a(u, e) + \frac{1}{2}\, a(e, e) - (p, u) + (p, e) \\ &= \Pi(u) \underbrace{-a(u, e) + (p, e)}_{(u,e)=0} + \frac{1}{2}\, \underbrace{a(e, e)}_{>0}\,, \end{aligned} \tag{9.54}$$

- and also

$$(p, u_h) = a(u_h, u_h) < a(u, u) = (p, u)\,. \tag{9.55}$$

- The potential energy of any $v_h \in \mathcal{V}_h$ depends on its "distance" from u [1],

$$\Pi(v_h) = \frac{1}{2}\,a(v_h, v_h) - (p, v_h) = \frac{1}{2}\,a(v_h - u, v_h - u) - \frac{1}{2}\,a(u, u)\,. \tag{9.56}$$

- A stiffness matrix is **singular** if the rigid-body motions u_0, i.e. the zeros of $a(u, u_0) = 0$ (for all u), lie in $\mathcal{V}_h$, since then $u_0(x) = \sum_i \bar{u}_i \varphi_i(x)$ has an address in $\mathcal{V}_h$, the vector $\bar{\boldsymbol{u}}$, and this vector is an eigenvector of $\boldsymbol{K}\bar{\boldsymbol{u}} = \boldsymbol{0}$ corresponding to the eigenvalue $\lambda = 0$.

In an LC d (support settlement) the function $w = 0$ is not in the solution space $\mathcal{S} = w_\Delta \oplus \mathcal{V}$, because $\mathcal{V}$ gets shifted by a fixed function w_Δ, which satisfies the inhomogeneous boundary condition at the support. Therefore $\mathcal{S}$ is not a vector space, but only a **manifold** (the sum $u + v$ of two functions in $\mathcal{S}$ is not in $\mathcal{S}$) and so one must be careful, when using the above formulas.

We also note: The **diagonal element** g_{ii} of the inverse is, up to the factor $(-1/2)$, the potential energy

$$\Pi(g_i) = \frac{1}{2}\,\boldsymbol{g}_i^T \boldsymbol{K} \boldsymbol{g}_i - \boldsymbol{g}_i^T \boldsymbol{e}_i = -\frac{1}{2}\,\text{eigenwork} = -\frac{1}{2} g_{ii} \cdot 1 \tag{9.57}$$

of the influence function $g_i(x) = \sum_j g_{ji}\,\varphi_j(x)$, since its nodal values $\boldsymbol{g}_i$ are the solution of $\boldsymbol{K}\boldsymbol{g}_i = \boldsymbol{e}_i$, ($g_i(x)$ is the influence function for u_i).

9.7 Eigenvalues and Eigenvectors

Stiffness matrices have eigenvalues and (orthogonal) eigenvectors. Let us look at the stiffness matrix $\boldsymbol{K}_{6\times 6}$ of a rope with 2 on the main diagonal and -1 on the two off diagonals. The **eigenvalues** are

$$\boldsymbol{\lambda} = \{0.1981, 0.7530, 1.5550, 2.4450, 3.2470, 3.8019\}^T\,, \tag{9.58}$$

and to the maximum eigenvalue λ_6 belongs the **eigenvector**

$$\boldsymbol{u}_6 = \{0.2319, -0.4179, 0.5211, -0.5211, 0.4179, -0.2319\}^T \qquad \|\boldsymbol{u}_6\| = 1\,, \tag{9.59}$$

so $\boldsymbol{K}\boldsymbol{u}_6 = 3.8019 \cdot \boldsymbol{u}_6$. If we multiply this from the left with $\boldsymbol{u}_6^T$, it gives

$$\boldsymbol{u}_6^T \boldsymbol{K} \boldsymbol{u}_6 = 3.8019 \cdot \boldsymbol{u}_6^T \boldsymbol{u}_6 = 3.8019 \cdot \|\boldsymbol{u}_6\|^2 = 3.8019 = \lambda_{max} \,, \tag{9.60}$$

which is, as can be shown (expand the vector $\boldsymbol{u} = \sum_i a_i \, \boldsymbol{u}_i$ with respect to the eigenvectors), the maximum possible energy if the maximum is searched among all $\|\boldsymbol{u}\| = 1$. While $\boldsymbol{u}_6$ is an *'up–down'* vector, which puts much strain on the rope, the eigenvector belonging to $\lambda_{min} = \lambda_1$

$$\boldsymbol{u}_1 = \{0.2319, 0.4179, 0.5211, 0.5211, 0.4179, 0.2319\}^T \tag{9.61}$$

is a quiet fellow in comparison. Interesting to note: If you iterate, i.e. you start with any vector $\boldsymbol{f}$, $\boldsymbol{u} = \boldsymbol{K}^{-1} \boldsymbol{f}$ and solve repeatedly for $\boldsymbol{f} = \boldsymbol{u}/\| \|\boldsymbol{u}\|$ (*power method* to determine λ_1) the rope converges to the first eigenvector $\boldsymbol{u}_1$.

The *up–down* is no surprise since the eigensolutions of ropes and fixed bars are **sinusoidal oscillations** and the one with the highest frequency generates the largest energy $a(u, u) = \boldsymbol{u}^T \boldsymbol{K} \, \boldsymbol{u}$.

The continuous beam in Fig. 9.3 has 27 degrees of freedom and 27 eigenvalues. The eigenvector of the lowest eigenvalue $\lambda_1 = 3.7 \cdot 10^3$ resembles a gentle swaying of the beam. The eigenvalue λ_{15} characterizes a more energetic eigenvector and the energy becomes really large, as in the case $\lambda_{22} = 2.2 \cdot 10^6$, if the beam works against the elastic supports or—here not shown, because it cannot be displayed—the supports move horizontally in opposite directions (like after a strike with a hammer on either end of the beam). This results in large axial strains, which typically belong to the eigenvalue $\lambda_{max} (= \lambda_{27})$.

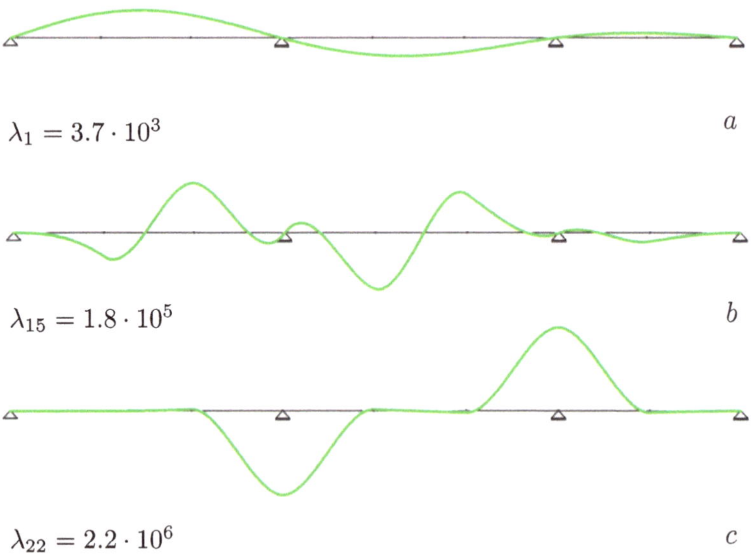

Fig. 9.3 Eigenvectors of a continuous beam on elastic supports

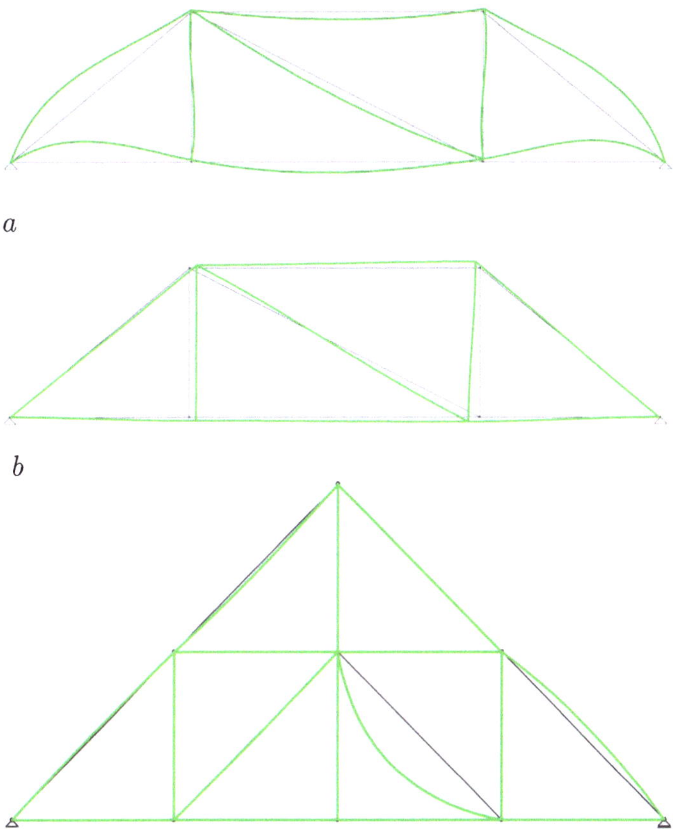

a

b

c

Fig. 9.4 Eigenvectors of trusses, **a–b** rigid joints, **c** pins

The lower eigenvectors of a frame are bending-dominated (soft) and the higher eigenvectors are normal-force dominated (stiff), see Fig. 9.4.

If you want to maximize the strain energy $\boldsymbol{u}^T \boldsymbol{K} \boldsymbol{u}$ in a structure, and the choice is among all $\|\boldsymbol{u}\| = 1$, you must opt for $\boldsymbol{f} = \lambda_{max} \boldsymbol{u}_e$ the eigenvector $\boldsymbol{u}_e$ of λ_{max}.

9.8 $\mathcal{V}_h$ and $\mathcal{V}_h^+$

The space $\mathcal{V}_h^+$ is the span of the shape functions of a mesh. The subset $\mathcal{V}_h \subset \mathcal{V}_h^+$ is $\mathcal{V}_h^+$ without the φ_i of the fixed nodes or rather fixed degrees of freedom u_i.

Since the φ_i in $\mathcal{V}_h^+$ form a ***partition of unity***

$$\sum_i \varphi_i(x) = 1 \qquad \text{at each point } x \text{ of the mesh}\,, \tag{9.62}$$

no load is lost, the sum $\sum_i f_i$ is equal to the resultant of the load p.

Since the matrix in $\boldsymbol{K}\boldsymbol{u} = \boldsymbol{f}$ is the reduced stiffness matrix,

$$k_{ij} = a(\varphi_i, \varphi_j) \qquad \varphi_i, \varphi_j \in \mathcal{V}_h\,, \tag{9.63}$$

one might assume the equivalence

$$f_{hi} = \int_0^l p_h\, \varphi_i\, dx = \int_0^l p\, \varphi_i\, dx = f_i \qquad \varphi_i \in \mathcal{V}_h \tag{9.64}$$

to only hold true with regard to the space $\mathcal{V}_h$, but no—it holds true for all $\varphi_i \in \mathcal{V}_h^+$.

After the system $\boldsymbol{K}\boldsymbol{u} = \boldsymbol{f}$ is solved, the support reactions f_{hi} at the fixed nodes are calculated with the original full system $\boldsymbol{K}_G\, \boldsymbol{u}_G = \boldsymbol{f}_G$ and these nodal forces correspond to the superposition of the FE-load p_h with the unlocked $\varphi_i \in \mathcal{V}_h^+$. The vector $\boldsymbol{u}_G$ is of course the vector $\boldsymbol{u}$ with the locked degrees of freedom, $u_i = 0$, added and $\boldsymbol{f}_G$ is the vector $\boldsymbol{f}$ with the nodal forces f_i at the supports added.

9.9 Galerkin

In Galerkin's method the FE-solution

$$u_h = \sum_i u_i\, \varphi_i(\boldsymbol{x}) \tag{9.65}$$

is the **projection** of the exact solution onto the trial space $\mathcal{V}_h$

$$a(u - u_h, \varphi_i) = 0 \qquad \text{for all } \varphi_i \in \mathcal{V}_h\,. \tag{9.66}$$

Since what is to be projected, u, is unknown, we replace with the help of Green's first identity

$$\mathcal{G}(w, \varphi_i) = \delta W_e - \delta W_i = 0 \tag{9.67}$$

the virtual internal energy $\delta W_i = a(u, \varphi_i)$ of the unknown solution with its virtual external work $\delta W_e = (p, \varphi_i) = f_i$ and we so arrive at the system

$$\boldsymbol{K}\boldsymbol{u} = \boldsymbol{f}\,. \tag{9.68}$$

The principle of minimum potential energy naturally leads to the same system, since $\boldsymbol{K}\boldsymbol{u} - \boldsymbol{f} = \boldsymbol{0}$ guarantees $\delta \Pi(\boldsymbol{u}) = \boldsymbol{K}\boldsymbol{u} - \boldsymbol{f} = \boldsymbol{0}$, i.e. $\Pi(\boldsymbol{u})$ has a horizontal tangent at the lowest point, and this in each direction u_i.

9.10 Weak Solution

Given the boundary value problem of a rope

$$-H\, w''(x) = p(x) \qquad w(0) = w(l) = 0\,, \tag{9.69}$$

we determine the FE-solution $w_h(x) = \sum_j w_j\, \varphi_j(x)$ by requiring

$$\int_0^l H\, w_h'\, \varphi_i'\, dx = \int_0^l p\, \varphi_i\, \varphi_i\, dx \qquad \text{for all } \varphi_i \in \mathcal{V}_h\,. \tag{9.70}$$

The solution of the boundary value problem (9.69) is called a ***strong solution*** and the solution of the variation problem (9.70) is called a ***weak solution***.

This distinction is usually explained by saying a weak solution does not have to be twice differentiable, as in $-H\, w'' = p$, but only once, as the w' in the strain energy product.

The following interpretation, though, seems more appropriate. In mathematics there is the concept of ***weak convergence***. It is an indirect proof for a sequence of functions $f_n(x)$ to converge to a target function $f(x)$. The test $f_n(x) \to f(x)$ is done against a set of control functions $\varphi_i(x)$, and the sequence $f_n(x)$ is said to converge *weakly* to $f(x)$ if

$$\lim_{n\to\infty} \int_0^l f_n(x)\, \varphi_i(x)\, dx = \int_0^l f(x)\, \varphi_i(x)\, dx \qquad \text{for all } \varphi_i\,. \tag{9.71}$$

Weak convergence is like the convergence of functionals. Every function $f_n(x)$ can be equated to a functional $J_n(.)$

$$J_n(\varphi_i) = \int_0^l f_n(x)\, \varphi_i(x)\, dx\,, \tag{9.72}$$

and weak convergence means, the functionals $J_n(.)$ converge towards the **target functional**

$$J(\varphi_i) = \int_0^l f(x)\, \varphi_i(x)\, dx\,, \tag{9.73}$$

in the sense of (9.71), the limit of the sequence $J_n(.)$ is “shake equivalent” to $J(.)$.

And this terminology fits exactly. The FE-solution is a weak solution since its agreement with the exact solution is not controlled by the differential equation, but it is controlled indirectly by $i = 1, 2, \ldots$ "shake tests"

$$\lim_{h \to 0} a(u_h, \varphi_i) = a(u, \varphi_i) \qquad \text{for all } \varphi_i \,. \tag{9.74}$$

Because of $\delta W_i = \delta W_e$ this is identical to

$$\lim_{h \to 0} \int_0^l p_h \, \varphi_i \, dx = \int_0^l p \, \varphi_i \, \varphi_i \, dx \qquad \text{for all } \varphi_i \tag{9.75}$$

i.e. the equivalence of the external virtual work. In practice, of course, the finite elements never reach the limit $h \to 0$ and only a finite number of tests is performed, since there are only a finite number of shape functions φ_i on a mesh.

The finite element method can be seen as a method of replacing a functional $J(\delta u) = (p, \delta u)$ with a functional $J_h(\delta u) = (p_h, \delta u)$, or, if you have infinite patience, $h \to 0$, with a sequence of functionals $J_h(\delta u) = (p_h, \delta u)$.

The practical significance of the concept of a weak solution becomes apparent when we watch a market woman, see Fig. 9.5, since she also draws conclusions indirectly. She has to solve the equation

$$P_l \cdot h_l = P_r \cdot h_r \,, \tag{9.76}$$

which means, as she knows, that at every turn $\delta\varphi$ of the balance beam, the work on the left and right side are the same

$$P_l \cdot h_l = P_r \cdot h_r \qquad \Rightarrow \qquad P_l \cdot h_l \cdot \tan \tan \delta\varphi = P_r \cdot h_r \cdot \tan \delta\varphi \,, \tag{9.77}$$

and so she concludes, by wiggling the balance, indirectly, arguing "backwards"

$$P_l \cdot h_l = P_r \cdot h_r \qquad \Leftarrow \qquad P_l \cdot h_l \cdot \tan \delta\varphi = P_r \cdot h_r \cdot \tan \delta\varphi \,. \tag{9.78}$$

The same is done by the toolmaker, who rolls a cylinder back and forth with his fingers, see Fig. 9.6, since he knows: If the cylinder has a perfect circular shape, the center of gravity does not change its height above the table when the cylinder is rotated. If the test fails, if his fingers feel a slight wobble, it is not a perfect cylinder.

The apprentice who is supposed to grind a cylinder out of a square iron, does it like the finite elements. At the start the square iron is equivalent to a cylinder with respect to all rotations $\delta\varphi$, which are a multiple of 90°, the center of gravity does not change its height. By grinding more and more edges (n) into the profile, the apprentice increases the test space, $\mathcal{V}_4 \to \mathcal{V}_8 \to \mathcal{V}_{16} \ldots$

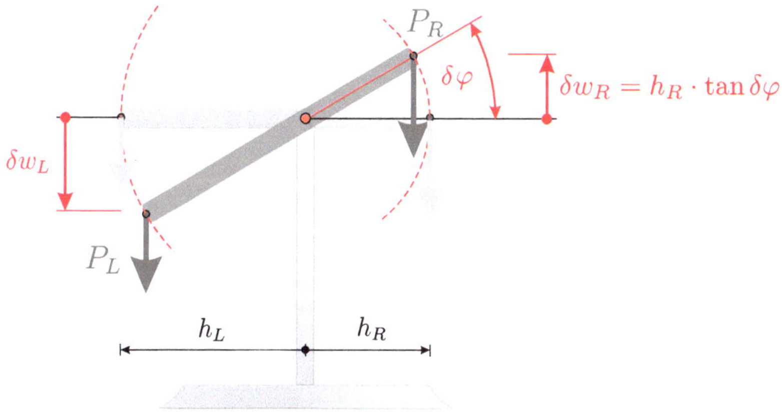

Fig. 9.5 The market woman checks the equilibrium with the principle of virtual displacements

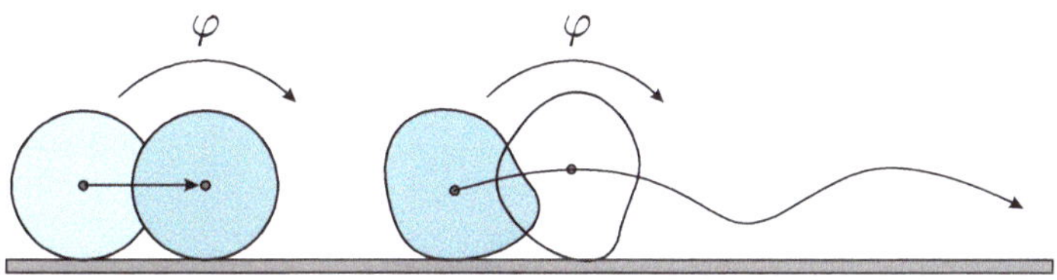

Fig. 9.6 The tool maker checks the excentricity by rolling the cylinder across the table

$$\mathcal{V}_n = \{\text{all multiples of } \delta\varphi = \frac{360}{n}\} \qquad n = 4, 8, 16 \ldots \text{edges} \tag{9.79}$$

and he so approximates the shape indirectly (via rotational equivalence), see Fig. 9.7 [2].

Equivalence is the key term in FE-analysis. The FEM does not solve the original load case but an equivalent load case. An equivalence relation means if $a \sim b$ and $b \sim c$ also $a \sim c$, and so

$$p \sim \varphi_i \quad \text{and} \quad p_h \sim \varphi_i \quad \Rightarrow \quad p \sim p_h \,. \tag{9.80}$$

In the finite element method this equivalence is "finite", is established only with respect to a finite set of test functions $\varphi_i, i = 1, 2, \ldots n$.

If we compare the length of two boards A and B via a folding rule, we use an equivalence relation. The boards have the same length, are equivalent, if they are in identical relations with the folding rule. Equivalence is indirect equality, is like weak convergence, and it leads to a true identity, $A \equiv B$, (all places after the decimal point are equal), if the relation passes all tests, and also the test with the Urmeter in Paris...

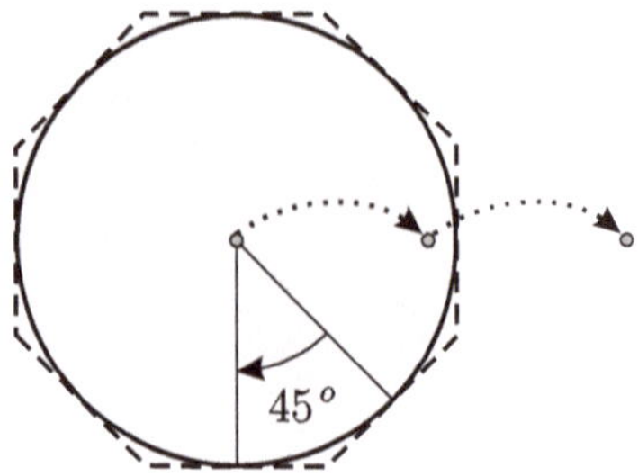

Fig. 9.7 The octogon is equivalent to a cylinder with respect to all rotations, which are a multiple of 45°

Remark 9.1 The concept of weak convergence also includes the difference between ***weak*** and ***strong boundary conditions***. Geometric boundary conditions are strong boundary conditions, like $w = 0$, since they are satisfied by all shape functions $\varphi_i \in \mathcal{V}_h$, while a static boundary condition like $v_n = 0$ (zero Kirchhoff shear) at a free edge of a plate, is guaranteed only in the integral mean—in the weak sense—$(v_n, \varphi_i) = 0$, but not pointwise. This is why static boundary conditions are called weak boundary conditions.

9.11 Variation and Green's First Identity

The potential energy of a bar, held fixed on the left, $u(0) = 0$, with a free end, $N(l) = 0$, and carrying a load p is

$$\Pi(u) = \frac{1}{2}\int_0^l \frac{N^2}{EA}\,dx - \int_0^l p\,u\,dx = \frac{1}{2}a(u,u) - (p,u)\,. \tag{9.81}$$

In the equilibrium position u the first variation of $\Pi(u)$ should be zero.

The value of Π at a neighboring point $u + \varepsilon\,\delta u$ is

$$\begin{aligned}\Pi(u+\varepsilon\delta u) &= \frac{1}{2}a(u+\varepsilon\delta u, u+\varepsilon\delta u) - (p, u+\varepsilon\delta u)\\ &= \frac{1}{2}a(u,u) + \varepsilon\cdot a(u,\delta u) + \varepsilon^2\cdot\frac{1}{2}a(\delta u,\delta u) - (p,u) - \varepsilon\cdot(p,\delta u)\end{aligned} \tag{9.82}$$

and so the first variation

$$\delta\Pi(u,\delta u) = \frac{d}{d\varepsilon}\Pi(u+\varepsilon\delta u)|_{\varepsilon=0} = a(u,\delta u) - (p,\delta u) \tag{9.83}$$

is identical to

$$\mathscr{G}(u, \delta u) = \int_0^l p\,\delta u(x)\,dx - \int_0^l \frac{N\,\delta N}{EA}\,dx = (p, \delta u) - a(u, \delta u) = 0 \quad (9.84)$$

since the work done on the boundary [...] is zero because of $N(l) = 0$ and $\delta u(0) = 0$.

The first variation of the potential energy is Green's first identity.

9.12 The Basic Functional (Hu-Washizu)

In [3] we introduced the basic functional Π_H and we showed how it can be derived from Green's second identity. We call it basic since all other functionals are modifications of this functional, in particular the **potential energy** and the **complementary potential energy**. We want to use this result to show that the potential energy, $\Pi(u) = 1/2\,a(u, u)$, is positive if a support settles.

To keep the algebra simple, we consider a bar. By integrating the last integral in Green's second identity by part

$$\mathscr{B}(u, \delta u) = \int_0^l -EAu''\delta u\,dx + [N\,\delta u - u\,\delta N]_0^l - \underbrace{\int_0^l u(-EA\delta u'')\,dx}_{int.\,by\,parts} = 0 \quad (9.85)$$

we can make a step back in the direction of $a(u, \delta u)$, and we thus obtain

$$\mathscr{V}(u, \delta u) = \left\{\int_0^l -EAu''\delta u\,dx + [N\,\delta u - u\,\delta N]_0^l\right\} + [u\delta N]_0^l - a(u, \delta u) = 0\,. \quad (9.86)$$

The curly brackets contain the "rest" of Betti, the untouched part.

Next, we replace all terms of u inside the curly brackets with their data as far as they appear in the boundary value problem, remove the brackets and delete all terms which cancel. The resulting expression is the first variation of the **basic functional** Π_H (H as in *Hu-Washizu*).

Consider for example the boundary value problem

$$-EAu'' = p \qquad u(0) = 0 \qquad u(l) = \Delta\,. \quad (9.87)$$

Applying the above scheme we get

$$\mathscr{V}(u, \delta u) = \int_0^l p\,\delta u\,dx + [N\delta u]_0^l - \Delta\,\delta N(l) + [u\,\delta N]_0^l - a(u, \delta u) = 0\,, \quad (9.88)$$

and this zero $\times(-1)$ is the first variation of the **basic functional**

$$\Pi_H = \frac{1}{2}a(u,u) - \int_0^l p\, dx + N(0)\, u(0) + (\Delta - u(l))\, N(l)\,. \tag{9.89}$$

If R_1 is the set of all u which satisfy the geometrical boundary conditions, inhomogeneous or not, the restriction of Π_H to R_1 is the **potential energy** and in the case of the boundary value problem (9.87), if we let $p = 0$ and $u(l) = \Delta$, this is

$$\Pi(u) = \frac{1}{2}\, a(u,u) \geq 0\,. \tag{9.90}$$

And if R_2 is the set of all u, which satisfy the static field equations ($-EAu'' = p$) and static boundary conditions (none in this problem), then—after some intermediate steps—the restriction of Π_H to R_2 is the **complementary potential energy**, see [3], p. 116,

$$\Pi_c(u) = -\frac{1}{2}\int_\Omega a(u,u) + N(l)\Delta\,. \tag{9.91}$$

The basic functional corresponds to the *Hu-Washizu* functional of elasticity theory, where u is not required to satisfy any field equations and any boundary conditions [4].

9.13 Force Method and Slope Deflection Method

In other disciplines we can also choose between a displacement formulation and a force method [1]. In electrical engineering the choice is between *Kirchhoff's Voltage Law*: Sum of the voltage drops in a loop is zero, (correspondingly the stress states of the X_i are equilibrium states), and *Kirchhoff's Current Law*: The sum of the currents at each node is zero, ("nodal equilibrium").

In *computational mechanics*, as *Gilbert Strang* noticed, the **nodal point methods** are preferred to the loop equations, *"Loop equations versus nodal equations"* [1], p. 158. The loop equations only have a chance as a hand method for small problems and are then even superior to the nodal methods. But one should not formulate them in matrix notation, this is cumbersome, and the result looks strange.

Irrespective of this, the force method is of course an excellent tool to learn statics!

9.14 The Adjoint Operator and Green's Function

Given an operator L and a scalar product the adjoint operator L^* is defined via $(Lu, v) = (u, L^*v)$ and to each (bounded) linear functional belongs a Green's function, which can also be a vector as in linear algebra.

Green's function g of the linear functional $J(u) = (j, u)$ is the solution of the adjoint equation $L^* g = j$.

To each symmetric (= self-adjoint) matrix $\boldsymbol{K}$ belongs an identity such as

$$\mathscr{B}(\boldsymbol{u}, \boldsymbol{v}) = \boldsymbol{v}^T \boldsymbol{K} \boldsymbol{u} - \boldsymbol{u}^T \boldsymbol{K} \boldsymbol{v} = 0\,. \tag{9.92}$$

If $\boldsymbol{J}(\boldsymbol{u}) = \boldsymbol{j}^T \boldsymbol{u}$ is a linear functional applied to the solution of $\boldsymbol{K}\boldsymbol{u} = \boldsymbol{f}$, then follows

$$J(\boldsymbol{u}) = \boldsymbol{j}^T \boldsymbol{u} = \boldsymbol{j}^T \boldsymbol{K}^{-1} \boldsymbol{f} = \boldsymbol{g}^T \boldsymbol{f}\,, \tag{9.93}$$

if $\boldsymbol{g}$ is the solution of the adjoint system $\boldsymbol{K}\boldsymbol{g} = \boldsymbol{j}$.

The operator $-EAu''$ in the boundary value problem

$$-EA\,u''(x) = p(x) \qquad u(0) = u(1) = 0 \tag{9.94}$$

is self-adjoint, since

$$\mathscr{B}(u, v) = \int_0^l -EA\,u''(x)\,v(x)\,dx + \cdots - \int_0^l u\,(-EA\,v'')\,dx = 0\,. \tag{9.95}$$

Given a linear functional

$$J(u) = \int_0^l j(x)\,u(x)\,dx \tag{9.96}$$

and the solution $g(x)$ to the adjoint boundary value problem

$$-EA\,g'''(x) = j(x) \qquad g(0) = g(1) = 0\,, \tag{9.97}$$

integration by parts applied twice, see (9.95), results in

$$J(u) = \int_0^l j(x)\,u(x)\,dx = \int_0^l -EA\,g''(x)\,u(x)\,dx = \int_0^l g(x)\,p(x)\,dx\,, \tag{9.98}$$

which corresponds to $\boldsymbol{J}(\boldsymbol{u}) = \boldsymbol{j}^T \boldsymbol{u} = \boldsymbol{g}^T \boldsymbol{f}$. So, the solution $g(x)$ of (9.97) is the Green's function of the functional $J(u)$.

Next, we look at the ***initial value problem***

$$u'(t) = f(t) \qquad u(0) = 0\,, \tag{9.99}$$

to which belongs the identity, ($T > 0$ arbitrary),

$$\mathscr{B}(u,v) = \int_0^T u'(t)\,v(t)\,dt - [u\,v]_0^T + \int_0^T u(t)\,v'(t)\,dt = 0\,. \tag{9.100}$$

The operator d/dt is not self-adjoint, since $(u', v) = (u, -v')$ (without boundary values).

Given a linear functional

$$J(u) = \int_0^T j(t)\,u(t)\,dt \tag{9.101}$$

and the solution $g(t)$ of the adjoint problem

$$-g'(t) = j(t) \qquad g(T) = 0\,, \tag{9.102}$$

integration by parts leads to

$$J(u) = \int_0^T j(t)\,u(t)\,dt = [g\,u]_0^T + \int_0^T g(t)\,u'(t)\,dt = \int_0^T g(t)\,f(t)\,dt\,. \tag{9.103}$$

Integration by parts has two limits, $(0, l)$ or $(0, T)$. In boundary value problems both limits are served, while in initial value problems only the lower limit is relevant and therefore Green's function has to be "grounded" at the endpoint T, which means $g(T) = 0$. Green's function runs "backwards".

If you look up a mountain range, the gradient is positive. If you look back from the top ($t = T$), the gradient is negative. This may explain the change in the sign[2]

$$u'(t) = f(t) \qquad\qquad -g'(t) = j(t)\,. \tag{9.104}$$

Example 9.1 We consider the initial value problem

$$u' = 1 - t \qquad u(0) = 0 \tag{9.105}$$

and let the functional $J(u)$ the integral of the solution $u(t) = t - 0.5\,t^2$

$$J(u) = \int_0^T 1 \cdot u(t)\,dt = \frac{T^2}{2} - \frac{T^3}{6} \qquad j(t) = 1\,. \tag{9.106}$$

The solution of the adjoint equation

$$-g' = 1 \qquad g(T) = 0 \tag{9.107}$$

[2]You can make (9.100) "self-adjoint" if you swap the integration limits in the second integral ($= \times(-1)$) and so motivate the retrograde.

is $g(t) = T - t$ and the scalar product of $g(t)$ and $f(t)$ is exactly the value in (9.106)

$$J(u) = \int_0^T g(t) f(t) dt = \int_0^T (T - t) \cdot (1 - t)\, dt = \frac{T^2}{2} - \frac{T^3}{6}\,. \tag{9.108}$$

9.15 Rope

Here, we want to briefly detail the formulation of the **rope equation** $-Hw'' = p$. At any point of the rope the tangential force S, the vertical force V and the horizontal prestress H form a right-angled triangle, $V/H = \tan\varphi = w'$, see Fig. 9.8. The equilibrium condition for a small piece Δx of the rope requires $V + \Delta V - V + p\,\Delta x = 0$ and given $-V' = p$ and $V = H\,w'$, we arrive at $-H\,w'' = p$.

To limit the sag of a rope under a uniform load p to a value f, the tensile force must be equal to

$$H = \frac{p\,l^2}{8\,f} \tag{9.109}$$

and the length L of the rope must not exceed

$$L = \int_0^l ds \simeq l \cdot (1 + \frac{8\,f^2}{3\,l^2})\,. \tag{9.110}$$

9.16 Filter

The *hat function* φ_i in Fig. 9.9 is a ***low-pass filter***—a sink, a "vacuum-cleaner"—and its derivative φ_i' is a ***high-pass filter***—it registers imbalances, it differentiates.

Let us consider a bar consisting of two elements, $[x_{i-1}, x_i] \cup [x_i, x_{i+1}]$. If the normal force $N(x)$ is continuous at the central node x_i, the two integrals

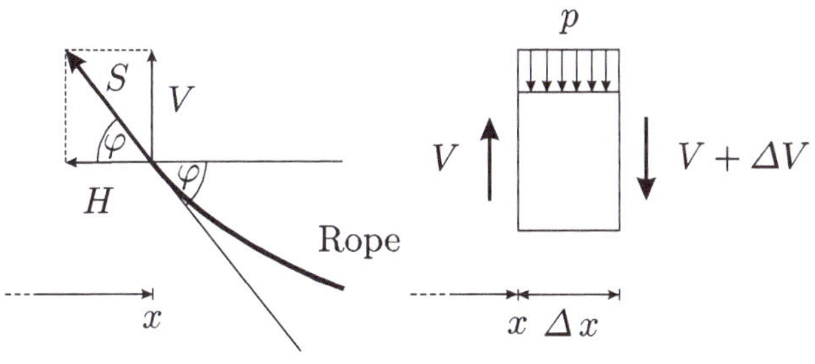

Fig. 9.8 Rope and section forces

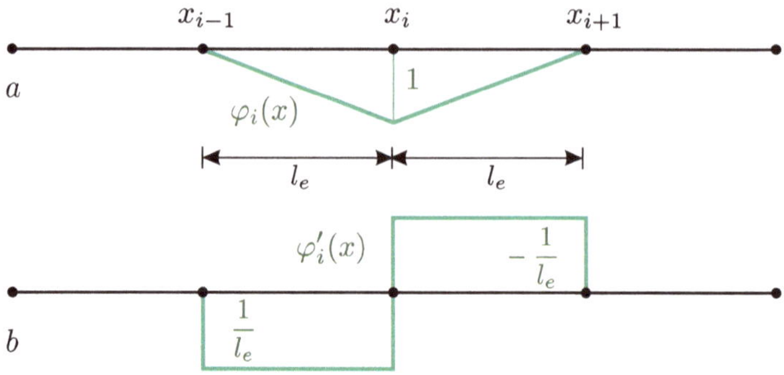

Fig. 9.9 Low-pass and high-pass filter

$$\int_{x_{i-1}}^{x_{i+1}} -N'\,\varphi_i(x)\,dx = \int_{x_{i-1}}^{x_{i+1}} N\,\varphi_i'(x)\,dx\,, \tag{9.111}$$

are the same and so the two filters φ_i and φ_i' applied to $p = -N'$ and N must provide the same result. If p is symmetric, N cannot be symmetric, because $sym \times anti = 0$, and if p is antisymmetric—which means the left side is zero, N must be symmetric with respect to the node x_i.

Repeating this with a complete bar, $-EAu'' = p$,

$$\int_0^l p\,\varphi_i\,dx = \int_0^l -N'\,\varphi_i(x)\,dx = \int_{x_{i-1}}^{x_{i+1}} N\,\varphi_i'(x)\,dx = a(u, \varphi_i)\,, \tag{9.112}$$

we recognize in this equation the statement $\delta W_e = \delta W_i$. So, the principle of virtual displacements is a consequence of the symmetry in the **filter law** (9.111). The derivative of a symmetric filter is an antisymmetric filter—with all the consequences, which this entails.

Integration by parts also means that we can write the L_2-scalar product of a derivative in "dual fashion"—the right side of (9.111).

In the limit, $\Delta x = x_{i+1} - x_{i-1} \to 0$, (9.111) becomes

$$-N'(x_i) = \lim_{\Delta x \to 0} -\frac{1}{\Delta x}(N(x_{i+1}) - N(x_{i-1}))\,. \tag{9.113}$$

So, if we differentiate something numerically, it is a *dual* formulation, is ***adjoint analysis***, the work outside, $\delta W_e = (-N', \varphi_i)$, is transformed into work inside, $\delta W_i = (N, \varphi_i') = a(u, \varphi_i)$.

9.17 Poisson Equation

The *Poisson equation* is the "inhomogeneous" Laplace equation, see Fig. 9.10,

$$-\Delta u(\boldsymbol{x}) = p(\boldsymbol{x})\,, \tag{9.114}$$

to which belongs the identity

$$\mathscr{G}(u, \delta u) = \int_\Omega -\Delta u(\boldsymbol{x})\, \delta u(\boldsymbol{x})\, d\Omega + \int_\Gamma \frac{\partial u(\boldsymbol{x})}{\partial n}\, \delta u(\boldsymbol{x})\, ds - a(u, \delta u) = 0 \tag{9.115}$$

where

$$a(u, \delta u) = \int_\Omega \nabla u(\boldsymbol{x}) \bullet \nabla\, \delta u(\boldsymbol{x})\, d\Omega = \int_\Omega (u,_{x_1}\, \delta u,_{x_1} + u,_{x_2}\, \delta u,_{x_2})\, d\Omega\,. \tag{9.116}$$

The relevance of the Poisson equation is based on its **fundamental solutions**

$$g(\boldsymbol{y}, \boldsymbol{x}) = -\frac{1}{2\pi} \ln r \qquad (2\text{-}D) \qquad g(\boldsymbol{y}, \boldsymbol{x}) = \frac{1}{4\pi}\frac{1}{r} \qquad (3\text{-}D)\,, \tag{9.117}$$

since any C^2 function over a domain Ω can be calculated from its boundary values u and $\partial u/\partial n$ and the second derivatives $\Delta u = u,_{11} + u,_{22}$ in the domain via the fundamental solutions

$$u(\boldsymbol{x}) = \int_\Gamma [g(\boldsymbol{y}, \boldsymbol{x})\, \frac{\partial u(\boldsymbol{y})}{\partial n} - \frac{\partial g(\boldsymbol{y}, \boldsymbol{x})}{\partial n}\, u(\boldsymbol{y})]\, ds_y + \int_\Omega g(\boldsymbol{y}, \boldsymbol{x})\, (-\Delta u(\boldsymbol{y}))\, d\Omega_y\,. \tag{9.118}$$

This equation is practically the extension of the integration by parts formula

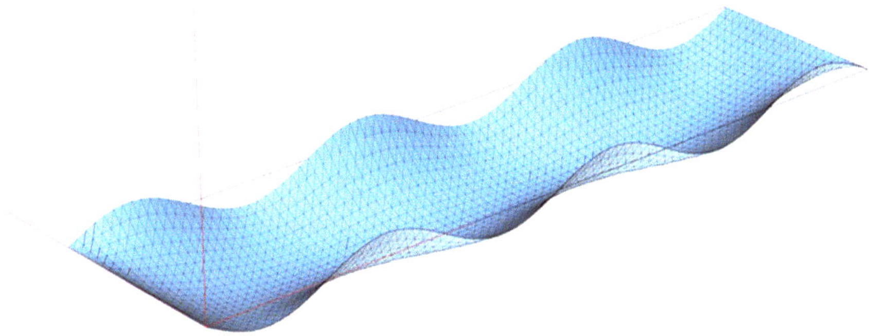

Fig. 9.10 Reconstructing the solution of a Poisson equation from its sinusoidal boundary values and the distributed load $-\Delta u = p$

$$w(x) = w(0) + \int_0^x w'(y)\,dy = \int_\Gamma \cdots + \int_\Omega \cdots\,, \tag{9.119}$$

to higher dimensions. We consider it the central equation of calculus. *A function is uniquely determined by its "trace" on the boundary, u and $\partial u/\partial n$, and its* "curvature" Δu in the domain.

The solutions of $-\Delta u(\boldsymbol{x}) = 0$ are called ***harmonic function***, since the value at a point $\boldsymbol{x}$ is the mean of the neighbors in the north, east, south and west. Each straight line, $-u'' = 0$, is a harmonic function.

9.18 Potentials and Potentials

The deflection $w(x)$ of a hinged beam minimizes the potential (energy)

$$\Pi(w) = \frac{1}{2}\int_0^l EI\,(w'')^2 dx - \int_0^l p\,w\,dx \tag{9.120}$$

and $w(x)$ is the superposition of the influence function with the load

$$w(x) = \int_0^l G(y, x)\,p(y)\,dy\,. \tag{9.121}$$

This too is a potential. Not in the sense of (9.120), but functions which are the superposition of an influence function with the load p are also called potentials [5].

The gravitational field of the Earth is a potential

$$u(\boldsymbol{x}) = \int_\Omega \frac{G}{r}\,m(\boldsymbol{y})\,d\Omega_y\,, \tag{9.122}$$

since it is the superposition of the mass $m(\boldsymbol{y})$ with the influence function G/r where $G = 6.7 \cdot 10^{-11}$ N m^2 kg^{-2} is the gravitational constant and the potential satisfies the equation $-\Delta u(\boldsymbol{x}) = 4\pi\,G\,m(\boldsymbol{x})$.

Linear static analysis and most of classical physics is potential theory since the fields which a distributed mass, the electrical charge in a wire, a line force on a plate generate are the superposition of the point solutions with the source.

9.19 Single Force Acting on a Plate

The following is an addendum to the remarks on page 364. The stress state generated by a single force in a plate Ω with an edge Γ can be split into two parts. A force acting on an infinitely long plate and a set of edge forces on Γ, which are so chosen that the sum of both load cases satisfies the boundary conditions at the edge Γ of the plate Ω.

The stresses of the second state are bounded at the source point, the integral of the traction vector $\boldsymbol{t}$ of this state over shrinking circles Γ_ε centered at the source point $\boldsymbol{x}$ is zero

$$\lim_{\varepsilon \to 0} \int_{\Gamma_\varepsilon} \boldsymbol{t}\, ds_y = 0\,, \tag{9.123}$$

and so we must only be concerned with the stresses of the singular state.

When we place a force $\boldsymbol{e}_i$ at a point $\boldsymbol{x}$ of an infinitely long plate, the stress tensor at a point $\boldsymbol{y}$ in the direction of a vector $\boldsymbol{\nu} = \{\nu_1, \nu_2\}^T$ has the components [6], (4.7),

$$\begin{aligned} T_{ij}(\boldsymbol{y}, \boldsymbol{x}) = &-\frac{1}{4\,\pi\,(1-\nu)\,r}\,[\frac{\partial r}{\partial \nu}((1-2\,\nu)\,\delta_{ij} + 2\,r,_i\; r,_j\,) \\ &- (1-2\,\nu)(r,_i\;\nu_j(\boldsymbol{y}) - r,_j\;\nu_i(\boldsymbol{y}))] \end{aligned} \tag{9.124}$$

where

$$r,_i := \frac{\partial r}{\partial y_i} = \frac{y_i - x_i}{r}\,. \tag{9.125}$$

If the point $\boldsymbol{y}$ lies on a circle with radius r and centered at $\boldsymbol{x}$, ν_i and $r,_i$ are identical

$$\nu_1 = r,_1 = \cos\,\varphi \qquad \nu_2 = r,_2 = \sin\,\varphi\,, \tag{9.126}$$

and so we have at any such point

$$\frac{\partial r}{\partial \nu} = \nabla r \bullet \boldsymbol{\nu} = \begin{bmatrix} \cos\,\varphi \\ \sin\,\varphi \end{bmatrix} \bullet \begin{bmatrix} \cos\,\varphi \\ \sin\,\varphi \end{bmatrix} = \cos^2\,\varphi + \sin^2\,\varphi = 1\,. \tag{9.127}$$

Since the force in Fig. 6.5 a points in x-direction, we let in (9.124) $i = 1$, and so the traction vector on the circle has the two components

$$t_x = T_{11} = -\frac{1}{4\,(1-\nu)\,\pi\,r} \cdot [(1-2\,\nu) + 2\,\cos^2\,\varphi] \tag{9.128}$$

$$t_y = T_{12} = -\frac{1}{4\,(1-\nu)\,\pi\,r} \cdot [2\,\cos\varphi\,\sin\,\varphi] \tag{9.129}$$

and the integrals are

$$\int_0^{2\pi} t_x\, d\varphi = -\frac{1}{r} \qquad \int_0^{2\pi} t_y\, d\varphi = 0\,. \tag{9.130}$$

9.20 Multipoles

Figure 9.11 and the following text is a supplement to page 129. The Taylor series of a curve $w(x)$

$$w(x) = w(0) + w'(0)\,x + \frac{1}{2}\,w''(0)\,x^2 + \cdots \tag{9.131}$$

leads to the representation

$$\begin{aligned} w(x) = &\int_0^l G_0(y,0)\,p(y)\,dy + \int_0^l G_1(y,0)\,p(y)\,dy \cdot x \\ &+ \frac{1}{2}\int_0^l G_2(y,0)\,p(y)\,dy \cdot x^2 + \cdots \end{aligned} \tag{9.132}$$

if we let G_1 and G_2 denote the influence functions for w' and w'' respectively.

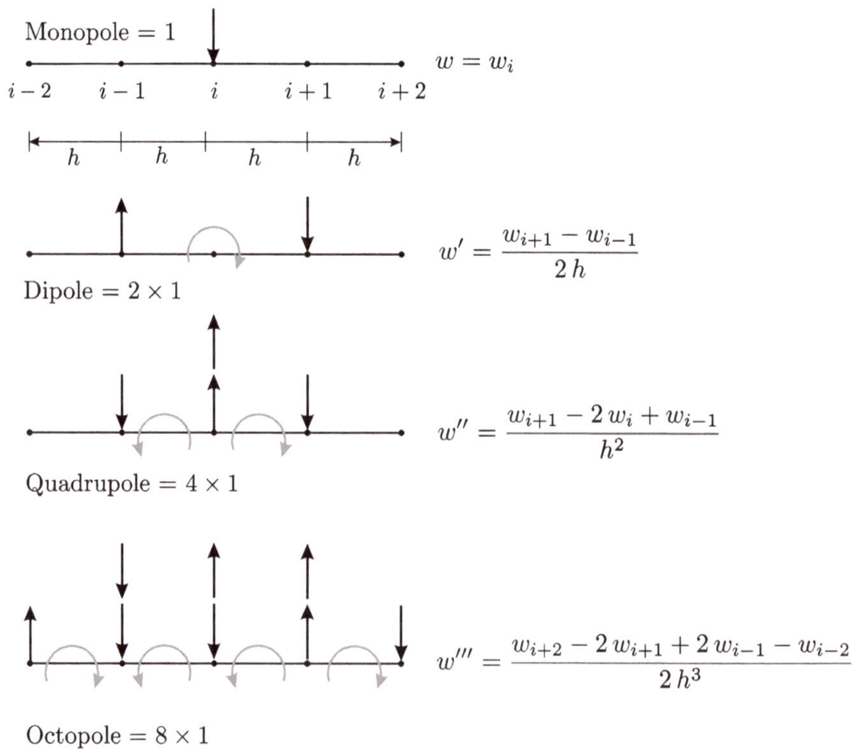

Fig. 9.11 Monopole—Dipole—Quadrupole— Octopole and finite differences

However, it is also possible to expand the kernel $G_0(y, x)$ into a Taylor series about the center of gravity y_s of the line load, which leads to

$$
\begin{aligned}
w(x) &= \int_0^l G_0(y, x)\, p(y)\, dy \\
&= G_0(y_s, x) \int_0^l p(y)\, dy + \frac{d}{dy} G_0(y_s, x) \int_0^l p(y)\,(y - y_s)\, dy \\
&+ \frac{1}{2} \frac{d^2}{dy^2} G_0(y_s, x) \int_0^l p(y)\,(y - y_s)^2\, dy + \cdots
\end{aligned}
\tag{9.133}
$$

or

$$
w(x) = G_0(y_s, x)\, R + \frac{d}{dy} G_0(y_s, x)\, M + \frac{1}{2} \frac{d^2}{dy^2} G_0(y_s, x)\, M_2 + \cdots \tag{9.134}
$$

if R is the resultant of the load, $M (= 0)$ is the moment of the load about the point y_s and M_2 is the "quadratic" moment of the load about y_s. This is also called the *multipole expansion* of $w(x)$.

The gravitational pull of the Earth works like this. If the Earth were a perfect, homogeneous sphere, we could concentrate the whole mass of the Earth at its center and the satellites would orbit the Earth on perfect circles. So, computers are needed to determine the orbits of the satellites.

With the help of (9.134) we can also assess the error if we would replace a distributed load p by its resultant or, more generally speaking, by the first terms of its Taylor expansion.

9.21 The Dimension of the f_i

An often-discussed question is, which dimension the f_i of a beam have?

$$
\frac{EI}{l^3} \begin{bmatrix} 12 & -6l & -12 & -6l \\ -6l & 4l^2 & 6l & 2l^2 \\ -12 & 6l & 12 & 6l \\ -6l & 2l^2 & 6l & 4l^2 \end{bmatrix} \begin{bmatrix} w_1 \\ w_2 \\ w_3 \\ w_4 \end{bmatrix} = \begin{bmatrix} f_1 \\ f_2 \\ f_3 \\ f_4 \end{bmatrix} . \tag{9.135}
$$

The answer is—it depends. If we calculate the entries k_{ij} in the stiffness matrix as in

$$
k_{ij} = a(\varphi_i, \varphi_j) = EI \int_0^l \varphi_i''\, \varphi_j''\, dx = [\mathrm{Nm}^2][\frac{1}{\mathrm{m}}][\frac{1}{\mathrm{m}}]\,[\mathrm{m}] = [\mathrm{Nm}]\,, \tag{9.136}
$$

the k_{ij} are energies, the u_i are dimensionless weights and the f_i are also energies, **energy quanta**.

To explain (9.136): Since each $\varphi_i(x)$ has the dimension meter, the derivatives have the dimension

$$\varphi_i(x)\ [\mathrm{m}] \qquad \varphi_i'\ [\ ] \qquad \varphi_i'' = [\frac{1}{\mathrm{m}}] \qquad \varphi_i''' = [\frac{1}{\mathrm{m}^2}]\,, \tag{9.137}$$

since with each d/dx we divide by meter.

But the matrix $\boldsymbol{K}$ can also be formulated by calculating the beam end forces of the unit displacements $\varphi_i(x)$ and placing these numbers into the respective column i. Done this way, the k_{ij} are forces per unit displacement $w_i = 1$ and $\boldsymbol{K}\boldsymbol{w} = \boldsymbol{f}$ are the beam end forces belonging to the beam end displacements $\boldsymbol{w}$.

9.22 Weak and Strong Influence Functions

We mentioned that there exist no weak influence functions for forces, say the moment $M(x)$ in a hinged beam, since

$$\lim_{\varepsilon \to 0} a(G_2, w)_{\Omega_\varepsilon} = 0\,. \tag{9.138}$$

It is possible, however, to define a sequence G_2^ε of smooth functions, which converge towards the triangular shape G_2, so that at the end the moment jumps out of the strain energy product, see [7], p. 67 *'A sequence which converges to* G_1 *'*, (there $N(x)$, here $M(x)$)

$$\lim_{\varepsilon \to 0} a(w, G_2^\varepsilon) = M(x)\,. \tag{9.139}$$

In (9.138) we start on $[0, x - \varepsilon] \cup [x + \varepsilon, l]$ and we let $\varepsilon \to 0$, while in (9.139) the ε is in the sequence $G_2^\varepsilon \to G_2$.

The first derivative of G_2 is a step function and the second derivative is a Dirac delta δ_0, and so the strain energy product yields the wished-for result.

$$\lim_{\varepsilon \to 0} a(w, G_2^\varepsilon) = \int_0^l M(y)\,\delta_0(y - x)\,dy = M(x)\,. \tag{9.140}$$

This is the way to go. The paradox however, if we want to call it so, is: When we substitute the target function $G_2(y, x)$ into (9.138) we get zero and not $M(x)$.

And one additional remark, concerning the observation that finite elements can (apparently) calculate forces with weak influence functions, see page 181. But in truth it is Betti and not Mohr, which provides for example the force $N_h(x)$

$$N_h(x) = \int_0^l G_1^h(y, x)\,p(y)\,dy = \int_0^l \sum_i g_i\,\varphi_i(y)\,p(y)\,dy = \boldsymbol{g}^T \boldsymbol{f}\,. \tag{9.141}$$

It is only, if we let $\boldsymbol{f} = \boldsymbol{K}\boldsymbol{u}$, that it looks like a weak formulation

$$N_h(x) = \boldsymbol{g}^T \boldsymbol{f} = \boldsymbol{g}^T \boldsymbol{K}\boldsymbol{u} \,. \tag{9.142}$$

9.23 How the Embedding Theorem Got Its Name

In the same way as the citizens of a city can be classified according to different criteria, *age, size*, ..., so the deflections $w(\boldsymbol{x})$ of a plate Ω can be classified in different ways. One of these possible scales is the so-called ***Sobolev norm***. The Sobolev norm of order $m = 2$ of a function $w(x, y)$ is the number

$$\|w\|_2 = \sqrt{\int_\Omega (w^2 + w,_x^2 + w,_y^2 + w,_{xx}^2 + w,_{xy}^2 + w,_{yy}^2)\, d\Omega} \,. \tag{9.143}$$

Following this pattern, we can define Sobolev norms $\|w\|_m$ to arbitrarily high order m: Square integrate the derivatives up to the order m and take the square root of the sum! The functions with a finite Sobolev norm of order m form the **Sobolev space** $H^m(\Omega)$.

Closely related to Sobolev spaces are the so-called ***energy spaces*** of functions with finite strain energy, $a(w, w) < \infty$, since the energy space of a Kirchhoff plate can be identified with the Sobolev space H^2, and the energy space of a plate with $\boldsymbol{H}^1 = H^1 \times H^1$ (horizontal displ. u_x and vertical displ. u_y).[3]

The Russian mathematician S. Sobolev has shown that the space $H^m(\Omega)$ is embedded into the space $C(\Omega)$ of the continuous functions over Ω,

$$H^m(\Omega) \subset C(\Omega) \,, \tag{9.144}$$

and the **embedding** is even continuous

$$\max |w| < c \cdot \|w\|_m \,, \tag{9.145}$$

if

$$m > \frac{n}{2} \,, \tag{9.146}$$

if the index m of the Sobolev space is greater than half the dimension of the space, which would be $n = 2$ for a Kirchhoff plate. The value $\max |w|$ is the norm of a function w in $C(\Omega)$, and (9.145) means: The norm $|w|$ of the projection is smaller

[3]Technically, the energy norm $\|w\|_E = \sqrt{a(w, w)}$ and the Sobolev norm $\|w\|_m$ are equivalent norms, if the rigid-body motions are excluded.

than the initial norm $\|w\|_m$ times a constant factor c, which only depends on the shape of Ω.

A small distance between two functions w_1 and w_2 in $H^m(\Omega)$ remains a small distance in $C(\Omega)$

$$\max |w_1 - w_2| < c \cdot \|w_1 - w_2\|_m \,, \tag{9.147}$$

—the neighborhood is preserved.

The interesting aspect of this result is the transition from an integral to a point. The Sobolev norm is an integral measure, but if the inequality (9.146) applies, as in the case of a Kirchhoff plate, $m = 2, n = 2$, the maximum deflection of the plate is bounded by the strain energy. If the strain energy tends to zero, the deflection of the plate also tends to zero—everywhere, at each point! ***An integral majorizes point values!***

In the case of a wall plate the situation is different. The energy space is $\boldsymbol{H}^1$, but since $m = 1$ and $n = 2$ do not satisfy the inequality (9.146), a small distance $\|\boldsymbol{u}_1 - \boldsymbol{u}_2\|_1$ between two displacement fields does not guarantee the maximum displacements to be equally close, which for a Kirchhoff plate is true

$$\max |w_1 - w_2| < c \cdot \|w_1 - w_2\|_2 \,. \tag{9.148}$$

If the reader has some patience left, we can pursue these ideas a little further.

People can be classified with respect to age (A) or weight (W). On the set A the functional "shoe size" is not constant, since a small distance in the age of two persons is no guarantee for similar shoe sizes. On the other hand, we (approximately) expect such a correlation to exist on the set W.

So, a functional is continuous if a small difference in the input guarantees a small distance in the output, if

$$|J(u_1) - J(u_2)| < c \cdot \|u_1 - u_2\| \,, \tag{9.149}$$

where c is a global constant, independent of the single u. Continuity therefore depends on how one measures distance in the input set and in the target set. Note, if you set $u_2 = 0$ such a bound implies $|J(u)| < c \cdot \|u\|$.

If a linear functional is continuous, it is also bounded and vice versa.

The point functional

$$J(w) = w(\boldsymbol{x}) \qquad \text{deflection at a point } \boldsymbol{x} \tag{9.150}$$

of a Kirchhoff plate is a continuous and bounded functional on $H^2(\Omega)$, because the continuous embedding (9.145) guarantees that the difference in two values is bounded by the difference in their energies

$$|J(w_1) - J(w_2)| = |w_1(\boldsymbol{x}) - w_2(\boldsymbol{x})| < c \cdot \|w_1 - w_2\|_2\,, \tag{9.151}$$

but this does not hold true for wall plates, since it cannot be guaranteed *for all (!)* $\boldsymbol{u} \in \boldsymbol{H}^1(\Omega)$ that the difference in the horizontal displacement of two displacement fields at a point $\boldsymbol{x}$

$$|J(\boldsymbol{u}) - J(\boldsymbol{v})| = |u_x(\boldsymbol{x}) - v_x(\boldsymbol{x})| \not< c \cdot \|\boldsymbol{u} - \boldsymbol{v}\|_1 \tag{9.152}$$

is bounded by the Sobolev norm $\|\boldsymbol{u} - \boldsymbol{v}\|_1$.

The emphasis is on *for all*. The deflection $u = -\ln(-\ln^{-1} r)$ of a membrane is infinite at the point $r = 0$, $J(u) = \infty$, but in a disk Ω with radius $R = 0.5$ and centered at the origin the H^1 norm is bounded, see [7], p. 98. So, there exists no bound c

$$J(u) = u(\boldsymbol{x}) = \infty < c \cdot \|u\|_1 \quad ? \tag{9.153}$$

and therefore $J(u)$ is not a continuous and bounded functional on $H^1(\Omega)$. One counterexample is enough to substantiate this claim.

Given a bar, $n = 1, m = 1$, the energy space is $H^1(0, l)$, and this means

$$J(u) = u(x) \qquad m - i = 1 - 0 = 1 > \frac{1}{2} \qquad \text{continuous} \tag{9.154}$$

$$J(u) = u'(x) \qquad m - i = 1 - 1 = 0 \not> \frac{1}{2} \qquad \text{not continuous} \tag{9.155}$$

The energy space of a beam, $n = 1, m = 2$, is $H^2(0, l)$, and so we have

$$\underbrace{J(w) = w(x) \qquad J(w) = w'(x)}_{continuous} \qquad \underbrace{J(w) = w''(x) \qquad J(w) = w'''(x)}_{not\ continuous}\,. \tag{9.156}$$

If a functional is continuous, the energy of the Green's function is finite, otherwise it is infinite and also the external work $W_e = displ. \times force$ *is infinite. One of the two quantities must be infinite at the source point.*

The Green's function of the (discontinuous) functional $J(u) = EA\,u'(x)$, the normal force $N(x)$ in a bar, is a dislocation and to produce such a gap requires infinitely large forces, which let the strain energy "overflow"

$$\int_0^l \frac{N^2}{EA}\,dx = \infty \qquad N = \text{normal force due to dislocation}\,. \tag{9.157}$$

Here, we follow the mathematician, who expands the dislocation into a Fourier series, see page 73, and not the engineer, who first installs a normal force hinge and then spreads the hinge.

Next, a comment on the inequality (1.268) from Chap. 1, which we repeat here

$$m - i > \frac{n}{2}. \tag{9.158}$$

If we differentiate a function in $H^m(\Omega)$, its derivative (possibly) lies one order of space lower, lies in $H^{m-1}(\Omega)$. This explains why we subtract from m the index $i = 0, 1, 2, 3$ of the Dirac delta δ_i, i.e. the signal how often the Dirac delta differentiates the function u

$$J(u) = \int_\Omega \delta_i(\boldsymbol{y} - \boldsymbol{x})\, u(\boldsymbol{y})\, d\Omega_{\boldsymbol{y}} \quad \sim \quad \frac{d^i u}{dx^i}. \tag{9.159}$$

Remark 9.2 Not continuous in (9.156) means: $J(w) = -EI\, w''(x) = M(x)$ is not a continuous ***functional*** on $H^2(0, l)$. There exists no constant c so that *all* functions $w \in H^2(0, l)$ satisfy the inequality

$$|J(w)| = |M(x)| < c \cdot \|w\|_2. \tag{9.160}$$

A curve w with a logarithmic singularity in the bending moment, $M(x) = \ln(x - x_0)^2$ at a point $x_0 \in (0, l)$ has $\|w\|_2 < \infty$, but $M(x_0) = \infty$.

We can read it also this way: The singular $M(x)$ is in $H^0(0, l)$, because we can integrate $M(x)^2$, but the norm on H^0, the integral $\|M\|_0 = (M, M)^{1/2}$, is not a bound for $\max |M| < c \cdot \|M\|_0$, since the embedding of $H^0(0, l)$ into $C^0(0, l)$ is not continuous, there are outliers and one of these is the singular moment. Even simpler:

That we can square-integrate a function $M(x)$ on $(0, l)$, does not imply that the function $M(x)$ is bounded on the interval $(0, l)$.

But if also the derivative $M'(x)$ is square-integrable, if $M(x)$ lies in $H^1(0, l)$, the conclusion is correct, since then the inequality $m > n/2$, set $m = 1, n = 1$, is satisfied.

But the derivative $M'(x) = 2/(x - x_0)$ of $M(x) = \ln(x - x_0)^2$ is not square-integrable, and so the loophole $M \in H^1(0, l)$ is not available.

Remark 9.3 The mathematical theory of finite elements is based on the concept of weak solutions on appropriate Hilbert spaces. With regard to influence functions the rule is: *An influence function*

$$J(u) = \int_0^l G(y, x)\, p(y)\, dy \tag{9.161}$$

exists if and only if the functional $J(u)$ *is linear and bounded,* $|J(u)| < c\, \|u\|$.

If one follows this *dictum*, there should be no influence function $G_2(y, x)$ for the moment in a beam, since the functional $J(w) = M(x)$ is unbounded on $H^2(0, l)$.

However, an engineer would tell you otherwise, since she or he simply places a kink at the source point and so manages to generate the exact influence for the bending moment.

But the above statement only means "no influence function with ***finite energy***", no influence function $G_2(y, x) \in H^2(0, l)$, which, if we calculate, see page 73, is confirmed: The triangular shape with the kink has an infinite energy, does not lie in $H^2(0, l)$. Now, of course, one could say: If mathematicians are so subtle and draw such narrow bounds, the engineer need not care.

But this is a weakness in the mathematical theory of finite elements according to which the FEM is an ***energy method***, and solutions which have infinite energy lie outside this theory. How do you minimize the distance to the exact influence function G in the energy,

$$a(G - G_h, G - G_h) = \|G - G_h\|^2 \qquad \rightarrow \text{ minimum}\,, \tag{9.162}$$

if the energy of the solution, $\|G\| = \infty$, is infinite?

This results in a curious situation: An FE-program calculates all section forces with influence functions which are approximate solutions of *ill posed problems*, ill posed, because the exact solutions do not lie in the solution space, in the energy space—all influence functions for internal forces have infinite energy.

A situation which every mathematician warns against: "Please make sure a solution exists, before you start the computer..."

9.24 Point Loads and Their Energy

Table 9.1 evaluates the inequality $m - i > n/2$, which must be true for the strain energy to be finite; it is m the order of energy

$m = 1$ Timoshenko beam, Reissner–Mindlin plate, 2-D and 3-D elasticity

$m = 2$ Euler–Bernoulli beam, Kirchhoff plates

The order of the energy corresponds to the highest derivative in the strain energy $a(u, u)$. It is always half the order $2\,m$ of the differential equation.

Figure 9.12 illustrates why it is better to convert point loads into short line loads.

9.25 Early Birds

We know two early papers, which deal with Green's functions and finite elements, *Kolář* (Brno, Czech) [8], and *Tottenham* (Southampton) [9], both from 1970. There arc probably other early papers. We would be grateful for hints.

Table 9.1 Finite (✓) and infinite (∞) energy

	dimension	$n=1$	$n=2$	$n=3$
	$m=1$	rope, bar, Timoshenko beam	2-D E-Th. Reissner–Mindlin plate	E-Th.
	point loads			
$i=0:$	↓	✓	∞	∞
$i=1:$		∞	∞	∞
	$m=2$	Euler–Bernoulli beam	Kirchhoff plate	
$i=0:$	↓	✓	✓	
$i=1:$		✓	∞	
$i=2:$		∞	∞	
$i=3:$		∞	∞	

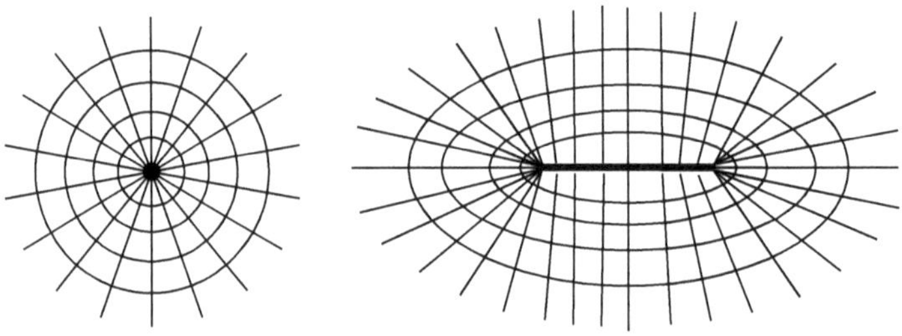

Fig. 9.12 Near the point load the lines of forces are so tightly packed that the material starts to flow, while a line load avoids the high stress concentration—the strain energy remains finite [7]

References

1. Strang G (2010) Computational science and engineering. Wellesley-Cambridge Press
2. Hartmann F, Katz C (2010) Structural analysis with finite elements, 2nd edn. Springer, Heidelberg
3. Hartmann F (1985) The mathematical foundation of structural mechanics. Springer, Heidelberg
4. Gurtin ME (1972) The linear theory of elasticity. Handbuch der Physik Band VIa/2 Festkörpermechanik II, ed. C. Truesdell, series ed. S. Flügge, Springer, Heidelberg
5. Kellogg OD (1967) Foundations of potential theory. Springer, Heidelberg (reprint)
6. Hartmann F (1989) Introduction to boundary elements. Springer, Heidelberg
7. Hartmann F (2013) Green's functions and finite elements. Springer, Heidelberg
8. Kolář V (1970) The influence functions in the finite element method. ZAMM 50:T 129–T 131
9. Tottenham H (1970) Basic principles. In: Tottenham H, Brebbia C (eds) Finite element techniques in structural mechanics. Southampton University Press

Chapter 10
Software

Four programs are offered for download:

BE-LAPLACE	http://simplel.ink/go/belaplacewinfem
BE-SLABS	http://simplel.ink/go/beSlabsWinfem
BE-PLATES	http://simplel.ink/go/bePlatesWinfem
BE-FRAMES	http://simplel.ink/go/beframes

The first three are based on the boundary element method. Many of the drawings in this book were created with these three programs, since influence functions and boundary elements are a good match, see Figs. 1.31, 2.42, 2.44, 2.45, 2.47, 2.48, 2.56, 3.3, 3.48, 3.63, 3.67, 3.68, 3.71, 3.72, 3.73, 2.51, 4.6, 4.7, 5.6, 5.19, 6.9, 6.10, 6.13, 6.15, 6.16, 6.17, 6.20, 6.21, 6.22, 7.2, 9.10.

BE-FRAMES is for the analysis of planar frames with an emphasis on influence functions and *one-click-reanalysis*, see section 5.24.

Manuals

BE-LAPLACE	http://simplel.ink/go/ManualBELAPLACE
BE-SLABS	http://simplel.ink/go/ManualBESLABS
BE-PLATES	http://simplel.ink/go/ManualBEPLATES
BE-FRAMES	http://simplel.ink/go/ManualBEFRAMES

For German versions of the programs and the book see http://www.be-statik.de

F. Hartmann and P. Jahn, *Statics and Influence Functions*,
Springer Series in Solid and Structural Mechanics 13,
https://doi.org/10.1007/978-3-030-55889-5_10

References

1. Altenbach H, Altenbach J, Kissing W (2004) Mechanics of composite structural elements. Springer, Heidelberg
2. Bathe K-J (1996) Finite element procedures. Prentice Hall, New Jersey
3. Blaauwendraad J (2010) Plates and FEM. Springer, Heidelberg
4. Blaauwendraad J, Hoefakker JH (2014) Structural shell analysis: understanding and application. Springer, Heidelberg
5. de Borst R, Crisfield MA (2012) Nonlinear finite element analysis of solids and structures. Wiley
6. Choi KK, Kim N-H (2005) Structural sensitivity analysis and optimization. Vol. 1, Linear Systems, Vol.2, Nonlinear systems and applications. Springer, New York
7. Clough RW, Wilson EL (1999) Early finite element research at Berkeley. Fifth U.S. National conference on computational mechanics, Aug. 4–6, 1999
8. Connor J (1976) Analysis of structural member systems. The Ronald Press Company
9. Cook RD, Malkus DS, Plesha ME (1989) Concepts and application of finite element analysis. John Wiley & Sons, New York
10. Dym CL, Willams HE (2012) Analytical estimates of structural behavior. CRC Press
11. Fish J, Belytschko T (2007) A first course in finite elements. Wiley
12. Gimsing NJ, Georgaki CT (2012) Cable supported bridges. Wiley
13. Ghali A (2009) Structural analysis: a unified classical and matrix approach. Taylor & Francis Ltd
14. Gopalakrishnan S, Ruzzene M, Hanagud S (2011) Computational techniques for structural health monitoring. Springer, Heidelberg
15. Grätsch T, Bathe KJ (2005) Influence functions and goal-oriented error estimation for finite element analysis of shell structures. Int J Num Methods Eng 63(5):631–788
16. Harsuijker C, Welleman JW (1999) Engineering mechanics. Springer, Heidelberg
17. Hibbeler RC (2017) Structural analysis. Prentice Hall, 9th edn
18. Hjelmstad KD (2005) Fundamentals of structural mechanics. Springer, New York. 2nd edn
19. Hughes TJR (2000) The finite element method. Dover
20. Karnovsky IA, Lebed O (2010) Advanced methods of structural analysis. Springer, Heidelberg
21. Kaveh A (2014) Computational structural analysis and finite element methods. Springer, Heidelberg
22. Kirsch U (2004) Design-oriented analysis of structures. Kluwer Academic Publishers
23. Kirsch U (2008) Reanalysis of structures. Springer, Heidelberg

F. Hartmann and P. Jahn, *Statics and Influence Functions*, Springer Series in Solid and Structural Mechanics 13, https://doi.org/10.1007/978-3-030-55889-5

24. Krenk S (2009) Non-linear modeling and analysis of solids and structures. Cambridge University Press
25. Krishnan S, Liao Y (2020) Geometric design of deployable spatial structures made ot three-dimensional angulated members. J Archit Eng 26(3):
26. Li G, Wong KKF (2014) Theory of nonlinear structural analysis. Wiley
27. Materna D, Barthold FJ (2008) On variational sensitivity analysis and configurational mechanics. Comput Mech 41(5):661–681
28. Melchers RE, Beck A (2018) Structural reliability analysis and predicition. Wiley 3rd edn
29. Nayfeh AH, Pai PF (2004) Linear and nonlinear structural mechanics. John Wiley & Sons
30. Nielson BG (2017) Structural analysis. John Wiley & Sons, Understanding Behavior
31. Oñate E (2013) Structural analysis with the finite element method. Linear Statics: Volume 1: Basis and Solids, Volume 2: Beams, Plates and Shells. Springer, Heidelberg
32. Pierce NA, Giles MB (2000) Adjoint recovery of superconvergent functionals from PDE approximations. SIAM Rev 42(2):247–264
33. Rugarli P (2010) Structural analysis with finite elements. Ice Publishing
34. Russell CH (2012) Structural analysis. Pearson
35. Rust W (2015) Non-linear finite element analysis in structural mechanics. Springer, Heidelberg
36. Saouma VE (2000) Structural engineering. University of Colorado, Design, Analysis
37. Schodek D, Bechthold M (2014) Structures. Pearson, Harlow, 7th edn
38. Stakgold, I (2011) Green's functions and boundary value problems. Wiley, 3rd edn
39. Strang G, Fix GJ (2008) An analysis of the finite element method. Wellesley-Cambridge Press, 2nd edn
40. Taranath BS (2019) Wind and earthquake resistant buildings. CRC Press
41. Wenzel H (2009) Health monitoring of bridges. Wiley
42. Williams A (2008) Structural analysis. Butterworth-Heinemann, In Theory and Practice
43. Xu XF, Chen X, Shen L (2009) A Green-function-based multiscale method for uncertainty quantification of finite body random heterogeneous materials. Comput Struct 87:1416–1426

Index

F. Hartmann and P. Jahn, *Statics and Influence Functions*, Springer Series in Solid and Structural Mechanics 13,
https://doi.org/10.1007/978-3-030-55889-5

MIX
Papier aus verantwortungsvollen Quellen
Paper from responsible sources
FSC® C105338

If you have any concerns about our products,
you can contact us on
ProductSafety@springernature.com

In case Publisher is established outside the EU,
the EU authorized representative is:
Springer Nature Customer Service Center GmbH
Europaplatz 3, 69115 Heidelberg, Germany

Printed by Libri Plureos GmbH
in Hamburg, Germany